Helmut Neunzert · Dieter Prätzel-Wolters
Editors

# Currents in Industrial Mathematics

## From Concepts to Research to Education

Translated from German by William Uber

Springer

*Editors*
Helmut Neunzert
Fraunhofer Institute for Industrial Mathematics ITWM
Kaiserslautern, Germany

Dieter Prätzel-Wolters
Fraunhofer Institute for Industrial Mathematics ITWM
Kaiserslautern, Germany

Translation of the German Edition: Mathematik im Fraunhofer-Institut. Problemgetrieben–Modellbezogen–Lösungsorientiert, published by Springer-Verlag GmbH 2014, © 2015. All rights reserved.

ISBN 978-3-662-48257-5 ISBN 978-3-662-48258-2 (eBook)
DOI 10.1007/978-3-662-48258-2

Library of Congress Control Number: 2015953325

Mathematics Subject Classification (2010): 00-02, 00A07, 00A35, 00A69, 01A74, 00A71, 34-XX, 35-XX, 60Gxx, 62-07, 93-XX, 76-XX, 91Gxx, 97-XX

Springer Heidelberg New York Dordrecht London


Editorial: Annika Denkert and Clemens Heine
Translator: William Uber
*Cover photo*: G. Ermel, Fraunhofer ITWM

Printed on acid-free paper

Springer is part of Springer Science+Business Media (www.springer.com)

Currents in Industrial Mathematics

# Foreword

The employees of a mathematical Fraunhofer Institute spend a great deal of their time discussing problems with industrial clients and then solving these problems with the help of mathematics and computers. The periods of reflection occasionally made possible by public projects and self-financed preliminary research are normally used to build the mathematical foundation necessary for finding our clients' solutions. Taking a step back to critically examine one's own activities and then to precisely and understandably articulate them requires great inspiration and an enormous commitment of time. Nonetheless, 18 employees and 5 other mathematicians closely connected to the Institute have ventured to report on their thoughts and actions in this book.

Our point of entry is represented by the four basic concepts that determine our work: modeling, computing, optimizing and analyzing data. What these terms mean to us is described in four relatively short concept chapters.

Next, five projects—perhaps better referred to as project groups—are presented as examples; and here, we take this business of presenting very seriously. First, we describe the non-mathematical problem and explain the deficiencies in the standard approaches for its solution. We also explain why the existing mathematics is often inadequate and describe how many preliminary works surrounding question clarification have already emerged from the Institute for Industrial Mathematics ITWM, in the form of doctoral theses, for example. The core of these five research chapters, however, is solid mathematics—the models and their numerical evaluation. Finally, we describe the "solution," that is, what the customer gets from us, which often includes software.

In a closing chapter, we describe in detail how this problem-driven, model-based, solution-oriented mathematics can be integrated into mathematics instruction in our schools, in order to emphasize its significance and to promote students' joy in learning mathematics.

In writing this book, we have kept quite diverse groups of readers in mind: First, there are the people in industry and business, to whom we wish to make clear that mathematicians don't just discuss or analyze problems, they also solve them. Second, there are the university mathematicians, whom we want to convince that this approach can also provide new impulses to mathematics. Third, there are university students, who want to know, and with good reason, what they will really be doing later in their professional lives—for only a small number of them will lecture at universities. And finally, there are those who want to become teachers or who already are; this group can read how mathematics instruction in the classroom can be revitalized.

Is a mathematical Fraunhofer Institute really entitled to claim that it can fulfill all these promises? There are more than 65 institutes in the Fraunhofer-Gesellschaft, and three of them are based on mathematical methods: The ITWM in Kaiserslautern, the SCAI (Institute for Algorithms and Scientific Computing) in Sankt Augustin, and MEVIS (Institute for Medical Image Computing) in Bremen. Among these, the ITWM today enjoys the highest industrial revenues and the most rapid growth. The Institute's fantastic growth over the nearly 20 years of its existence is shining evidence that mathematics really has become a key technology. For this reason, we believe that we can indeed turn all of our readers into fans of our kind of mathematics!

Kaiserslautern, Germany
June 2015

Dieter Prätzel-Wolters

Helmut Neunzert

# Contents

# Part I
# Introduction

# Problems Trump Methods: A Somewhat Different Mathematics from a Somewhat Different Institute

Dieter Prätzel-Wolters and Helmut Neunzert

This book is dedicated to mathematics-based topics that are driven by practical problems whose solutions generate innovation. The formulations of these problems have arisen in the context of projects carried out at the Fraunhofer Institute for Industrial Mathematics (ITWM), and the majority of the authors of this book are either employed at the Fraunhofer ITWM or are closely affiliated with it. Fraunhofer Institutes dedicate their work to problems in industry; a mathematical Fraunhofer Institute therefore makes "Industrial Mathematics".

The book's editors originally suggested "Fraunhofer Mathematics" as a title. This suggestion was discarded, however, since it found no consensus among the authors. A book about "Fraunhofer Mathematics" might have also had a polarizing effect. For many mathematicians, it would have also been a provocative title, one that generates confusion about what exactly "Fraunhofer Mathematics" refers to.

Mathematics is the science with the highest degree of abstraction; there is virtually one hundred percent agreement about what is recognized as mathematics; and mathematical results are highly objective, intrinsically verifiable, and formulated in a largely standardized language. Mathematics is divided into the categories of pure and applied, although making even this basic distinction is somewhat difficult. It also happens occasionally that the works of important mathematicians become identified with their originators, so that one then speaks, for example, of Hilbertian or Riemannian mathematics. There are also schools that have developed particular structural edifices of mathematical thought and whose works are then cited, for example, as Bourbaki or constructivist mathematics.

D. Prätzel-Wolters · H. Neunzert (✉)
Fraunhofer Institut für Techno- und Wirtschaftsmathematik ITWM, Kaiserslautern, Germany
e-mail: helmut.neunzert@itwm.fraunhofer.de



H. Neunzert, D. Prätzel-Wolters (eds.), *Currents in Industrial Mathematics*,
DOI 10.1007/978-3-662-48258-2_1

How then is Fraunhofer mathematics to be fitted into such a classification system? As the mathematics of Joseph von Fraunhofer, perhaps? Hardly. Although he too produced mathematically oriented works, Joseph von Fraunhofer (1787–1826) was not a mathematician. He was a very successful scientist who discovered the lines in the solar spectrum that were subsequently given his name and who was extremely well versed in physics and in the design of lenses and optical equipment. At the same time, he was a successful businessman who, at the tender age of 22, was made director of the glassworks in Benediktbeuren, which he then successfully managed (the telescope pictured below, manufactured for the University in Dorpat (today: Tartu, Estonia) was the largest and best of its day) (Fig. 1). Joseph von Fraunhofer also became the namesake of the Fraunhofer-Gesellschaft—after MIT, the second largest institution for applied science in the world.

The identity of Fraunhofer research is characterized by proximity to application, industrial relevance, and innovation. The Fraunhofer-Gesellschaft has recognized that research in applied mathematics not only serves as an aid to other scientific disciplines in the search for solutions to practical, in particular, technical and organizational problems. Mathematics also represents a discipline that is indispensible for maintaining economic competitiveness and meeting the challenges faced by society. It has evolved from being a key to

**Fig. 1** Joseph von Fraunhofer: researcher, inventor, and businessman (© Fraunhofer-Gesellschaft)

basic research and technology to being an enabling force for virtually every economically significant key technology.

The Fraunhofer-Gesellschaft, remaining cognizant of this evolutionary development, has added three new mathematics-based institutes to its ranks in the past decade:

- The Fraunhofer Institute for Industrial Mathematics ITWM, in Kaiserslautern,
- The Fraunhofer Institute for Algorithms and Computational Science SCAI, in Sankt Augustin, and
- The Fraunhofer Institute for Medical Image Computing MEVIS, in Bremen.

These institutes are dedicated, in terms of their research mission and focus, to application-oriented mathematics and the implementation of mathematics in society and industry.

Our book is dedicated to the mathematics practiced in the ITWM, whose spirit also prevails at the other institutes. It is problem-driven, model-based and solution-oriented. We will have more to say about this elsewhere in the book. If the goal is to highlight a unique feature associated with a particular "brand" of mathematics, then this is certainly the description "problem-driven, not method-driven." The style and structure of this book have been influenced by this attribute.

Beyond this, another motive was certainly to share the "success story" of the Fraunhofer ITWM. We want to illustrate how innovation in mathematics and the transfer of its results into the marketplace and society at large can be effectively carried out in a large research institute receiving relatively little basic funding. The success of the "ITWM model," as proven also by the formidable role played by mathematics in contemporary industry, might also serve as a motivating force for establishing similar institutions in other locations and other countries, adapted to the regional and national circumstances found there.

## 1 "Industrial Mathematics" Versus "Applied Mathematics"[1]

Many scientific disciplines profit from the solutions to practical problems developed through research in applied mathematics. As a rule, however, traditional, academically-oriented applied mathematics only examines and numerically treats problems that are also accessible to rigorous mathematical analysis; that is, problems for which existence and

[1] Portions of this introduction have been taken from the following publications:
H. Neunzert, U. Trottenberg: Mathematik ist Technologie – Ein Beitrag zur Innovations-Initiative aus Fraunhofer-Sicht, Fraunhofer ITWM und Fraunhofer SCAI, Kaiserslautern und Sankt Augustin, 2007
D. Prätzel-Wolters, U. Trottenberg: Rechnen für Fortschritt und Zukunft – Innovationen brauchen Mathematik, Jahresbericht der Fraunhofer-Gesellschaft 2007, S. 47ff., München 2008.

uniqueness statements for the solution and convergence statements for the applied numerical method, for example, can be proved. As a result, the problems treated in the mathematical literature are often highly idealized and not especially realistic.

The effective solution of large-scale, real-life problems only became the object of intensive mathematical research after technomathematics, economathematics, and computational science established themselves as new mathematical disciplines.

This practice-oriented mathematics, which further develops mathematical methods for the solution of specific problems and whose models and algorithms form the basis for simulating and optimizing complex products and processes, is at the heart of the mathematically oriented Fraunhofer Institutes. The fact that this research is far more than mere mathematics transfer is frequently underappreciated in the more academically-minded world found in universities. Here, one sometimes encounters the notion that such industrial-oriented mathematics is not "real" mathematics at all, or that the truly "new" mathematics is developed in universities—decoupled from practical application—and only after a time delay finds industrial application. The experience and expertise of the mathematically oriented Fraunhofer Institutes, gathered in extensive, long-term collaborations with industry, contradict these views.

Modeling and simulating the behavior of complex materials, for example, results in mathematically challenging problems involving the coupling of very diverse differential equations, such as those of fluid mechanics and Maxwell equations. This coupling represents a significant challenge, not only numerically, but also theoretically. The high-dimensional partial differential equations arising from the risk evaluation of financial securities require entirely new methods of numerical solution. The transition from smaller to larger scales can be tackled with homogenization methods, but only when the essential scales are well separated. When this is not the case—as happens in many practical applications, such as those involving turbulence or crack formation in materials under stress and in rocks—then there are currently only a few fruitful approaches for simplifying the models and/or the numerics. The digital interconnection of control systems demands new procedures for analyzing and synthesizing hybrid systems with continuous and discrete dynamics and logic based switching functions.

These few examples illustrate that substantial momentum for the development of "new" and "real" mathematics arises from treating complex, practical problems.

Nevertheless, the transfer of mathematics to the marketplace is a vital mission of the mathematically oriented Fraunhofer Institutes. Here, however, they don't restrict themselves to merely preparing general mathematical aids for the solution of practical problems, thus leaving the actual problem-solving to the users or to other technical software companies. Instead, they get involved themselves—in close cooperation with the users—to work towards a complete solution through the development of appropriate software modules. The goal of demonstrating a direct benefit to the economy, that is, of putting research results directly into practice with their industrial partners, is part of their identity and their mission. Here, it is accepted that the relevance of their research results is also reflected in the fact that the businesses making use of those results contribute substantially and directly to financing the costs of the research efforts. The Fraunhofer financing model assumes that an Institute will cover at least one third of its operating budget through business revenues.

However, to ensure long-term success in mathematics transfer, it is also essential to maintain contact with the frontlines of basic research and to actively pursue new mathematics oneself. Practical problems represent a wonderful source of new questions and methods that can then feed basic research in the Institutes.

In this context, the Institutes' joint ventures with other research institutions and universities, as well as with industrial partners in connection with projects publicly sponsored by the BMBF (Federal Ministry for Education and Research), the DFG (German Research Foundation), or the EU, for example, play a very significant role. They serve to build up new research areas and establish a trusting and cooperative working atmosphere with the participating institutions. The results of this research create innovation in economically and societally relevant fields of application and help finance the Institutes' knowledge-oriented basic research.

## 2 Problem-Driven or Method-Driven?

This view is reflected in the stereotype, still frequently encountered in public opinion, that mathematics is a difficult, dry, ivory-tower sort of endeavor. Mediocre or worse grades in school mathematics classes are accepted in society, where they are met with a shrug of the shoulders and commented upon sympathetically.

This attitude captures neither the fascination of mathematics as a playground for the mind nor its significance as a crucial instrument for shaping technological progress.

The mathematician himself is seen as a person who—cut off from the real world—performs his researches upon questions he has thought up himself, within the confines of his own system of thought. His research is driven by the methods and structures intrinsic to mathematics; solving practical problems doesn't interest him particularly. The ideal location for this endeavor is indeed the ivory tower, an intellectual refuge, inviolate and untouched by the world.

The ivory tower stands for the isolation of the scientist, who retreats from the events of the world and dedicates himself exclusively to pure research, paying no heed to either the practical uses or consequences of his investigations, but simply losing himself in his passionate pursuit of answers.

This image of the mathematician no longer fits into the research landscape of the 21st century. Applied mathematics has long-since abandoned the ivory tower, seized the computer as a tool of the trade, and addressed itself to the solution of practical, relevant problems. But it is a shortsighted view to assume that it was only through the computer that mathematics was finally rendered able and willing to solve practical problems.

Mathematics was always both: It was problem driven and it was method driven. It helped to solve practical problems, and it created culture, by following its own evolutionary path.

For the active participants in this process—mostly mathematicians—the past hundred years were dominated by the continued development of methods. This happened either in the pursuit of answers to questions that arose within mathematics—as in pure mathematics, such as with algebraic geometry—or in the pursuit of solutions to problems that typically manifest when dealing with practical questions—as in applied mathematics, such as with inverse problems. University mathematicians had, and still have, the privilege of being able to deeply immerse themselves for long periods of time in a particular class of mathematical problems.

Things were once different, however. Earlier, one's income depended on the successful treatment of problems posed from outside. Typical examples are the fluid dynamic problems that Euler needed to solve or the geodetic problems tackled by Gauss. And it is again different today; mathematics, with the aid of its tool, the computer, has become a technology in its own right, and a host of practical problems are standing in line, so to speak, outside its office door.

The doors of the Fraunhofer Institutes are opened wide to receive such problems, and the mathematics practiced there is driven very significantly by the need to solve them. This means that the focus of research is not on the further development of existing mathematical methods, but on the development of new methods for formulating and solving problems or the adaptation of known methods to the particular problem being addressed. The goal of solving the problem determines the direction in which the methods are developed and extended.

## 3 Model-Based and Solution-Oriented

Efficient mathematical treatment of practical problems calls for the preparation of "economical" mathematical models, as well as the development of efficient algorithms. A model is "economical" when it is as complex as necessary and as simple as possible. Often, the simplicity is also imposed by a desire for real-time simulations or because the simulations calculate the values of objective function(s) for an optimization task. Algorithms are efficient when they achieve maximal exactness on the computers at hand in the limited processing time available.

For most problems confronted in industrial practice, physics provides models. These are frequently continuum mechanical, thermodynamic, or electromagnetic equations, which very precisely describe the manufacturing processes of industrial goods or their behavior. Naturally, it is possible to describe the behavior of thousands of polymer fibers in the transition from fluid to solid phase in turbulent airflow. Or one can model very precisely at the particle scale the flow of a gas and the absorption of entrained particles by a porous medium.

However, even using high performance computers and the most modern algorithms, it is not possible to arrive at even a rough solution for these very complex equations. Presumably, this will not be possible decades from now either. But this isn't necessary, since one can simplify and reduce the models and still meet the specified precision requirements. The algorithms then have to be adapted to the model reductions, and vice versa: the first approximations in iterative solvers may work with simpler models; then, as the precision increases, the models themselves also become more precise. This interplay between model and algorithm is especially important for optimization tasks. Model reductions often deal with asymptotic analysis or multi-scale approaches, where small parameters are replaced by the limit value zero. Or they rely upon projection methods on lower-dimensional subspaces. It is also quite possible, however, that entirely new models based on a different mathematical theory are employed, for example, the use of stochastic models for very complex, deterministic behavior.

Because it is important when dealing with "real-life" problems to find usable solutions, the development of efficient algorithms, as already mentioned, also comes into consideration. Thus, multi-core approaches from modern computer architecture fit well together with multi-grid approaches, which, in turn, are often coupled with multi-scale models. Parallel algorithms also currently represent an important field. All of this, however, as is usually the case elsewhere, is not to be understood as "method for method's sake"—we repeat it once again here—but as problem-driven.

## 4 Mathematics as a Motor for Innovation in Technology and Society

The potential for applying mathematics is enormous. The scope of the mathematics that has found its way into industrial practice has grown explosively over the past 40 years. This can be explained for the most part by the fact that work with real models has been replaced by simulations, that is, by work with mathematical models. This development has been accompanied by the automation of work processes, sensory perceptions, and experiences in the form of algorithms, computer programs, or expert systems. Mathematics has become a key technology, one that can and should be mentioned in the same breath as nanotechnology or biotechnology.

At first blush, this may appear a rather audacious statement. It requires, at least, an explanation. To be sure, for thousands of years, natural scientists have used mathematics as a resource and as a language in which to formulate their theories, and it has formed the basis for the computations of the world's engineers. Thus, it is at least a raw material—the raw material of models that are then converted into technology. But simply being a raw material is not enough to qualify as a key technology. It is the computer that has elevated mathematics to the rank of technology. In a certain respect, the computer is the purest form of mathematics-turned-technology. Mathematics has taken on a material form in the guise of the computer, and it represents the intellect behind every computer simulation. Simulations need models and algorithms to evaluate and visualize their results. On closer inspection, it is always mathematics serving as the basis, as the "source code," so to speak, for these critical work steps.

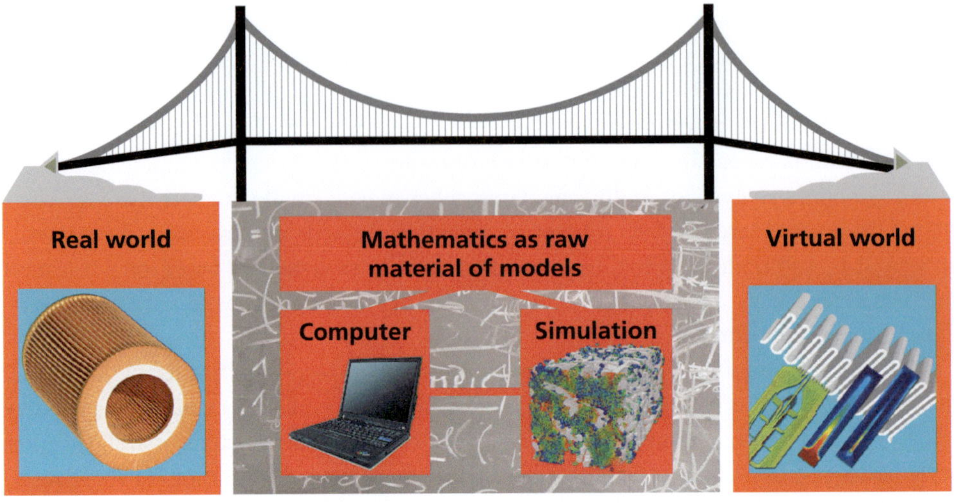


**Fig. 2** Mathematics is a key technology (Graphic: S. Grützner, Fraunhofer ITWM)

The computer has altered our world. In the view of the cultural philosopher Ivan Illich (1926–2002), it has become a universal and convivial tool. Computer simulations—and thus also mathematics—represent the essential tool for shaping and optimizing products and work processes.

Real models are being replaced by virtual models. Mathematics, as raw material and key technology, forms the foundation for a bridge to this second world—the world of virtual simulations—which has found a foothold in almost every area of our society and economy (Fig. 2).

## 5 Mathematics Is Universally Applicable, Because It Traverses Boundaries

This universal applicability stems from the fact that mathematical methods and tools developed for one sphere of reality or science can also be made useful in other areas of application, either directly or in an analogous form. Mathematical models fit horizontally into a landscape of scientific disciplines and technological applications that are arranged vertically. This transverse quality of mathematics makes it a "generic" technology.

The ideas developed in one area can bear fruit in others. In keeping with this motto, mathematics creates cross-links between disciplines and makes comprehensive insights possible. "Out-of-the-box thinking," as a characteristic of the mathematical approach to work, creates innovation by layering different levels of reference.

Mathematical models are in demand; solutions require simulations. As a rule, there is not just one solution, and mathematical optimization is also required to find the best ones. The abbreviation for this triad of Model-Simulation-Optimization is MSO. MSO is anchored in the research and development departments of today's large technology companies as its own field of competence, and is occasionally even part of the organizational structure of such companies. In practically all mathematics-based, practically-oriented research projects, MSO is an integral component of project work.

All in all, one may easily speak of a quantum leap over the past decades in the visibility of mathematics as an engine for innovation in technology and society.

There is a great deal of evidence for this development—a development that, in the interim, has become entrenched in the fields of politics, science, and industry.

### 5.1 Committee for Mathematical Modeling, Simulation, and Optimization (KoMSO)

This committee was established in connection with the "Strategic Dialog on Mathematics," an initiative of the German Ministry for Education and Research (BMBF). Its goal is...

> "*...to anchor the triad of mathematical modeling, simulation, and optimization in research and development as a new field of technology, in order to strengthen the innovative power of the technology nation Germany. Research and innovation are the foundation of prosperity for all of society. Therefore, the potential of MSO, which has remained undiscovered or only partially exploited up to this time, must be tapped into and made visible.*"

And, as also found in the strategy paper of the BMBF Strategy Commission:

> "*Improved mathematical methods and continuously improving computer performance make increasingly complex physical-technical, economic, or medical questions accessible to description with mathematical modeling, virtual simulation in computers, and optimization relative to a given technological goal. In this way, the most diverse simulation techniques have become as thoroughly established—as a third pillar of knowledge acquisition—as theory and experiment for optimizing automation and decision-making in an increasingly complex and interconnected world.*"

### 5.2 "Mathematics—Engine of the Economy"

The book "Mathematik – Motor der Wirtschaft"[2], published in the Year of Mathematics in 2008, and produced in close cooperation with the Oberwolfach Stiftung (Oberwolfach Foundation) and the Mathematisches Forschungsinstitut Oberwolfach (Oberwolfach Research Institute), contains, among other things, a series of contributions from prominent

[2] *G.-M. Greuel, R. Remmert, G. Rupprecht (Eds.): Mathematik – Motor der Wirtschaft, Springer-Verlag, Berlin, Heidelberg, 2008.*

representatives of German industry. The book illustrates that mathematics is today of great significance in virtually all branches, in all areas of industry, business, and finance. For example, Peter Löscher, former chairman of the board of Siemens, Inc., writes:

> *"Mathematics—this is the language of science and technology. This makes it a driving force behind all high technologies and, thus, a key discipline for industrial nations. Without mathematics, there is no progress and no technical innovation."*

Or, to quote Dieter Zetsche, Chairman of the Board of Daimler, Inc.:

> *"As does no other science, mathematics helps us in our branch to solve the most varied sorts of problems—and it is exactly this universal applicability that makes it the royal discipline."*

Of course, not all the companies surveyed by the Oberwolfach Stiftung responded. Nor was the vast economic sector of small and medium-sized businesses included, although these companies are responsible for the lion's share of German economic power. Nevertheless, there can be absolutely no doubt about the general validity of these statements.[3]

There are indeed other studies that confirm them completely: For example, "MACSI-net roadmap," published in 2004 by ECMI (European Consortium for Mathematics in Industry); "Mathematics: Giving Industry the Edge," published in 2002 by the Smith Institute at Oxford; and "Forward Look: Mathematics in Industry," a report prepared in 2010 by the European Science Foundation in cooperation with the EMS (European Mathematical Society). The experience of the mathematical Fraunhofer Institutes, whose mission is, after all, research cooperation with industrial partners, also lends support to the argument.

### 5.3 ECMI

Since 1986, the "European Consortium for Mathematics in Industry ECMI," to which many European institutions belong—including groups in Barcelona (E), Dresden (D), Eindhoven (NL), Florence (I), Glasgow (GB), Gothenburg (S), Graz (A), Grenoble (F), Kaiserslautern (D), Lappeenranta (FIN), Limerick (IRL), Linz (A), Lund (S), Lyngby (DK), Madrid (E), Milan (I), Oxford (GB), Sofia (BG), Trondheim (N), and Wroclaw (PL)—has endeavored to emphasize the significance of mathematics for European industry and organize the training and cooperation of European "industrial mathematicians."

Germany's applied mathematics enjoys an outstanding position internationally; it is one of the few areas in which Germany ranks globally among the top three nations. In industrial mathematics, Europe as a whole and Germany in particular are also at the forefront; the

[3]Cf. H. Neunzert: Mathematik ist überall – Anmerkungen eines Mathematikers zu den Beiträgen der Wirtschaftsunternehmen in G.-M. Greuel, R. Remmert, G. Rupprecht (Eds.): Mathematik – Motor der Wirtschaft, Springer-Verlag, Berlin, Heidelberg, 2008.

USA and Asia orient themselves for the most part on European examples. Here, once again, there is ample evidence to back up this claim.

The DFG sponsors a collection of graduate schools, memberships in excellence clusters, and collaborative research centers that have a strong link to applied mathematics. Applied mathematics is also strongly represented in the BMBF's large flagship projects, such as the Leading-edge Clusters and the Research Campus Program.

## 5.4 Berlin

In the past few decades, Berlin has evolved into a nationally and internationally recognized center of excellence in the area of applied mathematics. The Konrad-Zuse-Zentrum für Informationstechnik Berlin (ZIB) is one of the most successful institutes in scientific computing and has an excellent global network. ZIB is also home to the only mathematically oriented research campus "MODAL—Mathematical Optimization and Data Analysis Laboratories." Along with the graduate school "Stochastic Analysis with Applications in Biology, Finance and Physics" and the "Berlin Mathematical School," the DFG Research Center "Matheon—Mathematics for Key Technologies: Modeling, Simulation, and Optimization of Real Processes" is the German applied mathematics center having the widest international recognition. Matheon is supported by the mathematics institutes of the Technische Universität Berlin (TU Berlin), the Humboldt-Universität zu Berlin (HU Berlin), and the Freie Universität Berlin (FU Berlin), as well as by the ZIB and the Weierstrass Institute for Applied Analysis and Stochastics (WIAS) (see, also[4]).

Since 2010, Berlin, with the WIAS, has also been the permanent headquarters of the Internationale Mathematische Union (IMU), an umbrella organization for 77 national mathematical societies. Among other activities, it supports education and research in developing countries and organizes the International Congress of Mathematicians (ICM), the largest conference in the field of mathematics and venue for awarding the Fields Medals.

## 5.5 Kaiserslautern

The mathematics department at the TU Kaiserslautern (Technical University of Kaiserslautern) has acquired an outstanding global reputation by virtue of its research activities in theoretical and practical mathematics and its innovations in education. The curriculum of technomathematics was "invented" and conceived in Kaiserslautern, and the department here was one of the first in Germany, after Ulm, to introduce the economathematics

[4] P. Deuflhard, M. Grötschel, D. Hömberg, U. Horst, J. Kramer, V. Mehrmann, K. Polthier, F. Schmidt, C. Schütte, M. Skutella, J. Sprekels (Eds.): MATHEON—Mathematics for Key Technologies; EMS Series in Industrial and Applied Mathematics 1, European Mathematical Society Publishing House, Zürich 2014.

curriculum. Both fields of study have become successful curricula within Germany and developed into especially strong focal points in Kaiserslautern. The DFG has been a past sponsor of two graduate schools in mathematics in Kaiserslautern and a third, "Stochastic Models for Innovations in the Engineering Sciences," has just been approved.

With regard to its mathematics programs, the TU Kaiserslautern is among Germany's elite universities. This is evidenced by the university rankings, compiled by the CHE (Center for Higher Education Development) and the magazines Focus, Stern, Spiegel, and Zeit since 2003, in which mathematics in Kaiserslautern has always been placed in the top group.

Over the past five years, in connection with a mathematics initiative sponsored by the State of Rheinland-Pfalz, the TU Kaiserslautern, and the Fraunhofer ITWM, urgently needed specialists in differential-algebraic equations, image processing, biomathematics, and stochastic algorithms have been brought to Kaiserslautern.

The Fraunhofer ITWM emerged from the technomathematics working group and was the first mathematics institute to join the Fraunhofer-Gesellschaft. Today, with its yearly industrial revenues of more than 20 million euros and some 260 full-time employees and doctoral students, it is one of the largest applied mathematics institutes in the world.

The Institute is continually receiving new impulses for innovation from its cooperative efforts within the mathematical department of the TU Kaiserslautern. By the same token, the department is closely affiliated with the ITWM by virtue of third-party projects and doctoral programs, and research within the department is stimulated by the project-driven topics of the ITWM. Unfortunately, this close affiliation is not always perceived publicly and we encounter the misconception that there is a mathematics department very nicely situated in basic research and a Fraunhofer Institute that successfully transfers mathematics to industry, but the two have little to do with each other. The supposed separation into basic research within the department and mathematics transfer at the ITWM does not correspond to reality. The ITWM performs its own basic research within applied mathematics on a large scale. Between 2000 and 2013, for example, 150 PhDs and habilitations were successfully completed in the Institute and its immediate environment. Naturally, these degrees were granted by the TU Kaiserslautern.

In order to further strengthen the connection between the mathematical department and the ITWM, the "Felix-Klein-Zentrum für Mathematik" (FKZM) was founded in late 2008, in connection with the Rheinland-Pfalz "Mathematics Initiative." The center was named after the important mathematician and scientific promoter Felix Klein (1849–1925). This name was selected, because Felix Klein united Germany's pure and applied mathematics like no other mathematician in history, reorganized mathematics in Germany 100 years ago, built a solid bridge to industry, linked academic and school mathematics, and celebrated and promoted the history of science—all activities that have served and still serve as pole stars for Kaiserslautern's mathematicians. Therefore, the FKZM offers a platform and provides infrastructure for joint research projects, guest programs, scholarships, and school outreach activities. Last, but not least, the FKZM represents a forum for cooperation with other departments and industry.

## 5.6 Other Activities in Germany

It would exceed the scope of this introduction to offer detailed descriptions of all the sites in Germany at which applied mathematics plays a prominent role.

Heidelberg is certainly another exceptional location for applied mathematics, where great emphasis is placed on cooperation with industry. Along with the Heidelberg Graduate School of Mathematical and Computational Methods for the Sciences (HGS MathComp), the Interdisciplinary Center for Scientific Computing (IWR)—a research institute of the Ruprecht-Karls-Universität Heidelberg—is among the world's largest university-based centers for scientific computing. The previously mentioned Fraunhofer Institutes for Algorithms and Computational Science SCAI, in Sankt Augustin, and for Medical Image Computing MEVIS, in Bremen, along with the Max Planck Institute for Mathematics in the Sciences, in Leipzig, are all German centers of applied mathematics. In addition, there are numerous locations with well-funded chairs, state institutions, and special research areas that have also helped carve out Germany's applied mathematics landscape. Some examples are Bremen, Paderborn, Munich, Erlangen, Bonn, Stuttgart, Freiburg, Saarbrücken, Wuppertal, and Dresden—and this list is far from complete.

To conclude this introduction, we would like to offer the reader a bit of information about the design of this book and the various areas of focus in the individual chapters.

# 6 The Design of This Book

In structuring our book, we have kept in mind various groups of potential readers:

- Practitioners and interested laypeople who want to inform themselves—without having to dive into the technical details—about what today's mathematics can offer toward the solution of practical problems.
- Professional mathematicians and university-level mathematics students who want to understand the mathematics developed at the ITWM.
- Teachers, younger students of mathematics, and instructors or tutors who want to understand how to integrate the new image of mathematics into their school systems.

The triad "problem-driven–model-based–solution-oriented," has determined this book's design. The section entitled "The Concepts" (Part 2) is dedicated to the following superordinated topics:

- Mathematical modeling
- Computation
- Data analysis
- Optimization processes

These chapters serve to provide an overview of the essential questions, methodological approaches, strengths, and potentials—along with the weaknesses and limitations—of each topic. They are addressed to both practitioners and interested laypeople, as well as to professional mathematicians and university-level mathematics students. The aim here is not to offer a mathematical representation of specific models or algorithms. Instead, the chapters comment upon and give structure to the work of the Institute, work that has culminated in the research descriptions found later in the book. For this reason, these chapters tend to be written in more of a "prose" style.

This approach can be attributed to the fact that the mathematics of the ITWM is problem-driven, which means that the reality described by our models is much more complex than that forming the foundation of academic works. There are more complicated boundary conditions, the materials are non-homogeneous, the objective functions are not immediately clear, and the models must be simplified in order to make their application really practicable. All these aspects are discussed in the overview chapters. In addition, important models or algorithms that don't happen to appear in the "research" chapters presented later are also addressed briefly.

Significant results from the mathematics originating in the various ITWM departments are then introduced in the following five chapters under the rubric "The Research." These department-related chapters serve as prototypes of the model-based, problem-driven, and results-oriented mathematical research of the ITWM. They come nowhere near to providing a complete overview of the projects and results achieved during the past 20 years of research at the ITWM. Rather, they serve as examples from working areas that are especially suited to illustrate the unique flavor of industrial mathematics. All five chapters are structured in a similar fashion. The first three sections of each chapter are written so as to also be understandable to interested laypeople having no pertinent knowledge of mathematics.

In contrast, the fourth section is aimed principally at mathematicians. It comprises a "self-contained," compact mathematical presentation for one or two problem areas, addresses the mathematical challenges, and describes the significant results, including their relevance for the "problem solution." The remaining sections discuss simulations based on the previous results and round off each research chapter with descriptions of specific examples arising from "practice" that have been addressed in the joint projects.

The various chapter sections focus on the following questions:

**Basic structure of the research chapters**

1. Why is the industrial partner coming to us?
   - What are the industrial problems and challenges found in a particular area of focus in the department?
2. What are the mathematical challenges?
   - Which mathematical methods are needed and which results are available for solving these problems?
   - Why is the existing mathematics very often insufficient; i.e., why is it not simply a question of mathematics transfer?
3. What was achieved in the department?
   - What are the primary topics focused on in the department and what results were achieved?
   - What is the impact of doctoral dissertations and graduate theses and who are the visible cooperation partners and customers?
4. What problem-oriented mathematical results were achieved?
   - What results were achieved and to what extent are they relevant for the "problem solution"?
   - What works and what doesn't work?
5. How do the results apply in actual practice?
   - What is handed over to the customer in the end for his specific problems?
   - Are there simulation tools offered?

The final part, "The Training," containing the chapter entitled "Applied mathematics in schools—made in Kaiserslautern," is aimed primarily at high school teachers and students. As mentioned at length previously, recent decades have seen a quantum leap in the visibility of mathematics as an engine for innovation in technology and society. Unfortunately, the new role of mathematics as a key technology has not yet been recognized in our school systems. Mathematical modeling, computing for the solution of existing, practical problem, and interdisciplinary projects are hardly ever found in schools. Of course, one does find so-called "word problems," but these very rarely describe authentic problems whose relevance is clear to learners and who might thus be excited about finding solutions. Algorithms are introduced in schools, granted, but hardly any that have been developed in recent years to tackle large-scale challenges.

The MINT subjects (from German M = Mathematics, I = Informatics, N = Natural Sciences, T = Technology; in English maybe better STEM) are not sufficiently integrated into either the curriculum or into classroom practice. Learners perceive lessons in the various MINT subjects as sequences of contents and tools that often fail to make clear overarching relationships—even within a given subject. The linking of the subjects with each other happens even less often.

One reason for this is the way teachers are trained: to date, applied mathematics has been assigned a rather humble position in the teacher training curriculum. Neither modeling nor work with algorithms—which, in practice, have widely replaced the use of complicated formulas—play a role in school. Similarly, within the education curriculum, the interdisciplinary interplay of mathematics, computer science, the natural sciences, and technology is neither discussed nor trained adequately.

In the final chapter of our book, we want to present some ways and means for reforming instruction, both in our schools and in the training and continuing education programs for teachers, as they have been practiced for several years in Kaiserslautern.

After a short, application-oriented introduction into mathematical modeling, we will point out which measures can be adopted to bring learners into closer contact with applied mathematics and interdisciplinary work. Here, we will present both intracurricular and extracurricular events.

Activities such as "modeling week," "modeling day," and competitions can be used to offer pupils the opportunity, within the framework of a compact project, to become more closely acquainted with the role of mathematics, to actively and creatively practice mathematics, and to witness interdisciplinary connections. The sample problems serve as invitations to interested teachers to integrate modeling into their lessons. In the "Junior Engineer Academy" and the nation-wide "Fraunhofer MINT-EC Math Talents" program, participants have a chance to experience, over a longer period of time, a new philosophy of linking education with practical application.

We also offer pointers for the education and continuing education of future teachers that will help them to structure their lessons and additional intracurricular activities accordingly. For this purpose, prepared lesson material is far less important than the necessary technical training and a positive attitude towards using new methods and ways to address questions that have no clear-cut right or wrong answers. The information about the didactical integration of new instruction methods is designed to explain the impact and point out ways to connect new with traditional instruction.

## 7 A Brief Portrait of the Fraunhofer Institute for Industrial Mathematics ITWM

The Fraunhofer ITWM was founded by the working group "Technomathematics" from the University of Kaiserslautern. As a research institution belonging to the State of Rheinland-Pfalz, it was, from the beginning, under Fraunhofer administration. After a successful evaluation in 1999, it advanced to the status of the first mathematical research institute of the Fraunhofer-Gesellschaft, thus, becoming part of one of the world's largest and most successful research organizations (Fig. 3).

**Fig. 3** Institute building of the ITWM at the Fraunhofer Center in Kaiserslautern (Photo: G. Ermel, Fraunhofer ITWM)

As a mathematics institute, the ITWM has remained committed to one of civilization's oldest sciences while, at the same time, developing into one of the most successful institutes in the Fraunhofer-Gesellschaft, as measured by its economic revenues. The basis for this balancing act has been the previously mentioned, dramatic increase in the relevance of mathematics for all production, service, and communication processes in modern industry.

After 20 years of effort, the vision with which the ITWM began—to transport mathematics out of the ivory towers and cathedrals of pure science and transform it into a key technology for innovation in technology and business—has become realized to a significant extent. This vision was not always universally applauded. Hardly anyone would have believed at the time of the Institute's founding that, in so short a span of time, such a large and successful Fraunhofer Institute of Mathematics would develop out of the seeds of technomathematics and economathematics from the University of Kaiserslautern.

The warnings often had the ring of "modern technology needs mathematics, but not mathematicians; it remains the domain of engineers and scientists." In the interim, a reversal in thinking has taken place here.

In the past 30 years, the scope of the mathematics that has found its way into industrial practice has grown exponentially. The essential reason for this is that work on real models has been replaced by simulations, that is, by work on mathematical models. Augmenting this development has been the automation of work processes, cognitive capabilities, sense perceptions, and experiences in the form of algorithms, computer programs, and expert systems. The materialization of mathematics in computers and software programs has also played a role. As a raw material for models and the core of every simulation program, mathematics serves as a key technology and forms the foundation of the bridge to the world of simulations—a world based on the highly efficient assistance of the computer, a tool that has gained a foothold in nearly every sphere of our society and economy.

Research and development projects with industry, preparation of customized software solutions and systems, and support with the use of high performance computing technology are integral building blocks of this transformation. The projects of the ITWM reflect a broad range of clients, from low tech to high tech companies, from small and mid-sized companies to industrial heavyweights, from regional businesses to customers throughout Europe and overseas. Industry appreciates and needs the Institute's modeling competence, its algorithms, and its software products. Significant economic revenues, coupled with a strong emphasis on research—62 doctoral students are working on their dissertations at the Institute in 2014—form the basis for sustainable success and continuous growth.

Since its founding in late 1995, the ITWM has attracted more than 81 million euros' worth of industrial projects and almost 51 million euros' worth of publicly sponsored projects. In the past three years alone, more than 700 industrial projects have been successfully completed.

This is proof that there is a great demand on the part of industry for innovative mathematics and, simultaneously, that industrial problems can serve as a driving force for developing innovative mathematical methods and tools.

The ITWM budget has grown continuously since the Institute's founding and reached a total of more than 22 million euros in 2014; almost half of that is financed by industrial projects. This establishes the ITWM as one of the world's largest institutes in the area of applied and industry-oriented mathematics.

One quarter of ITWM business revenues come from contracts with small and mid-sized businesses. One third of ITWM business projects are contracted with regional businesses, and a further third with companies outside of Germany.

Analysis of ITWM's industrial projects reveals several trends that, in our view, are not attributable to local or regional effects, but have a general validity:

- Mathematical modeling, simulation, and optimization are in demand by large companies in all business sectors.
- The use of mathematical methods is also a significant innovation factor for small and mid-sized companies.
- The transfer of mathematics to industry is subject to globalization.
- Regional companies represent a large customer potential.
- Small batch sizes predominate in the projects.

The ITWM boasts a broad customer spectrum: the main sectors involved are plant and machine construction; the automobile industry; the plastics, metal, and mineral processing industries; information and communication technology; the wood, paper, and printing industries; microelectronics; medical technology; the pharmaceutical industry; the chemical industry; technical textiles; banks; and the insurance industry. Many of the projects involve large companies traded on the German Stock Exchange (DAX). In the automobile industry, the ITWM cooperates with all of the domestic companies and many foreign manufacturers as well. The ITWM works with a problem-oriented approach in a project landscape that encompasses the most varied business sectors and that allows, due to the cross-linking character of mathematics, an efficient transfer of methods. This results in structural stability and makes the Institute resilient in the face of economic downturns in any given industrial branch.

Many small and mid-sized companies are subject to enormous competitive pressures and take advantage of the ITWM's modeling and simulation competence to help them cope. The vanguards in this development have profited in the marketplace by using simulations as proof of innovation and quality assurance in their products. The fact that computing power can be purchased more cheaply every year has helped small and mid-sized companies, who have more limited financial resources. Here, it is not investments in computers, but in the relatively expensive software, that is the bottleneck. Moreover, technically qualified personnel must also be hired to support the ever more complex software programs. Because small and mid-sized companies often have little or no in-house R and D, the use of simulations frequently means hiring additional staff, which results in permanent costs. Along with this economic factor, the psychological challenge of giving up tried and tested, mainly experiment-based procedures—where one can always see and measure the results—and replacing them with simulations—where one must put faith in the computer and software tools—still occasionally impedes progress in project business. However, when implemented correctly, simulations are extraordinarily reliable. This, along with their almost limitless flexibility, will sooner or later convince everyone, which means that the potential for cooperative ventures is enormous.

Businesses located here and in the surrounding region use the ITWM's competences. In 2013, almost one third of the projects were carried out with cooperation partners from Kaiserslautern and environs, although these were predominantly small and mid-sized companies. This shows that a mathematics-based research institute can also significantly support the regional economy in the field of R and D and promote innovation.

The globalization of the economy is reflected in the ITWM's contracting partners. The portion of industrial revenues from projects with foreign partners has grown to more than one third. Many customers are based in Europe, but cooperative ventures with companies in the USA and Asia are becoming more and more significant.

With respect to the associated marketing efforts, the long-term planning of staff utilization and competence development, and the minimization of administrative costs, the ITWM's ideal partner is one who signs a multi-year contract with us that covers the execution of individual projects. Existing customers that meet this profile are very valuable to the Institute.

Small and mid-sized companies usually contract for single, smaller projects. Many companies which, when taken together, provide us with a large volume of contracts, have various R and D departments that sign separate contracts with the ITWM for their specific projects. New customers tend to first test the competence and capabilities of the Institute by means of smaller feasibility studies and computational jobs. As a result, the number of projects being processed yearly at the ITWM has grown to more than 250, and the average economic volume of the projects in 2013 was just under 40 000 euros. The large number of follow-up projects is a clear sign of the quality of our project work and a source of great satisfaction.

## 7.1 Which Competences and Structures Are Needed to Successfully Transfer Mathematics to the Market Place?

The cornerstones for successfully transferring mathematics are the classical disciplines of applied mathematics: numerics, optimization, stochastics and statistics, differential equations, and mathematical modeling. These are augmented by such strongly mathematically-oriented theoretical fields as 3D differential geometry, continuum mechanics, electrodynamics, system and control theory, financial mathematics, inverse problems, and image and signal processing, which have evolved into boundary fields between mathematics and technology over the past decades (Fig. 4). They are indispensible constituents for successfully carrying out application projects.

The ITWM's main field of activity consists of transforming mathematics that is applicable into mathematics that is actually applied: We adapt theorems and algorithms to models that come from actual practice and we convert optimal solutions that exist in theory into practicable solutions that can exist in reality. However, this transformation requires specific competences above and beyond the aforementioned cornerstones in order to build actual bridges to the virtual world. In relation to the processing of available experimental and observational data, they consist of setting up the mathematical model; transforming the mathematical solution of the problem into numerical algorithms; combining data, models, and algorithms in simulation programs; optimizing solutions in interaction with the simulation; and, finally, visualizing the simulation runs in the form of images and graphics. The

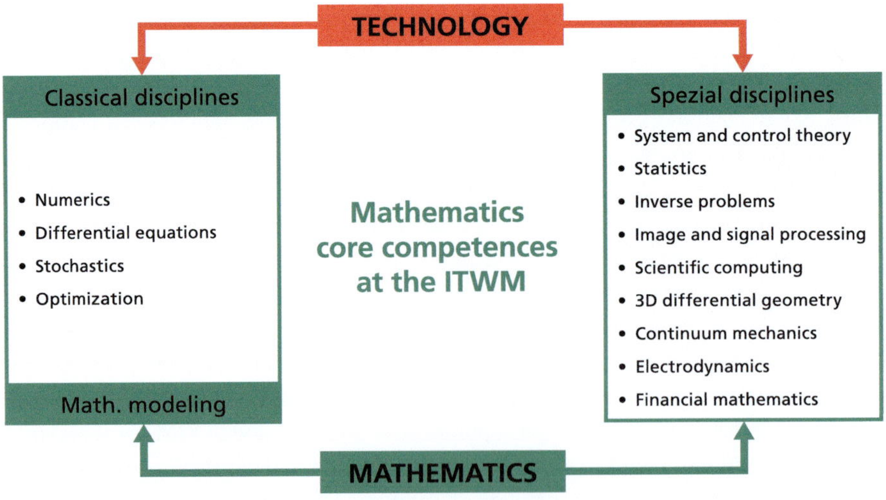


**Fig. 4** Mathematics core competences at the ITWM (Graphic: S. Grützner, Fraunhofer ITWM)

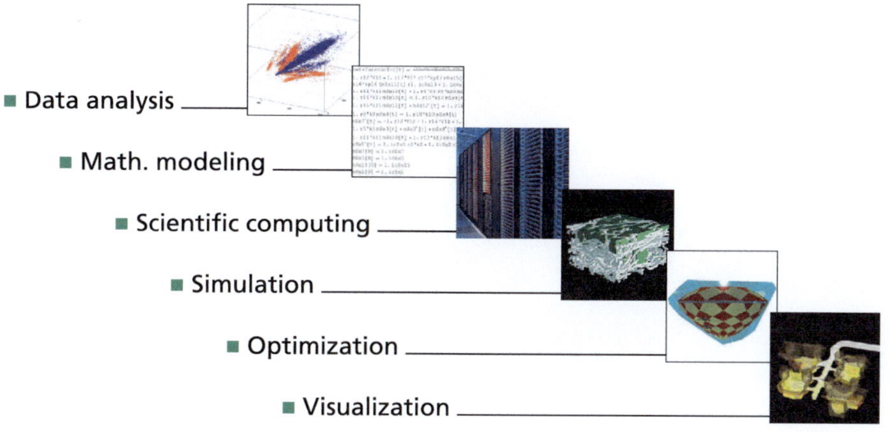


**Fig. 5** Process chain at the ITWM (Graphic: S. Grützner, Fraunhofer ITWM)

competences needed to build this process chain represent the ITWM's core competences (Fig. 5).

This entire process chain is frequently subsumed under the term "numerical simulation." In the past 20 years, the increasing performance capacity of computers has opened up entirely new possibilities for industrial simulation tasks. More and more, computer networks are achieving central significance. There has been a dramatic paradigm change with regard to generating the highest computational performance for industrial applications. PC clusters, multi-core systems, and cloud computing are replacing super computers. Parallel computing systems, which just a few years ago were found only in a few meteorological research centers, have now made their way into industrial settings. Adapting numerical algorithms to these rapid changes in hardware configuration is still a troublesome bottleneck in the complete realization of the performance potential of these new computing systems.

The full process chain is illustrated in very many ITWM projects. One of the Institute's great advantages is that all these competences are available in-house and their utilization in projects can be centrally planned. The original team of 34 scientists, PhD students, and staff in the centralized areas has grown into the current team of 260 employees. All in all, 170 scientists, most of them with doctorates in mathematics, but also coming from the fields of physics, engineering, and informatics, process a multitude of topics and develop simulation software (cf. Appendix: The Fraunhofer Institute in numbers). In contrast to project execution in university settings, there is no need for coordination and reconciliation of content and timing between working groups from different chairs. Zones of responsibility and authority, along with schedules and delivery of work packages, are already clearly defined during bid preparation.

At the same time, in order to maintain contact with the frontlines of research and remain competitive with other research institutions in the marketplace, it is necessary to continuously reflect on how our own focal points, ideas, and goals mirror events in research and development outside the ITWM microcosm (Fig. 6).

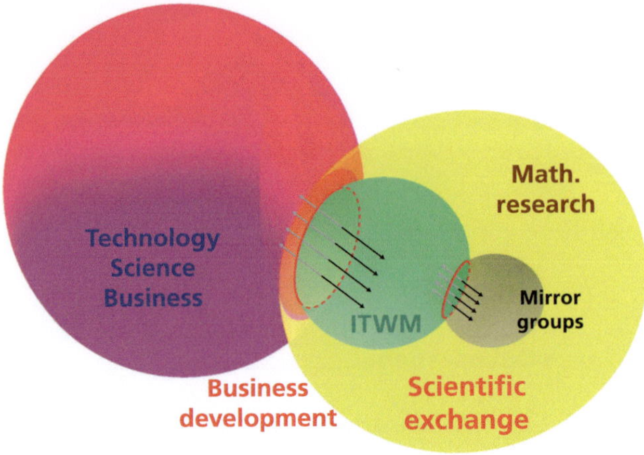


**Fig. 6** Scientific exchange (Graphic: S. Grützner, Fraunhofer ITWM)

The research at the ITWM is very tightly integrated with the research in the TU Kaiserslautern Department of Mathematics. At the University, there are counterparts to the groups working in the Institute's primary areas of focus. The University also participates in the State's research focus area CM2 and the graduate school "Stochastic Models for Innovations in the Engineering Sciences." Beyond this, there are cooperative projects with many chairs in the Informatics, Mechanical and Process Engineering, Civil Engineering, Electrical Engineering, and Information Technology Departments, including, for example, projects in the innovation center "Applied System Modeling" and in the Kaiserslautern "Science Alliance."

The bridging technology of mathematics is also reflected in a multitude of cooperative projects between the ITWM and other Fraunhofer Institutes. The ITWM is one of the most profitable members in the Fraunhofer ICT Group (Information and Communication Technology) and also enjoys the status of a guest institute in the Fraunhofer Group for Materials. Moreover, the Institute is a member of the Fraunhofer Alliances Automobile Production, Batteries, Big Data, Cloud Computing, Lightweight Construction, Numerical Simulation of Products and Processes, Transportation, and Vision, as well as of the Fraunhofer Innovation Cluster "Digital Commercial Vehicle Technology."

In association with other institutes, the ITWM participates in a series of larger in-house Fraunhofer cooperative projects. Here, we contribute our mathematically oriented competences, which, as a rule, complement those of the partner institutes. All told, the ITWM is one of the best-connected institutes in the Fraunhofer-Gesellschaft.

The ITWM's international network manifests itself also in the current research cooperative ventures with many foreign universities and research institutions, in the numerous foreign guest scientists working here, and in the extensive participation of ITWM scientists in scientific committees and in the publication of technical journals.

## 7.2 Departments, Business Areas, and Customers

The departments serve to structure the Institute's business areas, not always with perfectly sharp divisions, but with sufficient specificity for differentiation purposes. The matrix structure found in many institutes was consciously avoided in order to have few hierarchic levels in the ITWM and to minimize the internal coordination processes necessary in business development and project work. As a rule, the departments have at their disposal the relevant competences needed to serve the business areas they address.

It is beyond the scope of this introductory chapter to offer detailed descriptions of the competence and customer profiles of the various departments. Five departments have made significant contributions to this book, and the chapters in "The Research" that were prepared by these departments offer a glimpse into the work they have conducted. The ITWM's pallet of customers is also far too extensive to offer a complete accounting.

From 2009 to 2013, the ITWM processed 1070 industrial projects. The following short overview illustrates, using 2013 as an example, the Institute's diverse branch and customer pallet.

- Business sectors:
  Vehicle industry, general mechanical engineering, energy and raw materials, chemicals, financials, manual trades, information and communication technology, medical technology, and textiles.
- Customers:
  Accenture CAS GmbH, Assyst GmbH, AUDI AG, AUTEFA (A), BASF SE, BMW Group, BPW Bergische Achsen Kommanditgesellschaft, ClusterVision (NL), Daimler AG, DZ-Bank (L), ebm papst, FLSmidth Wadgassen GmbH, Freudenberg Filtration Technologies, Görlitz AG, IBS FILTRAN GmbH, John Deere, Johns Manville Europe GmbH, K + S Kali, Klinikum Essen, Liebherr, LONZA Group AG (CH), Lundin (N), M + W Process Industries GmbH, Marathon Oil (USA), Math2Market GmbH, MTU Aero Engines GmbH, Paul Wild OHG, proALPHA Software AG, Procter & Gamble (USA), Progress Rail inspection & information systems, Repsol (USA), Robert Bosch GmbH, Seismic Image Processing Ltd (GB), SGL Carbon, SIEDA GmbH, Siemens AG, Statoil (N), Teckpro AG, Voith Hydro, Volkswagen AG, Volvo CE (S), Woltz GmbH.

## 7.3 Cooperation with the Fraunhofer-Chalmers Center for Industrial Mathematics FCC

The ITWM was one of the first Fraunhofer Institutes to implement the recommendation of the Fraunhofer Board to promote internationalization in Europe. In Gothenburg, Sweden, the Fraunhofer-Chalmers Center for Industrial Mathematics FCC was successfully established as a joint venture between the Chalmers Technical University and the ITWM

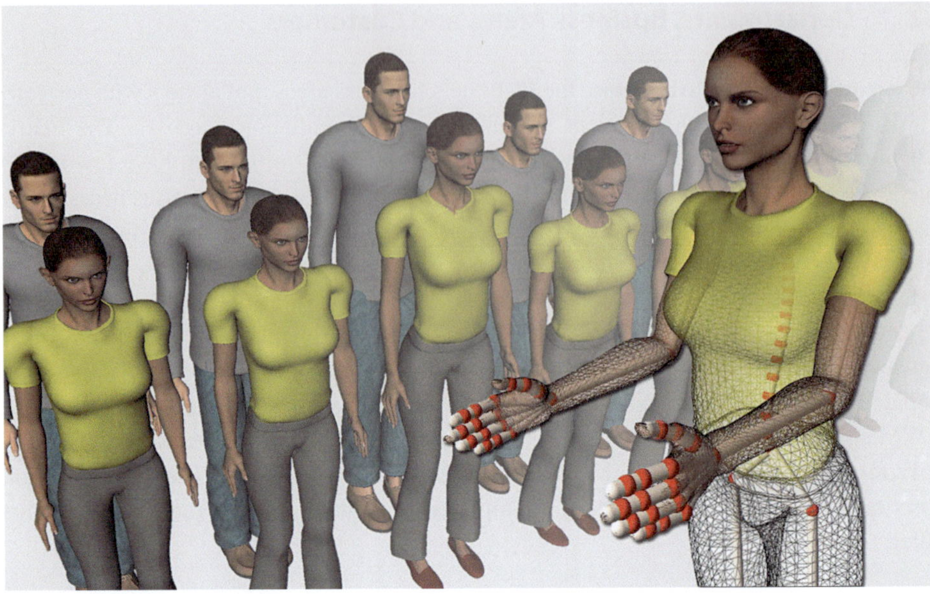

**Fig. 7** The Software IMMA$^{TM}$—Intelligently Moving Manikins—utilizes families of manikins in order to accommodate the majority of the population. The manikins are used for evaluating assembly ergonomics (Graphic: FCC, geometry mesh: Poser®)

(Fig. 7). Today, 51 employees generate an operating budget of almost 41 million Swedish kronor (approx. 4.3 Million euros). The Institute, with its departments

- Geometry and Motion Planning
- Computational Engineering and Design
- Systems and Data Analysis

was founded in 2001, and since that time has developed into one of Sweden's most renowned centers for "industrial mathematics."

## 8 Summing up the ITWM

Today, the ITWM already numbers among the largest institutes in the field of applied and industry-oriented mathematics. Its mission is to be the spearhead of mathematics in industry, with particular focus on small and mid-sized companies.

It will strengthen and enlarge this position and continue to contribute its part to making mathematics a key technology in industry and business. The outstanding connection with the TU Kaiserslautern in research and education guarantees proximity to current research topics, particularly (but not only) in applied mathematics and represents an important resource for attracting talented, young scientists. The ITWM's integration in the

Fraunhofer-Gesellschaft, its participation in a number of international cooperative ventures, and the close collaboration with its affiliated institute FCC in Gothenburg are also among the Institute's strengths.

The horizontal structures, with autonomous departments and a small, efficient administration, allow for simple operational procedures, operational flexibility without complex matrix structures, and direct coupling of ITWM competences with customers. A good working environment, a minimum of hierarchical friction, and a climate of mutual respect and appreciation contribute significantly to our employees' high level of commitment to and identification with their work and the Institute as a whole. Last, but not least, our straightforward dealings with our cooperation partners, based on the motto "promise only those things that you can really deliver," is an important element in the Institute's on-going economic success.

This is not to say that there is no room for improvement. We want to increase our cooperation with top national and international researchers in applied mathematics in order to ensure the quality of our research and to further develop and add to our competences. The publication activities in the departments vary widely; overall, an increase here is desirable, both for its own sake and to strengthen the Institute's visibility within the scientific community. The Institute addresses numerous application topics in almost all branches. This provides a degree of structural stability and helps ensure that economic downturns in individual branches have only a modest impact on the Institute's revenues. On the other hand, this high level of diversification is frequently associated with small project size, and the ITWM is the premium partner for MSO in only a few branches, such as commercial vehicles and the oil and gas industry. Moreover, there is a potential for further focusing, for example, in process technology, the energy sector, or the IT industry, which we want to promote more strongly in the future.

In addition, the Institute operates in a competitive environment: there are engineering offices offering companies R and D consulting with commercial software; there are software companies who are members of the ITWM's contract research pool offering commercial solutions for problems; there are university chairs pushing their way into the marketplace in response to the increasing market and third-party-funding orientation in academia; and there are also other Fraunhofer Institutes expanding their own modeling and simulation competences in their particular application domains. Of course, this competition is also directed toward attracting the best minds available in the employment marketplace. Naturally, the restrictions imposed by the TVöD (public service wage agreement) represent a competitive disadvantage. Attracting highly qualified new personnel and maintaining high employee motivation levels, while the team is increasing in age, will be one of the Institute's biggest challenges in the coming years.

The ITWM participates in many BMBF (Federal Ministry for Education and Research) projects as a partner for MSO. Although the innovation initiatives in Germany and the EU, when compared globally, may be viewed as providing a positive overall framework, it must nonetheless be admitted that mathematics does not fit squarely into the BMBF's funding channels. The BMBF mathematics program is certainly an important resource for applied mathematics in Germany, and mathematics funding also has a high priority for the DFG, as evidenced by its inclusion in all DFG subsidy programs. However, mathematics, as an independent technology, still has no funding program of its own, and the financial support of the BMBF program is exceedingly modest in comparison to the funding provided to other key technologies. The significance of applied mathematics as a driver of innovation is still not taken seriously in political circles. Thus, mathematically oriented research institutes and university chairs must continually rely on successfully docking their competences onto domain-oriented projects. However, they receive little dedicated funding for developing methodologies and expanding their core competences. There is no funding program for larger network projects with industry, in which methodological development, oriented on industrial needs, is expedited under the consortium management of the mathematics partners, and in which the companies themselves can also receive funding.

We do not, however, wish to conclude this introduction with what needs to be improved upon. The fact is that applied mathematics has experienced great growth in Germany within the past decades and has become a "motor of innovation," firmly anchored in the economy and society. The ITWM has made an important contribution to this development and is among the most renowned institutes of applied mathematics today. A significant part of the Institute's success is owed to the authors of this book. They have contributed to the project with great enthusiasm and attempted to identify the main elements of a problem-driven, model-based, and solution-oriented mathematics in the context of the Fraunhofer ITWM. Whether they have succeeded, we leave it for you to decide, dear reader. In any case, we wish you an interesting journey through this book, and look forward to receiving both positive feedback and constructive criticism.

## Appendix: The Fraunhofer ITWM in Numbers

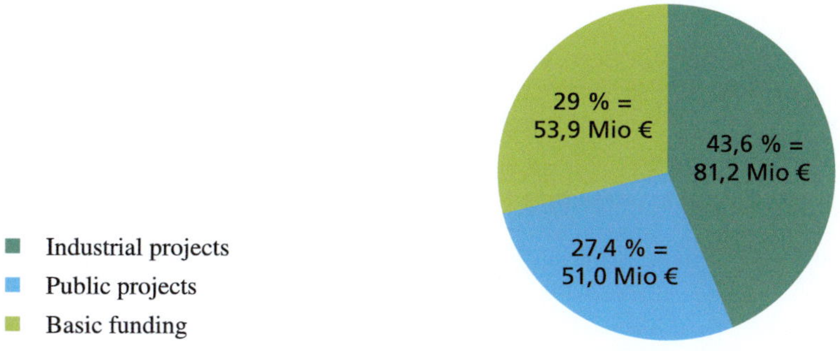


**Fig. 8** Total operating budget 1996–2013

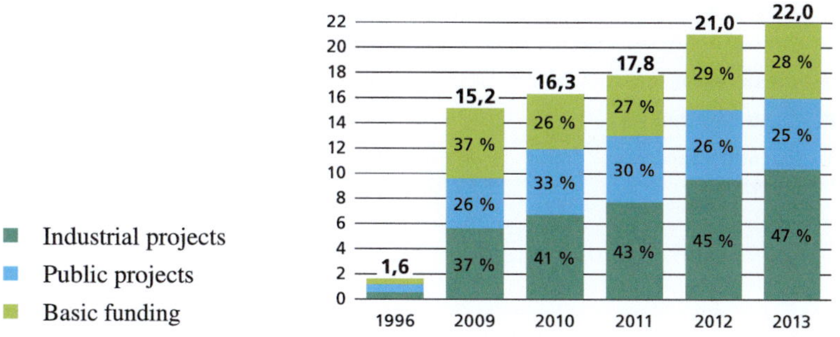


**Fig. 9** Development of operating budget in millions of euros

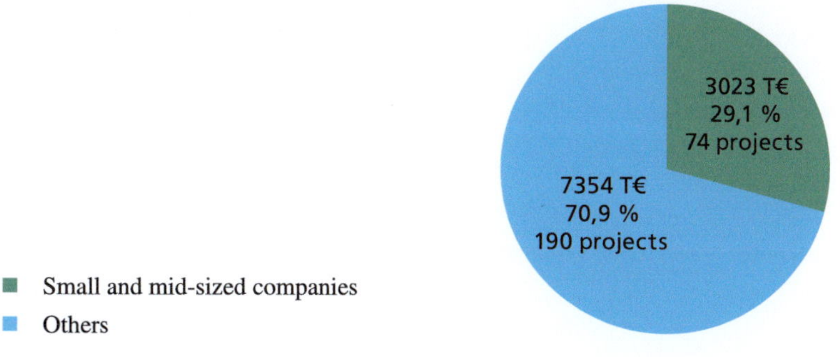


**Fig. 10** Breakdown of industrial revenues, 2013: small and mid-sized companies portion

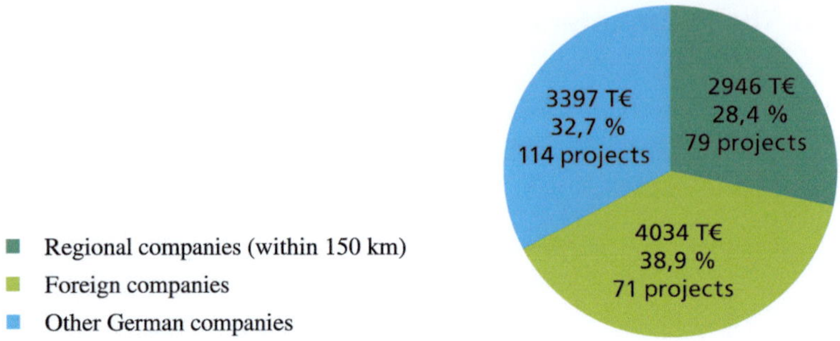


**Fig. 11** Breakdown of industrial revenues, 2013: distribution of industrial customers

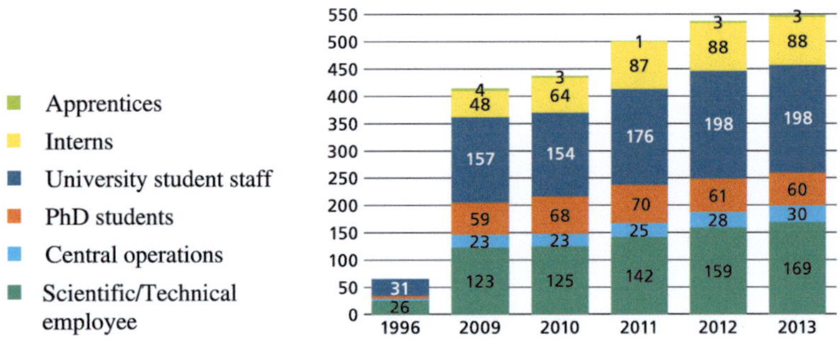


**Fig. 12** Staff levels 1996–2013

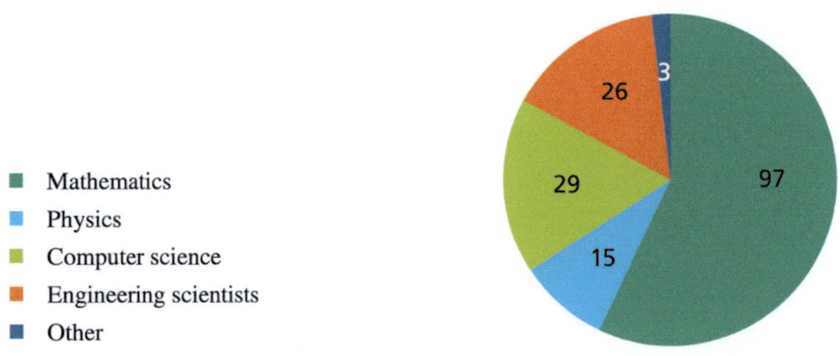


**Fig. 13** Breakdown of scientific/technical staff acc. to field, 2013

# Part II
# Concepts

# Modeling

Helmut Neunzert

Without doubt, "models and modeling" represents one of the most important core competences in industrial mathematics. Our primary purpose here is to clarify what we mean by models and modeling, for there are few terms in applied mathematics—perhaps few in all of the natural sciences—that have a wider variety of meanings and specialized uses. This is true, despite their also being key terms in natural science research, which is composed of the interplay between modeling and measuring.

So, let us begin with a definition of terms.

## 1 What Is a Model and what Is Modeling?

The literature is full of more or less original answers to this question. Here, for example, is an almost poetic entry: "Models describe our beliefs about how the world functions." And, further: "With mathematical modeling, we translate the contents of our beliefs into the language of mathematics" [4].

One might also say that we form hypotheses and pictures of our beliefs about how the world functions—or at least a part of the world. There is a note of caution in such sentences. We don't know how the world really functions, but we work with certain hypotheses about it until these hypotheses have been falsified. One hears the voice of Karl Popper here. These hypotheses have, at least, clear boundaries regarding the scope of their validity—and this is an important message, one that we place almost at the very beginning—especially when "world" means the world of industrial practice. Whatever is delivered in the form of solutions to industrial problems, one must never forget when applying these solutions that they were arrived at by virtue of simplified conditions that must

H. Neunzert (✉)
Fraunhofer-Institut für Techno- und Wirtschaftsmathematik ITWM, Kaiserslautern, Germany
e-mail: helmut.neunzert@itwm.fraunhofer.de



H. Neunzert, D. Prätzel-Wolters (eds.), *Currents in Industrial Mathematics*,
DOI 10.1007/978-3-662-48258-2_2

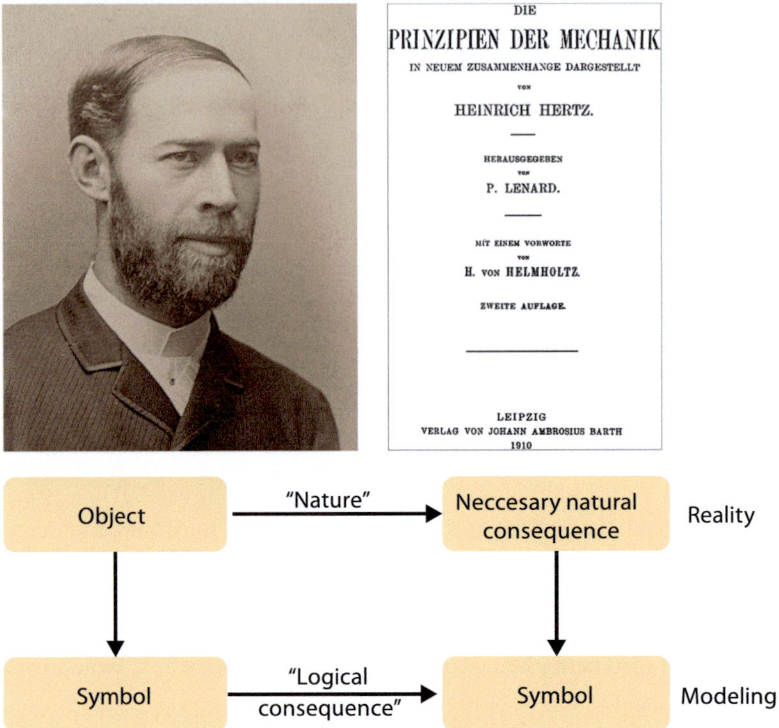


**Fig. 1** Heinrich Hertz (1857–1894, Photo: Robert Krewaldt)

always be taken into account. This caveat has often been forgotten–in financial mathematics, for example—but also with regard to technical problems. The unreliability of the results was not the fault of the mathematicians, but of the blithe utilizers of the results.

> However, it is also true that, within the boundaries of their validity, models can reflect the world with surprising accuracy—and this, above all, is what we want to discuss.

But let us remain for a while with the defining of terms. A less poetic definition than the first, but more useful from a scientific perspective, is that of the physicist Heinrich Hertz (Fig. 1), proposed in 1896 in his "Principles of Mechanics." To quote Mr. Hertz:

**Fig. 2** Systems: (**a**) natural ecosystem, (**b**) technical system (© Pressefoto BASF), (**c**) economic system (© Deutsche Börse)

> *"We create for ourselves internal replicas or symbols of external objects, and we create them in such a way that the logically necessary consequences of the pictures are always the pictures of the naturally necessary consequences of the replicated objects."*

In Fig. 1, we see a graphical representation of the above sentence. When we replace "internal replica or symbol" simply with model, then we understand that Hertz considered modeling to be the actual core of scientific research, for the above quote relates to the actual practice of science. We must determine the "logically necessary consequences of the pictures." This works best when the pictures are "made up of mathematics." Or, to put it another way: the raw material of the models under consideration here is mathematics. This corresponds to many other definitions of modeling also: "Mathematical modeling is the use of mathematics to describe phenomena from the real world" [7]. Moreover, the author goes on to say that modeling "investigates questions about the observable world, explains the phenomena, tests ideas, and makes predictions about the real world."

Wikipedia puts it even more simply. Here, a model is the description of a system by means of mathematical concepts and language. A model can help to explain a system and study the influence of various components, as well as help to make predictions about its behavior. Here too, science and modeling are closely coupled: "The quality of a scientific field depends on how well the mathematical models developed on the theoretical side agree with results of repeatable experiments."

A more thorough introduction to the most important terminology is provided by Velten [6]. He gives very formal definitions that are indeed correct, but not always helpful. The most comprehensible is still his Definition 1.2.1, which he adopts from Minski [3]: "To an observer B, an object A* is a model of an object A to the extent the B can use A* to answer questions that interest him about A." Is it now clear to everyone what modeling means? Another basic interest of modeling and modelers is objects. In most texts—and in [6] as well—these objects are quickly made more specific: one is interested in systems: natural systems, such as lakes; technical systems, such as installations and motors; economic systems, such as the stock market; and virtual systems, such as computer games (Fig. 2).

**Fig. 3** Model and reality: which is which? (© iStockphoto (*left*), Photo and montage: G. Ermel, Fraunhofer ITWM)

There are also more and less formal definitions for the term "system," but we accept the word as it is commonly understood. Without exception, the work of the ITWM also deals with systems of the kind mentioned above: spinning installations, grinding equipment, filter systems, generators, stock markets, etc. And although we prefer Heinrich Hertz's definition for our work and for this book, we too refrain from defining what a "picture" or replica is. We can say, however, that it illustrates to the investigator the essential qualities of a system; consequently, it excludes the non-essential qualities. In other words, a picture abstracts. Perhaps the word "caricature" would be more accurate than picture. Or perhaps the photograph in Fig. 3 says it best, entirely without words.

For us, then, a model is a picture of a system. The picture is composed of mathematics and reflects with satisfactory precision certain characteristics of the system that are of interest to the investigator. The model has clear boundaries for its validity, although these boundaries depend on the degree of precision that is desired. There are often parameters in the models that can be determined directly by measurement or extracted indirectly from measurement data via parameter identification. The precision requirements of the model must correspond to the precision of the data; it makes no sense to incorporate the tiniest phenomena into the model, when the parameters that belong to them can not be measured or can only be roughly ascertained—a problem that appears particularly often in biology. A cell's metabolism is extremely complex. On the basis of the structure model shown in Fig. 4, a mathematical model can easily be constructed using a system of ordinary differential equations, but the many transfer coefficients are neither measurable nor identifiable.

Modeling is a significant part of scientific practice. Physics, for example, consists of modeling and measuring. Newton's mechanics, Einstein's relativity theory, and Schrödinger's quantum physics are just as much models as Navier–Stokes or Euler equations, as Darcy's law of flow in porous media, as Cosserat's solid body theory, or as the Maxwell equations. In our projects, classical physics predominates—particularly continuum mechanics, thermodynamics, and electrodynamics—since the temporal and spatial magnitudes dealt with in industry are generally well-suited to it.

The supply of models in biology or the social sciences is not nearly so plentiful as in physics, and one often has to invent them anew. This is a challenge and sometimes a joy as well, but here, one is not as firmly rooted in solid ground as in physics.

Ultimately, I believe that modeling, as we have outlined it here, is essential for all problem solving, and thus represents a fundamental human activity. For, as Karl Popper reflected: "Life is problem solving."

## 2 Why Do We Model?

This question becomes superfluous once one has grasped what a model is. Nevertheless, the importance of mathematical models—and with it, the importance of mathematics—has increased greatly for industry, since computers have made it possible to utilize even more complex models. When one uses a computer to numerically evaluate a model that reflects a particular system, then one obtains in the computer a virtual picture of the system's behavior.

> We simulate the system. A simulation thus arises by means of the numerical evaluation of models, generally with the help of a computer. A simulation allows the behavior of a system to be predicted; one can investigate how system changes impact behavior and one can also optimize systems using a computer. Thus, models and simulations serve as important supports for decision-making: tactical decisions in the case of managers, and strategic decisions in the case of planners.

In the days of Heinrich Hertz, most models could not be evaluated. One had to simplify them–for example, by reducing the number of dimensions from three to two, or even one, or by using perturbation methods–in order to be able to then "solve" the simplified models analytically. The solutions of such simplified models often help one to better understand the system: Which parameters are important? Are there bifurcations? Can the system become unstable? And so on. However, if one wants to quantitatively predict system behavior in real, three-dimensional systems—and technical systems are mostly three dimensional—then such simplifications are not acceptable, and one must try to find at least an approximate evaluation of the original, complex model (Fig. 5).

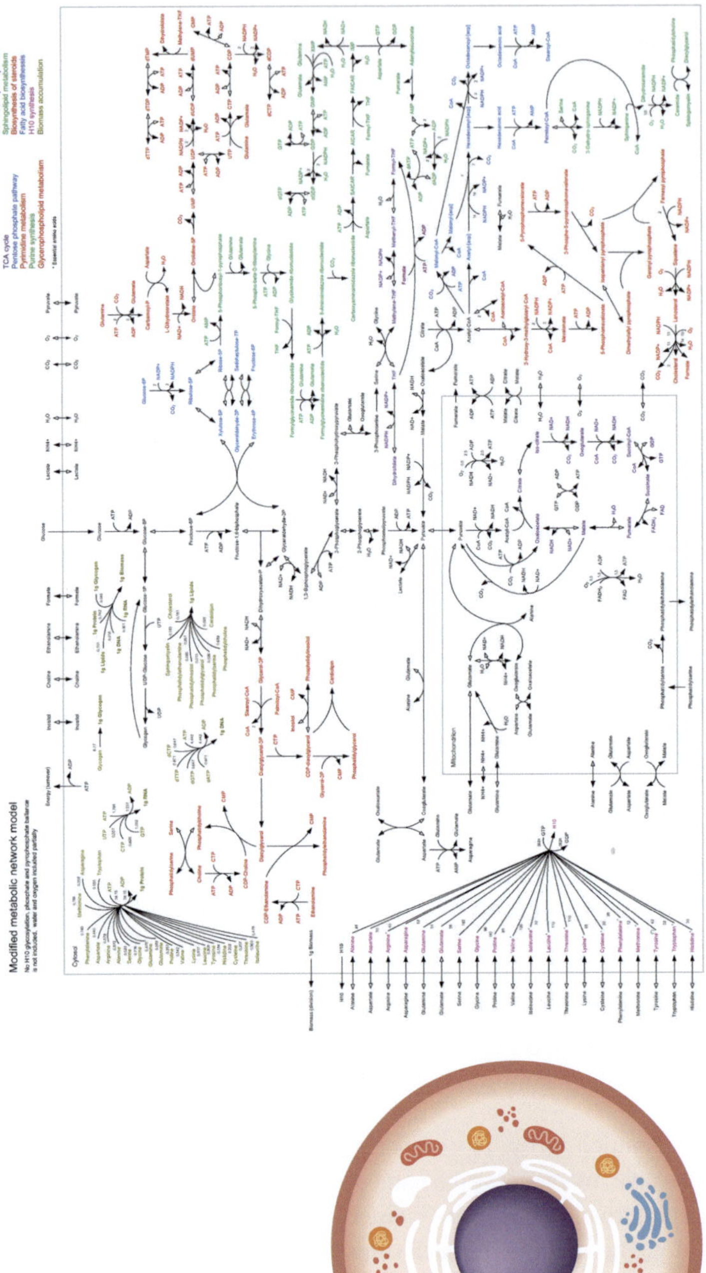

**Fig. 4** Sketch of a cell (© iStockphoto) and the structural model of a cell metabolic network (A. Shumilina, Fraunhofer ITWM)

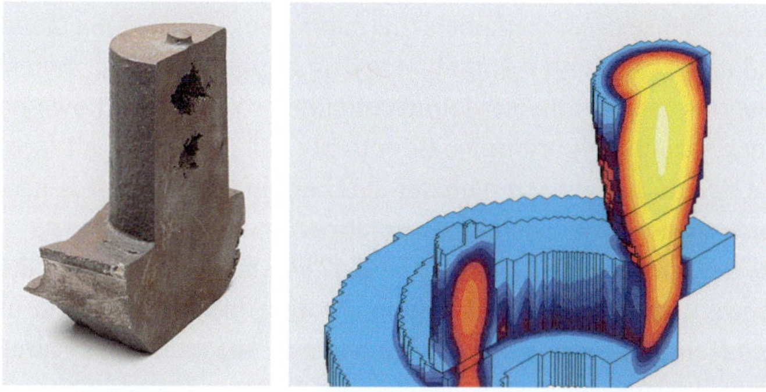

**Fig. 5** Real and virtual systems: a cast object and a solidification simulation (Photo: G. Ermel, Fraunhofer ITWM; simulation: Fraunhofer ITWM)

In industrial practice, one almost always wants relatively exact quantitative predictions. A purely qualitative understanding is indeed useful, but usually insufficient. This question also distinguishes between various groups offering "industrial mathematics" as a university or research institute topic. The "Study groups with industry" founded in Oxford some 45 years ago, for example, bear down for an entire week on industrial problems using mathematical methods and deliver interesting analyses, but rarely quantitative predictions. The Fraunhofer ITWM, in contrast, strives to ultimately provide the client with software for simulating, optimizing, or controlling the systems. The two approaches also call for different working tactics. At the Fraunhofer, the models should be "as simple as possible," but no simpler. A "small parameter" that is allowed to approach zero in Oxford in order to permit further investigations of the models is, at the Fraunhofer, often not small enough to cancel out without causing substantial quantitative errors.

Moreover, right from the start, when setting up the models, it is necessary to keep in mind that one must be able to efficiently evaluate them. Modeling and computation go hand in hand; artists of pure modeling and computation virtuosos, one without the other, are often inadequate to the real demands of industry. This calls for a genuine balancing act, for there are also "number crunchers" who will resort to faster algorithms, better computers, and sometimes coarser grids in their desire to evaluate the most complex of models—sometimes paying the price of large quantitative inaccuracies. Or of prohibitively expensive computing times. Modeling and computation specialists should form a team from the very beginning if they want to deliver reliable software to the client in the end. However, the idea of starting with the development of so-called "computer models," that is, with models formulated directly in the language of finite elements (FEM), for example, is untenable in our opinion. Numerical methods, such as FEM, help with the evaluation of differential equations, which in turn represent models from continuum mechanics or electromagnetism. The models are one thing; their evaluation algorithms are another. There might indeed be, for example, more efficient algorithms than FEM, and one loses out on the chance of using them.

Nor does it make sense to set up models that cannot be evaluated. Such models were occasionally found in the past with system biology problems, where the systems of ordinary differential equations used for the modeling contained so many unknown parameters that no parameter identification algorithm could reliably calculate them all (see Fig. 4). Here as well, close cooperation between modeler and computation expert is needed to ensure that optimal use is made of the existing information.

So, to repeat the question: Why do we model? We model so that, in the end, we can simulate, optimize, and control a real system—within the virtual world of a computer. The picture from Heinrich Hertz comes to mind again. The simulation should replace real experiments, since it is simpler, faster, and cheaper. Imagine the geometry of a production line, of a car, of a chemical reactor; how much easier it is to vary them in a computer than it is in reality! But always with the caveat, the simulation must also be reliable.

Optimization algorithms can only be executed in the virtual world—the raw gemstone must be "virtualized" before it can be optimally cut or ground. Gerda de Vries [7] has one more argument: "Experimental scientists are very good at taking apart the real world and studying small components. Since the real world is nonlinear, fitting the components together is a much harder puzzle. Mathematical modeling allows us to do just that."

Modeling and simulating is problem solving. We are always doing this, wherever we may find ourselves—though we are not always doing it consciously. However, this fact should be made conscious at an early age, while we are still in school. The models don't have to be differential equations; counting and adding can suffice for model evaluation purposes. It represents great progress that modeling is included as a permanent topic in school instruction plans in some federal states of Germany. This book also includes a chapter that reports on the valuable experience with modeling that Kaiserslautern's mathematicians have gained in schools (cf. "The training"). Just how deeply this look at modeling can penetrate into our daily life was made clear to me when taking leave of a Burmese student after her completion of a two-year Masters in "Industrial Mathematics." "I cannot open a refrigerator any more without thinking about how I can model the cooling loss and change the controls so that the energy consumption is lowered," she said laughing, and full of pride.

## 3 There Is Never Just One Model. How Can We Find the Right One?

Naturally, there is never just one model, not even when the questions about the system under consideration have been very clearly and unambiguously formulated.

For one thing, the model will depend very strongly on the modeler's previous knowledge and experience. Perhaps the modeler only finds problems that are reconcilable with his existing knowledge base. I used to poke fun at how often my Oxford colleagues managed to discover "free boundary value problems, until I found myself discovering an astonishing number of problems that fit into the even more exotic area of kinetic equations. Of course, that should come as no surprise. "Very many, perhaps the majority, of people, in order to find something, must first know that it's there," said G. Ch. Lichtenberg in his Sudelbücher book of aphorisms.

Naturally, only university mathematicians can afford such a luxury; when they search out their modeling problems, their search is "method driven." The mathematician at Bosch doesn't have this option. He has to optimize the transmission, regardless of what method fits. The Fraunhofer ITWM also takes on all problems that are mathematically interesting (and which lie within the Institute's competence). Granted, a problem may be transferred to the department that has the appropriate method for it. To be perfectly honest, however, departments usually attract the problems that suit them, which makes such problems transferals relatively rare.

Even when the modeler's methodological competence is not the determining factor, the model of choice is not unambiguous. Again, we see varying degrees of complexity. One begins with the "full physics" (models of first principles)—for example, the full compressible Navier Stokes equations—and, since these are not utilizable for the given parameters, ends with Prandtl boundary layer equations or with simpler turbulence models. This is the true art of industrial mathematics: how far can I drive the simplifications without violating my precision requirements for the simulation? Naturally, asymptotic analyses yield an error on the order of $(\epsilon^k)$ and a numerical error on the order of $(h^p)$. My $\epsilon$ and $h$ are small, but not zero, and these orders tell me absolutely nothing about the size of my error for a given $\epsilon$ and $h$. And the constants in the order estimations are much too rough to be usable.

One must validate the models and simulations in order to know what is really meant by "as simple as possible, as complex as necessary." We return to this point in the section "How do we construct the correct model?"

Here, however, we must still discuss the structure of the system somewhat. Actually, these are always input-output systems. They take input data, such as the environmental conditions or the tributaries to a lake, the control values of a machine, the trading data on a stock market, or the use of topography or solar irradiation in a solar farm, and convert it to output data, such as the lake's algae growth, the performance or consumption of the machine, the stock market quotation, or the daily energy production of the solar farm (see Fig. 6).

The system is the "piece in the middle" that transforms the input into the output. What this transformation looks like depends of course on the "state" of the system. Along with state variables that describe the system's instantaneous condition, there are also parameters that distinguish the system from other, similarly structured, systems. With an engine, for example, the geometry and fuel are described by parameters, while the temperature, pressure, piston position, etc. are state variables. The model's job is to describe their changes over time for a given input. The output is then usually a direct function of the state.

When there are natural laws that describe these state changes, such as the flow dynamic equations and the equations governing the combustion process in an engine, then all parameters have a geometric, physical, or chemical significance, and one can measure them. The model is then built from these equations, which often come from different areas of physics—one speaks then of "multi-physics"—and the measured parameters are

**Fig. 6** Another example of an input-output system: a solar collector farm (© Siemens)

inserted. When the equations can be solved numerically with enough accuracy, then the input-output system can be simulated, and the output can be calculated and predicted for each input, without experimentation. Even better, one can change the parameters—say, the geometry or the materials—and then calculate how the output changes. In this fashion, one can try out ways to improve the output, for example, to reduce fuel consumption and harmful emissions in the engine and/or increase performance. And still better, one can optimize these criteria by varying the parameters; one can develop an "optimal engine." Here, however, one should proceed cautiously. There are usually several criteria to be minimized or maximized; that is, one almost always has a "multi-criteria" problem. This will be discussed further in "The Concepts—Optimization Processes."

There was a time when many companies used optimization algorithms to help them "calculate" the form of an auto body. This led to autos that all looked the same, once their decorative elements were stripped away. Today, one often foregoes the absolute optimum in favor of a little individuality.

Optimization algorithms require many evaluations of the target function(s), and each evaluation requires a simulation. Consequently, one must simulate the system many, many times, which means that the individual simulation runs cannot take too long. Model simplifications are called for here, at least for the initial optimization steps. The clever coupling of optimization and model/computation is an important, modern research area, about which we will likewise report elsewhere in this book (cf. "The Concepts—Optimization Processes" and "The Research—Maximum Material Yield for Gemstone Exploitation.")

Optimization is an important reason for wanting a simplified model. The coupling of different simulations can also make it necessary to perform faster, although perhaps less precise, individual simulations. The simplified models sometimes contain parameters that are not measurable. These can be determined by means of a computation with the complex model. A good example of this can be found in "The Research—Virtual Production of Filaments and Fleeces."

Models that are based completely on natural laws and contain measurable parameters, that is, in which the system is completely "understood," are also referred to as white box models. The box, that is, the system between the input and output, is "white," in other words, transparent. These models stand in contrast to input-output systems in which the system is observable, but not really understood, which are referred to as black box models. The latter represent the best choice when one has many observations of system inputs and their associated outputs, but no theoretical knowledge of the system.

With black box models, one makes an approach for the transformation input $\rightarrow$ output that is as general as possible—one that has many free, that is, not directly measurable parameters—and then tries to determine these parameters from the measurement series. Good examples for such approaches are dynamic systems with an input $u(t)$, output $y(t)$, and a system

$$\begin{aligned} \dot{x}(t) &= f\big(t, x(t), u(t)\big) \\ y(t) &= g\big(t, x(t), u(t)\big), \end{aligned}$$

where the states $x(t)$ are from $\mathbb{R}^n$, and the dimension $n$ reflects the complexity of the system. The functions $f$ and $g$ are yet to be selected and are often assumed to be linear and even time-invariant

$$\begin{aligned} \dot{x} &= \mathbf{A}x + \mathbf{B}u \\ y &= \mathbf{C}x + \mathbf{D}u, \end{aligned}$$

where **A**, **B**, **C**, **D** denote matrices whose dimensions are given by the dimensions of the state space and the input-output space. In this case, **A** alone contains $n^2$ parameters that must also be identified by inserting observed $\tilde{u}(t)$ and $\tilde{y}(t)$. Similar, but non-linear, cases are handled with neural networks, which approximate general input $\rightarrow$ output transformations especially well. These are black box models, which require no theoretical knowledge and whose parameter identification algorithms can be taken "off the shelf." For this reason, they are quite popular. They enjoy a prestige, in fact, that is further increased through the notion that "neural networks" function according to the model of the human brain, in which the neurons "fire." Whether this prestige is justified is a matter of opinion.

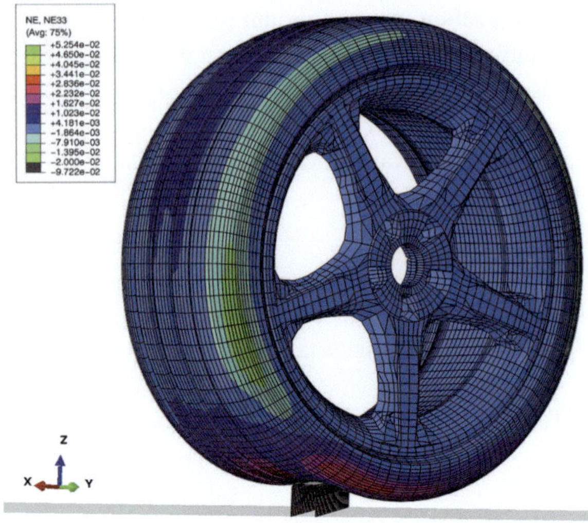


**Fig. 7** An already very finely-grained FEM model of a tire rolling over a threshold (Simulation: Fraunhofer ITWM, department MDF)

Certainly, however, black box models also have significant disadvantages. Because the coefficients of the matrices, that is, the parameters, have no relationship to the natural world, one never knows how changing them will affect the system. Thus, only for a given observed system can one identify the parameters and predict its behavior. Changes and improvements in the system are not possible. Therefore, one tries to reserve black box models for simulations of biological or ecological systems that are resistant to change; for technical systems, one tries to avoid them. For want of theory, however, black box models are also commonly used in economics, despite the frequent changes such systems undergo.

There are also intermediate stages between white box and black box. For example, one has theory that establishes the model's substructures, but one also has terms that are selected for mathematical reasons and contain non-measurable parameters. These so-called grey box models are found much more frequently in practical work than in the mathematical theory of the natural sciences. Grey box models are found, for example, in the deposition model used in "The Research—Virtual Production of Filaments and Fleeces." The models for car tires developed at the ITWM are also grey box models. A detailed resolution of the structure of a modern tire, that is, a white box model, is feasible in principle, but it cannot be evaluated. The tremendously fine-grained structure would make the elements of an FEM so microscopically small that an evaluation would be prohibitively expensive (cf. Fig. 7). One therefore aggregates the tire into larger compartments; the material values that characterize them are non-linear averages and thus non-measurable. These parameter values must be identified by means of tire tests; for example, recordings of tire deformation during traversal of a threshold—still a numerically delicate task for which the mathematics must be invented. The tire model shown here resolves many of the details, but not all of them. The nylon filaments in the tire are not individually reproduced. So, we have here indeed a grey box model, albeit one that is rather light grey.

This type of grey box model often arises in so-called multi-scale modeling, to which we will return when we speak of model reductions.

So, there are always many ways to arrive at a model. The reputation and success of an institute seeking to solve real-world problems surely depend primarily on finding good, suitable models. This means, models that predict system behavior with the desired precision and do so with modest computing effort. Such criteria are not the guiding motives of a university mathematician, but their use contributes significantly to the prestige of mathematics as a practical science, and indeed, as a technology. "Technology is the application of scientific knowledge to the practical aims of human life," according to the Encyclopedia Britannica. This application of knowledge takes place through the use of mathematical models, which thus makes modeling *the* key technology.

## 4 Avenues to Model Reduction

How does one make a mathematical model? In the classroom, one starts perhaps with quite a simple model and then advances to more complex ones (see "The Training"). In practice, however, one usually starts with the complex models. Here, the natural laws are known and the materials can be described in detail. One therefore has to reduce the complexity to arrive at simpler models, since evaluating the complex ones is too expensive. The process of systematic simplification is called "model reduction." Here, there are various approaches that can be used, initially, for genuine white box models.

### 4.1 Methods of Asymptotic Analysis, Perturbation Theory, and Up-scaling

The art of finding small parameters by non-dimensionalizing a model—which should always be the first step, for otherwise, one cannot say what are the "large" and what are the "small" terms in an equation—and then letting them approach zero is called asymptotic analysis. This art, which is still especially cultivated in Great Britain, is what practitioners there mean when they speak of "modeling." It is learned very nicely by studying Barenblatt's book [1]. In the research articles in this book, one also finds examples that demonstrate how difficult it sometimes is to find the "right" small parameter.

A special case arises when the medium is periodically inhomogeneous and the periods of these inhomogeneities are very small compared to the size of the total system. Here, one can skillfully apply a two-scale approach, for example, by letting the "period length approach zero"—which is called "homogenizing"—in order to obtain models whose inhomogeneities are no longer so finely scaled. One thus obtains models that only capture large-scale effects, but still maintain a memory of the micro-scales. One therefore speaks of up-scaling, which can also be attained–granted, in a somewhat "more robust" manner—by means of averaging in numerical methods (numerical up-scaling) (Fig. 8).

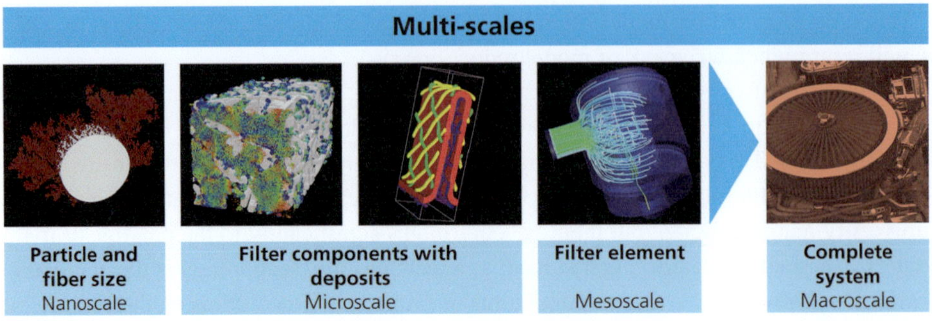


**Fig. 8** Filter simulation at various scales (Graphic: S. Grützner, Fraunhofer ITWM, simulations: Fraunhofer ITWM, department SMS, photo: iStockphoto)

This multi-scale modeling is closely related to the multi-grid methods of numerics. In "The Research—Modeling and Simulation of Filtration Processes," up-scaling will be examined for the simulation of filters (which indeed exhibit a crucial microstructure), along with the interplay between multi-scale and multi-grid.

Asymptotic analysis, perturbation theory, homonogenization, etc. are important analytical methods for the reduction of white box or grey box models.

## 4.2 Model Order Reduction (MOR) and Projection Methods

The simplification of models using projection methods, such as principal component analysis, balanced truncation, and proper orthogonal decomposition (POD), comes from system and control theory. These methods, which arose in statistical problems in the context of the Karhunen–Loève expansion, are based on the fundamental assumption that the relevant effects or temporal evolutions of the sought-after quantities play out in subspaces of the entire state space, so that projections onto these subspaces are possible without the resulting errors violating the accuracy requirements. This method of reducing dimensions is well established for linear systems [2]. The article "The Research–Robust State Estimations of Complex Systems" may be referred to in this regard.

The manner in which the subspaces are found varies from method to method. POD, for instance, uses information from representative snapshots of the solution, which has been obtained, for example, through elaborate FEM calculations: If $u(t)$ from $\mathbb{R}^N$ is a spatially discretized solution, then one observes snapshots of it, that is, $(u(t_1), \ldots, u(t_k))$, and tries to find the subspace $W$ with the smaller dimension $n$, onto which the subspace $U$ created from the snapshots can be best projected, that is, which minimizes $\|U - \text{Projection of } U \text{ onto } W\|$. Thus, one starts with the already discretized model in order to arrive in finite dimensional spaces, such as $\mathbb{R}^N$, and one discretizes time. $W$ is then found using an eigenvalue problem from the correlation matrix. POD delivers good results for simple flow or diffusion problems, but usually breaks down on the choice of snapshots for rapid and stiff transport processes.

Methods for nonlinear, parameter-dependent models are particularly attractive for applications and represent a current research field that is also being actively pursued at the Fraunhofer ITWM [5]. Parametric model reduction methods allow reduced models to be determined for changed parameters without repeated observation and simplification of the original model. For this reason, they are popular for use in parameter studies. The methods use interpolation of the transfer function, for example, or of the projection spaces, or even of the entire solution. This last example is known as the reduced basis method and is motivated by classical error estimators for Galerkin approximations for partial differential equations, whereas the empirical interpolation approaches are oriented toward the approximation of dynamic behavior.

## 5 Summary

In the standard texts on modeling, distinctions are often made between deterministic and stochastic, discrete and continuous, linear and nonlinear, explicit and implicit, and static and dynamic models. I suspect that, in each case, this is a function of the existing knowledge base of the modeler. For us, however, stochastic models are sometimes simplifications of too complex deterministic models; discrete are sometimes discretizations of continuous models; linear are sometimes desperate attempts at simplifying what are in reality nonlinear models (The world is not linear!). Such a system of classification has little value for us—the problem determines the model, not our knowledge or lack thereof. Black box models, however, represent the method of last resort, to be used only when we have no theory, and only observations, to work with. In reality, however, almost everything is "grey."

We repeat here once again the steps involved in modeling:

(a) We check to see if there are theories which, when appropriately compiled, describe our problem. Often, we must amend these theories, for example, by setting up the right boundary values. Here, we must be clear as to the questions we want to answer about the problem: which quantities do we want to predict, and with what accuracy? We have to thoroughly consider not only the desired output, but also the input: What belongs to the state of our system; what is the input, that is, what is the environmental data that must be entered; and what can we control? How often will we have to repeat the simulation? What aids (computers, toolboxes) do we have at our disposal?

(b) We have to simplify the complex "complete" model enough so that we can evaluate it. Here, model simplification and numerics work hand in hand. After non-dimensionalizing, we must investigate the remaining parameters. Above all, we have to identify the non-measurable parameters (and there are such parameters in grey box models). We must find algorithms and estimate their precision, and here, the standard order estimates are of little help. We have to implement the whole algorithm, while paying attention to our computer architecture. The coupling of multi-core, multi-grid, and multi-scale is becoming more and more important. In the end, we have a simulation program.

(c) We must test this program with the user, the problem provider, the client. Here, we will often note that we have not understood him correctly; he will perhaps only at this point really understand his problem himself. And then, we start again at the beginning: What is the desired output? What can we control? How exact must everything be? Etc.

(d) Finally, we hand over the program to the client and collect our fee. And, if the work was well done, he will soon come knocking at our door again with more requests: "I'd like to change this; I'd like to know this more precisely; this should be optimized..." Marvelous! For in this way, science and practice—and our Institute as well—all make progress together.

## References

1. Barenblatt, G.: Scaling, Self-Similarity, and Intermediate Asymptotics. Cambridge Texts in Applied Mathematics, vol. 14. Cambridge University Press, Cambridge (2009)
2. Benner, P., Mehrmann, V., Sorensen, D.C. (eds.): Dimension Reduction of Large-Scale Systems. Lecture Notes in Computational Science and Engineering, vol. 45. Springer, Heidelberg (2005)
3. Minsky, M.L.: Matter, minds and models. Proc. IFIP Congress **1**, 45–49 (1965)
4. Open University Course Team: Introduction to Mathematical Modelling. Open University Worldwide (2005)
5. Panzer, H., Mohring, J., Eid, R., Lohmann, B.: Parametric model order reduction by matrix interpolation. Automatisierungstechnik **58**(8), 475–484 (2010), Oldenbourg
6. Velten, K.: Mathematical Modelling and Simulation: Introduction for Scientists and Engineers. Wiley, New York (2008)
7. Vries de, G.: What is Mathematical Modelling. University of Alberta. Lecture Notes (2001)

# Numerics

Oleg Iliev, Konrad Steiner, and Oliver Wirjadi

We now turn to the development and use of numerical algorithms and software for industrial problems, once again, guided by the philosophy and experience of the ITWM. After we recapitulate a few basic ideas, we elaborate on the specifics of Fraunhofer research in the field of computational mathematics and formulate some of the essential requirements and criteria of such research. We begin with grid generation and discretization. Because of its important role in the Institute, the so-called Lattice-Boltzmann Method, LBM, will be treated separately. Several departments work on multi-scale problems, and so we address this topic next. Efficient methods for problems in image processing will be briefly mentioned. Here as well, we will continue to point out the differences between research at the Fraunhofer ITWM and academic research. Finally, we reflect a bit upon the topic "validation" of models and algorithms.

## 1 The Fundamentals

Most of the models discussed in this book use partial differential equations (PDE). Ordinary differential equations (ODE), differential algebraic equations (DAE), and integro-differential equations also occur, albeit less frequently. These models describe deterministic systems. For stochastic cases, of course, stochastic ordinary or partial differential equations (SDE and SPDE) must also be analyzed and solved. Some examples of scalar PDE are non-steady-state or steady-state diffusion or heat conduction and, for systems of PDE, the Navier–Stokes equations or the equations of linear or nonlinear elasticity. Examples of DAE are found in models of two-phase flow in porous media and in incompressible Navier–Stokes equations. We find SDE in models describing the transport of

O. Iliev · K. Steiner (✉) · O. Wirjadi
Kaiserslautern, Germany
e-mail: konrad.steiner@itwm.fraunhofer.de



H. Neunzert, D. Prätzel-Wolters (eds.), *Currents in Industrial Mathematics*,
DOI 10.1007/978-3-662-48258-2_3

nanoparticles in cases where the Brownian motion of the particles plays an important role and, naturally, in financial mathematics as well. We find SPDE in single or multi-phase flows in porous media with random permeability and in the extrusion of fibers. And, finally, integro-differential equations are used to describe viscoelastic fluids or to represent kinetic equations.

Of course, one only very rarely finds analytical solutions to these equations for initial boundary value problems, and we need numerical algorithms to compute approximate solutions. These algorithms will be discussed extensively in the case of PDE; elsewhere, we will only touch upon them briefly.

Almost all numerical algorithms for solving PDE belong to one of the following two large classes: the grid-based methods or the grid-free methods. Spectral methods play a smaller role for us and are not discussed here. With grid-based methods, one must generate a regular or irregular grid in the computational domain in order to then use it for discretizing the continuous problem. The most popular methods for this are the finite difference method (FDM), the finite volume method (FVM), and the finite element method (FEM). FDM is the easiest to understand, since its basic idea is to replace partial derivatives with finite differences. The method works very well with Cartesian grids but is much more difficult to manage with complex domain shapes and irregular grids. FDM used to be applied mainly to scalar equations, such as heat conduction equation, and also to fluid dynamics equations. In contrast, it is rarely used in structural mechanics. As already mentioned, the advantage of FDM is its simplicity; among its drawbacks, one should mention the complexity when working with complicated domains and, in many cases, the high smoothness requirements for the solutions. FVM is especially suited to fluid dynamics problems. Fluid dynamics models are usually based on conservation laws (conservation of mass, momentum, and energy). Such conservation laws are often written in integral form for small sub-domains (finite volume) and yield the differential equations when one lets the size of these partial domains suitably approach zero. The latter limit is not taken for FVM. Instead, the conservation equations are discretized over the finite volumes directly from the integral form of the conservation laws. Here, the union of the finite volumes overlaps the computation domain. The significant advantage of FVM is that the conservation laws are fulfilled locally also at the discrete level. This is especially important in industrial problems, where, for example, loss of mass in the computations is not tolerated. FVM, like FDM, is used predominantly with scalar problems and fluid dynamics problems, but hardly ever with problems in structural mechanics.

Finite element methods are very suitable for solving problems in irregular domains. Unlike the FDM, which works with the so-called strong formulation, i.e., directly with the PDEs, the FEM is a variational method, and it works with the weak (integral) form of the governing equations. In this case, the solution is sought as a linear combination of a finite number of basis functions, with each basis function having local support (hence the name, finite elements). The standard procedure, for example, for scalar elliptic equations is to replace the continuous solution in the equation, which belongs to infinitely dimensional functional space, with the best approximation from a properly selected finite dimensional space (i.e., the span of the basis functions). Next, one multiplies the equation by test functions and integrates over the computational domain. There is a large variety of finite element methods: Galerkin or Petrov–Galerkin, conforming or non-conforming, discontinuous Galerkin, etc. The basis functions are usually polynomial functions with local support. Originally, the FEM was derived for the needs of structural mechanics. Today, however, it is widely used in other areas as well.

The result of all approximation methods, whether FDM, FVM, or FEM, is a large system of linear equations (one must also sometimes linearize nonlinear equations or systems of PDEs; this process will not be discussed here, however). Solving these systems with direct methods generally consumes much time and memory. Therefore, iterative methods are used. Toward this end, geometric or algebraic multi-grid iterative methods usually show the best performance. These are best used as pre-conditioners in Krylov subspace methods.

In our research in computational mathematics, we are not interested merely in developing numerical algorithms for PDEs, ODEs, etc.; we are mainly interested in providing solutions to industrial problems by using proper numerical algorithms and performing and analyzing computer simulations. This distinguishes our work from pure academic research in computational mathematics.

Several factors play a crucial role in developing numerical algorithms and performing computer simulations when the goal is to provide solutions to industry:

- As emphasized in the previous chapter, during the discussion of modeling, the algorithms cannot be separated from the models, because modeling error and computational error must be balanced out.
- The same applies to the input data; the expected accuracy of the solution should correspond to the accuracy with which the input data is given.
- When developing algorithms to solve industrial problems, the foremost goal must be to achieve a solution using the resources available with a reasonable amount of computational time. Algorithms with theoretical, but no practical, value are not considered here.
- Similarly, algorithms that cannot be implemented with existing programming languages or whose complexity exceeds the capabilities of the available computers are not (yet) relevant for the solution of industrial problems. In numerical analysis, the convergence behavior when the grid-size parameter approaches zero plays a crucial role. We keep this in mind, while at the same time remembering that we solve problems on realistic (and not on asymptotic) grids. In certain cases, methods that are preferable with respect to the asymptotic properties might not perform well on relatively coarse grids. In practice often we can solve the problem only on one grid with a fixed size.
- Often the numerical algorithms must be developed in a way that preserves as many of the physical properties of the model as possible (e.g., conservation of mass and momentum, monotonicity). Industrial clients might not accept a solution with a non-physical sink or source of mass, for example, and/or with non-physical oscillations of the solution, even if there are theoretical estimates proving that the method is asymptotically conservative.

- In projects carried out directly with industry, the work usually needs to be executed in steps, with each step being of several months' duration. After each step, one must deliver results proving that a correct strategy is being used. Further on in such projects, the final project solution generally has to be delivered before a specified deadline. If a solution is not provided on time, the project may be terminated. Consequently, we mainly develop algorithms that can be modularized. This allows the modules to be developed, implemented, and tested in a relatively short time; in the ideal case, they can be used to solve simplified industrial problems.
- Because the work at the ITWM is problem driven (and not method driven), we find ourselves subjected to a large variety of problems that need solving, which means that we also need a large number of algorithms and software tools to solve them.

## 2 New Development or Adaptation of Known Algorithms?

There is no simple answer to this question. One has a better feel for the strengths and weaknesses of algorithms that one has developed oneself, but in-house development takes time. With public projects, therefore, we develop our own algorithms. These then also find application in industrial projects, just as do the algorithms in the literature that are adapted to a given problem. It must be mentioned, however, that the process of adaptation is not always faster. Many researchers working in the area of computational mathematics, as well as researchers active in "computational X," where X stands for mathematics, physics, chemistry, biology, etc., develop algorithms that they consider to be especially good for solving practical problems. However, it often happens that these algorithms don't converge for the industrial parameters, or they converge to a false solution, or they yield results that exhibit non-physical behavior (oscillations, loss of mass, etc.). When developing and adapting algorithms, we often have to pay attention to their robustness and—as emphasized earlier—make sure that computational, model, and data errors are of comparable size.

**In-House Software or Commercial Software?** Actually, the answer to this question requires a multi-criterion optimization. Using commercial software is a good idea when it has already demonstrated its ability to solve models of the form being currently considered. Thus, we are also confronted with problems that we tackle using ANSYS, FLUENT, Comsol, etc. And we do indeed take this software right up to the limits of its capabilities. However, if it cannot provide a usable answer—and this is not so seldom an occurrence—then we develop new software ourselves. The pros and cons of commercial software are well known. We would like to briefly address the counter arguments. Comsol, for example, is outstanding for training purposes. It also claims to be able to quickly solve every industrial problem. Experience has shown us, however, that this is only true when the same or very similar problems have already been solved in the past using Comsol, and are therefore contained in the library. If a completely new and complex problem must be solved quickly, then this software, in our experience, is not always reliable. Naturally, the developers at this software firm keep very close track of these cases and, after a certain period of time, release a new version that redresses these deficiencies. Unfortunately, we can rarely wait, and therefore need our own developments. The situation is similar with Open Source software tools, which, on top of the aforementioned problem, offer even less support. The developers of FLUENT take great pains to ensure that their software is as robust as possible. One sometimes pays a price for this robustness, however, in terms of precision.

Our own software development has many facets. The algorithms are often implemented in MATLAB for quick testing or the quick computation of a solution. We often couple our own developments with commercial tools by processing only those parts of a composite problem ourselves for which the commercial software is inadequate to our purposes. For example, in the chapter "Virtual Production of Filaments and Fleeces," a situation is described in which the software for simulating fiber production is composed of a FLUENT simulation of turbulent flow and an in-house development for simulating filament dynamics.

Our software tools are not only used within our Institute; they are also released to our customers. Here, the software is usually coded in C++, equipped with GUI, accompanied by handbooks, etc. Sometimes, these in-house developments attain a degree of sophistication that allows them to be marketed in spin-off firms. This book points out several such cases.

## 3 Grid Generation

Of course, there are numerous commercial and academic tools for generating grids. Nonetheless, this is still today a very active research area. In many industrial problems having complex geometry, the effort involved in generating a grid of acceptable quality can be comparable to the entire effort involved in solving the problem. It can, in fact, be the determining factor (see Fig. 1). Sometimes one needs a tremendous amount of experience to design a good grid. In structural mechanics, where FEM dominates, grid generation is unavoidable.

In addition, we pursue two other options for flow simulations: the generation of Cartesian grids (and voxel-based grids, in particular) or the use of finite point set methods in connection with grid-free methods. Our use of Cartesian grids is motivated by the following reasons:

- One can generate them quickly, simply, and reliably.
- The error that arises when the domain is approximated by such a grid is often comparable to the error of the input data.
- This error can be controlled, particularly for laminar, viscous flows or diffusion type problems.

**Fig. 1** Generating a good quality unstructured grid for simulation of flow through this structure would be a real challenge (grid generation: I. Shklyar, Fraunhofer ITWM)

- Both, our comparisons and examples from the literature, show that comparable precision can be attained for voxel-based and unstructured grids for certain classes of problems (e.g., computations of permeability of porous media or the effective mechanical characteristics of composite materials).
- Not much time is required to develop the algorithms and test the implementation.
- Computation domains arising from three-dimensional computer tomography are a priori only available in voxel format.

The mesh-free approach (see also the next section) is especially suitable when the computation domain varies very quickly over time and, therefore, grid generation and the interpolation that it necessitates takes up time and also represents an error source. At the ITWM, the "Flow and Material Simulation" and "Image Processing" Departments, in particular, use Cartesian access. For image processing, this is quite natural, since images are based on either pixels or voxels. Grid-free methods are focused on in the "Transport Processes" Department.

## 4 Discretization Approaches

FEM is generally used for structural mechanics problems. On occasion, however, plate or rod models are used instead. For these cases, mimetic type finite difference methods have been developed in the TV and MDF Departments of the Fraunhofer ITWM. Mimetic type methods aim to also preserve conservation properties or momentum at the discrete level, when they are preserved at the continuous level. With few exceptions, in the area of FEM, we work mainly on modeling or optimization and more rarely on developing entirely new FEM algorithms.

Lately, the ITWM has been developing FEM algorithms to simulate lithium-ion batteries. The ion concentration and potential are discontinuous on the boundary surfaces between the solid particles and the electrolytes (the electrodes have a porous structure that is resolved by the grid), whereas the temperature is continuous there. Therefore, one must approximate these boundary surfaces very carefully. The situation is quite different with flow problems. Historically, FDM and FVM dominated "computational fluid dynamics (CFD), and they are also implemented in most commercial packages. In the past decade, however, FEM has also been developed for this application. Here, too, there is now software available on the market. Nonetheless, we still prefer FVM. Here are our arguments in its favor:

- FVM provides discretizations that locally fulfill the conservation laws, a factor that is important, particularly for flow problems with discontinuous coefficients.
- The monotonicity of the solution can be easily checked when FVM (which has a lower order) is used. Thus, for example, the probability of finding non-physical oscillations in the numerical solutions of convection problems is markedly higher with FEM than with FVM. This is related to the fact that the stability of FEM is only ensured in the weak (integral) sense, but not locally. Appreciable progress has been made in the development of stabilization methods for FEM discretizations for flow problems, but the reliability and robustness that have been reached are not yet adequate for industrial problems. It bears repeating: these methods function when they are applied to the classes of problems, geometries, and flow zones for which they were developed. Adapting them to other geometries, flow zones or modified equations can require a great deal of effort.
- FVM discretizations are indeed of a lower order, but our input data are also often inexact.
- In the past few decades, efficient solutions have been developed for the linear equation systems arising in FVM discretizations and have reached an adequate degree of maturity. The same cannot be said for the new FEM discretizations. Today, for example, iso-geometric methods are very much in vogue. However, along with other challenges, the development of robust and efficient multi-grid solvers for these discretizations still remains to be achieved. The situation is similar for the discontinuous Galerkin method (DG).

FEM is indeed a powerful method, but new problems may demand new elements. Let us illustrate this with a few examples. FEM made a breakthrough for flow problems when "bubble finite elements" were invented and LBB stability conditions for the pairs of finite elements were analyzed and understood. The need to use special finite elements for elasticity problems with almost incompressible material is also known. In the past 5 to 7 years, there has been very intensive research directed at finding finite elements suitable for a robust solution of the Brinkman equation (Brinkman is a Stokes equation perturbed by a Darcy term). These examples show that the solution of new models (or of known models applied to new flow regimes) may require extensive investigations into new basis functions, stability, etc. before applying FEM. Such investigations cannot be undertaken with

**Fig. 2** A further example for the application of the FPM method: a car driving through water; experiment and associated FPM simulation (photo and simulation: Volkswagen, Inc., as part of a cooperative project with the Fraunhofer ITWM)

industrial projects running on tight schedules, and we therefore in general rely on FVM discretizations for the new models. Looking ahead, once the behavior of the solutions is better known and academic researchers have made more progress, then we will also use FEM more extensively, just as we already do now with battery simulations.

In our FVM developments, we place special emphasis on the conservation characteristics, the monotonicity of the solutions, and the accurate treatment of the discontinuity surfaces of coefficients. For example, when solving the Navier–Stokes–Brinkman equations needed to simulate flow in filtration processes in plain and porous media, we have suggested discretizing the interface conditions. This correctly captures the linear pressure gradient in porous media, although only one layer of cell-centered finite volumes is used.

Over the past 20 years, the ITWM has developed a grid-free method, which we have named the "finite point-set method" (FPM). This successful tool comes from the "Transport Processes" Department. In contrast with the other methods discussed above, it is not based on a fixed grid. Rather, the (grid) points are free to float, and often move with the flow in the manner of Lagrange. FPM is especially apt for situations in which the computation domains vary very quickly; the classic example is the simulation of an airbag. Instead of continually varying the grid ("re-meshing"), as would be required by FDM, FVM, and FEM, FPM only has to monitor the density of the points (Fig. 2).

## 5 Microstructure Simulation (Voxel-Based Methods)

In this section, we will use the topic of microstructure modeling and simulation to illustrate how problem-oriented modeling (see "The Concepts—Modeling") and the application of numerical methods cross-fertilize each other in the mathematical research of the ITWM. Moreover, the mathematical research is also driven forward—here, in particular, voxel-based numerical solution procedures for PDE. In the early years of the Institute, calculations were made for flows in porous media for design simulations of diapers or filter components. Here, it soon became evident very difficult it is in industrial applications

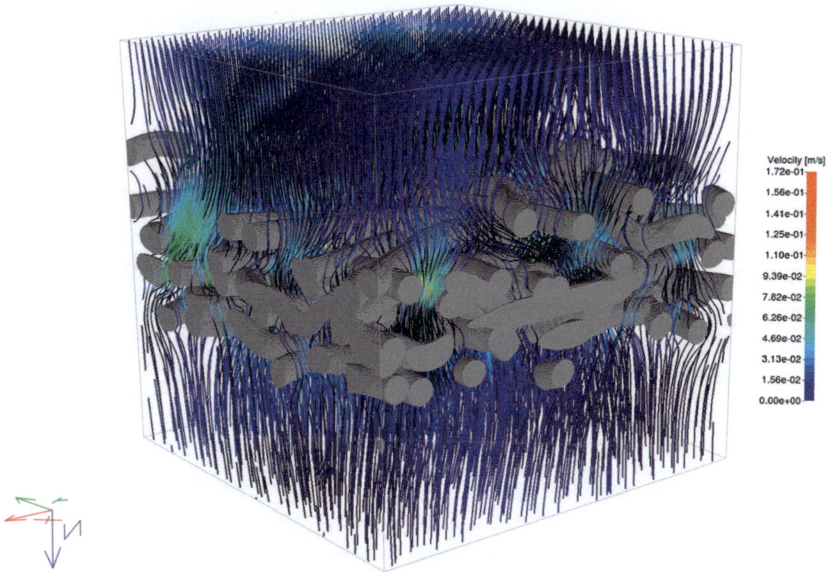


**Fig. 3** Visualization of streamlines through a microstructure on a voxel-based grid (simulation: S. Rief, Fraunhofer ITWM)

to experimentally determine material parameters such as permeability and capillarity for soft, porous materials (fleeces, filter papers). Consequently, the idea quickly arose of pursuing microstructure simulation, which is based on the homogenization theory and which numerically calculates the material characteristics as a solution of the cell problem for the most realistic, porous microstructure geometries possible (in 2001, the ITWM received its first Fraunhofer prize for this development).

Here, realistic, high-resolution, three-dimensional images of the porous structure are needed, since they are significant ingredients of the microstructure simulation. These are customarily delivered by micro-computer tomography in a 3-D image format (voxel). The material characteristic calculation is then generated as a solution of the fluid dynamic equations (Stokes or Navier–Stokes) in the pore space, that is, in the complex microstructure geometry. Because the given discretization of the image data has a large number of uniform voxels (generally $1000 \times 1000 \times 1000$), lattice Boltzmann methods (LBM) provide themselves to this application as a very specific, but also very adequate, solution procedure and were developed further at the ITWM.

Lattice Boltzmann methods solve a discrete and linearized Boltzmann equation on the voxel network by means of a time-explicit procedure and take advantage of the fact that, with an appropriate limit value of the discretization parameter, the Navier–Stokes equations are approximated. The aligned discretization of the velocity model with the space discretization (voxel grid) allows for an exact time integration of the transport step. By implementing a simple data exchange, the no-slip conditions on the geometric walls can also be taken into consideration (Fig. 3). The linear collision step does indeed take more effort and require many computation operations, but these are completely local, which means that the entire procedure can be very simply and efficiently parallelized. Through the use of clever extensions, LBM make possible the solution of single and multiple phase flows for both Newtonian and non-Newtonian fluids in arbitrarily complex microstructures, and thus have become established as a suitable tool for micro-simulating porous media.

Unfortunately, lattice Boltzmann methods cannot be directly applied to elliptic problems, such as the ones for determining effective thermal conductivity or mechanical stiffness. Here as well, however, specialized procedures have emerged from the ITWM, thus stimulating research into techniques that possess advantages similar to those of the lattice Boltzmann methods.

These are numerical methods that use voxel partition directly as discretization, work without matrices, and can be efficiently parallelized. These voxel-based methods rely on the solution of the associated integral equations in the perturbed form (Lippmann–Schwinger formulation) and the highly efficient solution in Fourier space utilizing the explicit form of the Green's function by means of fast Fourier transformation. Generalized boundary element procedures, such as the explicit-jump immersed interface methods, are related approaches that also allow for more exact boundary approximations. Just like the lattice Boltzmann methods, these voxel-based methods have existed for several decades. Their industrial utility, as robust microstructure solvers for arbitrary geometries, in particular for arbitrary material contrasts, has been significantly improved at the ITWM.

Today, at the Institute, microstructure simulation is a significant tool for determining the anisotropic characteristics of heterogeneous composite materials. Because the numerics are so effective, it is also used in industry for designing and optimizing virtual material structures and as a high-resolution material model in multi-scale simulations.

## 6 Numerics for Multi-Scale Problems

Multi-scale problems are of special interest to mathematicians, not least because they are very important in many areas of the natural sciences and industry. There has been much progress made in the past decades with problems that permit a clear separation of fine and coarse scales (e.g., periodic microstructures; media that are heterogeneous on the fine scale, but can be considered homogeneous on the coarser scale) (Fig. 4). In this case, the ratio of the characteristic lengths of the two scales can play the role of a small parameter, so that asymptotic analysis can be applied. Thus, rigorous results were obtained in the area of asymptotic homogenization, and it was possible to prove that the problems on the two scales can often be decoupled. In this case, the multi-scale problem then reduces to a two-step procedure: (a) one solves the periodicity "cell problem" at the fine scale and uses this solution to determine the coefficient of the macro-problem, and (b) one solves the problem on the coarse scale with the "scaled-up" coefficients, i.e., the coefficients determined in step 1.

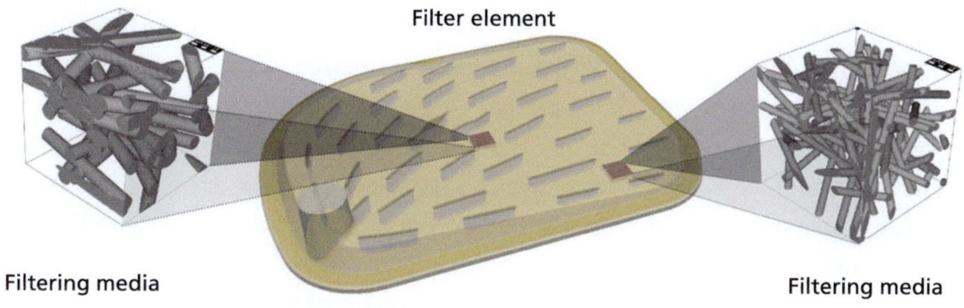


**Fig. 4** Illustration of micro-scale (filter medium, enlarged) and macro-scale (filter element) in filtration (graphics: G. Printsypar, KAUST)

The homogenization theory delivers everything needed to solve the overall problem: the operators that link both scales, the equations of the coarse scale (whose type can differ from that of the fine scale), estimations for asymptotic solutions, etc.

However, a clear separation of the scales is not always possible. Many problems are heterogeneous on several scales, and one cannot define any small parameters. Here, the problems of the different scales are strongly coupled and the two-step procedure mentioned above is no longer applicable. Instead, the coupled problems must be solved iteratively. The most active mathematical research in the field of numerical upscaling is currently being carried out in three directions: upscaling based on multigrid methods, upscaling based on multiscale finite element method and related approaches, and the application of upscaling for solving multiscale industrial problems.

The ITWM is active in the last two directions; in the first direction, there have only been isolated developments concerning the linear elasticity of composite materials. It is obvious that calculating the effective characteristics of composite materials or porous media from their microstructure is of great importance for industry. This was the point of departure for the development of numerous algorithms and software tools designed to help solve the cell problem derived in homogenization theory. Here, the focus was on the efficiency of the methods, so as to make the underlying "cells," that is, the microscopic sub-domains being treated, as large as possible. We have, for example, already reported on microstructure simulation in this section.

We now want to return to the case of non-separable scales and examine it more closely. Many successful developments have been based on the aforementioned multi-scale finite element method (MSFEM). Here too, one resolves the microstructure into sub-domains belonging, in turn, to elements of a coarser scale. This solution then feeds into the design of customized basis functions for the coarser scale. Where needed, an outer iteration across the various scales is applied, and the result is a "coupled global–local upscaling." MSFVM and a variational multi-scale method (VMSM) are both related to MSFEM. Another member of this family is a two-level domain decomposition method for equations with oscillating coefficients, where upscaled equations are used for the coarse grid correction. In their classical formulation, all members of this family are quite expensive, computationally, since they "see and touch" every unknown of the finer scale when they calculate the basis functions. A variant of MSFEM, the heterogeneous multi-scale method (HMM), sometimes seems to have a broader scope of application. With HMM, the subdomains for the fine scale cell problems do not cover the entire computational domain. Instead, one selects small sub-domains around the integration points of the coarser scale elements, and the micro-scale problems are solved only in these small sub-domains. The purpose of this procedure, of course, is to obtain macroscopic coefficients to be used for the macroscopic problems. The method is appreciably faster than MSFEM, since only a small part of the domain must be processed on the fine scale. On the other hand, this fact restricts the use of this method to problems in which the scales are "almost" separable. One can also carry out HMM with iterations between the scales, and one must do so in cases of non-linear problems. In the field of mechanics, the method is then known as $FE^2$. HMM is often viewed as a general framework for solving multi-scale problems, and not just as a special method. This framework also fits well when one considers the micro-problems discretely ("atomically," so to speak), while the macro-problems are treated continuously, that is, as PDE. Engineers have been using these approaches for many years, and HMM is an attempt to give them a suitable mathematical formulation.

The mathematical solution of multi-scale problems is often especially helpful for consortia of several industrial partners, in cases where the partners consider one and the same process, but some are more interested in micro-scale characteristics (e.g., filter media manufacturers, in the case of filtration), while others are more interested in macroscopic characteristics (e.g., filter element manufacturers). However, multi-scale approaches are also used by companies that generally work on a fine or coarse scale, but understand that their processes are, in reality, multi-scaled, and who therefore try to improve their products by taking the multi-scale interactions into consideration. Thus, manufacturers of filter media are usually interested in investigations of filtering processes on the pore scale, and they often simplify the conditions in the proximity of the actual filter medium (by assuming, for example, uniform flow velocity). But these companies sell their media to filter manufacturers. For this reason, it is advantageous if they also understand how their materials behave under the actual working conditions of the filter. For their part, the filter manufacturers seek to understand how their products behave on the scale of the filters themselves and, to do so, they assume they are dealing with ideal filter material. They too must know how filter media behave under operating conditions in order to be able to specify and utilize a suitable design. In summary: both kinds of firms profit from a true multi-scale simulation.

## 7 Image Analysis

The analysis of two-dimensional and three-dimensional image data in industrial environments is another area of activity at the Fraunhofer ITWM. The spectrum of applications runs from quality assurance in production via optical systems to statistical evaluation of

**Fig. 5** Visualization of a glass fiber-reinforced polymer, imaged using X-ray computed tomography (visualization: H. Riedel, Fraunhofer ITWM, department "Image processing")

three-dimensional microstructures in engineering and materials science. PDE were discussed at the start of this chapter. In the sense defined there, PDE are also used in image processing, for example, with so-called diffusion filters (operators for adaptive image smoothing). Various linear and nonlinear image-processing filters are described in the literature. Frequently, however, these are not directly applicable for industrial challenges. For example, it is often necessary to process enormous quantities of image data in order to attain statistically representative results with regard to the three-dimensional microstructures of materials such as composites, fleeces, or insulation materials. In such cases, standard algorithms frequently require a long run time, much storage space, or both. Adapting these algorithms to specific industrial challenges offers a way out of this dilemma. In this section, we describe two examples of efficient numerics for linear and nonlinear image-processing filters that we have developed in years past.

The first example comes from the application area of image analysis for fiber-reinforced polymers. These composites are currently used extensively in lightweight construction. In order to be able to analyze such materials—above all, their fiber systems–ITWM utilizes three-dimensional imaging by means of X-ray computed tomography. This technique achieves a spatial resolution down to the scale of a few microns (Fig. 5). But how can such image data be efficiently evaluated so as to process and analyze the largely anisotropic fiber system in fiber composites?

One possible way is to apply with an anisotropic Gaussian convolution filter. These linear image filter operators can, among other things, assume an elongated shape in 3D, so that they offer a good fit to the local shape of a fiber. Efficient implementations of these operators for two-dimensional images are indeed found in the literature (so-called "separable filters"). But how could a corresponding implementation be realized for three-dimensional data? To answer this question, we took a closer look, together with a colleague from the German Research Center for Artificial Intelligence (DFKI), at the underlying mathematical principles. The result of this work was not only a very efficient algorithm for this anisotropic filter operation, but also a fundamental geometric insight: a separated anisotropic Gauss smoothing corresponds to a shearing of the voxel grid that is independent of the dimension of the data. The geometric characteristics of this efficient filter implementation had previously been unknown, even in 2D. This fundamental understanding thus allowed fibers in fiber-reinforced polymers to be precisely and efficiently analyzed.

A second example of numerics in image processing comes from the field of nonlinear smoothing filters. Nonlinear filters are filter operations for which the result of the filtering in a voxel cannot be represented as a linear function of the voxels in the original image. The median filter is a very versatile nonlinear filter. It can suppress certain types of noise without damaging important information, such as the edges depicted in the data set. It has been shown in the literature that this filter operation can be implemented with a fixed number of calculation steps per voxel. When viewed from a theoretical standpoint, this means that it was hardly possible to find a more efficient algorithm.

However, at ITWM, we routinely use these median filters for very large three-dimensional image data sets ($2000^3$ voxels, or more). In these cases, the above-mentioned filter, with its fixed number of calculation steps per voxel, is indeed fast. At the same time, however, it occupies a very large memory block. The implementation described in the literature, for example, requires 8 GB of additional memory when filtering an image with $2000^3$ voxels, and 31 GB for an image with $4000^3$ voxels (when using a $7 \times 7 \times 7$ voxel mask). This storage capacity is indeed available on modern computers, but not on a normal desktop PC. Moreover, the storage problem will only intensify with every new generation of still higher-resolution CT scanners.

For these reasons, a novel median filter algorithm was developed at the ITWM. This new algorithm makes it possible to use the median filter for image data in three or more dimensions with a substantially lower memory requirement (e.g., 8 MB, instead of 31 GB, for the $4000^3$ voxel example mentioned above). This algorithm sacrifices the theoretical runtime optimality of the algorithm described in the literature for the sake of a reduced memory overhead. We achieve this by not executing the procedure described in the literature algorithm for each of the d dimensions of a data set (e.g., $d = 3$, for volume images). Instead, we interrupt after $d - c$ dimensions with a parameter $c$ $(0 < c < d)$, and manage the remaining calculations using a simpler algorithm. Surprisingly, we have found that, for many practically relevant situations, this theoretically slower algorithm is even faster than the theoretically optimal algorithm from the literature. This can be explained by the memory management overhead (allocation, reading, and writing of data). This overhead makes the theoretically faster algorithm, in practice, slower.

One can say, in summary, that the use of custom-made numerics in image processing makes it possible for us to work efficiently with large image data sets. When a technical challenge can be recognized and its underlying mathematical structure analyzed, one is often rewarded with innovative, practicable solutions.

## 8 Validation and Verification

When one develops new models—and the new software that often comes with them—both must be carefully validated. For cases in which there are no known analytical solutions for the models, then the validations must take place using numerical solutions. Here, one must be aware of various sources of error: modeling error, discretization error, rounding error, measurement error, and data error.

The first three sources of error are well known from a mathematical perspective and are treated in every lecture on modeling and numerical analysis. The last two error sources originate more in engineering practice, however, and are generally not mentioned in lectures delivered to mathematicians. Here, one often assumes perfect measurement accuracy. In reality, however, this is unachievable. Consequently, from a mathematical perspective, a total evaluation is not simple.

For industrial problems in particular, the validation process presents some challenges, which can arise for the following reasons:

Exact measurement results are often not available. Moreover, the question arises as to which portion of the results can be used. For example, a measurement of the pressure difference between the upstream and downstream sides of a filter contains no information about the velocity and pressure distribution within the filter. If one wants to validate a model that contains these quantities, then one needs detailed measurements. Or, one must treat the problem of identifying these internal quantities from the pressure difference as an ill-posed problem. In this case, measurement errors are especially critical. This is particularly risky when new parameter ranges are being processed and the measurement devices have not yet been calibrated for these values.

The communication between those taking the measurements and those doing the computational work, that is, between the engineers and the mathematicians, is often difficult, for they speak different languages. Great care must be taken here to ensure clear understanding. The human factor must be kept in mind.

As these reasons show, it is also very important for mathematicians to understand how engineers think and to learn something about their sphere of activity. Here as well, the difference between our work at the ITWM and that of academic institutions is made clear.

Even the phrase "exact result" has different meanings for engineers from different fields. To put it in plain terms: a 0.01 % relative error is considered an exact result in the aviation industry, as is 5 % in the filter industry, and up to 30 % in CFD for chemical process technology. It is also necessary to bear this in mind.

The world of research and the world of industry are different; each has its own language. But when the inhabitants of these two worlds meet, when they work together in a spirit of genuine cooperation, then both sides profit from their efforts: with innovations and, equally important, with enjoyment and gratification for a job well done!

# Data Analysis

Patrick Lang and Jürgen Franke

## 1 Data Sources

Today, due to the continuously advancing digitalization of production and business processes, data is being produced and often archived in an amount that only a few years ago would have been hardly imaginable. The drivers of this trend are the availability of numerous new sensor technologies and higher-performance data storage equipment. For many production processes in large industry, all potentially relevant adjustment and equipment parameters are now being recorded at high temporal resolution and then stored. Moreover, implementation of the Industry 4.0 concept, in which diverse, context-specific communication is to flow between production goods and production equipment, and between one production step and another, will lead to numerous additional data streams and, consequently, to a further significant increase in data volume.

The availability of ever more complex and precise measurement and analysis procedures also leads to the generation of larger quantities of data. One can think, for example, of the Next Generation Sequencing Procedure for genome analysis in the context of personalized medicine. Here, data on the order of terabytes can easily accrue with each analysis.

A further source for this flood of data is the increased networking of our world. One only has to consider the many data streams in the Internet, such as real-time stock market index updating; numerous social media with their own news channels; on-line service providers, such as eBay and Amazon, with their movements of customer data; or locally-resolved meteorological data streams. Moreover, in addition to current data, for

P. Lang (✉) · J. Franke
Kaiserslautern, Germany
e-mail: patrick.lang@itwm.fraunhofer.de



H. Neunzert, D. Prätzel-Wolters (eds.), *Currents in Industrial Mathematics*,
DOI 10.1007/978-3-662-48258-2_4

almost any question that can be asked, there exists a data base with corresponding historical data to answer it. Not only is the quantity of data increasing, but the opportunities of the individual for utilizing the publicly accessible flood of data are increasing as well.

Data, as it is generally understood, is not necessarily a structured combination of numerical values in the form of vectors, matrices, or time series; it can also refer to semi-structured or unstructured information, such as a simple piece of text. Due to its nature, the latter is not directly accessible to mathematical processing. Instead, it must first be prepared appropriately. The methods of information retrieval and text mining deal with this topic.

Media reports also currently feature the problems associated with "Big Data," which is typically characterized by the three "V's": volume, velocity, and variety. Volume refers simply to the size of such data sets, and velocity, to the speed with which streaming services can supply new data. Variety describes the heterogeneity of the data that might appear together in a common context. This brief description outlines the challenges facing the data analysis procedures that will be needed in the future.

## 2 Data Quality and Informational Content

The enormous amounts of existing and newly arising data remain relatively useless, unless we succeed in discovering new connections and knowledge within it. This is the main task of data mining and statistical learning theory, fields that have provided a multitude of algorithms for diverse scenarios (see [1] and [14]). Despite the existence of these methods and the software tools that accompany them, their use in the context of industrial production processes, for example, has not yet caught on widely. As shown in a joint project entitled "Supporting Decisions in Production Using Data Mining Tools," carried out by a consortium consisting of the ITWM, other Fraunhofer institutes, and representatives from the manufacturing industry, the disproportionately large adaptation efforts required for heterogeneous production domains and communication structures often cause significant difficulties. The lack of real-time capability for many of the analysis procedures also plays an important role here.

Generally speaking, especially in the context of dynamic systems, not all arbitrarily measured combinations of system inputs and outputs contain enough information in and of themselves to allow for complete identification of the system dynamics and generation of a corresponding system model. Discussions with customers from the manufacturing industry have consistently revealed that, although the adjustment and equipment parameters, for example, may indeed be highly temporally resolved, the product qualities to which they are assigned are only sampled randomly on a coarse time schedule. And there is another factor. Because the determination of these quality characteristics is often not automated, but performed manually in the lab, there are also long time delays before the data becomes available. Taken as a whole, this often means that the potential of high-resolution input data can only be realized in a limited way for modeling product quality.

For successful, data-based system identification, it is also crucial to have data from different operating points and/or different dynamic excitation states. Otherwise, the resulting system models are only valid within a very limited area and are usually not suitable for use in subsequent optimization or control approaches. The most informative generation of process data is methodologically supported by the design of experiments (DOE) framework, which seeks to achieve the largest possible variance reduction in the model parameters being estimated by means of the smallest possible number of suitably chosen measurement points. In our projects, however, we regularly run up against technical or economic limits regarding specifications in the experimental design about the amount of data to be collected and the selected process points. The insertion of appropriate filters to protect against technically impossible parameter combinations is very helpful, but, for reasons of complexity, is usually only partly feasible. It should also be noted that the experimental design only delivers explicit formulas for determining the system input settings for models that are linearly dependent on their parameters. For nonlinear dependencies, no generally valid formulas can be specified in advance. Instead, the DOE plans themselves depend on the results of the executed measurements and are of an iterative nature.

In the life sciences–for example, when considering the expression patterns of the more than 20 000 human genes–there is also often a multitude of potential influencing factors that might explain a specific disease. However, one has only a small number of patients available who have been classified and analyzed.

Another crucial point in the evaluation of data quality is the proportion of disturbances contaminating the observed data. Particularly with measurement data, there is always contamination of this kind caused by the measurement-principle-dependent characteristics of the sensors being used. If the characteristics of the processes generating the disturbances are known with sufficient precision, then they can be modeled explicitly, and this model can be used to correct the data for the impact of the disturbances. In practice, however, one is often dealing with the simultaneous overlapping of several disturbance sources, and the resulting complexity often makes mechanistic modeling impossible. Instead, one describes the disturbances as the result of stochastic processes, which can be characterized by the appropriate distribution information. The frequently made assumption that this data follows a normal distribution can indeed be justified in many situations, due to the law of large numbers. There are, however, very many technical and biological questions for which this assumption is false. Nonetheless, many well-established procedures presume a normal distribution, along with the linearity of the underlying data-producing process dynamics. If one generalizes these assumptions, for example, in the field of state and parameter estimation, one then moves from the well-known Kalman filter based methods to the sequential Monte Carlo approach. This is a method that has been actively pursued for several years in the System Analysis, Prognosis, and Control Department in its work with particle filters (see also "The Research—Robust State Estimations of Complex Systems").

In many application cases, it is not just disturbances in the data that cause difficulties. Often, the observed data sets are also incomplete, that is, some entries are missing. The values of some data sets may also be many times higher than the level of comparable data sets. The correct treatment of these defects and outliers, which can be caused by damaged sensors, for example, plays a decisive role in dealing with industrial data sources.

## 3 Data Integration and Pre-processing

The selection and allocation of suitable information-bearing quantities is crucial for the successful use of data analysis methods. In many industrial cases, this data is not to be found initially all together in some data warehouse, easily accessible to analysis. It is more likely to be distributed across different sources. Here, the spectrum runs from diverse databases to ASCII and/or Excel files to other, application-specific data formats. Occasionally, it still happens that certain data is only available in paper form and must first be digitalized. One initially looks for opportunities to extract the relevant data from all sources and bring it together into a higher-level data structure. Here, there are often problems in correctly assigning the data sets. In addition to solving these problems, one must also find suitable treatments for other incompatibilities, such as differing sampling rates among sensor data. Organizational challenges can arise when the needed data is distributed among different spheres of responsibility within different departments of a company.

As mentioned in the previous section, data sets are generally incomplete. One can almost always count on finding discontinuities and outliers. There are various procedures for identifying and adequately managing such problem cases, which must be chosen and executed according to the situation.

Along with integrating the data, one generally also subjects it to a normalization process and, possibly, a disturbance correction. Here as well, there are many procedures available for these work steps. In general, however, our project experience has taught us that, from the perspective of data analysis, it is desirable to retain as much control as possible over the entire chain of data processing steps. In accordance with this goal, one should always try to obtain data from project partners in its "rawest" form.

Moreover, to optimally select the next processing steps, it helps to first gain an overview of the data distribution. Especially for highly dimensional problems, one will make decisions on dimension reduction on the basis of the data's correlation structure and remove strongly correlated quantities from further consideration. In many cases, it also makes sense to execute the subsequent modeling steps not on the basis of the original data, but to draw upon compressed features instead. A well-known example of this is the principal component analysis, in which the original data is projected onto those sub-spaces that explain the largest portion of variance in the data. If the corresponding background information is available, one attempts in this step—in the manner of grey box modeling—to transfer this knowledge into a set of appropriate features. For more on this topic, see Sect. 5.

## 4 Data-Based Modeling

In almost all mathematical modeling questions arising from practical applications, the existence of an adequate amount of real, measured data plays a decisive role in the success of the model design. Depending on the type of modeling, however, the requirements for the quantity and information content of the needed data fluctuate markedly. With so-called white box modeling, in which the model design is strongly guided by the explicit implementation of physical, biological, or economic laws, the data requirements are rather moderate and serve primarily scaling and calibration purposes. In contrast, so-called black box approaches assume purely data-driven modeling, with correspondingly high requirements on the quantity and information content of the available data. With so-called grey box modeling, a hybrid form of knowledge-driven and data-driven modeling, the data requirements lie somewhere in between. For the remainder of this chapter, we will be concerned primarily with questions of purely data-driven modeling. For further discussion of white box and black box modeling, refer also to "The Concepts—Modeling."

Data-driven modeling approaches come into consideration primarily when sufficiently informative measurement data is available and the interrelations and dynamics of the observed systems or processes resist explicit description due to their complexity. Two examples here are the extrusion of plastic components, including variation in the material recipe, and the crash behavior of carbon-fiber composite materials.

In general, data mining includes procedures with which relevant information can be extracted from complex data. Here, statistical learning methods model the data as results from random experiments. This perspective makes it possible to derive, verify, and better understand procedures for gaining information on the basis of statistical theory and intuition.

Statistical learning has a great deal in common with machine learning. With complex data, statistics must rely on appropriate, computationally intensive learning algorithms. Conversely, the statistical perspective in machine learning often allows one to understand when and why data analysis algorithms function and how they can be extended.

An important distinction of data mining problems lies in the type of data being observed. So-called structure-describing procedures, such as regression and classification, are normally confronted with the problem of approximating a target quantity $Y$ (output, dependent variable) as accurately as possible using a function of the input quantity $U$ (input, independent variable, predictor). The data forms a *random sampling* or *training set* $(U_1, Y_1), \dots, (U_N, Y_N)$ of input variables $U_j$, together with the output variables $Y_j$. When learning the connections between input and output, one can therefore judge and optimize the system's performance on the basis of correct, observed values $Y_j$. In this case, one speaks of *supervised* learning.

With so-called structuring problems, in contrast, one has only input data $U_1, \dots, U_N$, in which one wishes to identify structures such as clusters or low-dimensionality. Because there are no output variables that can serve as starting points for correcting errors in the learning results, this is also described as *unsupervised* learning. The features $U_j$ are generally high-dimensional, and their structures usually cannot be simply visualized. Graphically representable projections onto two or three coordinate dimensions do not typically show the structures of interest. To make cluster formation or low dimensionality graphically visible, one must identify the most informative projections possible for this data.

## 5 Unsupervised Learning

With unsupervised learning, the focus is on characterizing the distribution and structure of the existing data. Along with observing standard quantities from the descriptive statistics, one is especially interested in discovering clusters and low-dimensional structures in the data. Here, there is also a strong overlap with the goals of data pre-processing, and unsupervised learning is therefore often used as a preparatory step in supervised learning problems.

One class of structuring problems arising in practice contains so-called variant management problems. Here, the input data describes the composition of complex products, such as commercial vehicles, for example, on the basis of their structural components. The goal is to find a sensible way to structure the product space, as defined by the customers of the associated company by means of the purchased products.

Here, the space should be approximated by the smallest possible number of representative products. This then allows one, in a subsequent step, to derive a plan for revising and reducing the necessary component spectrum and thus, decreasing inventory costs. The so-called cluster analysis is one method suitable for working on this question.

## 5.1 Cluster Analysis

One considers a finite set $U$ of objects, each of which is described by the characteristics $U_1, \ldots, U_m$ of a number of attributes. The central prerequisite for the grouping of data is the existence of a dissimilarity or distance measure $d : U \times U \to R^{\geq 0}$, which permits measurement of the similarity between two objects; the larger the value $d(U_i, U_j)$, the more dissimilar are the objects $U_i$ and $U_j$. In the cluster analysis, the goal is now to decompose the finite set $U$ into pairwise disjoint groups or clusters $C_1, \ldots, C_r$ :

$$U = \bigcup_{i=1}^{r} C_i, \quad C_i \bigcap C_j = \emptyset, \text{ for } i \neq j.$$

Such a decomposition is also called a partition of $U$. Each two objects within a cluster should be as similar as possible, whereas two objects from different clusters should be highly dissimilar. There are numerous algorithms for determining an optimal partition of $U$, which differ in search strategy and in the data types permissible for the features. The algorithms themselves frequently need specifications for the values of control parameters, such as the number of clusters to be sought, the minimal number of elements in a cluster, or the minimum dissimilarity between the objects of different clusters. Some algorithms also assume the specification of a start partition. This multitude of choices militates in favor of an external evaluation of the result partitions (in contrast to an evaluation within the algorithm regarding optimality) [8]. By comparing the results of a cluster algorithm for different parameter settings or start partitions, one can draw conclusions about, among other things, the stability of a result partition, the optimal number of clusters, and the coarse structure of the similarity space $(U, d)$. The comparison of partitions can itself be accomplished by means of a distance measure

$$D : P(U) \times P(U) \to R^{\geq 0}$$

which is defined on the set $P(U)$ of all partitions of the set $U$. Such measures have been used for many years in biology and the social sciences. One possibility for comparing partitions is the information variation introduced in [11], which represents a metric based on an entropy approach.

## 5.2 Feature Selection

During the process of preparing a data-based regression model, the choice of which features one uses to build up the model is crucial. In our experience, this decision is significantly more important for successful modeling than the choice of a special model class. Although individual input quantities can be used as features, in many cases, one relies instead on the functional linking of different input quantities. Clues as to how one arrives at the definition of the most information-rich features often come in the form of problem-specific expertise. Our project experience has shown us that these clues should definitely be followed. This helps to turn the original black box modeling at least partially grey.

Particularly in cases where there are no application-specific clues about feature definition, there is indeed in many applications the problem of a disparity between the high dimension of the input space and the relatively small number of existing input-output data pairs. Here, a dimension reduction is necessary, and one often carries out a principal component analysis of the input data. Restricting oneself to the principal components assigned to the largest singular values then delivers a corresponding subspace that is defined by the selected principal components. Another advantage of this approach is that the transformed data is uncorrelated and thus, in the case of normally distributed data, is even independent. Data that is given as a linear mixture of independent, arbitrarily distributed data sources can be decomposed into independent individual components using entropy based methods, such as independent component analysis (ICA) [9]. Entropy-based measures for quantifying the independence of two random variables, such as Mutual Information, are also often suitable for evaluating the explanatory power of a feature or a collection of features with regard to a given output quantity. On the basis of corresponding ranking criteria, one can then derive a variety of selection strategies for building up information-rich feature sets.

## 6 Supervised Learning

In the remainder of this section, we will consider supervised learning on the basis of input-output pairs $(U_j, Y_j)$, $j = 1, \ldots, N$, which are modeled as independent and identically distributed (i.i.d.) realizations of random variables. For the sake of simplicity, we will only look at the case in which $Y_j$ is one-dimensional. In contrast, the features $U_j$ used to predict $Y_j$ are typically highly dimensional in data mining. $(U, Y)$ stands for a representative input-output pair that has the same distribution and is independent from the observed data.

The goal of the learning is to find a mapping $f$, so that $f(u)$ approximates or predicts "as well as possible" a new value $Y$, when the associated input value $U = u$ is known. In order to refine this, a loss function $L_f(u, y)$ is specified that measures the quality of the approximation. The most widely used loss function for regression problems is the quadratic forecasting error $L_f(u, y) = (y - f(u))^2$.

Statistical learning now attempts to find a classification or prediction function $f(u)$ that delivers a good approximation on average, i.e., for which the expectation value $R(f) = \mathrm{E}L_f(U, Y)$ is as small as possible. For regression problems with quadratic loss functions, the optimal prediction is

$$f(u) = m(u) = \mathrm{E}\{Y \mid U = u\}$$

of the *conditional expectation value* of $Y$, given that $U = u$ is known. Because the distribution of the data $(U_j, Y_j)$ is unknown and can be quite arbitrary, the conditional expectation value $m(u)$, in practice, cannot be calculated. The goal of statistical learning is therefore to use the data to calculate approximations or *estimators* for this optimal function.

One demanding regression problem from the area of production is the quality prognosis of extruded plastic components. During extrusion, a mixture of plastic granules and other raw materials is melted in an extruder under the influence of temperature and pressure and, with the help of the extruder screw, pressed through an application-dependent mold. Such processes are used to manufacture window frame stock and insulation sheets, for example. Here, one is interested in the functional dependency of the thermal conductivity coefficient and the compressive strength of the extruded insulation sheets on the starting recipe and the settings of the equipment parameters, as well as on the various temperature zones along the extruder and the rotation speed of the extruder screw. Due to the complexity of the dependencies, an explicit modeling of the interactions is futile, and one resorts instead to historical production data and regression methods. The identified transformations then serve as the starting point for a subsequent process optimization by means of suitable Pareto optimization methods. For more, see "The Concepts—Optimization Processes".

A further example of a complicated regression problem from business economics is the calculation of the expected residual value of a leasing vehicle according to the specified duration of the contract. The value depends on numerous predictor variables, such as distance driven, vehicle model, engine, color, diverse equipment options, vehicle age, etc. If one knows the dependence of the residual value on this vehicle data, then one can estimate the capital value of the leasing inventory, plan future equipment packages so as to optimize the residual value, and so on. A similar regression problem is estimating the value of a house as a function of square footage, lot size, roof style, location, number of separate apartments, age and condition of the house, etc. What we are looking for is a forecasting function that predicts the price obtainable on the market as a function of all this data. In addition to providing support for specific purchase and sales decisions, this value information also plays an important role in appraising and mortgaging larger real estate projects.

In addition to regression problems for which the target quantity is continuous, so-called classification problems are also of practical importance. Here the $Y_j$ only assume values in a finite set $\mathcal{K}$, which, for the sake of simplicity, correspond to the numbers $1, \ldots, K$ of the $K$ classes. Figure 1 shows an example of a classification problem for two classes that is not separable by a linear classifier, but by a nonlinear one. Classification problems can be represented mathematically as special regression problems and thus will not be treated as a topic in their own right in the following discussion.

A challenging classification problem from economics is automatic fraud detection within the very large number of invoices that contracting firms submit to a company. On the basis of extensive information about the accounting data, such as amount and scope of the individual items, the identity of the invoicing party, etc., one uses statistical learning to decide whether there are any grounds for suspecting fraud and whether the invoice must therefore be examined more closely. An everyday example for the use of statistical-learning-based classification procedures are the spam filters in email accounts, which decide, on the basis of a large set of features, whether an incoming item is spam or a genuine email.

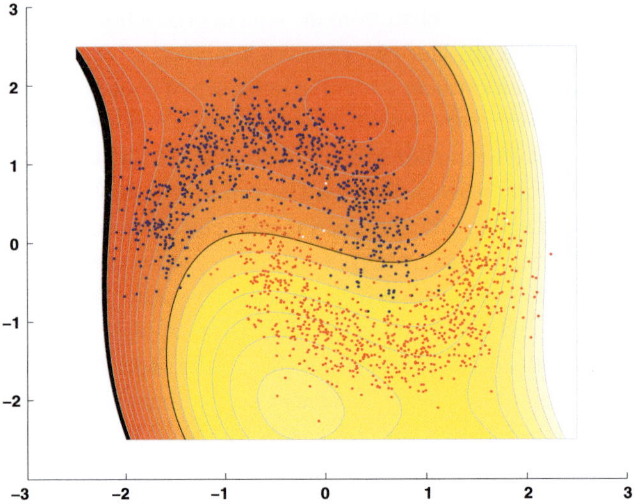


**Fig. 1** Nonlinear classification problem: class 1: *blue*, class 2: *red*

A representative classification problem from the field of bio-informatics is identifying a so-called biomarker for a particular disease within a set of gene expression data. In other words, one searches for genes whose common expression pattern is characteristic for the presence and severity of the disease in question. If such a biomarker is found, then it can be used to manufacture disease-specific test kits, which allow one to quickly verify the presence of the disease.

## 6.1 Non-parametric Regression

If one defines the residuals $\varepsilon_j = Y_j - m(U_j)$, $j = 1, \dots, N$, then they have conditional expectation values $\mathrm{E}\{\varepsilon_j \mid U_j = u\} = 0$. That is, $U_j$ contains no information about which average value $\varepsilon_j$ will assume. One usually assumes that the $\varepsilon_j$ are i.i.d., which means that the following standard model of non-parametrical regression [3, 6] applies for the data:

$$Y_j = m(U_j) + \varepsilon_j, \quad j = 1, \dots, N, \ \mathrm{E}\varepsilon_j = 0, \tag{1}$$

where $U_1, \dots, U_N$ are i.i.d. and independent from the likewise i.i.d. $\varepsilon_1, \dots, \varepsilon_N$. Moreover, one also usually assumes that the residuals possess a finite variance: $\operatorname{var} \varepsilon_j < \infty$.

In contrast to classical regression analysis, where the regression function $m(u)$ is assumed to be known except for a few parameters, non-parametric regression, and thus statistical learning as well, does not need these restrictive pre-requisites. Weak regularity assumptions about $m(u)$, such as twice continuous differentiability or quadratic integrability with respect to the distribution of $U_j$, are sufficient. The estimation procedure makes it possible to use the data to "learn" a predictive function that is largely unknown at the start.

Non-parametric regression approaches are not restricted to the standard model (1). For example, the residuals $\varepsilon_j$ can also depend on the independent variables $U_j$. One example is the heteroscedastic regression model

$$Y_j = m(U_j) + \varepsilon_j = m(U_j) + \sigma(U_j)\eta_j, \quad j = 1, \ldots, N, \tag{2}$$

with i.i.d. $\eta_j$, for which $\mathrm{E}\eta_j = 0$ and var $\eta_j = 1$. Here, it is not only the average, but also the variability of $Y_j$ that depends on $U_j$. The term $\sigma^2(u)$ is the conditional variance $\mathrm{var}\{Y_j \mid U_j = u\}$ of $Y_j$, given that $U_j = u$, and it can also be estimated using the same procedures as for $m(u)$.

An important class of problems that one repeatedly encounters in practice is characterized by dynamic developments in the target quantity over time. The above methods are also used in the corresponding non-parametrical time series analysis; one merely abandons the assumption that the $U_j$ are independent. For example, if one sets $U_j = (Y_{j-1}, \ldots, Y_{j-p})$, then the result is a non-parametrical auto-regression model

$$Y_j = m(Y_{j-1}, \ldots, Y_{j-p}) + \varepsilon_j, \quad j = 1, \ldots, N, \ \varepsilon_1, \ldots, \varepsilon_N \text{ i.i.d. with } \mathrm{E}\varepsilon_j = 0.$$

In this case, the auto-regression function $m$ delivers the best prediction of the value $Y_j$ of the time series at time $j$, using the last $p$ observations $Y_{j-1}, \ldots, Y_{j-p}$, inasmuch as the average quadratic prediction error is minimized. Correspondingly, one obtains non-parametrical versions of the ARCH models from (2), which play an important role in risk measurement in financial statistics.

## 6.2 Empirical Risk Minimization

The predictive function $m(u) = \mathrm{E}\{Y_j \mid U_j = u\}$ minimizes the expected loss $R(f) = \mathrm{E}(Y - f(U))^2$ (also known as *risk*) relative to $f$. With empirical risk minimization, in order to estimate $m(u)$, the risk is first estimated from the data, taking reference here to the law of large numbers, by

$$\widehat{R}(f) = \frac{1}{N}\sum_{j=1}^{N}(Y_j - f(U_j))^2. \tag{3}$$

Depending on the application, other loss functions might be more suitable, such as the $L^1$-risk, as defined by adding the absolute deviations of the amounts. Particularly for multi-dimensional target quantities, the search for an optimal loss function is commensurately complex. One must also consider that many prominent learning algorithms take advantage of the special characteristics of a quadratic loss function, in particular for the derivative formation. Thus, one must assume that there are significantly fewer suitable learning algorithms for more general loss functions. Particularly with classification problems, one is also often dealing with the kinds of problems for which the costs caused by a misclassification depend on the original class affiliation; that is, they are often particularly non-symmetric. Let us consider here a healthy person who is incorrectly classified as sick, and a sick person who is classified as healthy. While in the former case, a superfluous therapy is prescribed that is possibly accompanied by quite unpleasant side effects and unnecessary monetary costs, in the latter case, a possibly life-saving treatment is withheld from a sick person who requires it to survive. Arriving at a loss function that accurately reflects the characteristics of the problem under investigation and that can also be efficiently minimized is, in many cases, a key milestone in a successful data-based modeling endeavor.

One then attains an estimator for $m$ by minimizing the empirical risk $\widehat{R}(f)$. Minimizing across all measurable functions, or even merely across all twice continuously differentiable functions, leads to a function $\hat{f}$, however, that interpolates the data, that is, $Y_j = \hat{f}(U_j)$, $j = 1, \ldots, N$. Such a solution is unserviceable for use in predicting future data, since it models exactly the random disturbances $\varepsilon_j$ in the collected random samples, instead of adequately reflecting the general form of dependency between the random quantities $U$ and $Y$.

There are three strategies that allow empirical risk minimization to circumvent this problem:

- Localization, that is, restricting the averaging in the empirical risk to those $U_j$ lying in the neighborhood of that point $u$, at which one wants to estimate $m(u)$;
- Regularization, that is, imposing variation limitations on $f$ that rule out interpolating solutions;
- Restricting the set of functions across which (3) is minimized, which leads to the class of *sieve estimators*.

In the following sections, we will discuss important further aspects and implementations of these strategies.

## 6.3 Local Smoothing and Regularization

The idea of local smoothing for the estimation of a largely arbitrary regression function $m(x)$ can be derived directly from the law of large numbers: when $Y_1, \ldots, Y_N$ i.i.d., with expectation value $\mathrm{E}Y_j = m_0$, then the random sample average for $N \to \infty$ converges almost surely toward $m_0$:

$$\frac{1}{N}\sum_{j=1}^{N} Y_j \underset{\text{a.s.}}{\longrightarrow} m_0.$$

If, in regression model (1), $m(u)$ is smooth–for example, twice continuously differentiable—then $m$ is approximately constant within a small neighborhood around $u$. This means that, for small $h > 0$

$$m(z) \approx m(u), \quad \text{when } \|z - u\| < h. \tag{4}$$

If one now averages only those observations $Y_j$ in the neighborhood of $u$, that is, with $\|U_j - u\| < h$, then, for all $EY_j \approx m(u)$, that is, for large $N$,

$$\hat{m}(u,h) = \frac{1}{N(u,h)} \sum_{j=1}^{N} 1_h\big(\|U_j - u\|\big)Y_j \approx m(u), \quad \text{with } N(u,h) = \sum_{j=1}^{N} 1_h\big(\|U_j - u\|\big) \tag{5}$$

in which $1_h(z) = 1$ for $-h \leq z \leq h$, and $= 0$ otherwise. $N(u,h)$ is the number of observations in the neighborhood of $u$. Local smoothing of the data, that is, averaging of the data in the neighborhood of $u$, delivers a convenient estimator for $m(u)$. One obtains a convergence of $\hat{m}(u,h)$ towards $m(u)$ for one-dimensional $U_j$, for example, for $N \to \infty$, $h \to 0$ and $Nh \to \infty$.

The local averaging is based on assumption (4), for $z = U_j$, an assumption that becomes better and better as the distance between $U_j$ and $u$ decreases. This suggests therefore the idea of weighting the contribution of $Y_j$ to the local averaging according to how closely $U_j$ lies to $u$. Instead of a simple average, one then obtains a weighted local average. One example of this is the *kernel estimator*, in which the weights are generated by a function $K(u)$ known as a *kernel*. Typical choices for $K$ are probability densities, that is, $K(u) \geq 0$ and $\int K(u)du = 1$.

With a simple local average (5) and, in general, with kernel estimators, the bandwidth $h$ determines the size of the area used for local averaging. This leads to problems in estimating $m(u)$ when there are only a few observations $U_j$ in the neighborhood of $u$. Therefore, drawing on the same insight, *k-nearest-neighbor estimators* do not average across a fixed neighborhood surrounding $u$. Instead, they average across a fixed number $k$ of data points. Those data points $Y_j$ are chosen for which $U_j$ lies closest to $u$, that is, the averaging is performed across the $k$ nearest neighbors to $u$.

At first glance, *regularization estimators* appear to follow an entirely different approach than localized smoothing procedures for ruling out interpolation when minimizing an empirical risk. In (3), $\widehat{R}(f)$ measures how well the function values $f(U_j)$ fit to the observations $Y_j$. In order to avoid over-fitting, an auxiliary condition $r(f) \leq c$ is placed on the minimization of $\widehat{R}(f)$, where $r(f)$ is a measure for the variation of the function $f$. As a result, when $N$ is large, the strongly-fluctuating interpolating functions or nearly interpolating functions are ruled out as solutions. For some regularization estimators, an asymptotic equivalence to special kernel estimators can be shown (see [10] and [13]). Because the latter allow for a simple asymptotic theory, corresponding distribution approximations can be transferred to the regularization estimators and used for hypothesis tests and the calculation of confidence intervals and quantiles.

A recognized problem with the use of local smoothing procedures, one that, unfortunately, arises frequently with applications relevant to practice, is the so-called "curse of dimensionality." When these procedures are applied in the direct form described here for input spaces $U$ with high dimension $d$, then, except for extremely large random samples, the neighborhoods determined by $h$ are almost empty for many values of $u$. As a result, the random error is not averaged out. With k-nearest-neighbor estimators, the design does indeed ensure that averaging is always performed across $k$ values. But here, when the number of dimensions is large, the adaptively selected neighborhoods are necessarily very large. This corresponds to the choice of a very large bandwidth $h$ for kernel estimators, which leads to a systematic distortion of the estimator.

Especially when working on attractors of nonlinear dynamic systems that have been reconstructed using phase space methods, the above next neighbor methods can often be used successfully. Indeed, here, we have relevant project experience in connection with the risk evaluation of electrocardiogram data. In this context, the dimensions $d$ being observed are small to medium in size, and there is a relatively large data set. Nevertheless, it is definitely advisable to use efficient procedures when searching for each of the nearest neighbors; a naive implementation quickly reaches its performance limits. See [7], also.

## 6.4 Sieve Estimators

Sieve estimators dispense with localization or regularization as a means of avoiding over-adaptation or, worse, interpolation of the data. Instead, they achieve this by restricting the function class across which the empirical risk (3) is to be minimized. In order to still achieve the necessary flexibility and avoid limiting assumptions about the estimated function $m(x)$, the function class $\mathcal{F}_N$ being considered here grows with the random sample size $N$. A sieve estimator therefore solves the minimization problem

$$\min_{f \in \mathcal{F}_N} \widehat{R}(f).$$

To ensure that the resulting function estimator $\hat{m}_N(x)$ converges to $m(x)$, the function classes $\mathcal{F}_1 \subset \mathcal{F}_2 \subset \cdots$ must possess a *universal approximation characteristic*. That is, for each regression function $m$ being considered, there must be a suitable $N$ and an $m_N \in \mathcal{F}_N$, so that $m_N$ approximates the function $m$ with sufficient accuracy. There are various possibilities for refining this requirement. The function classes $\mathcal{F}_N$ are typically parametric, that is, they contain only functions that have been specified, except for a single parameter $\theta \in R^p$. Actually, just as in classical statistics, one adapts a parametric model to the data, but allows the model to be mis-specified. That is, one allows the function

$m$ being estimated to lie outside of $\mathcal{F}_N$. The non-parametric consistency of the procedure is achieved by allowing the parameter dimension $p = p(N)$ to grow as the number $N$ of data grows. Next, we will briefly discuss the three most important function classes.

As a starting point for designing the function classes $\mathcal{F}_N$, one often resorts to *series expansions* relative to orthogonal basis functions. Accordingly, the number of summands then depends on $N$. Sieve estimators can also be derived for non-orthogonal functions, provided that a universal approximation characteristic applies. In order to guarantee the stability of the estimator, one usually carries out an additional regularization of the coefficients of the series expansion. For corresponding convergence results, see [3].

The starting point for *partition estimators* is a disjoint decomposition of the domain of the input variables. Each of the estimators is then constant on each set of this partition, and the corresponding values are calculated as the average of the observations lying within the set. If the partitions become finer and finer as $N$ grows, then $\mathcal{F}_N$ possesses the universal approximation characteristic. A data-adaptive choice of partitions is advantageous. In many cases, tree-based methods are used here, and the corresponding estimators are then called *classification* or *regression trees*. See [2]. These approaches are useful for practical applications requiring the estimator to be interpretable, such as is almost always the case in medical applications, for example. Here, in very rare cases, one accepts a black box whose decisions may indeed be correct, but cannot necessarily be explained or argued satisfactorily. In particular, rule bases for decision-making can also be derived directly from the classification trees. This allows the plausibility of this procedure to then be evaluated in discussions with experts in the application domain.

*Neural networks* (see [4, 5], and [12]), originally developed as models for signal processing in the brain, represent an important class of sieve estimators. The best known of these are the feed-forward networks. In addition to the input and output layers, these networks possess at least one nonlinear, hidden layer of so-called neurons. These lead with the *activation function* $\psi$ to the following class of functions:

$$\mathcal{F}_N = \left\{ f(x) = v_0 + \sum_{k=1}^{H} v_k \psi \left( w_{0k} + \sum_{\ell=1}^{d} w_{\ell k} x_\ell \right); v_k, w_{\ell k} \in R \right\}$$

with the parameter $\theta = (v_0, \ldots, v_H, w_{01}, \ldots, w_{dH})' \in R^{(d+2)H+1}$. The classes $\mathcal{F}_N$ of output functions of feed-forward networks possess the universal approximation characteristic when the number $H$ of neurons grows as a function of $N$. The practical success of neural networks is the result of the existence of fast algorithms, particularly the back propagation algorithm and suitable modifications [15], which allow the network parameters to be learned within an acceptable time, even for large data sets $N$. An important point for successfully learning the underlying dependencies in the given data is the selection of a neural network whose size is adapted to the informational content of the data. The next section describes approaches for doing this.

## 7 Data-Adaptive Complexity Selection

All non-parametrical regression estimators contain tuning parameters with which the variation or complexity of the function can be controlled. They are utilized to force the estimation procedure to adapt to an adequate description of the actual dependency structure between input $U_j$ and output $Y_j$, instead of reproducing irrelevant random effects that are inconsequential for the prediction of future data. With kernel estimators, the tuning parameter is the bandwidth $h$; with next-neighbor estimators, it is the number $k$ of neighbors; and with sieve estimators, it is basically the number of free parameters of the function class $\mathcal{F}_N$. There is a variety of procedures that allow for data-adaptive selection of these tuning parameters.

The choice of tuning parameters is closely connected with the bias-variance dilemma and the problem of finding a balance between over-adaptation (overfitting) to the data and insufficient adaptation (underfitting) to the data. If the estimator is allowed too much freedom, overfitting will result; the estimator $\hat{m}$ adapts itself not only to the desired function $m$, but also tries to model parts of the random error $\varepsilon_j$. Conversely, if the estimator is allowed too little freedom, the result is underfitting. Here, the variability of the function estimator $\hat{m}$ is indeed small, but it deviates systematically from the function $m$ being estimated, since the bias $\mathrm{E}\hat{m}(u, h) - m(u)$ is large. Accordingly, it is also unsuitable for predicting future data.

The goal of the data-adaptive selection of tuning parameters is an estimator of the function $m$ that is as good as possible and that delivers optimal predictions. The average estimation error should be as small as possible, but is unknown. Therefore, one generally proceeds by splitting the data into training data and validation data; the training data is then used to calculate the estimator and the validation data is used to compare different estimators with different tuning parameters or complexity. When there is only a small amount of data available, and the estimation quality suffers significantly because some of the data must be put aside for validation purposes instead of being used for the estimation, then the cross-validation approach can be used [6]. This approach uses the data more efficiently, but at the cost of appreciably higher computation time.

## 8 Concluding Remarks

Our experience with industrial data analysis questions shows that an application-specific problem formulation, combined with the selection of suitable data sources and the features derived from them, plays the central role. Here, as much expertise as possible from every application domain should be brought to bear on the modeling process. The success of the endeavor generally depends more on this expertise than on the choice of a special machine-learning procedure.

Nonetheless, in all cases, the quality and informational content of a given data set also implicitly set an upper limit to the maximum attainable quality for learning a dependency structure based on the data. Here, it is very important to suitably adapt the complexity of

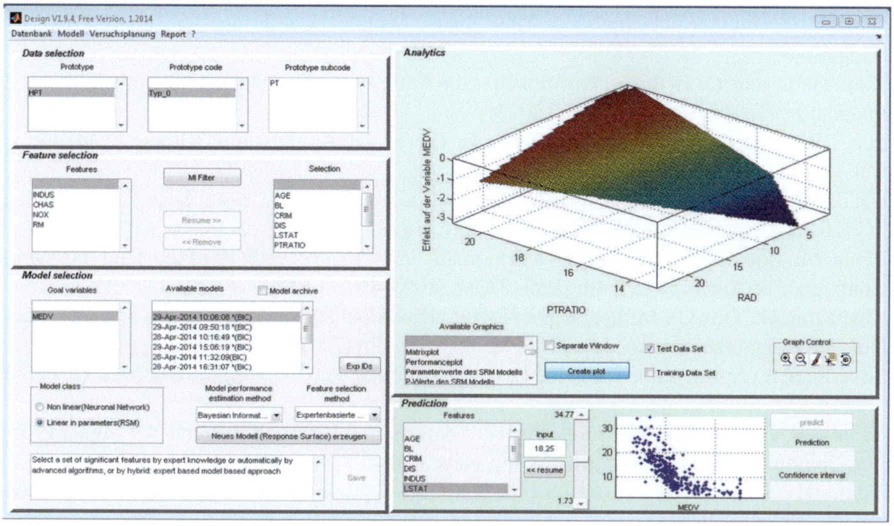


**Fig. 2** Data mining platform "Design"

the chosen model approach to this informational content. Acknowledging and integrating any additionally available expertise and domain-specific knowledge is always beneficial.

To promote acceptance of data mining procedures in industry, it is important, on the one hand, to supply high-performance algorithms that take into account the corresponding requirements and restrictions regarding run-time or data volume. At the same time, it is also crucial to support the user in selecting procedure parameters and interpreting and evaluating the results. Toward this end, we in the System Analysis, Prognosis, and Control Department have developed the analysis platform "Design" (Fig. 2). It can be easily adapted to diverse application contexts and data structures, and it contains a selection of effective machine learning algorithms. At the same time, however, it relieves the user of much of the work of setting critical procedure parameters.

## References


1. Bishop, C.M.: Pattern Recognition and Machine Learning. Information Science and Statistics. Springer, Berlin (2008)
2. Breiman, L., Friedman, J., Stone, C.J., Olshen, R.A.: Classification and Regression Trees. CRC Press, Boca Raton (1984)
3. Györfy, L., Kohler, M., Krzyżak, A., Walk, H.: A Distribution-Free Theory of Nonparametric Regression. Springer, Berlin (2002)
4. Hastie, T., Tibshirani, R., Friedman, J.: The Elements of Statistical Learning, Data Mining, Inference, and Prediction. 2nd edn. Springer, Berlin (2008)
5. Haykin, S.: Neural Networks: A Comprehensive Foundation. Prentice Hall, New York (1999)
6. Härdle, W.: Applied Nonparametric Regression. Cambridge University Press, Cambridge (1990)

7. Kantz, H., Schreiber, T.: Nonlinear Time Series Analysis. Cambridge University Press, Cambridge (1997)
8. Knaf, H.: Distanzen zwischen Partitionen – zur Anwendung und Theorie. Technical Report 226, Fraunhofer ITWM, Kaiserslautern (2013)
9. Lee, T.W.: Independent Component Analysis: Theory and Applications. Kluwer Academic, Norwell (2000)
10. Linton, O., Härdle, W.: Nonparametric regression. In: Banks, D., Kotz, S. (eds.) Encyclopedia of Statistical Science, vol. X. Wiley, New York (1998)
11. Meila, M.: Comparing clusterings – an axiomatic view. In: Proceedings of the 22nd International Conference on Machine Learning, vol. 84, pp. 1003–1013 (2005)
12. Montavon, G., Orr, G., Müller, K.R.: Neural Networks: Tricks of the Trade. 2nd edn. Lecture Notes in Computer Science, vol. 7700. Springer, Berlin (2012)
13. Silverman, B.: Spline smoothing: the equivalent variable kernel method. Ann. Stat. **12**, 898–916 (1984)
14. Vapnik, V.N.: Statistical Learning Theory. Adaptive and Learning Systems for Signal Processing, Communications and Control. Wiley, New York (1998)
15. White, H.: Some asymptotic results for learning in single hidden-layer feedforward network models. J. Am. Stat. Assoc. **84**, 1003–1013 (1989)

# Optimization Processes in Practice

Karl-Heinz Küfer

## 1 *Improve?* or Perhaps even *Optimize?*

"We are always working to optimize our processes!" or "Our products are continually optimized!" are sentences often heard at press conferences or at the kickoff events of large business projects. *Optimize* has a positive connotation and is a word that is heard and spoken gladly, due to its positive associations. It is often used together with *always* or *continuously* or *ongoing*, words that imply "we are relentless in our efforts" and suggest a permanent process.

But what does *optimize* actually mean? Upon asking, one usually hears that optimize means to improve, to make something better than it was before. To put it abstractly: one starts with a current status, and one strives to change the current status so that something is made better. The colloquial expression, *make something better*, means that one attains more favorable values or evaluations for at least one objectively measurable benchmark. For example, that something can be produced more cheaply with the same quality, or that the consumption characteristics of a product are improved.

Actually, however, when one reflects on the meaning of the word *optimal* in light of its Latin origins, then it becomes clear that optimal means *best possible*, and not simply *better*. Thus, when optimizing, one is striving to choose the best alternative from the multiplicity of alternatives defined by a set of boundary conditions, according to an objective benchmark. To do so, one needs not only a starting situation and a target situation with comparison criteria, as when one merely wants *to improve*. One needs instead a description of every sensible alternative, out of which then *the best one* is chosen. Optimizing is therefore more than just improving, or, to return to Latin, *meliorating*.

K.-H. Küfer (✉)
Kaiserslautern, Germany
e-mail: kuefer@itwm.fhg.de



H. Neunzert, D. Prätzel-Wolters (eds.), *Currents in Industrial Mathematics*,
DOI 10.1007/978-3-662-48258-2_5

Was everyone daydreaming in Latin class? Why do not people speak of meliorating instead of optimizing when they want to say they are improving something? Or do they in fact really want to optimize, that is, to find the best possible alternative? "Can this even be done?" one is inclined to ask.

## 2 The Mathematical Optimization Task

For mathematicians, the world needs to be well defined: there are many layout alternatives within a *design space* that are exactly defined, either explicitly or implicitly. *Exactly defined* means it is absolutely clear whether a particular alternative is permissible or not. "That depends", the typical colloquial response to the question "Can we do that?" does not exist. Along with these many *permissible solutions*, one also needs at least one, or even several, usually numerical—also known as scalar—valuations or grades. These so-called *objective functions* or *target quantities* help with the comparison of alternatives. One always assumes that such evaluations are possible for all options and that each alternative is completely rankable with regard to each of the individual evaluations. The evaluation quantities then create a *decision space* of one or more dimensions in which the evaluated options can be selected.

One thing is clear: if the decision space is one-dimensional, that is, if there is but one objective function, then it is simple to characterize the best possible choice, at least when there is a finite number of options to choose from. However, when there are several dimensions or objective functions in the decision space, as is usually the case in everyday life, then it is a bigger challenge. Take price and quality, for instance. The least expensive alternative is rarely the one with the best quality. Should this indeed be the case, then we have the happy circumstance that one alternative *dominates* all others. The more likely situation, however, is that one must make compromises. For example, one establishes a price limit and then selects the best quality. Quality is frequently evaluated using several target quantities, however. What do we do then?

For this situation, mathematicians have defined the term *Pareto optimal solutions*, or options, named after the Swiss economist and sociologist Vilfredo Pareto. A solution is Pareto optimal when it is impossible to find an alternative that is better in at least one objective function while remaining at least as good in all the others. Pareto optimal solutions are thus alternatives that are not completely dominated by at least one other solution. Usually there are numerous such Pareto optimal solutions from which decision-makers have to select a compromise that appears favorable to them.

If one looks at university lecture catalogs and the relevant mathematical literature, one finds a multitude of lectures, books, and technical articles under the heading *optimization*, which address themselves to the following topics:

**Questions from the field of mathematical optimization**
Given a set of alternatives in a well-defined design space and a set of well-defined scalar objective functions that establish a decision space.

- Are there optimal solutions and are they unambiguously described? If so, then find a complete description or characterization of these optimal solutions?
- How does one find the optimal solutions? Which algorithmic concepts exist to do so?
- How much time and effort will it take to find the optimal solution(s), or at least to come close to them?
- What is the best algorithm for finding, or at least approximating, the solution set, taking into consideration the computation time and storage complexity?

Along with the various disciplines of mathematical optimization, one should also mention decision-theory. This field seeks to shape the finding and selecting of best-possible alternatives into an (optimizable) process of its own. It also tries to help decision-makers orient themselves quickly and efficiently in the decision space, offers them rules-of-thumb for decision processes, and helps them evaluate decisions made in the context of uncertainty and partial information. A complete overview of mathematical optimization can be found in the compendium [12].

## 3 Are There Mathematical Optimization Problems in Practice?

Before we reflect upon the practical benefits of the mathematical optimization concept just introduced, it makes sense to first critically examine the principal assumptions of the mathematical optimization task. Is the mathematician's picture of an optimization task with well-defined alternative sets and explicitly known, completely described target quantities at all relevant in practice?

If one comes to industry as an applied mathematician with expertise in optimization and introduces oneself in this manner, one is welcomed with open arms. *Optimization* is a very familiar term in industry, although it is perhaps used in a slightly different way there. It has, as mentioned previously, a positive connotation, which makes starting a conversation quite easy. Practitioners frequently even have experience with optimization software and, thus, also with mathematical optimization approaches in various contexts. Optimization components are readily bought and are therefore important to many software suppliers. Optimization routines are often viewed as intelligent extensions to administrative software or simulation tools, and they generate interest at trade fairs and software demonstrations.

However, if one looks at the daily realities of businesses and inquires as to which processes, which daily optimization tasks, are actually supported with the help of the optimization routines from the purchased software, the picture is quite sobering. Unfortunately, automatic optimization routines are very seldom used. And this, despite the fact that they are, by definition, designed to simplify and accelerate the tiresome process of finding good or best solutions. That is, despite the fact that they promise an inherent, generic benefit.

Why is this? Why isn't better use made of these often very expensive systems? Are the algorithms bad? Are the routines too slow? Are they poor at finding or approximating good solutions? Generally speaking, these are not the explanations. In fact, two cardinal problems frequently lead to rejection of optimization routines:

*Cardinal problem A—an inflexible model:* routines are not used because the modeling of the optimization task, that is, the defining of the objective functions and the feasibility of the alternatives, is not flexible enough. Or, the user himself must first prepare or modify an optimization mode—a very time-consuming prospect. The latter is typical of *generic* optimization packages, in which the mathematical method or algorithm is the focal point, and not the particular application.

*Cardinal problem B—no interaction:* in optimization packages, genuine interaction with the model is often impossible. All boundary conditions for the set of feasible solutions and exact objective criteria must be fixed *a priori*. Then, on the basis of this rigid model, solutions are found. This forces the user to accept the calculated solutions just as they were found by the optimizer. *What if scenarios* cannot usually be considered *a posteriori* without a completely new computation using a suitably modified model. This often makes it impossible to fix partial aspects of the initial solution in the design space and then let the optimizer make further improvements around these fixed aspects. Moreover, it is also impossible to simultaneously analyze solutions in the design and decision spaces. This is a fatal shortcoming, since practitioners love to think and plan in design aspects; to them, target quantities are merely dependent auxiliary variables.

Let us now illustrate these observations using an example taken from the Fraunhofer ITWM's past experience. The research and development department of a power plant builder contracted with the Fraunhofer ITWM to evaluate planning software they had prepared themselves for photovoltaic power plants. During a visit to coordinate the evaluation work it became clear that there was a wide gap between the wishes of the management team and the ideas of the planning engineers. The managers had the notion that a comprehensive set of rules describing how a photovoltaic power plant should be erected on a pre-selected building site could be used to generate a well-defined proposal, after entering a few parameters. This could then be applied to the piece of land in question and, with the help of an algorithm, used to design a power plant yielding the highest expected profit for the investor. Quality assurance of the technical design, cost modeling, and return-on-investment simulation were to be components of the package. The goal was to use this software to generate, within a few minutes, valid planning recommendations that could be used as a basis for preparing a bid for the customer. The planning engineers, for their part,

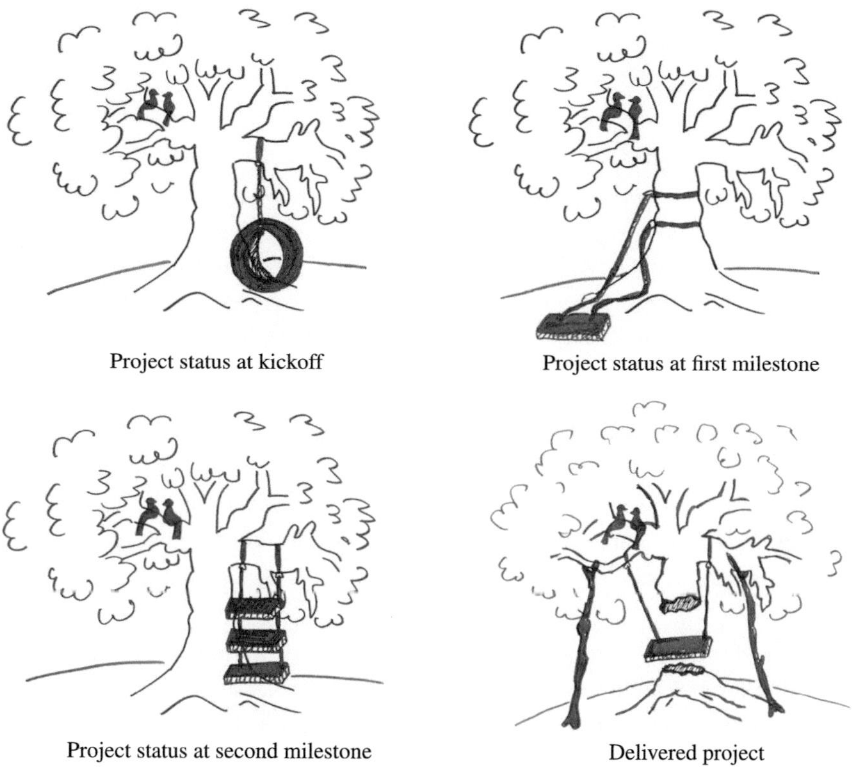


**Fig. 1** Practial project work requires flexibility!

viewed the idea of such a "turn-key solution" with great skepticism. Initial tests with the new system showed that one could fail for the widest variety of reasons in one's search for good recommendations. The engineers favored, instead, a simple modular system, which could be used to prepare a recommendation in an extended CAD system without built-in "optimization intelligence". Planning a new installation in this way would take several days.

So, the contract with the customer called for us to analyze the model and solution approaches used in the software and, where needed, to offer suggestions for improvements. At the very least, however, we were to help objectify the points of contention. Under these circumstances, one felt a bit like one had been beamed into the world of Fig. 1, in which equally large discrepancies in the objective targets are apparent.

A first examination of the software and several long discussions with potential users from the circle of planning engineers revealed evidence of both cardinal problems. The model of the photovoltaic plant recommended by the software was much too rigid and was partially over-specified, which led to many good solutions being absent from the alternative set. The lack of interaction capabilities prevented specifications that were necessary in this context, but could not be captured in the model, from being entered manually. Thus, it was impossible in even simple cases to use the system to generate viable solutions. On the other hand, it was just as clear that a complete customized planning, as desired by the engineers, would greatly hinder a comparison of solutions or an orientation on generally valid standards. In the end, the Fraunhofer ITWM was able to deliver a solution concept for the contracted task that led to a larger, still on-going, research and development project.

For mathematicians, some old questions of principle may heave into view here: Doesn't mathematics always need a rigid or well-defined model? When there is no such model, is this not due to the insufficient diligence and underdeveloped mathematical discipline of the practitioner? Of what use is interactivity when one can fix exact constraints and target quantities in advance so that only the best possible solutions can appear?

## 4 Flexible Optimization Concepts for Practical Use

On the basis of many years' experience in practical project work, the scientists at the Fraunhofer ITWM have come to the basic conviction that both cardinal problems can be solved with the help of mathematics.

**Parametrical Boundary Conditions as a Lever for Interaction** From out of a well-defined multiplicity of alternatives, a mathematical optimization process uses the light of target quantities to illuminate the best possible proposals. The definition of the alternatives is carried out by means of constraining conditions, which can be formulated mathematically as equality or inequality constraints. In some cases, one uses combinatorial constraints or stochastic bounds.

Some of the specified constraints result from natural laws, others from legal stipulations. In both cases, one must accept them as *hard constraints*, since they lie outside the defining powers of the modeler. In practice, however, one often observes that a far larger number of constraints can be traced back to the arbitrary defaults and guidelines of planners and decision-makers. Such constraints are often revisited when their acceptance results in optimization solutions that don't fulfill one's wishes or expectations. Here, one then hears phrases like "If we can, we ought to relax this constraint a bit!" Mathematically speaking, this means that one should change the constraints and once again compute and optimize. But how then should the constraints be determined? Should one ask how to shape the constraints on the basis of a target number for a target quantity? If so, one might then want to couple the constraint parameter itself to the corresponding target quantity.

But this might then compromise another target quantity. This means that treating a *soft constraint* in the same manner as the hard constraints frequently leads to a sequential solution and evaluation of parametrically altered optimization tasks. Hence, using the iterative trade-off of model selection and target evaluation, one seeks the best possible model for the optimization, into which one would like to feed knowledge gained from the objective function relations *a posteriori*. How can such an iterative modeling process be avoided?

There are basically two variants in the optimization process for treating soft constraints. As mentioned above, one can integrate constraint parameters *a priori* into the objective function(s). This assumes that one has at least a rough overview of their functional relationships. Otherwise, one runs again into the aforementioned iteration dilemma. Alternatively, one can solve a host of optimization problems that are dependent on the constraint parameters of the soft constraints; store the collection of solutions; and then illustrate to the decision-maker the *a posteriori* unknown functional relationships by means of an interactive *navigation* through the solution landscape. Navigating the solution landscape means considering the parametrical solution diversity with the help of a computer-supported depiction. One can thus contemplate the solution landscape in real time, as it is altered by the constraint parameters, and study the dependencies between the constraint parameters and the achievable target quantity or quantities. The extra overhead accruing from the solution process of such a parametric task is compensated for by avoiding the unsatisfying iterative process of model modification and optimization run.

In engineering practice, this technique is known as a parameter study. However, engineers usually carry out such studies merely for simulation purposes; the evaluation of simulation results using target quantities is done manually, and optimization runs for finding favorable parameters don't normally take place. The approach described above goes further, however. Instead of considering the results of a simulated parameter study, one analyzes interactively the solution landscape of a host of optimization problems with parametrical constraints. An overview of parametrical optimization is found in [2].

**Multi-criteria Models for Improving Target Function Flexibility** Even more difficult than describing the permissible alternatives for an optimization problem is accurately evaluating the alternatives using objective functions that ought to cover all aspects of an intrinsic value. In practice, one frequently observes the attempt to postulate a single objective function that integrates all these target values. This is usually done by coupling all relevant objective values, such as those for cost and quality, with the help of artificial weights. These weights typically add up to one. Thus, the weight of an evaluation aspect indicates its significance for the whole in the form of a percentage. What is the most sensible way to choose such weights?

Actually, such weights represent translation rates between the various target quantities. So, how can quality be translated into cost?

Managers usually offer a simple business solution: "We convert all aspects of an evaluation into currency and then maximize our margin!" This sounds simple and convincing, but is it really possible? Let's assume we are considering a suggestion for a new product, and we have to translate the benefit of a quality aspect into money. To do so, we have to know how the increase in quality will translate into additional units sold on the market, and thus, into additional revenues. This brings us to prognosis-based, expected revenues, which we then have to discount by the expected costs. The stochastic comes from the unknown sales figures of a new or altered product. Management will put a question mark behind these figures, since sales estimations for a new product are notoriously risky. One will have trouble finding good weights that are valid for all alternative solutions.

The situation is even more difficult when we must weigh the benefits of a medical therapy against its *costs*. *Costs*, in this case, might refer to the actual costs of the therapy, such as those for an expensive chemotherapy for a gravely ill patient. Is it fair to burden the community of insurees with the costs of a very expensive treatment that increases one individual's life expectancy by a few weeks or months? *Costs* might also refer, however, to the side effects or post-treatment complications afflicting a basically curable patient. How does one translate the hours of life of an individual into costs for a community of insurees? How does one balance the healing of an illness against subsequent complaints that last a lifetime?

This makes weighting a very complex matter, if only due to ethical considerations. In practice, such weights are indeed set; they are rarely talked about, however. Is it necessary to work with weights? How do the solutions depend on the weights selected? What happens if one sets the wrong weights?

Let's assume initially that one has set the weights according to past experience, or simply made them all equal to start with. One then computes solutions and considers the relationships of the objective values. If these are unsatisfactory or unbalanced, then one tries adjusting the weights. In a minimization problem, for example, if an aspect is unacceptable, its weight is increased. The weights for the other criteria are then decreased accordingly. One solves the optimization task again, and then iterates again across the setting of the weights, just as with the soft constraints, until one arrives at a solution that is more or less balanced in all relevant aspects. How can such a "human iteration loop" be avoided?

Here too, it makes sense to let aspects that are non-comparable or at least non-translatable remain independent of each other. Instead of computing the solutions of a one-criterion task, one then computes the solution landscape of a multi-criteria task, that is, the Pareto solutions. *A posteriori*, once the Pareto set has been computed, the decision-maker can determine by means of a navigation process which target criteria are in conflict with which others, how an assumed constraint for one quantity impacts the achievability of another, and so on. Thus, here too, one can first work with a more generally described task, whose more complex solution landscape can be studied and used to gain insights into implicit dependencies and to find a balanced alternative that is appropriate for the context. An overview of the treatment of multi-criteria tasks and the interactive use of their solution sets is found in the monograph [22].

Figure 2, for example, shows the main screen of a software program developed by the ITWM for planning radiation therapies. The screen is divided basically into two halves: the right half shows a single radiation therapy plan, depicted by means of dosage distribution in color wash on the 3 classic 2D-CT body cross-sections. The contours of the relevant organs and target volumes (violet and pink), along with the healthy risk structures (other colors), provide orientation as to which dosage is to be expected in which volume. An additional dose-volume histogram shows what percentage of the relevant structures (target volumes and risk structures) are being radiated with what dosage. This plan, shown on the right side, is a Pareto optimal alternative within a set of feasible solutions specified in

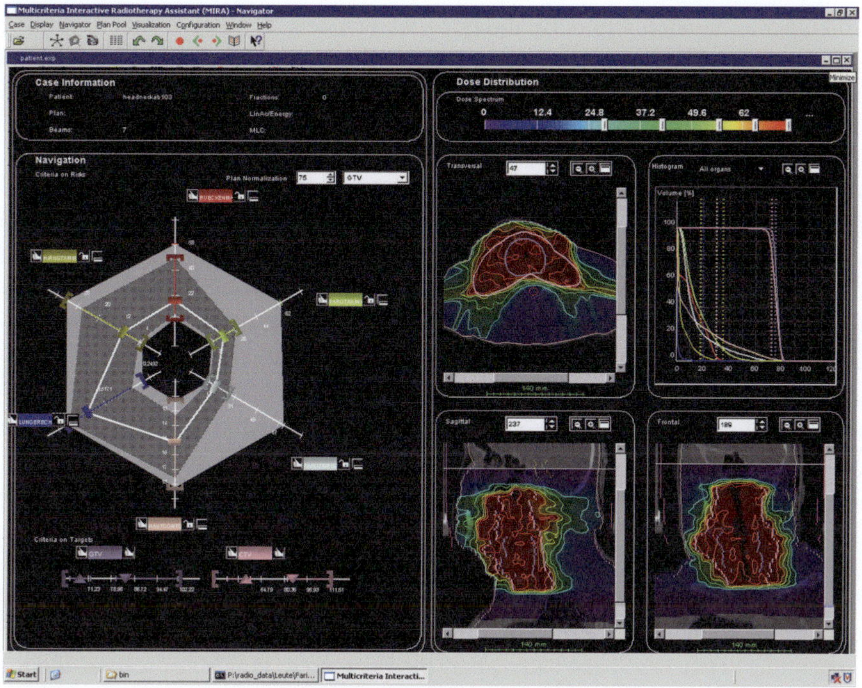


**Fig. 2** Navigation screen in the ITWM radiotherapy planning software

advance. The target criteria here are the target dosage averages that are to be achieved as a minimum or maximum, and the maximum dosage averages that are not to be exceeded in the risk structures. The specific method of averaging dosages is selected for the risk structures on an organ-by-organ basis.

The left side of the figure shows an overview of all the Pareto alternatives. The radar chart with six evaluation axes in the upper section shows the possibilities for the risk structures, and the two linear axes below show the situation for the 2 target volumes. The evaluations on the axes are specifically-selected dosage average values for each of the structures of interest. The area covered by the Pareto solutions is bounded by brackets. The intervals thus marked out are joined together two-dimensionally in the radar chart. The alternative selected for the right side, with each of its dosage averages, is depicted in the radar chart as a polygon. The currently shown solution for the target volumes is marked in each case by a lower dosage value and an upper dosage value, each of which is depicted in the graphic by a triangle.

There are two options available to help in selecting a desired solution from the Pareto solutions:

**Restriction** The intervals in the alternative landscape can be limited by dragging the brackets. This is primarily used to interactively exclude undesirable alternatives. Because the evaluations on the axes depend implicitly on each other, such restrictions render certain areas on other axes inaccessible. By dragging the brackets, these no longer accessible areas, visualized by surfaces on the radar chart in real-time, project forward in high contrast tones. Thus, the decision-maker immediately sees the effect of the desired constraint. If the limitation goes too far, for example, one can adjust the bracket accordingly.

**Selection** Likewise, one can change the currently depicted solution using the tags (the polygon corners) on the axes. If a different dosage average value is desired, one drags the corresponding tag of the desired structure, thus changing the solution. Generally, this is possible as long as one remains within the active interval of the Pareto set. However, this action causes the values on the other axes to change as well. This is because all the depicted solutions are Pareto optimal. If one wants to improve such an alternative with regard to one target quantity, then at least one other will worsen. The selection mechanism works so that the burdens that result for other structures are distributed onto the other axes as uniformly as possible, in order to keep each of those changes as small as possible. Obviously, this only happens within the active intervals, as they are defined by the bracket settings. Here too, the decision-maker can see the effect of his changes immediately in real-time and respond appropriately.

Restriction and selection are two examples of interaction mechanisms on Pareto sets that can change bounds *a posteriori* or create a desired solution ratio between the various target quantities. They have been patented by the Fraunhofer ITWM for several application domains.

Since 2001, the multi-criteria optimization of radiation therapy planning has been prepared and improved for clinical use at the Fraunhofer ITWM in several interdisciplinary, sequentially coordinated projects with international partners in research and industry (cf. [10, 14, 18–20, 25, 30, 37] and [35]). The first clinical evaluations are presented in [36, 38, 40] and [11]. A mathematics oriented presentation of this can be found in the dissertations [24] and [31]. In addition to radiation therapy planning, similar studies on radiofrequency ablation were carried out in the medical therapy planning field. This minimally invasive technique uses heatable applicators to remove tumors or metastases by thermal ablation. See [13] and [34] for more on this topic.

Multi-criteria optimization is helpful not only in medical therapy planning, but also in the design of complex technical systems. At the Fraunhofer ITWM, work has been done on the design of cooling channels in injection and pressure casting molds (see [21]), on the design of photovoltaic installations and power plants (see [6] and [5]), and on the planning of chemical production processes (cf. [8] and [9]).

Along with the above-mentioned physically modeled applications, research has also been done on organizational tasks, such as improving the connection reliability in local public transportation networks (see [28] and [29]).

**Integration of Modular Elements and Optimization Using Design Rules** In the previous sections, we have illustrated how to fulfill the postulate of an optimization task—mathematical rigor regarding the alternative set and the target function(s)—by appropriately relaxing soft constraints and using multiple objective concepts. This presupposes, however, that constraints and target quantities are essentially known. In practice, however, it is not uncommon that the context of an optimization task plays an important role. In other words, constraints or changes must sometimes be implemented due to requirements that are not contained in the model.

In such cases, is it still possible to make productive use of mathematical optimization?

Yes, it is possible. Provided one is willing to relinquish a fundamental paradigm of mathematical optimization. Mathematical optimization concepts always assume that the design space or the alternative set is completely known, and that the attention of the decision-maker can thus be completely directed to the result space. This attitude is essentially foreign to the engineer, however. For him, the design space is "where the action is"; the target quantities are simply dependent indicators for evaluating the alternatives. For him, the changeable objects in the design space are often context dependent, and he wants to be able to work with them.

One can also gratify this wish by means of interaction alternatives. Here, one presents the design space to the decision-maker as a modular system, so that he can piece together alternatives for himself. One also provides a rule-checker with which these alternatives can be verified against known constraints. One uses the objective function(s) as evaluation quantities and then places the alternatives thus found in the context of an automatically computed solution landscape. This procedure creates the possibility of integrating context, while at the same time offering the decision-maker a chance to evaluate the quality of his own solutions *vis a vis* a mathematically optimized solution landscape that satisfies all known conditions.

A further step for simultaneously using the design space as a construction kit and a decision space is an alternating sequence of setting binding specifications in the design space and then automatically optimizing according to these specifications. This allows one to build a seamless bridge between a completely automatic optimization with a previously known, complete set of design rules and a totally manual planning process, in which target quantities serve only the purpose of orientation.

This last concept is accepted by most practitioners without all too much resistance, since it can be flexibly implemented by all parties. The fan of modular elements can stick to his accustomed way of working, just as the fan of the "one button solution" can stick to his. One disadvantage of this concept, however, is the increased difficulty of comparing solutions when the design rules are only partially binding.

## 5 Integrating Simulation and Optimization—The Curse of Complexity?

With many layout questions concerning technology, it is necessary to also draw on physical simulation models, which are used to help create a virtual representation of the product or process of interest and to study its behavior. Evaluation dimensions for quality and cost are often functions taken from simulation results. Depending on the physics used and the model granularity, simulation runs can require a good deal of computation time. For their part, optimization algorithms frequently retrieve target function evaluations. Now, if one needs a complete simulation run for each target function retrieval, then it is often impossible to simply and sensibly integrate optimization and simulation algorithms, since this would lead to absolutely unacceptable computation times.

Let's look, for example, at radiation therapy planning in medical physics. Using fixed settings with the physical therapy set-up, a single run on the basis of Monte Carlo simulations lasts at least several minutes, since imitating the complex physics of collision processes and particles in non-homogeneous body tissues is very time-consuming.

When confronted with such a challenge, how can one nonetheless integrate simulation and optimization? The classical approach is to simulate a few constellations that are composed according to past experience. In this case, mathematical optimization algorithms are not used. This approach is widely used in practice and frequently leads to results that are far from being "the best possible".

**Reduced Models and Hierarchical Concepts** One way out of the incompatibility between simulation and optimization is to use simplified models, which in some domains are also referred to as short-cut models. See [4], for example, for process technology. Here, physical laws are initially simplified or left out entirely in order to achieve faster simulation run-times. It only makes sense to use such a reduced model, however, when the differences relative to a more refined and realistic physical model can be assessed with an error estimator, and the coarse and fine models can be used in a complementary fashion.

Using a coarser model with error estimators, one can initially make optimization runs and then, with the help of the error estimator, exclude zones in the design space that, under no circumstances, can contain good or optimal solutions. In the literature, so-called surrogates are also often used here for the objective functions (cf. [15])

In the most favorable case, the solutions found using the reduced model can even be verified by means of the more refined simulation run. It is standard practice in radiation therapy planning to approximate the discrete physics with a continuous reduced model, such as a pencil beam model; to optimize using the reduced model; and finally, to verify the solutions found using a Monte Carlo simulation run. The deviations of about 1 to 2 percentage points that occur here are typically smaller than the effects of the data uncertainty.

As an alternative to a verification run, one can also initiate the fine simulation in a post-processing step using the optima found with the reduced model, so as to improve the results. Often, only a few iterations are needed. This approach is familiar from numerics. The solution of linear equation systems, such as found in machine arithmetic with single precision, is then iterated again with double precision, in order to reduce the size of the residual.

In more complex cases, one works not just with a coarse and a fine model. Instead, a hierarchy of models is used, each of which can be compared to the others with error estimators. Examples of this approach can be found in [23]. At the Fraunhofer ITWM, hierarchic concepts were used in radiotherapy planning (cf. [1, 16] and [32]) and photovoltaic installation planning (see [5, 6]).

**Optimization-Driven, Adaptive Simulation Granularity** As an alternative to the model hierarchies described above, with their interplay of coarser and finer simulation models, one can also steer granularities within a fixed simulation model using the optimization algorithm. Within discretization models, for example, one can use increment controls to obtain faster or slower simulation runs. Increment control is frequently used with discretization schemes to ensure error monitoring during simulation runs. Normally, one verifies that a comparable simulation error can be maintained across the entire simulation result. If discretization is used in the context of optimization, one can relax this procedure, since one merely needs to monitor whether the discretization errors strongly compromise the values of the objective function(s). In the end, one can usually discretize coarsely where the objective function(s) tend toward flat gradients, and more finely where the gradients are steeper.

This means, with such an optimization-driven discretization, one normally manages with significantly smaller discretization patterns than with a simulation-driven discretization. The integration of simulation and optimization can be achieved with particular efficiency by using model hierarchies with error estimators across all models to supplement the optimization-driven discretizations performed within the models (cf. [26, 27, 31, 33] and [39]).

## 6 Optimizing with Uncertainties

In practice, optimization processes are usually influenced by a variety of uncertainties. Besides the already described model uncertainties, there is also imprecision in the simulation and in the quality of the available data. In order to obtain good solutions, it is essential to address these fundamental problems and reconcile the optimization and simulation models in an appropriate fashion.

**Impacts of Data Uncertainty and Simulation Error** In practice, optimization processes are usually influenced by a variety of uncertainties. Besides the already described model uncertainties, there is also imprecision in the simulation and in the quality of the available data. In order to obtain good solutions, it is essential to address these fundamental problems and reconcile the optimization and simulation models in an appropriate fashion.

**Impacts of Data Uncertainty and Simulation Error** The following is an important principle in numerics: when there are known errors in the data, the method and simulation errors must be kept in a healthy proportion to the data errors in order to achieve a favorable total error. An analog to this principle also applies to optimization procedures, although optimizations, due to damping effects, often behave more graciously than pure simulations.

Data flows into optimization problems at many points. It is frequently inhomogeneous, due to its heterogeneous origins. It is also stochastic, since it arises from inherently random processes or because it is afflicted with a randomly scattering measurement error. Despite these realities, in practice, data is frequently accepted with "no questions asked" and put to use without any preparation or treatment. Ultimately, this leads to strong distortions in the results, which, in the worst case, can render an optimization process useless. For this reason, it is essential, wherever possible, to check for systematic errors using statistical methods and to treat the data before using it in an optimization process. To do this, it is useful to enlist data models expressly to assist with the analysis and reconciliation of data.

Let's once again look at the example of designing photovoltaic power plants. To evaluate the anticipated revenue, it is essential to have at one's disposal local weather data pertaining to solar radiation forecasts. The weather data available for purchase is frequently model data for a "typical" month, which has been developed from values measured over the previous years. This data is then scaled up to the amortization period for the power plant. If one is only interested in the expected value of the total solar radiation for the 20-year amortization period, then this procedure will lead to satisfactory results, due to the law of large numbers. However, if one wants to be certain, for example, that the revenue within each accounting quarter will lie within a specified interval, in order to be able to meet the loan installment payment deadlines, then this artificial "typical" month is inadequate to guarantee sufficient certainty. How should one deal with these uncertainties?

**Robust Optimization and Solution Sensitivity** In recent decades, much work has been done on the topic of robust optimization. Here, it is assumed that one has a set of possible scenarios for all uncertain model parameters, such as those from measured, often stochastic, data, and that they are known with sufficient accuracy. Optimization is performed using the scenario sets so that the solution for the worst scenario is as good as possible ("optimum for the worst-case scenario") or that the solution in the middle is as good as possible ("optimum for an expected scenario"). A good introduction to robust optimization can be found in [3].

These concepts of robust optimization depend significantly on the choice of a scenario set size. Here one can choose more conservatively or less so. At what point is one on the safe side? With which concept can one work best? These questions are similar to those of model uncertainty. Interactive methods can also be used for robust optimization, in order to be able to change assumptions about the scenario sets *a posteriori* and, at the same time, study the sensitivity of the result to the various influencing factors.

As an alternative to robust optimization using scenarios, one can also optimize with fixed specifications and introduce a sensitivity estimator as a further target quantity. These might be target function gradients, for example, with whose help one can estimate local changes in the result under conditions of uncertainty. A simple way to achieve a robust optimization is to only choose solutions whose sensitivity is constrained by a limit that has been set in advance. With this method, an advance definition of scenario sets is not necessary; one need only consider the influencing factors that may possibly arise as a result of changes. [7] is a standard work on sensitivity analysis.

Naturally, one can also mix the two concepts—the advanced definition of scenarios and the use of sensitivity estimators. One reasonable approach is to first clarify with sensitivity estimators which quantities are especially sensitive and then, as a safeguard, to pursue a scenario approach with just these sensitive quantities. At the Fraunhofer ITWM, one finds examples in the planning of radiotherapies [17] and chemical production facilities [9].

## 7 Concluding Remarks

So, what is special about optimization at the Fraunhofer ITWM? What do we excel at? What do we do differently?

Primarily, our work is dedicated to the question of how to make mathematical optimization of practical use. It is less concerned with answering classical questions, such as the existence and unambiguity of solutions, or with gaining and analyzing fundamental insights into solution concepts. It is important for us to think as the decision-maker thinks, and work from there. What makes a model suitable for optimization? The classical questions about the definition of alternative sets ("What is feasible?") and the objective quantities ("What is a good result?") are of tremendous relevance. Here, the observation that one cannot clarify everything *a priori* is very important. Design and decision spaces must be considered at the same time, and not consecutively. Otherwise, one runs into the dilemma of iterative models. Interactive models that help one to gain knowledge about the dependencies of constraints and to find solutions represent one way to meet this challenge.

Managing the complexity of optimization, particularly when integration of optimization and simulation algorithms is necessary, is a second, equally important subject. Simple, weak coupling of simulation and optimization frequently leads into a complexity trap. Here, the hierarchical integration of models via error estimators and optimization-adaptive discretization patterns can provide a way out.

A third relevant topic is data uncertainty and how to manage its impacts on the solution. Using sensitivity estimators as target quantities and interactive scenario selections for robust optimization are methods than can help one come to terms with model and data uncertainties.

In our efforts to optimize this chapter, we have excluded many other likewise relevant topics due to space constraints (Constraints are lurking everywhere, it seems). Dealing with NP-hard problems is worth mentioning briefly. Here, we frequently prefer explainable, hierarchical concepts with guaranteed partially-optimal solutions, as opposed to quick, non-explainable heuristics. Although these often yield excellent solutions for many cases, they can also deliver strongly sub-optimal solutions, and are therefore rejected by practitioners. Nor have we discussed how we deal with mixed integer problems, which we encounter frequently in practice. Here, a maximal decoupling of the discrete and non-discrete variables often makes sense, in order to ensure an appropriate algorithmic treatment. Dynamic optimization, as found in control problems, also accompanies the daily work of the ITWM. To adequately discuss it would require a whole chapter of its own.

In closing, we can state that the successful implementation of mathematical optimization can only be ensured by means of a good model. Solutions are good when they are considered good from the perspective of others. Only in very few instances is it acceptable to label solutions as the best possible, simply because they are the best possible found within a mathematical model. The practitioner continuously calls the model itself into question; he does not doubt the art of the mathematical optimizer. One must succeed in creating an optimization environment that can also help decision-makers to make modeling trade-offs. Only then can optimization do full justice to the needs of its users.

## References

1. Azizi Sultan, A.: Optimization of beam orientations in intensity modulated radiation therapy planning. Ph.D. thesis, Department of Mathematics, Technical University of Kaiserslautern, Germany (2006)
2. Bank, B., Guddat, J., Klatte, U., Kammer, B., Tammer, C.: Nonlinear Parametric Optimization, 2nd edn. Birkhäuser, Basel (2014)
3. Ben Tal, A., El Ghaoui, I., Nemirovski, A.: Robust Optimization. Princeton University Press, Princeton (2009)
4. Biegler, L.T., et al.: Systematic Methods of Chemical Process Design. Prentice Hall International, New Jersey (1997)
5. Bischoff, M., König, M., Schüle, I., Plociennik, K.: Utilizing full planning potential. PV-Mag. **2/2014**, 60–62 (2014)
6. Bischoff, M., Schüle, I.: Photovoltaik-Kraftwerke besser planen. Brennst. Wärme Kraft Energ. Fachmag. **6**, 22–24 (2012)
7. Bonnans, J., Shapiro, A.: Perturbation Analysis of Optimization Problems. Springer, New York (2000)
8. Bortz, M., Burger, J., Asprion, N., Blagov, S., Böttcher, R., Nowak, U., Scheithauer, A., Welke, R., Küfer, K., Hans, H.: Multi-criteria optimization in chemical process design and decision support by navigation on Pareto sets. Comput. Chem. Eng. **60**, 354–363 (2014)
9. Burger, J., Asprion, N., Blagov, S., Böttcher, R., Nowak, U., Bortz, M., Welke, R., Küfer, K., Hasse, H.: Multi-objective optimization and decision support in process engineering—implementation and application. Chem. Ing. Tech. **86**, 1065–1072 (2014)
10. Craft, D., Monz, M.: Simultaneous navigation of multiple Pareto surfaces, with an application to multicriteria imrt planning with multiple beam angle configurations. Med. Phys. **37**(2), 736–741 (2010)
11. Craft, D., Süss, P., Bortfeld, T.: The tradeoff between treatment plan quality and required number of monitor units in intensity-modulated radiotherapy. Int. J. Radiat. Oncol. Biol. Phys. **67**, 1596–1605 (2007)
12. Floudas, C., Pardalos, P. (eds.): Encyclopedia of Optimization, 2nd edn. Springer, New York (2009)
13. Haase, S., Süss, P., Schwientek, J., Teichert, K., Preusser, T.: Radiofrequency ablation planning: an application of semi-infinite modelling techniques. Eur. J. Oper. Res. **218**, 856–864 (2012)
14. Hamacher, H., Küfer, K.: Inverse radiation therapy planning—a multiple objective optimization approach. Discrete Appl. Math. **118**, 145–161 (2002)

15. Koziel, S., Leifsson, L.: Surrogate-Based Modelling and Optimization. Springer, New York (2013)
16. Kratt, K., Scherrer, A.: The integration of dvh-based planning aspects into a convex intensity modulated radiation therapy optimization framework. Phys. Med. Biol. **12**, 239–246 (2009)
17. Krause, M., Scherrer, A., Thieke, C.: On the role of modeling parameters in imrt plan optimization. Phys. Med. Biol. **58**, 4907–4926 (2008)
18. Küfer, K., Monz, M., Scherrer, A., Alonso, F., Trinkaus, H., Bortfeld, T., Thieke, C.: Real-time inverse planning using a precomputed multi-criteria plan database. Radiother. Oncol. **68**, 76 (2003)
19. Küfer, K., Monz, M., Scherrer, A., Süss, P., Alonso, F., Azizi Sultan, A., Bortfeld, T., Thieke, C.: Multicriteria optimization in intensity modulated radiotherapy planning. In: Pardalos, P., Romeijn, H. (eds.) Handbook of Optimization in Medicine, pp. 123–168. Kluwer Academic, Boca Raton (2009). Chap. 5
20. Küfer, K., Scherrer, A., Monz, M., Alonso, F., Trinkaus, H., Bortfeld, T., Thieke, C.: Intensity-modulated radiotherapy—a large scale multi-criteria programming problem. OR Spektrum **25**, 223–249 (2003)
21. Maag, V.: Multicriteria global optimization for the cooling system design of casting tools. Ph.D. thesis, TU Kaiserslautern (2010)
22. Miettinen, K.: Nonlinear Multiobjective Optimization. Springer, New York (1998)
23. Migdalas, A., Pardalos, P., Värbrand, P. (eds.): Multilevel Optimization: Algorithms and Applications. Kluwer Academic, Dordrecht (1998)
24. Monz, M.: Pareto navigation—interactive multiobjective optimisation and its application in radiotherapy planning. Ph.D. thesis, Technical University of Kaiserslautern, Department of Mathematics (2006)
25. Monz, M., Küfer, K., Bortfeld, T., Thieke, C.: Pareto navigation—algorithmic foundation of interactive multi-criteria imrt planning. Phys. Med. Biol. **53**, 985–998 (2008)
26. Scherrer, A.: Adaptive approximation of nonlinear minimization problems—the adaptive clustering method in inverse radiation therapy planning. Ph.D. thesis, Department of Mathematics, Technical University of Kaiserslautern, Germany (2006)
27. Scherrer, A., Küfer, K., Bortfeld, T., Monz, M., Alonso, F.: Imrt planning on adaptive volume structures—a decisive reduction in computational complexity. Phys. Med. Biol. **50**, 2033–2053 (2005)
28. Schröder, M., Schüle, I.: Interaktive mehrkriterielle Optimierung für die regionale Fahrplanabstimmung in Verkehrsverbünden. Straßenverkehrstechnik **6**, 332–340 (2008)
29. Schüle, I.: Rlt approaches to qsaps—applied to timetable synchronization in public transport. Ph.D. thesis, TU Kaiserslautern (2010)
30. Serna, J., Monz, M., Küfer, K., Thieke, C.: Trade-off bounds for the Pareto surface approximation in multi-criteria imrt planning. Phys. Med. Biol. **54**(20), 6299–6311 (2009)
31. Süss, P.: A primal-dual barrier algorithm for the imrt planning problem—an application of optimization-driven adaptive discretization. Ph.D. thesis, TU Kaiserslautern (2008)
32. Süss, P., Bortz, M., Küfer, K., Thieke, C.: The critical spot eraser—a method to interactively control the correction of local hot and cold spots in imrt planning. Phys. Med. Biol. **21**, 1855–1867 (2013)
33. Süss, P., Küfer, K.: Balancing control and simplicity: a variable aggregation method in intensity modulated radiation therapy planning. Linear Algebra Appl. **428**(5), 1388–1405 (2008)
34. Teichert, K.: A hyperboxing Pareto approximation method applied to radiofrequency ablation treatment planning. Ph.D. thesis, Technische Universität Kaiserslautern (2013)
35. Teichert, K., Süss, P., Serna, J., Monz, M., Küfer, K., Thieke, C.: Comparative analysis of Pareto surfaces in multi-criteria imrt planning. Phys. Med. Biol. **56**(12) (2011).

36. Thieke, C., Küfer, K., Monz, M., Scherrer, A., Alonso, F., Oelfke, U., Huber, P., Debus, J., Bortfeld, T.: A new concept for interactive radiotherapy planning with multicriteria optimization: first clinical evaluation. Radiother. Oncol. **85**(2), 292–298 (2007)
37. Thieke, C., Küfer, K., Monz, M., Scherrer, A., Alonso, F., Oelfke, U., Huber, P., Debus, J., Bortfeld, T.: A new concept for radiotherapy planning with fast automatic multicriteria optimization—a first clinical evaluation. Radiother. Oncol. **85**, 292–298 (2007)
38. Thieke, C., Spalke, T., Monz, M., Scherrer, A., Süss, P., Küfer, K., Nill, S., Bendl, R., Debus, J., Huber, P.: Prostate imrt planning using a new multicriterial interactive planning system. Int. J. Radiat. Oncol. Biol. Med. Phys. **69**, 371–372 (2007). Proceedings of the American Society for Therapeutic Radiology and Oncology (ASTRO) 49th Annual Meeting
39. Thieke, C., Süss, P., Grebe, T., Serna, I., Scherrer, A., Bortz, M., Küfer, K., Rhein, B., Nicolay, N., Debus, J.: A new planning tool for fast online dose optimization of imrt plans. Int. J. Radiat. Oncol. Biol. Phys. **84**, S785–S786 (2012)
40. Uhrig, M., Thieke, C., Alonso, F., Küfer, M., Monz, M., Scherrer, A., Oelfke, U.: First evaluation of a new multicriteria optimization tool—investigation of Pareto-surfaces for imrt prostate plans. Med. Phys. **32**, 1975 (2005)

# Part III
# Research

# Virtual Production of Filaments and Fleeces

Raimund Wegener, Nicole Marheineke, and Dietmar Hietel

## 1 Consistency out of Chaos—A Challenge for Production

Production processes for manufacturing continuous filaments and fleeces are on-line processes in which the individual process steps are highly coordinated with each other and integrated into a tightly linked chain. The process chain for fleeces formed from filaments consists of the operations melting, spinning, swirling, and deposition. Here, molten polymer exits an extruder via a tube and is distributed on a spinning plate, where it is pressed through capillary jets and spun to filaments by means of aerodynamic forces. The filaments are swirled in an open air jet, decelerated, and deposited on a moving conveyor belt. The overlapping of thousands of filaments produces a fleece, with its typically irregular and cloud-like structure. The application spectrum for fleeces is extremely broad and ranges from everyday products like diapers and vacuum cleaner bags to high-tech goods like battery separators and medical products. Naturally, filament spinning is also used in conjunction with further processing steps, in the production of technical yarn products or synthetic short-fibers, for example. Moreover, we include the production of fiber-like insulation materials, such as glass wool and mineral wool, in the category of filament production, since these processes are based on similar physical, albeit technically different, principles.

R. Wegener (✉) · D. Hietel
Fraunhofer-Institut für Techno- und Wirtschaftsmathematik, Fraunhofer-Platz 1,
67663 Kaiserslautern, Germany
e-mail: raimund.wegener@itwm.fhg.de

N. Marheineke
FAU Erlangen-Nürnberg, Lehrstuhl Angewandte Mathematik 1, Cauerstr. 11, 91058 Erlangen,
Germany

 
H. Neunzert, D. Prätzel-Wolters (eds.), *Currents in Industrial Mathematics*,
DOI 10.1007/978-3-662-48258-2_6

The fluctuating characteristics of filaments and fleeces—a consequence of the stochastic, and often turbulence-induced, impacts on production processes—can lead to problems in product quality. In the spinning processes, for example, such problems might come in the form of fluctuations in filament diameter and strength, due to an unsteady temperature history during cooling. These problems can frequently be traced back to the economic necessity of high machine throughput rates and tight filament bundling. Above and beyond the problems of the individual filaments, fleeces also exhibit problems with fluctuations in the weight and strength of the material. These latter arise on a sufficiently small scale from the production principle itself, since a chaotic, turbulence-driven overlapping of the filaments takes the place of an expensive weaving procedure. The bold challenge faced by production is therefore to create *consistency out of chaos*, a challenge that has already resulted in the development of astonishing installations and processes through decades of technical advances in machine engineering. The currently available and continuously improving instruments for simulating such complex processes, however, represent a qualitatively new opportunity for the simulation-supported design and control of these installations and processes. With their help, it is now possible to take the next step toward creating even *more consistency out of chaos*.

## 2 Simulatable, but only in Principle—A Challenge for Mathematics

Fundamentally, almost all the steps in the process chain outlined above—melting and spinning, for filaments, plus swirling and deposition, for fleeces—can be viewed as continuum-mechanical, multi-phase problems. Depending on the degree of cooling and the stage in the process, one is treating a viscous, viscoelastic, or elastic filament phase, coupled with turbulent airflow, in a complex machine geometry. Classical continuum mechanics offers the models for such multi-phase problems. There is an abundance of numerical discretization ideas, solution algorithms, and even ready-made software tools in the arsenal of applied mathematicians and engineers. In other words, the problems can indeed be simulated, in principle. Unfortunately, however, *only in principle*.

A closer look reveals, in fact, the hopelessness of such a monolithic approach: as our examples of fleece production (Sect. 6) and glass wool production (Sect. 7) show, the actual production processes demand the coupled filament flow simulation of thousands of filaments having diameters as small as 10 microns in highly turbulent flows across macroscopic scales on the order of meters. The mathematical challenge is therefore to use modeling strategies such as homogenization and asymptotics, along with the generation of surrogate models having a grey box character, to prepare adequate models for all the partial aspects and then to couple these aspects together. After a thoroughgoing analysis of these models, numerical algorithms must then be developed and adapted to the problem. Only in this way can one portray the processes so as to allow realistic application scenarios to be computed in an acceptable time and, thus, made accessible to optimization. The procedure requires, in particular, the compatibility between the various modeling approaches, the derivation of coupling conditions, and the identification of model parameters. Using this procedure, we want to avoid the trap of *simulatable in principle*, and achieve instead the state of *simulatable in practice*, which will allow us to contribute significantly to the design and optimization of production processes. By concentrating diverse approaches from various mathematical areas in a single application domain, the Fraunhofer ITWM has an outstanding opportunity to substantiate its claim to be a problem-solver, to make innovative contributions to existing research into applied mathematics, and to initiate the exploration of brand new thematic areas. Our contribution to this book is designed to document the current state of our work, but we hope that it also generates a host of new questions.

## 3 Studies in Filament Dynamics and Fleece Production at the Fraunhofer ITWM

The work in filament dynamics at the Fraunhofer ITWM has its origins in a project that has absolutely nothing to do with filaments and their production processes. In 1995, the year the Institute was founded, we began work on simulating the paper flight in a printing press. This was one of the first industrial projects in the Transport Processes Department, and the starting point for at least two thematic areas that are today pursued in force within the Department. The largely two-dimensionally characterized flow of paper in a printing press is a coupled fluid-structure interaction problem. Therefore, particle methods were tested for the flow domains below and above the sheet, which are time variant due to sheet movement. For the sheet dynamics, shell models from continuum mechanics were refurbished, which, in their two-dimensional variant, are mathematically equivalent to rod models for filament dynamics. The work on particle methods led to development of the ITWM software FPM (Finite Pointset Method), which is today one of the best-performing grid-free simulation tools available on the market for a wide and still continuously growing field of continuum mechanical problems. The work on sheet dynamics was the breeding ground for all subsequent research in the area of filament dynamics, which is the subject matter of this chapter. This short story illustrates the enormous power generated by problem-oriented research in industrial projects: the specific questions breed approaches, which then often grow far beyond the original field of investigation and the short-term concerns of daily business.

In 1998, concurrently with the above-mentioned industrial printing press project, our contact with the company Freudenberg, which dates back to before the founding of the ITWM, was revitalized in Kaiserslautern in connection with the topic of fleece production. It took a while, however, before the tender sprout would grow into a large-scale Institute activity, whose salient points we want to selectively outline here. Our work in this area received an initial impulse in 2003, in the form of a large, in-house Fraunhofer project on market-oriented preparatory research. An accompanying dissertation [27] laid the foundation for our turbulent force model in 2005 (Sect. 4.3 and Ref. [9, 16, 17]). The following year witnessed the first ideas for stochastic model analogies for deposition simulations (Sect. 4.4 and Ref. [5, 6]). At the same time, again on the basis of a dissertation [29], work commenced on the asymptotic derivation of viscous string models [7, 20]. All three of these thematic areas have been widely pursued and thematically extended up to the present date (see development and status for *turbulent force modeling* [19], for the *stochastic surrogate lay down models* [8, 11–13], and for *asymptotic rod and string models* [1, 4, 14, 18]). Likewise, as a consequence of the above-mentioned Fraunhofer project, there has been an enormous broadening of our industrial customer base. Johns Manville (2003) and Oerlikon Neumag (2004) are examples of a fleece manufacturer and a machine designer in the field of technical textiles. Both remain today steady customers of the Fraunhofer ITWM.

It was then two projects sponsored by the BMBF at the start of this decade that set long-term developments in motion: the project 'Nano-melt-blown fibers for filter media' (NaBlo, 2008–2011) set the stage for our current work on *turbulence reconstruction for filament dynamics* [10]. In the project 'Stochastic production processes for the manufacturing of filaments and fleeces' (ProFil, 2010–2013), a consortium project in the BMBF mathematics program under the leadership of the Fraunhofer ITWM, the complete production chain for filaments and fleeces was simulated for the first time. Several PhD projects resulted either directly from the project [22, 23, 25, 28] or were offshoots from it [26, 30, 31]. These represent an important foundation for further investigations in this thematic area. The project also forms the basis for the current status of the central ITWM software for filament dynamics, the FIDYST suite, with the software tools FIDYST (Fiber Dynamics Simulation Tool, Sect. 5.1) and SURRO (Surrogate Model, Sect. 5.2). On the industrial side, our contact with the company Woltz (2010) and the resulting, on-going cooperation have proven extremely fruitful. Here, we were able to couple the filament and flow dynamics in a complex production process for the first time, in connection with the manufacture of glass wool (Sect. 7 and Ref. [3, 15]). The simulation toolbox VISFID (Viscous Fiber Dynamics, Sect. 5.3) for *coupled flow-filament simulations in spinning processes* was conceived in projects involving this production process.

Although this chapter discusses many of the above-mentioned topics, it makes no attempt to offer a complete historical portrayal. Instead, it attempts to present a cohesive overview from our current perspective. We therefore dedicate some space to the presentation of a consistent and integrated modeling basis (Sect. 4), before we then show the performance status of the software tools available today at the Fraunhofer ITWM (Sect. 5) and demonstrate their capabilities using two typical industrial applications as examples: the production of fleeces in the spunbond process (Sect. 6) and the production of glass wool via rotational spinning (Sect. 7). To promote readability, we offer annotations at various points that summarize more detailed aspects of the work and illustrate how it fits into the framework of current international research. Readers interested primarily in the applications can also begin with Sects. 6 and 7, follow the references to the simulation tools being used (Sect. 5), and consult with the underlying models (found in grey boxes in Sect. 4) as desired.

In addition to the authors, substantial credit for the modeling ideas, software developments, and industrial projects that serve as the foundations for this chapter must be given to some of our current colleagues from the Transport Processes Department of the Fraunhofer ITWM (Sergey Antonov, Dr. Walter Arne, Dr. Christian Leithäuser, Dr. Robert Feßler, Dr. Simone Gramsch, Dr. Jan Mohring, Johannes Schnebele), as well as to some former colleagues (Dr. Daniel Burkhart, Dr. Marco Günther, Dr. Jalo Liljo, Dr. Ferdinand Olawsky). The past and current PhD projects mentioned here have been or are being supervised by Prof. Nicole Marheineke (FAU Erlangen-Nürnberg), Prof. Andreas Meister (Universität Kassel), and Prof. Hans Hagen, Prof. Axel Klar, Prof. Helmut Neunzert, Prof. Rene Pinnau, and Prof. Klaus Ritter (all from the TU Kaiserslautern).

## 4 Foundations of the Modeling

The *Cosserat rod theory* serves as the framework for the partial differential equation models considered here for filament dynamics. At their core are 1D balances for linear and angular momentum. These are complemented by *geometric models* for describing angular momentum, *material laws* for the emerging internal stress forces and moments, as well as models for the external forces acting on the system. In view of the target application, the *interaction of the filaments with the surrounding, often turbulent, airflow* is especially significant.

These Cosserat rod models can be used to successfully simulate single filaments in spinning and swirling processes. However, the significant computational effort prevents a virtual mapping of complete fleece deposition processes involving large numbers of filaments. Therefore, *surrogate models based on stochastic differential equations* (SODE) were developed and implemented at the Fraunhofer ITWM, which allow highly efficient simulations of the fleece deposition structure. The parameters of these surrogate models are identified using the Cosserat rod computations for single filaments.

**Folklore and Convention** We embed our continuum mechanical models in an abstract three-dimensional Euclidean space $\mathbb{E}^3$. In this space, we take $\{\mathbf{e}_1, \mathbf{e}_2, \mathbf{e}_3\}$ to be a fixed orthonormal basis (ONB). Such an ONB induces an isomorphism $i_e : \mathbb{E}^3 \to \mathbb{R}^3$, $\mathbf{a} \mapsto i_e(\mathbf{a}) = \bar{\mathsf{a}} = (\bar{a}_1, \bar{a}_2, \bar{a}_3)$ with $\bar{a}_j = \mathbf{a} \cdot \mathbf{e}_j$, $j = 1, 2, 3$. Because we are operating with different bases, it is important to us to always distinguish between the vectors $\mathbf{a} \in \mathbb{E}^3$ and their component tuples $\bar{\mathsf{a}} \in \mathbb{R}^3$ in the arbitrary, but fixed ONB $\{\mathbf{e}_1, \mathbf{e}_2, \mathbf{e}_3\}$. This is motivated largely by the fact that we also introduce, as a component of the Cosserat rod theory, a temporally and spatially (along the rod) varying director basis $\{\mathbf{d}_1, \mathbf{d}_2, \mathbf{d}_3\}$. The components of a vector $\mathbf{a}$ in this basis are denoted as $\mathsf{a} = (a_1, a_2, a_3)$. The canonical basis of $\mathbb{R}^3$ (that is, the component tuples of any ONB in relation to itself) is denoted by $\mathsf{e}_1$, $\mathsf{e}_2$, $\mathsf{e}_3$.

We use a tensor calculus that is oriented on the calculus of Antman [32]. That is, we consistently use the point $\cdot$ for scalar products and tensor-vector operations; we make no distinction between vectors of $\mathbb{E}^3$ and their adjoints; and, consequently, no distinction between row and column vectors of $\mathbb{R}^3$. In contrast to [32], however, we use $\otimes$ in place of a blank space for tensor products. $3 \times 3$-matrices are identified with tensors having values in $\mathbb{R}^3 \otimes \mathbb{R}^3$ and are frequently, with respect to a basis, the components of tensors with values in $\mathbb{E}^3 \otimes \mathbb{E}^3$. For all further details of our selected calculus, we refer the reader to [32]. We use a generalized summation convention in which Latin indices run between 1 and 3 and Greek indices, between 1 and 2.

Because we are mainly examining modeling aspects, we generally assume, for the needed manipulations, that there is sufficient differentiability and invertibility—as was just needed—and we do not usually critically reflect upon these points. This does not mean that we consider such reflections superfluous, or that all models we examine have classical solutions. Quantities are always introduced with their SI units, unless this is completely trivial (or forgotten!). Frequently, this clarifies their physical significance better than many words.

## 4.1 Cosserat Rod Theory

The Cosserat rod theory describes a filament as a spatial curve with oriented cross-sections. This results in a 1D manifold embedded in 3D, to which an element of the rotation group $SO(3)$ is differentiably assigned at each point. The theory is characterized by 1D balances for linear and angular momentum, which result from 3D continuum mechanics by averaging over the cross-sectional areas and restricting degrees of freedom. These restrictions mean that a re-orientation of the cross-sections can indeed be described, but not a genuine deformation that overcomes their planarity. We largely follow [32] in introducing the theory in a material parameterization (Lagrangian description), but we place a general and spatial variant (Eulerian description) on an equal footing alongside it. We take pains to present the theory as self-contained and reflect upon its embedding in 3D continuum mechanics as little as possible. Nevertheless, this embedding can be undertaken in order to thereby identify all elements of the theory in 3D continuum mechanics.

### 4.1.1 Material Description

**Reference State** A Cosserat rod or filament is described in its reference state by a curve $\mathbf{r}^\circ : (s_a, s_b) \to \mathbb{E}^3$ and two normalized, orthogonal vectors $\mathbf{d}^\circ_\alpha : (s_a, s_b) \to \mathbb{E}^3$, which are referred to as directors.

One also defines $\mathbf{d}^\circ_3 = \mathbf{d}^\circ_1 \times \mathbf{d}^\circ_2$, so that the directors form a right-handed orthonormal system. The reference state can be assumed for any given point in time, but this is not actually imperative. The interval $(s_a, s_b) \subset \mathbb{R}$ addresses the section of the filament whose dynamic is to be subsequently described. A parameter $s \in (s_a, s_b)$ addresses the materially-determined cross-section of the filament to be modeled. For our applications concerning filament dynamics, we require that $\mathbf{d}^\circ_3 = \partial_s \mathbf{r}^\circ$ and $\partial_s \mathbf{d}^\circ_i = \mathbf{0}$ for the reference state. The geometry and material models formulated in Sect. 4.2 are attuned to this reference state. More precisely, they ensure an absence of tension and torque in the reference state. With

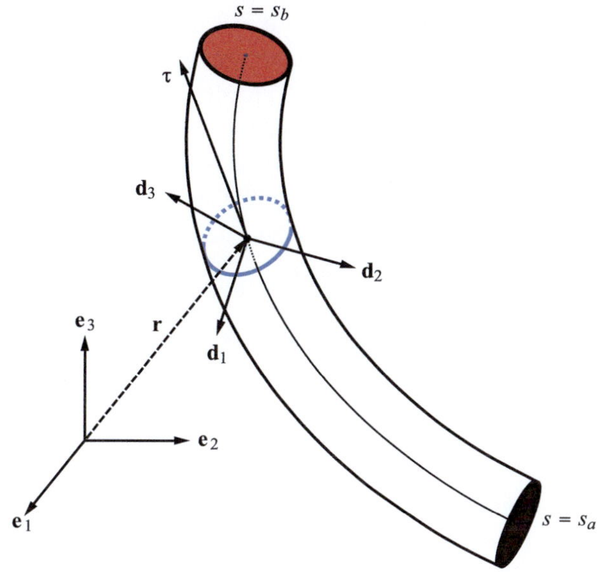


**Fig. 1** Cosserat rod, consisting of curve and director triad (Graphic: Steffen Grützner, Fraunhofer ITWM)

these assumptions, we select, in particular, an arc-length parameterization of the reference state, but only of the reference state.

**Kinematics** At an arbitrary point in time $t$, the state of the rod is defined by the curve $\mathbf{r}(\cdot, t)$ and the orthonormal directors $\mathbf{d}_\alpha(\cdot, t)$, where $\mathbf{d}_\alpha \cdot \mathbf{d}_\beta = \delta_{\alpha\beta}$. The curve describes the position and the directors describe the orientation of the cross-sections addressed by $s$ (Fig. 1). Using the consistently applied definition $\mathbf{d}_3 = \mathbf{d}_1 \times \mathbf{d}_2$, the directors form a right-handed orthonormal system at all times. Both the referential linking of $\mathbf{d}_3$ with the tangent $\partial_s \mathbf{r}$ and the arc-length parameterization, however, are generally not valid in a moving state.

The velocity and tangent of the rod are characterized by the vector fields

$$\partial_t \mathbf{r} = \mathbf{v}, \qquad \partial_s \mathbf{r} = \boldsymbol{\tau}.$$

Because the directors form a right-handed orthonormal system, there exist also unambiguous vector fields $\boldsymbol{\kappa}$ (1/m) (curvature) and $\boldsymbol{\omega}$ (1/s) (angular velocity), so that the equations

$$\partial_t \mathbf{d}_i = \boldsymbol{\omega} \times \mathbf{d}_i, \qquad \partial_s \mathbf{d}_i = \boldsymbol{\kappa} \times \mathbf{d}_i$$

are valid. These vector fields describe the change in the triad over time, and along the curve. By changing the order of the partial derivatives with respect to $t$ and $s$, one obtains the following compatibility relations:

$$\partial_t \boldsymbol{\tau} = \partial_s \mathbf{v}, \qquad \partial_t \boldsymbol{\kappa} = \partial_s \boldsymbol{\omega} + \boldsymbol{\omega} \times \boldsymbol{\kappa}.$$

In order to use the Cosserat rod theory in specific applications, it proves helpful to represent vector fields and model equations partially mixed in two basis systems (external basis and director basis). The change from the invariant formulation to a fixed external basis $\{\mathbf{e}_1, \mathbf{e}_2, \mathbf{e}_3\}$ is, in this instance, trivial. The transition from the external to the director basis $\{\mathbf{d}_1, \mathbf{d}_2, \mathbf{d}_3\}$ can always be accomplished technically using the following calculus. As agreed, for an arbitrary vector field $\mathbf{a} \in \mathbb{E}^3$ of the rod, $\bar{\mathsf{a}} \in \mathbb{R}^3$ and $\mathsf{a} \in \mathbb{R}^3$ denote the component tuples relative to the external basis or the director basis. The director basis is transformed with the rotation

$$\mathbf{D} = \mathbf{e}_i \otimes \mathbf{d}_i = D_{ij}\mathbf{e}_i \otimes \mathbf{e}_j \in \mathbb{E}^3 \otimes \mathbb{E}^3 \quad \text{with } D_{ij} = \mathbf{d}_i \cdot \mathbf{e}_j$$

into the external basis. The orthogonal matrix $\mathsf{D}$ is assigned to the components $D_{ij}$ of this rotation. If one now considers an arbitrary vector field of the rod, then

$$\mathsf{D} \cdot \bar{\mathsf{a}} = \mathsf{a}, \qquad \mathsf{D} \cdot \partial_t \bar{\mathsf{a}} = \partial_t \mathsf{a} + \omega \times \mathsf{a}, \qquad \mathsf{D} \cdot \partial_s \bar{\mathsf{a}} = \partial_s \mathsf{a} + \kappa \times \mathsf{a}.$$

Moreover, the kinematic base equations for the directors can be transformed into corresponding equations for the rotation matrix:

$$\partial_t \mathsf{D} = -\omega \times \mathsf{D}, \qquad \partial_s \mathsf{D} = -\kappa \times \mathsf{D}.$$

This formulation of the kinematics also serves as the starting point for other representations of the rotation group (Euler angles, unit quaternions, rotation vectors), each of which has its merits, depending on the application.

The fundamental deformation variables for the formulation of objective material laws are the component tuples $\tau$ and $\kappa$ of tangent and curvature in the director basis. More precisely, $\tau_1$ and $\tau_2$ quantify shearing, $\tau_3$, strain, $\kappa_1$ and $\kappa_2$, bending, and $\kappa_3$, torsion. Moreover, with

$$e = \|\boldsymbol{\tau}\|$$

we introduce a further strain measure that refers solely to the curve.

**Dynamics** Balancing linear and angular momentum (dynamic equations) for a rod leads to the following generalized forms:

$$(\rho A)\partial_t \mathbf{v} = \partial_s \mathbf{n} + \mathbf{k}, \qquad \partial_t \mathbf{h} = \partial_s \mathbf{m} + \boldsymbol{\tau} \times \mathbf{n} + \mathbf{l}.$$

The line density of the rod $(\rho A)$ (kg/m) in the reference state is traditionally designated using a slightly confusing symbol that suggests a product. When embedded in 3D continuum mechanics, it results in the integral of the density over the cross-section of the rod in the reference state, and is thus dependent on the filament parameter $s$, but not on the time $t$. The angular momentum line density $\mathbf{h}$ (kg m/s) is described as a function of the remaining base quantities of our theory, in particular, of the angular velocity (see geometric modeling, Sect. 4.2.1). The internal stress forces $\mathbf{n}$ (N) and torques $\mathbf{m}$ (N m) are modeled via suitable material laws as functions of the internal variables. Section 4.2.2 consists primarily of a discussion of two types of such material laws—elastic and viscous. In the dynamic equations, $\mathbf{k}$ (N/m) and $\mathbf{l}$ (N) denote line force density and line torque density for modeling the external force and torque effects on the rod. Each of these can depend on different internal variables and thus decisively impact the coupling of the dynamic and kinematic equations. In the following discussion, we generally restrict ourselves to models with no external moment effects; that is, $\mathbf{l} = \mathbf{0}$. Ultimately, geometric modeling, material laws, and external forces are the primary determinants of the type of PDE system.

### 4.1.2 General and Spatial Description

So far, the entire theory has been formulated in a Lagrangian description; that is, the parameter $s \in (s_a, s_b)$ addresses a material point (or cross-section) of the rod. Except for the orientation and a constant, the parameterization is then completely determined by requiring the arc-length parameterization of the reference state. This is not essential, but it simplifies much of the treatment. As we show below, a simple typing concept for the theory's base quantities can be used to formulate the model equations very easily in any other time-dependent parameterization. Without doubt, the most important application case is the spatial description (Eulerian description), in which, for all times, the transformation is made to an arc-length parameterization.

**Parametrizations** Suitable time-dependent re-parameterizations can be introduced with bijective transformations

$$\phi(\cdot, t) : (s_a, s_b) \to \big(\varphi_a(t), \varphi_b(t)\big), \quad s \mapsto \phi(s, t).$$

In order to define the transformation behavior of the different fields of the Cosserat rod theory, we introduce the term type-$n$-quantity. A type-$n$-quantity, $n \in \mathbb{Z}$, is transformed as follows:

$$j^n(s,t)\tilde{f}\big(\phi(s,t),t\big) = f(s,t), \quad j = \partial_s \phi.$$

Here, $f(s,t)$ characterizes a type-$n$-quantity in the material parameterization (Lagrangian description) and $\tilde{f}(\varphi, t)$ characterizes the associated field in the new parameterization. For the different fields of our theory, we specify that $\mathbf{r}$, $\mathbf{d}_i$, $\mathbf{v}$, $\boldsymbol{\omega}$, $\mathbf{n}$, $\mathbf{m}$ are to be treated as type-0-quantities and $\boldsymbol{\tau}$, $\boldsymbol{\kappa}$, $\mathbf{k}$, $\mathbf{l}$, $(\rho A)$, $\mathbf{h}$ as type-1-quantities. This specification allows the various quantities to retain their physical character and defining interrelationships (point-related observables, densities, derivatives, etc.) in the transformation. Time-independent re-parameterizations do not disturb the material character of the parameterization, nor do they change the form of the base equations. In contrast, time-dependent re-parameterizations bring convective terms into play. These have their origins in the scalar velocity $u(\varphi, t)$ (convective speed), which is defined by

$$\partial_t \phi(s,t) = u\big(\phi(s,t), t\big).$$

On the basis of this definition, the following applies for type-$n$-quantities:

$$\partial_t f(s,t) = j^n(s,t)(\partial_t \tilde{f} + n\partial_\varphi u \tilde{f} + u\partial_\varphi \tilde{f})\big(\phi(s,t), t\big).$$

The application of this rule leads directly to the base equations of the Cosserat rod theory (1) in an arbitrary parameterization, which are formulated below. The suitability of the selected typing of all quantities is demonstrated by the fact that (1) does not depend explicitly on the selected parameterization, but only on the associated convective speed. The definition of $u$ and $j$ yields

$$\partial_\varphi u\big(\phi(s,t), t\big) = \frac{\partial_t j}{j}(s,t).$$

A change in the sign of $j$ indicates a re-orientation in the parameterization. Without restricting ourselves significantly, we stipulate that $j > 0$.

Ultimately, every parameterization requires one to define the convection $u$. This definition can take place directly or indirectly. The simplest case, $u = 0$, corresponds to either our starting point of the Lagrangian description or one of the time-independent, and thus still material, re-parameterizations. Another special case, the Eulerian description, results from the arc-length parameterization requirement

$$\|\tilde{\boldsymbol{\tau}}\| = 1.$$

If one adds this requirement to the balance equations, then the convective speed $u$ is defined indirectly as Lagrange parameter to the constraint $\|\tilde{\boldsymbol{\tau}}\| = 1$. To be consistent, we treat the deformation measure $e = \|\boldsymbol{\tau}\|$ just as we do $\boldsymbol{\tau}$, that is, as a type-1-quantity. Thus, for the transition to the Eulerian description, $\tilde{e} = 1$ and $e = j$ are valid. Moreover, $\partial_t \phi(s,t) = u(\phi(s,t), t)$ is the rate of change in the arc-length $\phi(s,t)$ of the material point $s$, and $\partial_\varphi u(\phi(s,t), t) = \partial_t e / e(s,t)$ is the associated relative strain rate.

**Base Equations** With the formalism introduced above, the balance equations follow in an arbitrary parameterization. To streamline the appearance of the notation, we remove the marker $\tilde{}$ from the fields and also select $s$ instead of $\varphi$ as parameter in the general case.

*Kinematic and dynamic base equations for the Cosserat rod theory*

$$
\begin{aligned}
\partial_t \mathbf{r} &= \mathbf{v} - u\boldsymbol{\tau} \\
\partial_t \mathbf{d}_\alpha &= (\boldsymbol{\omega} - u\boldsymbol{\kappa}) \times \mathbf{d}_\alpha \\
\partial_s \mathbf{r} &= \boldsymbol{\tau} \\
\partial_s \mathbf{d}_\alpha &= \boldsymbol{\kappa} \times \mathbf{d}_\alpha \\
\partial_t(\rho A) + \partial_s\big(u(\rho A)\big) &= 0 \\
\partial_t\big((\rho A)\mathbf{v}\big) + \partial_s\big(u(\rho A)\mathbf{v}\big) &= \partial_s \mathbf{n} + \mathbf{k} \\
\partial_t \mathbf{h} + \partial_s(u\mathbf{h}) &= \partial_s \mathbf{m} + \boldsymbol{\tau} \times \mathbf{n} + \mathbf{l}.
\end{aligned} \tag{1}
$$

The equations for $\partial_t \mathbf{r}$ and $\partial_t \mathbf{d}_\alpha$, or $\partial_s \mathbf{r}$ and $\partial_s \mathbf{d}_\alpha$, can also be replaced by the compatibility relations

$$
\partial_t \boldsymbol{\tau} + \partial_s(u\boldsymbol{\tau}) = \partial_s \mathbf{v}, \qquad \partial_t \boldsymbol{\kappa} + \partial_s(u\boldsymbol{\kappa}) = \partial_s \boldsymbol{\omega} + \boldsymbol{\omega} \times \boldsymbol{\kappa}.
$$

As mentioned, the convective speed $u$ is a scalar degree of freedom in (1), which is to be defined by means of an additional condition. In the Lagrangian description, this is $u = 0$. The continuity equation $\partial_t(\rho A) + \partial_s(u(\rho A)) = 0$ then degenerates to the time constancy of $(\rho A)$, and is therefore generally not included as a balance equation. In the case of the Eulerian description, the convective speed $u$ is a system variable and is defined in the manner of a Lagrange parameter by the additional constraint $\|\boldsymbol{\tau}\| = 1$ of the arc-length parameterization. Provided nothing else is stated, we choose the Lagrangian description in the following discussion.

## 4.2 Geometry and Material Modeling

The considerations presented thus far establish the general framework for the Cosserat rod theory. Modeling the elements of geometric and material characteristics is important for completing the theory. On the one hand, these two steps deliver the angular momentum as a function of the angular velocity. On the other, they deliver the emerging internal stress forces and moments as functions of the fundamental deformation variables. Geometric constraints on the dynamics frequently replace material laws. These constraints usually reflect stiff material behavior, for example, inextensibility.

### 4.2.1 Geometric Modeling

When modeling the angular momentum $\mathbf{h}$ as a function of the angular velocity $\boldsymbol{\omega}$, the inertial tensor $(\rho\mathbf{J})$ (kg m) plays a central role. In contrast to the line density already discussed in connection with the dynamic equations, it is a function of time even in the Lagrangian description, via the director dynamics:

$$(\rho\mathbf{J})(s,t) = (\rho J)_{ij}(s)\mathbf{d}_i(s,t) \otimes \mathbf{d}_j(s,t).$$

Analogously to the notation introduced for the vectors, $(\rho\mathsf{J})$ designates the symmetrical $3 \times 3$-matrix formed from the components $(\rho J)_{ij}$. This results, when embedded in a 3D theory, from the area moments of inertia $(\rho J)^{\times}_{\alpha\beta}$ of the reference state

$$(\rho\mathsf{J}) = \begin{pmatrix} (\rho J)^{\times}_{22} & -(\rho J)^{\times}_{12} & 0 \\ -(\rho J)^{\times}_{12} & (\rho J)^{\times}_{11} & 0 \\ 0 & 0 & (\rho J)^{\times}_{11} + (\rho J)^{\times}_{22} \end{pmatrix}$$

and is thus, as a referential quantity, time-independent.

For a circular cross-section, $(\rho\mathsf{J})$ is defined by the polar area moment of inertia $(\rho I)$

$$(\rho\mathsf{J}) = (\rho I)\mathrm{diag}(1,1,2).$$

Provided this cross-section with referential surface $A^\circ$ exhibits a homogeneous referential mass distribution of the density $\rho^\circ$, then

$$(\rho A) = \rho^\circ A^\circ, \qquad (\rho I) = \frac{1}{4\pi}\rho^\circ A^{\circ 2}. \tag{2}$$

In general, however, just as in the case of line density $(\rho A)$, the matrix $(\rho\mathsf{J})$ is only to be understood as the symbol for a quantity, and not as a product. The relationships (2) are initially bound to the material description, since we have not yet defined the type of the new quantities.

**Inertia-Free Geometry Model** The simplest geometrical model results by neglecting all inertial terms in the angular momentum balance; that is,

$$\mathbf{h} = \mathbf{0}.$$

With this approach, the angular momentum balance in (1) degenerates to a quasi steady-state equation and can be used, in particular, to compute explicit expressions for the non-tangential components of the stress.

**Geometrical Standard Model** In the following discussion, we refer to the linear dependency of the angular momentum on the angular velocity

$$\mathbf{h} = (\rho\mathbf{J}) \cdot \boldsymbol{\omega}$$

which is given directly by $(\rho\mathbf{J})$, as the geometrical standard model. The basis for this model is the geometrical assumption that the cross-sections of the rod remain unchanged in their form and extent for the entire dynamics. The designation standard' is traced to the model's use in the field of elastic materials, which represents the original application domain of the Cosserat rod theory [32, 34]. For the transition to an arbitrary parameterization, we require a continuance of the above relationship, which characterizes the standard model. This forces one to treat the matrix $(\rho\mathsf{J})$ as a type-1-quantity, so that an additional balance equation

$$\partial_t(\rho\mathsf{J}) + \partial_s\big(u(\rho\mathsf{J})\big) = 0$$

in a general description corresponds to the time-independence of the matrix in the Lagrangian description.

This general treatment only makes use of $(\rho\mathsf{J})$ and initially avoids all deeper discussions of the geometry. For more detailed modeling of external forces, materials, and temperature effects, however, it is useful to introduce the density $\rho$ (kg/m$^3$) and the cross-sectional area $A$ as further type-0-quantities. Note that we distinguish—carefully, and as a function of the geometry model—between these quantities and their referential counterparts $\rho^\circ$ and $A^\circ$. For the standard model, we consider a rod with homogeneous, circular cross-sections of density $\rho = \rho^\circ/\tau_3$ and cross-sectional area $A = A^\circ$, so that (2), with $\tau_3 = \boldsymbol{\tau}\cdot\mathbf{d}_3$, leads to

$$(\rho A) = \tau_3\rho A, \qquad (\rho I) = \frac{1}{4\pi}\tau_3\rho A^2. \tag{3}$$

Due to the specified typing, these relationships are form-invariant during re-parameterization. In the Lagrangian description, the time-independence of $(\rho A)$ and $(\rho I)$ leads to

$$\partial_t(\tau_3\rho) = 0, \qquad \partial_t A = 0$$

and therefore, in an arbitrary parameterization,

$$\partial_t(\tau_3\rho) + \partial_s(u\tau_3\rho) = 0, \qquad \partial_t A + u\partial_s A = 0. \tag{4}$$

When dealing with the standard model in an arbitrary parameterization under the assumption of homogeneous, circular cross-sections, the above considerations allow one to use the definitions in relationship (3) to replace $(\rho A)$ and $(\rho I)$ with $\rho$ and $A$. In this case, (4) then replaces the associated balances for $(\rho A)$ and $(\rho J)$ or $(\rho I)$. Alternatively, one can retain $(\rho A)$ and $(\rho I)$ as base quantities and use (3) to calculate $\rho$ and $A$:

$$\rho = \frac{1}{4\pi}\frac{(\rho A)^2}{\tau_3(\rho I)}, \qquad A = 4\pi\frac{(\rho I)}{(\rho A)}.$$

**Incompressible Geometric Model** In contrast to the geometric standard model, the approach

$$\mathbf{h} = \frac{1}{\boldsymbol{\tau} \cdot \mathbf{d}_3} (\rho \mathbf{J}) \cdot \boldsymbol{\omega}$$

which is also linear in $\boldsymbol{\omega}$, but scaled with $\tau_3$, accounts for changes in the dimensions of the cross-sections, which are irreversible for incompressible material behavior. This model also begins with the assumption that the form of the cross-sections remains unchanged when the rod undergoes deformation [1]. As with the standard model, we require for the transition to another parameterization—especially the Eulerian description—that the above characteristic relationship is invariant in the face of re-parameterization. This then forces one to treat the matrix $(\rho \mathsf{J})$ as a type-2-quantity, unlike in the standard model. Consequently, in a general description, the additional balance equation

$$\partial_t (\rho \mathsf{J}) + \partial_s \big(u(\rho \mathsf{J})\big) = -(\rho \mathsf{J}) \partial_s u$$

corresponds to the time-independence of the matrix in the Lagrangian description. It becomes clear at this point that, although the typing of the quantities is indeed physically motivated, it remains ultimately a matter of definition.

In order to clarify for the incompressible model the relationship to the additionally introduced quantities $\rho$ and $A$, we once again consider a rod with homogeneous, circular cross-sections. This time, however, due to the incompressibility, the rod has density $\rho = \rho^\circ$ and cross-sectional area $A = A^\circ / \tau_3$. This leads to the following equations, which are form-invariant during re-parameterization:

$$(\rho A) = \tau_3 \rho A, \qquad (\rho I) = \frac{1}{4\pi} \tau_3^2 \rho A^2.$$

Rearrangement yields

$$\rho = \frac{1}{4\pi} \frac{(\rho A)^2}{(\rho I)}, \qquad A = 4\pi \frac{(\rho I)}{\tau_3 (\rho A)}.$$

If one wants to replace $(\rho A)$ and $(\rho I)$ with $\rho$ and $A$ in the incompressible model, the replacement balances then become

$$\partial_t \rho + u \partial_s \rho = 0, \qquad \partial_t (\tau_3 A) + \partial_s (u \tau_3 A) = 0. \tag{5}$$

### 4.2.2 Constraints and Material Laws

In the Cosserat rod theory, material laws describe the dependency of the internal stress forces $\mathsf{n}$ and moments $\mathsf{m}$ on the fundamental deformation variables $\tau$ and $\kappa$. As the tuple notation suggests, they are formulated in the director basis in order to ensure objectivity (invariance in the face of rigid-body movements). Keeping in mind the applications considered later, we limit ourselves here to linear-elastic and viscous material laws. Stiff material behavior is treated by means of geometric requirements on the dynamics, that is, by formulating appropriate constraints. This reduces the number of material laws to be formulated. The most important example in practice is the Kirchhoff constraint, which we will discuss in both its classical form and in a generalized variant.

**Classical Variant of the Kirchhoff Constraint** The requirement of a strain and shear free rod

$$\boldsymbol{\tau} = \mathbf{d}_3,$$

that is, $\tau = \mathsf{e}_3$ in the director basis, is referred to as the Kirchhoff constraint [32]. By fixing the strain and shear quantities, all stresses $\mathsf{n} = (n_1, n_2, n_3)$ become Lagrange parameters to the constraint; that is, they become variables in the theory, and all that remains is to formulate a material law for the moments $\mathsf{m} = (m_1, m_2, m_3)$. The classical Kirchhoff constraint has its greatest significance in the area of elastic materials. It is not form-invariant in the face of re-parameterization ($\boldsymbol{\tau}$ is a type-1-quantity and $\mathbf{d}_3$, a type-0-quantity). Models using the classical Kirchhoff constraint are therefore treated exclusively in the Lagrangian description. Because the constraint ensures the arc-length parameterization, this restriction is not practically relevant with regard to the Eulerian description.

**Generalized Variant of the Kirchhoff Constraint** For viscous and viscoelastic material behavior, shear effects frequently play a subordinated role as well. Strain effects, however, may not be ignored. This situation is handled by means of a generalization of the Kirchhoff constraint

$$\boldsymbol{\tau} = e\mathbf{d}_3$$

or $\tau = e\mathsf{e}_3$ in the director basis. Using this weaker requirement, the normal stress force components $n_1$ and $n_2$, as Lagrange parameters to the constraint, become variables in the theory, and the task remains to formulate material laws for the tangential stress force $n_3$ and the moment $\mathsf{m}$. The generalized Kirchhoff constraint is form-invariant in the face of re-parameterization.

**Elastic Filaments—Bernoulli–Euler** Among the numerous elastic variants of the Cosserat rod theory, the Bernoulli–Euler model dominates the filament applications. Here, a rod is considered using an inertia-free geometry model. The material model in the Lagrangian description consists of a combination of the classical Kirchhoff constraint and a linear moment-curvature relation:

$$\tau = \mathsf{e}_3, \qquad \mathsf{m} = (EI)\mathrm{diag}\big(1, 1, 1/(1+\nu)\big) \cdot \kappa.$$

Here, $\nu$ is the Poisson's ratio and $(EI)$ ($\mathrm{N\,m^2}$) is the bending stiffness of the rod.

In accordance with the discussion of the classical Kirchhoff constraint, we formulate the model only in the Lagrangian description. In the director basis, we obtain the following:

$$\mathsf{D} \cdot \partial_t \bar{\mathsf{r}} = \mathsf{v}, \qquad \partial_t \mathsf{D} = -\omega \times \mathsf{D}, \qquad \mathsf{D} \cdot \partial_s \bar{\mathsf{r}} = \mathsf{e}_3, \qquad \partial_s \mathsf{D} = -\kappa \times \mathsf{D}$$
$$(\rho A)\partial_t \mathsf{v} = \partial_s \mathsf{n} + \kappa \times \mathsf{n} + \mathsf{D} \cdot \bar{\mathsf{k}} + (\rho A)\mathsf{v} \times \omega, \qquad 0 = \partial_s \mathsf{m} + \kappa \times \mathsf{m} + \mathsf{e}_3 \times \mathsf{n} \quad (6)$$

(see [32, 34] also). As an alternative to the inertia-free geometry model, the standard model is also used for the angular momentum. However, the degenerate angular momentum balance in (6), in interplay with the simple linear material model, permits, by means of several algebraic rearrangements, a radical simplification of the model. Here, the torsion module $M = m_3$ proves to be constant, and $N = n_3 - (EI)\|\partial_{ss}\mathbf{r}\|^2$ replaces the tangential stress force $n_3$ as a variable [11].

*Elastic Bernoulli–Euler model, Kirchhoff rod*

$$(\rho A)\partial_{tt}\mathbf{r} = \partial_s(N\partial_s\mathbf{r}) - \partial_{ss}\big((EI)\partial_{ss}\mathbf{r}\big) + M\partial_s\mathbf{r} \times \partial_{sss}\mathbf{r} + \mathbf{k}, \quad \|\partial_s\mathbf{r}\| = 1. \quad (7)$$

The system (7) shows a wave-like behavior, with an elliptic regularization governed via the bending stiffness and the constraint of inextensibility, with $N$ as associated Lagrange parameter. Provided one filament end is stress-free, then $M = 0$, and the torsion elements of the momentum balance disappear completely. The simplified formulation (7) forms the core of the ITWM software tool FIDYST (Fiber Dynamics Simulation Tool) for simulating cured filaments.

*Remark 1* (Elastic rods) Elastic Cosserat rods represent a very old, extremely comprehensive, but still current, field of research. We mention here only a few of the key points that are important for our work and refer to the existing literature for further information. The foundations of the theory outlined here can be traced back to Bernoulli, Kirchhoff [72], the Cosserat brothers [41], and Love [78], among others. From today's perspective, [96] and [32] offer a comprehensive overview. The works [80, 81, 111] explore analytical aspects, such as solution theory and stability. For the basics of Lagrange-based discretization strategies (geometrically exact approach, discrete differential geometry), we refer to [101, 102] or the more recent [70], and for computer graphic considerations, [34]. Finally, [82] is of particular interest for the deposition behavior of filaments.

**Viscous Filaments—Ribe** Ribe proposed modeling viscous jets on the basis of the Cosserat rod theory. This was initially formulated for steady state [93] and later generalized for transient systems [94]. In our system of classification, Ribe's model is a Cosserat rod model with incompressible geometry for homogeneous, circular cross-sections of density $\rho$. The material model is based on a generalized Kirchhoff constraint with specification of the tangential stress forces $n_3$ and moments $\mathsf{m}$. In the Lagrangian description [1], it becomes

$$\tau = e\mathsf{e}_3, \qquad n_3 = 3\mu \frac{A}{e} \partial_t e, \qquad \mathsf{m} = \frac{3\mu}{4\pi} \frac{A^2}{e} \mathrm{diag}(1,1,2/3) \cdot \partial_t \kappa,$$

where $\mu$ (Pa s) refers to the dynamic viscosity of the jet. Note that the formulated material laws are form-invariant in the face of re-parameterization. This allows us to formulate the complete Cosserat rod model for viscous jets in a generalized description with use of the director basis.

*Viscous rod model*

$$\begin{aligned}
\mathsf{D} \cdot \partial_t \bar{\mathsf{r}} &= \mathsf{v} - u e \mathsf{e}_3 \\
\partial_t \mathsf{D} &= -(\omega - u\kappa) \times \mathsf{D} \\
\mathsf{D} \cdot \partial_s \bar{\mathsf{r}} &= e \mathsf{e}_3 \\
\partial_s \mathsf{D} &= -\kappa \times \mathsf{D} \\
\partial_t (eA) + \partial_s (ueA) &= 0 \\
\partial_t (eA\mathsf{v}) + \partial_s (ueA\mathsf{v}) &= \frac{1}{\rho}(\partial_s \mathsf{n} + \kappa \times \mathsf{n} + \mathsf{D} \cdot \bar{\mathsf{k}}) + eA\mathsf{v} \times \omega \\
\mathsf{P}_2 \cdot \left(\partial_t \left(eA^2\omega\right) + \partial_s \left(ueA^2\omega\right)\right) &= \frac{4\pi}{\rho}(\partial_s \mathsf{m} + \kappa \times \mathsf{m} + e\mathsf{e}_3 \times \mathsf{n}) + eA^2(\mathsf{P}_2 \cdot \omega) \times \omega \\
\partial_t e + \partial_s (ue) &= \frac{1}{3\mu} \frac{e}{A} \mathsf{n} \cdot \mathsf{e}_3 \\
\partial_t \kappa + \partial_s (u\kappa) &= \frac{4\pi}{3\mu} \frac{e}{A^2} \mathsf{P}_{3/2} \cdot \mathsf{m}.
\end{aligned} \tag{8}$$

Here, $\mathsf{P}_c = \mathrm{diag}(1,1,c)$ for $c \in \mathbb{R}$. The model is completed by means of an additional equation for selecting the description: for example, $u = 0$, for the Lagrangian description and $e = 1$, for the Eulerian description. If the referential density is not constant, then a convection equation for $\rho$ must be added (see (5)). The system (8) has a hyperbolic character, with additional ordinary differential equations for the curve and rotation group in place of evolution equations for $\mathsf{n}$ and $\mathsf{m}$. Due to their structural closeness to the material laws, the compatibility relations

$$\partial_t e\mathsf{e}_3 + \partial_s (ue)\mathsf{e}_3 = \partial_s \mathsf{v} + \kappa \times \mathsf{v} + e\mathsf{e}_3 \times \omega, \qquad \partial_t \kappa + \partial_s (u\kappa) = \partial_s \omega + \kappa \times \omega$$

allow diverse re-formulations of (8). In practice, the model presented here forms the descriptive foundation for spinning processes.

*Remark 2* (Slender-body asymptotics) The Cosserat rod models do not arise—as one might perhaps suppose—as asymptotic limits $\varepsilon \to 0$ of 3D continuum mechanics for the slenderness parameter $\varepsilon$ (ratio of typical filament thickness to filament length). In such limits, one obtains instead string models, which neglect angular momentum effects and, in the linear momentum balance, only account for the tangential stress forces. The Cosserat rod models also include terms of the order $\varepsilon^2$ via the angular momentum balance. As demonstrated using the example of elastic Kirchhoff rods (7), inertia-free geometry models allow the remaining angular momentum effects to be integrated into the linear momentum balance. We then speak of generalized strings. Particularly for the viscous case, there are asymptotically strict derivations of string models based on boundary value problems. For uni-axial jets based on Stokes equations, see [45], for example; for those based on Navier–Stokes equations, see [69]. Curved jets are handled in a corresponding manner in [18, 20] (Navier–Stokes without and with surface tension). The solvability of these string models relative to the underlying parameters is restricted by the occurrence of singularities [4, 7], [67]. By a comparative consideration of the associated globally-applicable rod models as a regularization of the string, we were able to completely analyze this problem [4]. The viscous rod can also be transferred into a generalized string with an inertia-free geometry model. However, the structure of the equations remains complex compared to the elastic case [2, 33].

*Remark 3* (Viscous and viscoelastic strings and rods) Fundamental works on viscous filaments are [88], for strings, and [51, 93, 112], for rods, pursuant to Remark 2. There have been many theoretical and numerical investigations, and the contributions mentioned below are merely representative examples that may serve as starting points for the interested reader. For spinning, see [43, 45, 87, 110]; for break-offs and drops, [48, 49, 69, 103]; for the deposition of highly viscous filaments, [40, 93–95]; and for instabilities, [56, 59, 85, 99, 100, 113]. The dynamics of non-Newtonian and viscoelastic strings, as well as crystallization aspects, are discussed in [14, 24, 37, 53, 54, 63, 92].

### 4.2.3 Energy Balance

Up to this point, we have presented the classical Cosserat rod theory, which only includes mass, linear momentum, and angular momentum balances and, consequently, initially excludes thermal effects. Particularly in the cooling phase of spinning processes, however, the temperature $T$ (K) plays an important role: it determines the choice of an adequate rheological model. Here, material coefficients such as the viscosity or bending strength are functions of temperature. An accompanying energy balance is therefore frequently included in the applications. This can be constructed from various base quantities (internal energy, enthalpy, entropy), much as in 3D thermo-mechanics. In the framework of this chapter, we consider the specific enthalpy $h$ (m$^2$/s$^2$) as a function of the temperature $T$. The associated derivative $c_p = \partial_T h$ (J/(kg K)) is the specific heat capacity of the filament material. $h$, $T$ and $c_p$ are introduced as type-0-quantities. In this treatment, as a type-1-quantity, the enthalpy line density $(\rho A)h$ (J/m) is the actual energetic balance quantity.

Using the definition of $c_p$ and the conservation equation for $(\rho A)$, the general form of the energy balance becomes:

*Energy balance*

$$c_p\big(\partial_t\big((\rho A)T\big) + \partial_s\big(u(\rho A)T\big)\big) = q.$$

By means of the source term $q$ (W/m), both thermal conduction along the filament and various other warming and cooling effects can be included. We restrict ourselves here to the thermal exchange with a surrounding airstream of temperature $T_a$, which is described with the help of a heat transfer coefficient $\alpha$ (W/(m$^2$ K)). Thermal exchange takes place across the circumference $\pi d$, with diameter $d = (2/\sqrt{\pi})\sqrt{A}$, of the filament of cross-sectional area $A$. Thus, for this type of model, a geometric model is needed that uses $(\rho A)$ to define the density $\rho$ and the cross-sectional area $A$, pursuant to (4) and/or (5). The energy balance for the viscous rod model (8) with an incompressible geometry model and generalized Kirchhoff constraint $\tau_3 = e$ then becomes:

*Convective air cooling of a viscous rod*

$$c_p\rho\big(\partial_t(eAT) + \partial_s(ueAT)\big) = q, \quad q = -e\pi d\alpha(T - T_a). \tag{9}$$

## 4.3 Filaments in Airflows

The targeted applications are characterized by interactions between filaments and airflows. This holds for the spinning-and-cooling phase, as well as for the turbulent swirling phase. The geometric reconstruction of the Cosserat rods in a 3D flow is one possible approach, but it leads to resource-consuming interface problems. Therefore, we design simplified models for the line force density $\mathbf{k}$ and the heat source $q$ as functions of the relevant filament quantities $\Psi$ and the flow fields $\Psi_a$:

$$\Psi = (\mathbf{r}, \mathbf{v}, \boldsymbol{\tau}, d), \qquad \Psi_a = (\mathbf{u}_a, T_a, \rho_a, \nu_a, \lambda_a, c_{p,a}, k_a, \epsilon_a).$$

Simulations of the Navier–Stokes equations (NSE) yield the flow velocity $\mathbf{u}_a$, the temperature $T_a$ (K), the density $\rho_a$ (kg/m$^3$), the kinematic viscosity $\nu_a$ (m$^2$/s), the thermal conductivity $\lambda_a$ (W/(m K)), and the specific heat capacity $c_{p,a}$ (J/(kg K)) of the air. To the extent that turbulence effects must be included, we use the Reynolds-averaged Navier–Stokes equations (RANS) with the statistical $k$-$\epsilon$-turbulence model [52], since these still represent the industrial standard in the complex machine geometry of the applications. Here, $k_a$ (m$^2$/s$^2$) is the turbulent kinetic energy and $\epsilon_a$ (m$^2$/s$^3$) is the dissipation rate. To date, the alternative of a description based on large-eddy simulations (LES) has not been pursued. Our concept allows for a one-sided consideration of the flow effect on the filaments by using a filament-free flow as a basis, but also for the complete coupling of filament and aerodynamics via additional source terms in the flow equations.

While the general framework of the Cosserat rod theory in Sect. 4.1 exhibits an axiomatic character, the inclusion of specific material and geometry models in Sect. 4.2 already begins to bring phenomenological elements into consideration. For the outlined inclusion of air effects, we draw upon a variegated mixture of experimental results, simulations, symmetry considerations, and asymptotic reflections. It is precisely this mixture that permits the formulation of a complete, simulatable model and which is a characteristic feature of industrial mathematics.

### 4.3.1 Drag Coefficients and Heat Transfer Coefficients

In order to describe the effect of air on the filaments, it is necessary to specify the associated external force $\mathbf{k} = \mathbf{k}_{air}$ and the heat transfer coefficient $\alpha = \alpha_{air}$ in the base equations. Both models are based on the treatment of a cylindrical incident flow, which is locally evaluated with the relevant fields for filament $\Psi = \Psi(s,t)$ and flow $\Psi_a = \Psi_a(\mathbf{r}(s,t),t)$. Here, the normalized tangents of the filament and the relative velocity between flow and filament are identified with the orientation and incident velocity in the idealized cylinder configuration.

*Models for aerodynamic force and heat transfer*

$$\mathbf{k}_{air}(\Psi, \Psi_a) = e\mathbf{f}\left(\frac{\boldsymbol{\tau}}{e}, \mathbf{u}_a - \mathbf{v}, d, \rho_a, \nu_a\right)$$

$$\alpha_{air}(\Psi, \Psi_a) = \frac{\lambda_a}{(\pi/2)d} Nu\left(\frac{\boldsymbol{\tau}}{e} \cdot \frac{\mathbf{u}_a - \mathbf{v}}{\|\mathbf{u}_a - \mathbf{v}\|}, \frac{(\pi/2)d}{\nu_a}\|\mathbf{u}_a - \mathbf{v}\|, \frac{\rho_a \nu_a c_{p,a}}{\lambda_a}\right). \quad (10)$$

What follows is a discussion of the functional dependencies of the line force density $\mathbf{f}$ and the Nusselt number $Nu$, the latter of which is the basis for the heat transfer model. Whereas $\mathbf{k}_{air}$ is referenced to the filament parameter, note that the line force density is referenced to length—which implies the pre-factor $e$ in (10).

**Drag Coefficients** We consider the line force density $\mathbf{f}$ exerted on a straight cylinder with orientation $\mathbf{t}$, $\|\mathbf{t}\| = 1$, and diameter $d$ by a homogeneous, steady-state airflow of density $\rho_a$ and kinematic viscosity $\nu_a$, and constant incident velocity in the distance field $\mathbf{w}$ (Fig. 2). Assuming a functional relationship among these quantities, a dimensional analysis necessarily results in the existence of a dimensionless function $\mathbf{f}^\star$, which depends only on $\mathbf{t}$ and the dimensionless velocity $\mathbf{w}^\star = d/\nu_a \mathbf{w}$, such that

$$\mathbf{f}(\mathbf{t}, \mathbf{w}, d, \rho_a, \nu_a) = \frac{\rho_a \nu_a^2}{d} \mathbf{f}^\star\left(\mathbf{t}, \frac{d}{\nu_a}\mathbf{w}\right).$$

**Fig. 2** Cylinder incident flow (Graphic: Steffen Grützner, Fraunhofer ITWM)

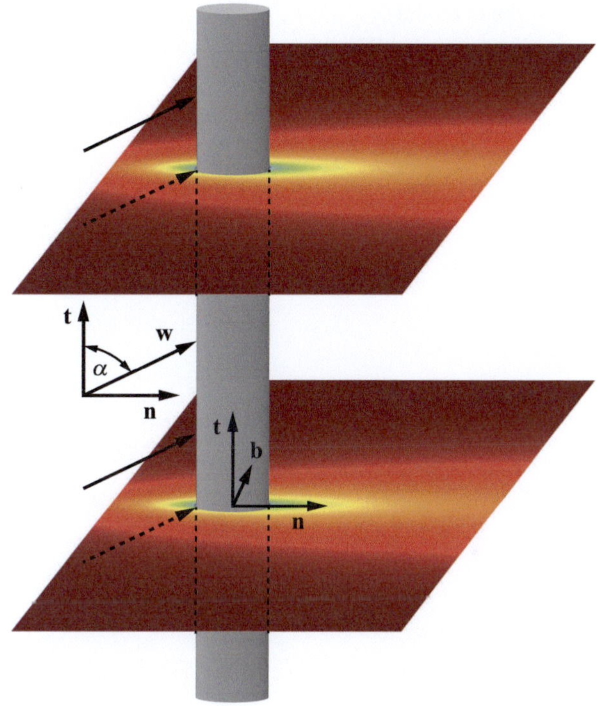


For steady-state flow around the cylinder, the line force density $\mathbf{f}^\star$ is given by the normal and tangential resistance coefficients $r_n$ and $r_t$, which, for their part, depend solely on the normal component of $\mathbf{w}^\star$ (Independence principle [68]. For a formal proof, see [19], for example.)

$$\mathbf{f}^\star\big(\mathbf{t}, \mathbf{w}^\star\big) = r_n(w_n)\mathbf{w}_n + r_t(w_n)\mathbf{w}_t$$

$$\mathbf{w}_t = \big(\mathbf{w}^\star \cdot \mathbf{t}\big)\mathbf{t}, \qquad \mathbf{w}_n = \mathbf{w}^\star - \mathbf{w}_t, \qquad w_n = \|\mathbf{w}_n\|.$$

Along with the resistance coefficients, the drag coefficients $r_n/w_n$ and $r_t/w_n$ are discussed in the literature as alternatives. For the normal drag coefficient, that is, for a vertical incident flow, there are several analytical results for creeping flows. See [76] for an infinitely long cylinder and [36, 42, 71] for finitely elongated objects. For rapid flows as well, [98, 104, 114] provide an overview of the numerous numerical and experimental investigations. Our experimentally validated model [19] for the normal and tangential resistance coefficients $r_n, r_t \in \mathscr{C}^1(\mathbb{R}_0^+)$ comprises piecewise the asymptotic Oseen theory [107, 108] with the auxiliary function $S(w_n) = 2.00 - \ln w_n$, an exponential spline approximation of our own numerical simulations, and the heuristic Taylor model [105] for the parameter $\gamma = 2$ (see Fig. 3):

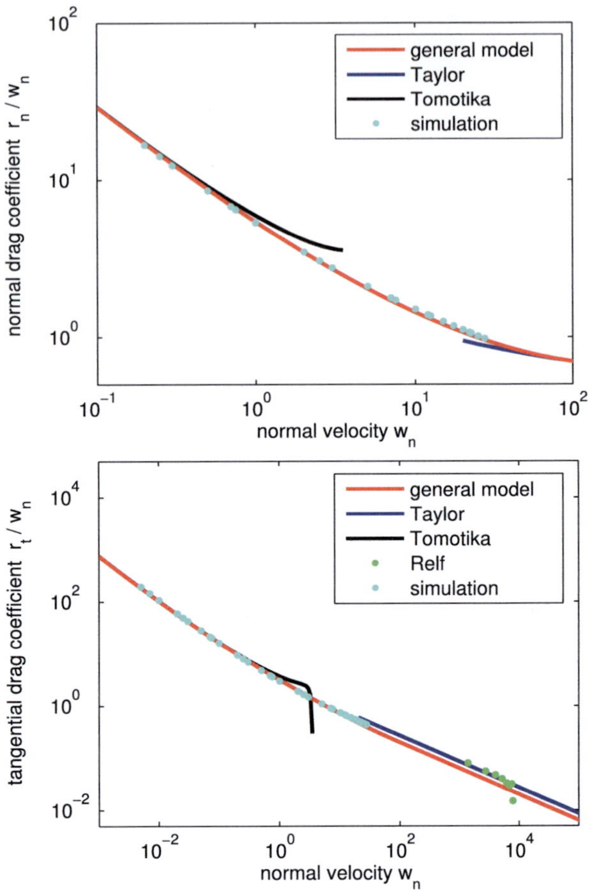


**Fig. 3** Model for drag coefficients based on literature data (see [105] for Taylor, [107, 108] for Tomotika, and [98] for Relf)

$$r_n(w_n) = \begin{cases} \sum_{j=0}^{3} q_{n,j} w_n^j & : \quad w_n < w_0 \\ 4\pi/S(1 - \frac{S^2 - S/2 + 5/16}{32S} w_n^2) & : \quad w_0 \le w_n < w_1 \\ \exp(\sum_{j=0}^{3} p_{n,j} \ln^j w_n) w_n & : \quad w_1 \le w_n < w_2 \\ 2\sqrt{w_n} + 0.5 w_n & : \quad w_2 \le w_n \end{cases}$$

$$r_t(w_n) = \begin{cases} \sum_{j=0}^{3} q_{t,j} w_n^j & : \quad w_n < w_0 \\ 4\pi/(2S - 1)(1 - \frac{2S^2 - 2S + 1}{16(2S-1)} w_n^2) & : \quad w_0 \le w_n < w_1 \\ \exp(\sum_{j=0}^{3} p_{t,j} \ln^j w_n) w_n & : \quad w_1 \le w_n < w_2 \\ \gamma\sqrt{w_n} & : \quad w_2 \le w_n. \end{cases}$$

The result of the Oseen theory (creeping flow) is restricted here to $w_0 \le w_n$, since, in the Oseen theory, both resistance coefficients disappear for $w_n \to 0$. This well-known result is traceable to the consideration of an infinitely long cylinder as an object in a flow. To generate realistic results for finite objects, the domain $w_n < w_0$ is therefore used, so that for

$w_n \to 0$, one runs smoothly into a Stokes expansion $r_n^S = (4\pi \ln(4/\delta) - \pi)/\ln^2(4/\delta)$ and $r_t^S = (2\pi \ln(4/\delta) + \pi/2)/\ln^2(4/\delta)$, with the regularization parameter $\delta < 3.5 \cdot 10^{-2}$. The transition points of the model, which are adjusted to measurements and simulations, are $w_0 = 2(\exp(2.00) - 4\pi/r_n^S)$, $w_1 = 0.1$, $w_2 = 100$. The $\mathscr{C}^1$-smoothness is guaranteed for $i = n, t$ by $q_{i,0} = r_i^S$, $q_{i,1} = 0$, $q_{i,2} = (3r_i(w_0) - w_0 r_i'(w_0) - 3r_i^S)/w_0^2$, $q_{i,3} = (-2r_i(w_0) + w_0 r_i'(w_0) + 2r_i^S)/w_0^3$ as well as $p_{n,0} = 1.69$, $p_{n,1} = -6.72 \cdot 10^{-1}$, $p_{n,2} = 3.33 \cdot 10^{-2}$, $p_{n,3} = 3.50 \cdot 10^{-3}$ and $p_{t,0} = 1.16$, $p_{t,1} = -6.85 \cdot 10^{-1}$, $p_{t,2} = 1.49 \cdot 10^{-2}$, $p_{t,3} = 7.50 \cdot 10^{-4}$. Here, $r_i'$ is to be understood as the right-hand limit from the zone $[w_0, w_1)$. The model's zone of application is limited to $w_n < 3 \cdot 10^5$, that is, to values below the so-called drag crisis [97, 98]. Due to vortex shedding, steady flows are not realizable for the zone $w_n > 40$. The model is to be understood here in the sense of time-averaged resistance coefficients.

**Heat Transfer Coefficient** For the Nusselt number $Nu$, we use a heuristic based model that was initially formulated in [109] for a vertical cylinder incident flow and then, on the basis of experimental data, modified in [3] for an arbitrary incident flow direction. In this model, the Nusselt number is a function of the cosine of the incident flow angle $c \in [-1, 1]$, the Reynolds number $Re$, and the Prandtl number $Pr$. The value $(\pi/2)d$ is chosen as a typical length,

$$c = \mathbf{t} \cdot \frac{\mathbf{w}}{\|\mathbf{w}\|}, \qquad Re = \frac{(\pi/2)d}{\nu_a}\|\mathbf{w}\|, \qquad Pr = \frac{\rho_a \nu_a c_{p,a}}{\lambda_a}.$$

In an analogous manner to the drag coefficients, the appropriate regularization with the associated parameter $\delta_h \ll 1$ guarantees a smooth transition to the case of parallel incident flow. The complete model becomes:

$$\begin{aligned}
Nu(c, Re, Pr) &= \left(1 - 0.5h^2(c, Re)\right)\left(0.3 + \sqrt{Nu_{lam}^2(Re, Pr) + Nu_{turb}^2(Re, Pr)}\right)\\
Nu_{lam}(Re, Pr) &= 0.664 Re^{1/2} Pr^{1/3}, \qquad Nu_{turb}(Re, Pr) = \frac{0.037 Re^{0.9} Pr}{Re^{0.1} + 2.443(Pr^{2/3} - 1)}\\
h(c, Re) &= \begin{cases} 1 - Re/\delta_h + cRe/\delta_h & : \ Re < \delta_h \\ c & : \ Re \geq \delta_h. \end{cases}
\end{aligned} \tag{11}$$

### 4.3.2 Turbulence Effects

Turbulent flows are characterized by a broad spectrum of variously sized vortices, whose resolution with the help of direct numerical simulation (DNS) is limited to moderate Reynolds numbers. Large-eddy simulations (LES) and statistical turbulence models represent alternatives. Whereas LES filters out turbulent structures below the grid resolution and accounts for their effects by means of a grid-dependent correction of the viscosity, the statistical models assume the Reynolds-averaged Navier Stokes equations, characterize the fluctuations by means of additional fields with associated transport equations, and introduce heuristically a turbulent viscosity that is a function of these fields (Boussinesq approximation) [52]. The treatment that follows is based on the $k$-$\epsilon$-model with turbulent kinetic energy $k_a$ and dissipation rate $\epsilon_a$, which are introduced for model justification as expectation values dependent on the velocity fluctuations $\mathbf{u}'$:

$$k_a = \frac{1}{2}\mathbb{E}(\mathbf{u}'^2), \qquad \epsilon_a = \nu_a \mathbb{E}(\nabla \mathbf{u}' : \nabla \mathbf{u}').$$

Turbulence effects enter into the model of heat transfer coefficients (11) via the turbulent Nusselt number $Nu_{turb}$, but must be discussed in more detail for the aerodynamic forces. We pursue the strategy of reconstructing the turbulent flow fluctuations $\mathbf{u}'$ as a random field from the $k$-$\epsilon$-model, taking them into account in the proposed force model, and, where necessary, transitioning to white noise. The starting point here is a complete quantitative description of homogeneous, isotropic turbulence.

**Homogeneous Isotropic Turbulence** Homogeneous, isotropic, advection-driven turbulent flow, pursuant to the $k$-$\epsilon$-model [77], is characterized by constant values of $\mathbf{u}_a$, $\nu_a$, $k_a$, and $\epsilon_a$. The velocity fluctuations $\mathbf{u}'$ can therefore be described by the dimensionless function $\mathbf{u}'^{\star} = \mathbf{u}'/k_a^{1/2}$, with dimensionless position $\mathbf{x}^{\star} = \epsilon_a/k_a^{3/2}\mathbf{x}$ and time $t^{\star} = \epsilon_a/k_a t$, which is then parametrically dependent only on the turbulence-strength-scaled mean velocity $\boldsymbol{\upsilon} = \mathbf{u}_a/k_a^{1/2}$ and the ratio of the small and large turbulent length scales $\zeta = \epsilon_a \nu_a / k_a^2$:

$$\mathbf{u}'(\mathbf{x}, t) = k_a^{1/2} \mathbf{u}'^{\star}\left(\frac{\epsilon_a}{k_a^{3/2}}\mathbf{x}, \frac{\epsilon_a}{k_a}t; \frac{1}{k_a^{1/2}}\mathbf{u}_a, \frac{\epsilon_a \nu_a}{k_a^2}\right). \tag{12}$$

We model the fluctuations as a centered, differentiable, stochastic Gaussian random field [16, 19]. This is clearly defined by its correlations (covariance function), to which we apply a product approach in space and time, taking the advection into account:

$$\mathbf{K}(\mathbf{x}^{\star} + \mathbf{y}^{\star}, t^{\star} + \tau^{\star}, \mathbf{y}^{\star}, \tau^{\star}; \boldsymbol{\upsilon}, \zeta) = \boldsymbol{\gamma}(\mathbf{x}^{\star} - \boldsymbol{\upsilon} t^{\star}; \zeta)\varphi(t^{\star}). \tag{13}$$

Assuming isotropy and taking advantage of an incompressibility argument, the spatial correlation across its Fourier-transform $\mathcal{F}_{\boldsymbol{\gamma}}$ is defined by the energy spectrum $E$ [55]:

$$\mathcal{F}_{\boldsymbol{\gamma}}(\boldsymbol{\kappa}; \zeta) = \frac{1}{4\pi} \frac{E(\|\boldsymbol{\kappa}\|; \zeta)}{\|\boldsymbol{\kappa}\|^2}\left(\mathbf{I} - \frac{1}{\|\boldsymbol{\kappa}\|^2}\boldsymbol{\kappa} \otimes \boldsymbol{\kappa}\right).$$

Thus, the modeling task is reduced to the specification of two scalar functions for the energy spectrum $E$ and the time correlation $\varphi$.

Our proposal for a differentiable energy spectrum $E \in \mathscr{C}^2(\mathbb{R}_0^+)$ is in agreement with Kolmogorov's 4/5-law [55] and the $k$-$\epsilon$-model:

$$E(\kappa;\zeta) = C_K \begin{cases} \kappa_1^{-5/3} \sum_{j=4}^{6} a_j (\frac{\kappa}{\kappa_1})^j & : \ \kappa < \kappa_1 \\ \kappa^{-5/3} & : \ \kappa_1 \le \kappa \le \kappa_2 \\ \kappa_2^{-5/3} \sum_{j=7}^{9} b_j (\frac{\kappa}{\kappa_2})^{-j} & : \ \kappa_2 < \kappa \end{cases}$$

with Kolmogorov constant $C_K = 1/2$. For the dimensionless formulation selected here, the significance of the turbulence parameters $k$ and $\epsilon$ is reflected in the integral terms

$$\int_0^\infty E(\kappa;\zeta)\mathrm{d}\kappa = 1, \qquad \int_0^\infty \kappa^2 E(\kappa;\zeta)\mathrm{d}\kappa = \frac{1}{2\zeta},$$

which define the transition wave numbers $\kappa_1$ and $\kappa_2$ as functions of the parameter $\zeta$. The regularity requirement is satisfied by $a_4 = 230/9$, $a_5 = -391/9$, $a_6 = 170/9$, $b_7 = 209/9$, $b_8 = -352/9$, and $b_9 = 152/9$. For the time correlation $\varphi \in \mathscr{C}^\infty(\mathbb{R}_0^+)$, exponential decay behavior is plausible (for the time scale $t_T = 0.212$, see [79, 91]):

$$\varphi(t^\star) = \exp\left(\frac{-t^{\star 2}}{2t_T^2}\right).$$

**Correlated Stochastic Aerodynamic Force in Turbulent Flows** In turbulent flow, the aerodynamic force is primarily defined by the mean flow velocity and the fluctuations. As mentioned previously, we follow the approach of using the turbulence information to reconstruct the fluctuations as a stochastic random field. We then superimpose the fluctuations additively on the force model (10) of the main flow. The force itself thus becomes a correlated random field and the associated Cosserat rod models become randomized PDEs:

$$\mathbf{k}_{air}(\Psi, \Psi_a) = e\mathbf{f}\left(\frac{\boldsymbol{\tau}}{e}, \mathbf{u}_a + \mathbf{u}'_a - \mathbf{v}, d, \rho_a, \nu_a\right). \tag{14}$$

To reconstruct $\mathbf{u}'_a$ for an arbitrary turbulent flow (inhomogeneous, anisotropic), we draw on the homogeneous isotropic model (12) by localizing the non-dimensionalizing scales and parameters [10]:

$$\mathbf{u}'_a(\mathbf{x},t) = k_a^{1/2}(\mathbf{p})\mathbf{u}'^\star\left(\frac{\epsilon_a}{k_a^{3/2}}(\mathbf{p})\mathbf{x}, \frac{\epsilon_a}{k_a}(\mathbf{p})t; \frac{1}{k_a^{1/2}}\mathbf{u}_a(\mathbf{p}), \frac{\epsilon_a \nu_a}{k_a^2}(\mathbf{p})\right)\Bigg|_{\mathbf{p}=(\mathbf{x},t)}.$$

It proves extremely advantageous for the numerical implementation that, in the applications, the turbulent length ratio $\zeta = \epsilon_a \nu_a / k_a^2$ generally satisfies $\zeta \ll 1$, so that the asymptotic limit $\zeta \to 0$ is justified (see [10] for our algorithms, which are based on the ideas and strategies concerning sampling found in [50, 74, 83]). However, for observation scales significantly larger than the correlation scales, the resolution of spatial and temporal turbulence is just as inefficient as it is unnecessary; the asymptotic transition to an uncorrelated aerodynamic force model (white noise limit) [16, 17] can be used instead.

**Uncorrelated Stochastic Aerodynamic Force** Starting from the assumption of a linear approximation, we decompose the aerodynamic force into deterministic and stochastic elements. The deterministic force $\mathbf{f}$, with its derivative $\partial_{\mathbf{w}}\mathbf{f}$, is yielded by (10). The correlated fluctuations $\mathbf{u}'_a$ are approximated asymptotically along the filament and, in time, by a Gaussian white noise with turbulence-dependent amplitude $\mathbf{A}$. Here, $\mathbf{W}$ characterizes a vector-valued Wiener process in the filament parameter and in time. For the appropriate Cosserat rod models, this leads to stochastic partial differential equations (SPDE) with the following force terms in integral notation (Riemann and Ito integrals):

$$
\begin{aligned}
&\mathbf{k}_{air}(\Psi, \Psi_a)\mathrm{d}s\mathrm{d}t \\
&\quad = e\mathbf{f}\left(\frac{\tau}{e}, \mathbf{u}_a - \mathbf{v}, d, \rho_a, \nu_a\right)\mathrm{d}s\mathrm{d}t \\
&\quad + e\partial_{\mathbf{w}}\mathbf{f}\left(\frac{\tau}{e}, \mathbf{u}_a - \mathbf{v}, d, \rho_a, \nu_a\right) \cdot \mathbf{A}\left(\frac{\tau}{e}, \mathbf{u}_a - \mathbf{v}, d, \nu_a, k_a, \epsilon_a\right) \cdot \mathrm{d}\mathbf{W}_{s,t}. \qquad (15)
\end{aligned}
$$

The amplitude $\mathbf{A}$ represents the cumulative effects of the localized velocity fluctuations and is calculated, after the associated non-dimensionalization, by integrating across the homogeneous, isotropic correlation tensor (13):

$$
\begin{aligned}
\mathbf{A}(\mathbf{t}, \mathbf{w}, \nu_a, k_a, \epsilon_a) &= \frac{k_a^{7/4}}{\epsilon_a}\mathbf{A}^\star\left(\mathbf{t}, \frac{1}{k_a^{1/2}}\mathbf{w}, \frac{\epsilon_a \nu_a}{k_a^2}\right) \\
\mathbf{A}^{\star 2}(\mathbf{t}, \mathbf{w}^\star, \zeta) &= \int_{\mathbb{R}^2} \boldsymbol{\gamma}(s\mathbf{t} - t\mathbf{w}^\star; \zeta)\varphi(t)\mathrm{d}s\mathrm{d}t.
\end{aligned}
$$

Analogously to the cylinder incident flow (Fig. 2), we decompose $\mathbf{A}^\star$ with respect to the orthonormal basis $\{\mathbf{n}, \mathbf{b}, \mathbf{t}\}$. Thus, the calculation reduces finally to two integrals across the energy spectrum and the Fourier transforms of the time correlation function ($a_n^2 + a_b^2 = a_t^2$), which are numerically evaluated (Fig. 4):

$$
\begin{aligned}
\mathbf{A}^\star(\mathbf{t}, \mathbf{w}^\star, \zeta) &= a_n(w_n, \zeta)\mathbf{n}\otimes\mathbf{n} + a_b(w_n, \zeta)\mathbf{b}\otimes\mathbf{b} + a_t(w_n, \zeta)\mathbf{t}\otimes\mathbf{t} \\
a_{n,b,t}^2(w_n, \zeta) &= 4\pi \int_0^\infty \frac{E(\kappa; \zeta)}{\kappa} \int_0^{\pi/2} \{\sin^2\beta, \cos^2\beta, 1\}\mathscr{F}_\varphi(w_n \kappa\cos\beta)\mathrm{d}\beta\mathrm{d}\kappa.
\end{aligned}
$$

As in the correlated case, the simplification $\zeta = 0$ normally makes sense here as well, so that the model is completely represented by two functions. Associated data can be filed in a look-up table.

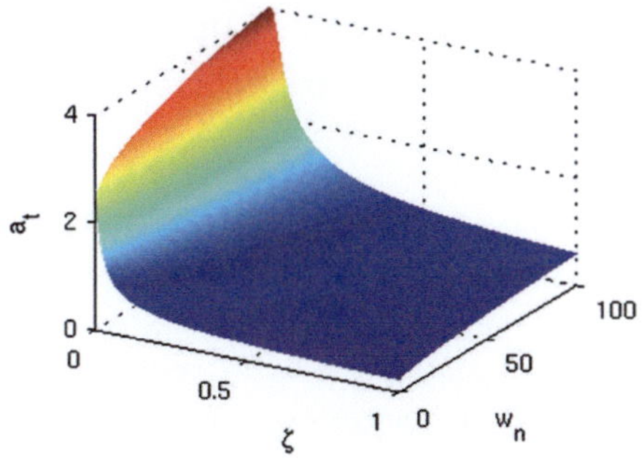


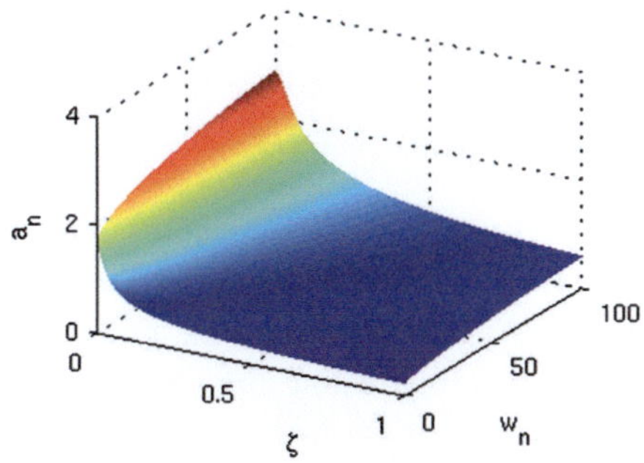


**Fig. 4** Amplitude of the uncorrelated aerodynamic force

### 4.3.3 Coupled Filament and Flow Dynamics

In continuum mechanics, the coupling of multiple phases, such as structure and flow, is accomplished by means of interface conditions, which ensure the conservation of momentum and energy [65]. In view of the necessary numerical resolution, it makes sense in the asymptotic context of the Cosserat rod model to represent the interaction with an action-reaction principle via source terms in the conservation equations of the filaments and the airflow. Here, the force and heat sources described in the Cosserat rod theory can be enlisted and incorporated in the flow equations. However, the source terms of the Cosserat rod theory are line-based, whereas the flow equations expect volume sources. A distributional approach of the form

$$\mathbf{k}_{rod}(\Psi, \Psi_a)(\mathbf{x}, t) = -\int_{s_a}^{s_b} \delta\big(\mathbf{x} - \mathbf{r}(s, t)\big)\mathbf{k}_{air}(\Psi, \Psi_a)\mathrm{d}s$$

on the basis of the deterministic force model (10) suggests itself for the treatment of single filaments—and, analogously, for the heat sources $q_{rod}(\Psi, \Psi_a)$ [3]. However, the models presented here lead to non-removable singularities, since the flow data is directly evaluated locally on the filaments. Making modifications via suitable averaging strategies represents one remedy; here, however, we follow another path: the inclusion of feedback effects is unavoidable, particularly for high filament densities. For this reason, we turn to a homogenization strategy in which a continuous filament length density $\sigma$ ($1/\mathrm{m}^2$) is defined for the volume occupied by filaments. This is to be scaled with $e$ in the source terms, since $\mathbf{k}_{air}$ was introduced as the line force density in relation to the parameterization. Thus, we arrive at

$$\mathbf{k}_{rod}(\Psi, \Psi_a) = -\frac{1}{e}\sigma\mathbf{k}_{air}(\Psi, \Psi_a). \tag{16}$$

And once again, we have the analogous form for the heat sources. The dynamics of the filament length density $\sigma$ result directly from the dynamics of representative single filaments, which, for the purposes of numerical resolution, are to be arranged with sufficient density.

Both modeling approaches lead to a fully coupled, dynamic system of all state variables $\Psi$ and $\Psi_a$ (see Sect. 5.3 for an algorithmic solution proposal; a turbulent extension of this treatment has not yet been developed).

*Remark 4* The classical approach of continuum mechanics to fluid-structure interactions (FSI) with spatially resolved phases for fluid and structure causes significant discretization problems for the flow, due to the time-changing computation domain. There are indeed some methods for treating this problem: for example, fictitious domain [58], immersed boundary [89], and mesh-free methods, such as SPH [86] and FPM [21, 106] (see Sect. 3 also). All of these methods, however, reach their limits for high particle or filament loads, since the structures must be resolved. For particles, kinetic model approaches [35, 57] that lead to coupled Navier–Stokes/Fokker–Planck systems have proven effective. Here, the feedback is accomplished via an extra stress tensor in the momentum balance of the flow. For filaments, however, these methods are limited to short objects that allow for a local description of the orientation.

## 4.4 Stochastic Surrogate Models for the Deposition of Fleeces

In a typical fleece production process, swirled filaments are deposited on a conveyor belt. Overlapping produces the stochastic microstructure typical of fleeces. The swirling and deposition phases of the process can be described on the basis of elastic Kirchhoff rods (7) in turbulent flow, which enters into the model either reconstructed or in the asymptotic limit as white noise. However, the full simulation of the Cosserat rod dynamics is extremely complex and can therefore only be carried out at an acceptable level of effort for individual filaments. In order to nonetheless represent the entire microstructure of a fleece fabric through a simulation that incorporates the chosen production parameters, we developed stochastic surrogate models that describe the deposition structure of a single filament. These models are based on stochastic (ordinary) differential equations (SODE), can be very efficiently simulated, and allow for the generation of the entire microstructure by superimposing repeated runs. Moreover, the parameters of the surrogate models are identified from the outlined full simulation of single representative filaments. It helps here that, in typical production situations, many and sometimes all of the filaments run through process conditions that are identical except for stochastic fluctuations caused primarily by turbulence. We will discuss the 2D standard lay down model in detail here, and then offer a brief overview of various model extensions.

### 4.4.1 Standard Lay down Model

We consider a deposited filament to be an arc-length parameterized curve on a flat conveyor belt, which we model as the stochastic process $(\boldsymbol{\eta}_s)_{s\in\mathbb{R}_0^+} \in \mathbb{E}^2$. Under real production conditions, the filament curve $\boldsymbol{\eta}$ fluctuates around a deterministic, process-specific reference curve $\boldsymbol{\gamma}$. For the simplest process, in which $v_0$ filament length per time is generated and deposited on a conveyor belt with speed $v_{belt}\mathbf{e}_1$, it is $\boldsymbol{\gamma}_s = -v_{belt}/v_0 s\mathbf{e}_1$. In

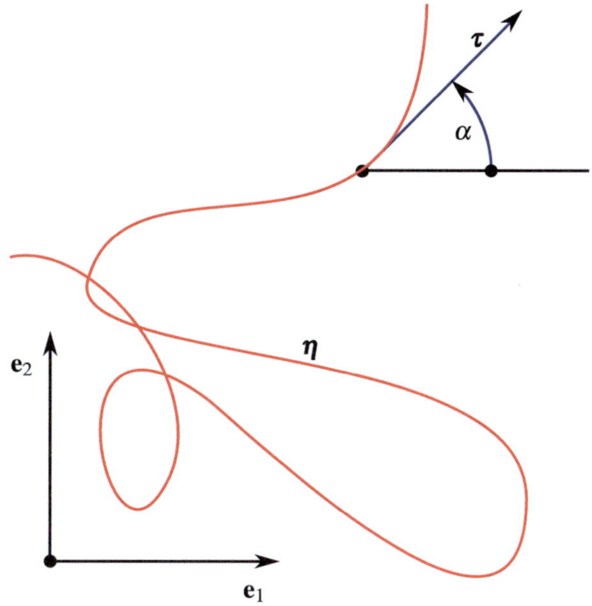


**Fig. 5** Basic quantities and notation of the deposition models

practice, however, more complex reference curves are also found in processes, such as those produced by rotating spinning positions or oscillating flows. We use $\boldsymbol{\xi} = \boldsymbol{\eta} - \boldsymbol{\gamma}$ to refer to the $\boldsymbol{\gamma}$-related curve (withdrawn process) and $\alpha = \angle(\mathbf{e}_1, \boldsymbol{\tau})$ to denote the angle between the production direction $\mathbf{e}_1$ and the tangent $\boldsymbol{\tau}$ to the filament curve $\boldsymbol{\eta}$; that is, $\boldsymbol{\tau}(\alpha) = \cos\alpha\mathbf{e}_1 + \sin\alpha\mathbf{e}_2$ and $\boldsymbol{\tau}^{\perp}(\alpha) = -\sin\alpha\mathbf{e}_1 + \cos\alpha\mathbf{e}_2$ (see Fig. 5).

**SODE Model** The model for the filament curve $\boldsymbol{\eta}$ is formulated in $\boldsymbol{\xi}$ and $\alpha$ as a stochastic differential equation [5, 6]. Here, the arc-length $s$ takes on the role of time in dynamic systems.

*Stochastic surrogate model for the deposition of the filament*

$$\mathrm{d}\boldsymbol{\xi}_s = \boldsymbol{\tau}(\alpha_s)\mathrm{d}s - \mathrm{d}\boldsymbol{\gamma}_s, \qquad \mathrm{d}\alpha_s = -\nabla B(\boldsymbol{\xi}_s) \cdot \boldsymbol{\tau}^{\perp}(\alpha_s)\mathrm{d}s + A\mathrm{d}W_s. \tag{17}$$

The typical process behavior of a filament is modeled by means of the potential $B$, which depends solely on $\boldsymbol{\xi}$. The fluctuations of the process are taken into consideration by means of an additive noise with amplitude $A$ (1/m$^{1/2}$) to the scalar-valued Wiener process $(W_s)_{s\in\mathbb{R}_0^+}$. In (17), $(\boldsymbol{\xi}, \alpha)$ represents a degenerated diffusion process.

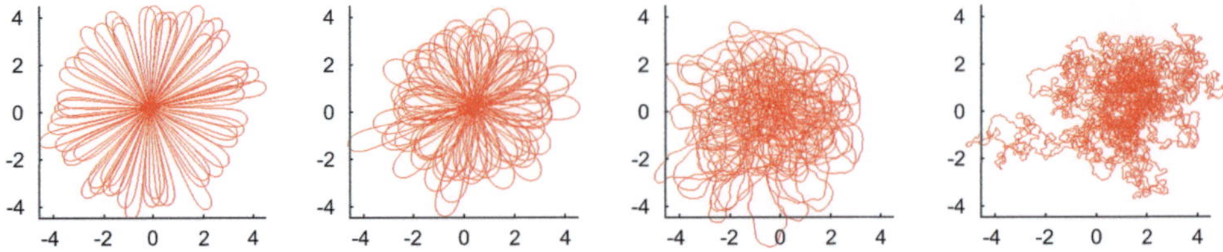


**Fig. 6** Effect of noise amplitude on the filament curve for a fixed process position; model with $\mathbf{C} = \lambda^2 \mathbf{I}$. *From left to right*: $A\sqrt{\lambda} = 0.001, 0.025, 1, 4$. Spatial depiction in units $\lambda$ (Graphics: Simone Gramsch, Fraunhofer ITWM)

**Fokker–Planck Equation and Steady State** The Fokker–Planck equation belonging to the SODE model (17) describes the probability density $p : \mathbb{E}^2 \times \mathbb{R} \times \mathbb{R}_0^+ \to \mathbb{R}_0^+$, $(\boldsymbol{\xi}, \alpha, s) \mapsto p(\boldsymbol{\xi}, \alpha, s)$ in the state space of the stochastic process

$$\partial_s p + \big(\boldsymbol{\tau}(\alpha) + \partial_s \boldsymbol{\gamma}\big) \cdot \nabla_{\boldsymbol{\xi}} p - \partial_\alpha \big(\nabla B(\boldsymbol{\xi}) \cdot \boldsymbol{\tau}^\perp(\alpha) p\big) = \frac{A^2}{2} \partial_{\alpha\alpha} p.$$

As can be easily shown, in the case of a fixed process position, i.e., for a constant $\boldsymbol{\gamma} = \boldsymbol{\gamma}_0$,

$$p_S(\boldsymbol{\xi}) = c \exp\big(-B(\boldsymbol{\xi})\big) \tag{18}$$

represents the steady-state, that is, the parameter $s$-independent solution to the Fokker–Planck equation (equilibrium solution), with $c > 0$ as the normalizing constant. Remarkably, this stationary solution is independent of the angle $\alpha$ and permits an immediate interpretation of the potential $B$. As the standard approach in practice,

$$B(\boldsymbol{\xi}) = \frac{1}{2} \boldsymbol{\xi} \cdot \mathbf{C}^{-1} \cdot \boldsymbol{\xi} \tag{19}$$

with the positive definite tensor $\mathbf{C}$ has proved reliable. With this choice, the position vectors $\boldsymbol{\xi}$ relative to spinning position $\boldsymbol{\gamma}_0$ are static and normally distributed, with expected value zero and two-dimensional covariance matrix $\mathbf{C}$, whose eigenvalues we designate as $\lambda_i^2$, $i = 1, 2$. The term 'throwing range' has therefore become established for the standard deviation $\lambda_i$. One must keep in mind, however, that this interpretation is only strictly valid for a conveyor belt at rest; it is approximately valid for a weakly varying reference curve $\boldsymbol{\gamma}$, that is, in the simplest case, for $v_{belt}/v_0 \ll 1$, since for arbitrary reference curves (18), there is no steady-state solution for the Fokker–Planck equation. To observe the interactions of the various parameters, see Fig. 6.

**Parameter Identification** The parameters appearing in the surrogate model, that is, the reference curve $\boldsymbol{\gamma}$, the potential $B$, and the noise amplitude $A$, are identified using deposition images from Cosserat rod calculations of representative filament dynamics [11].

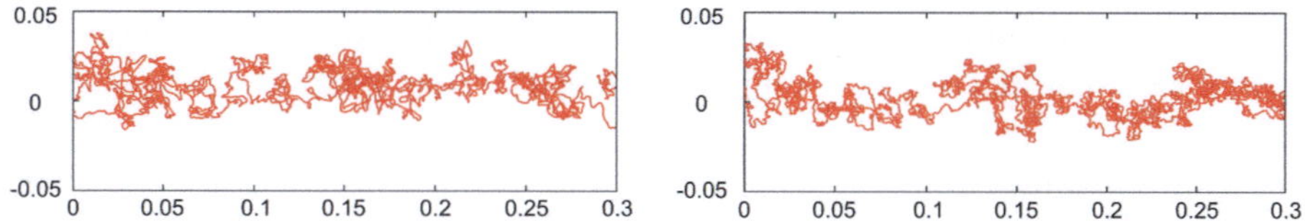


**Fig. 7** Comparison of filament curves from Cosserat rod simulation (*left*) and surrogate model with identified parameters (*right*) (Graphics: Simone Gramsch, Fraunhofer ITWM)

There are established approaches for estimating parameters in stochastic differential equations [75]. We choose a simple but very robust procedure. We also consider the approximation of an arc-length parameterized deposition curve by means of $N$ equally distributed (by $\Delta s$) supporting points $(\boldsymbol{\eta}_i)_{i=1,\ldots,N}$, which are acquired from a full simulation of a suitable Cosserat rod model. In the first step of our parameter identification, one selects a suitable reference curve $\boldsymbol{\gamma}$, based on the existing data set and further information about the process (spin speed, conveyor belt speed, and any oscillation and rotation frequencies). Using subtraction by $\boldsymbol{\gamma}$ and numerical differentiation, one can determine the data set $\mathbf{D}_{rod} = (\boldsymbol{\xi}_i, \alpha_i)_{i=1,\ldots,N}$ used for identification. Moreover, a suitable parametrical approach is made for the potential $B$. Here, the quadratic forms from (19) with the tensor $\mathbf{C}$ perform nicely. The valuation should always be checked at the end of the procedure, however. What remains is to identify the parameters $\mathbf{P} = (\mathbf{C}, A)$ to which we assign the data $\mathbf{D}_{surro}(\mathbf{P})$, toward the end of executing the surrogate process with fixed random numbers. We formulate the identification task as a minimization problem with the choice of a suitable function $\mathscr{F}$ across the data set:

$$\hat{\mathbf{P}} = \operatorname{argmin}_{\mathbf{P}} \left\| \mathscr{F}\big(\mathbf{D}_{surro}(\mathbf{P})\big) - \mathscr{F}(\mathbf{D}_{rod}) \right\|^2. \tag{20}$$

As the function, we choose

$$\mathscr{F}(\mathbf{D}) = \left( \frac{1}{N} \sum_{i=1}^{N} \boldsymbol{\xi}_i \otimes \boldsymbol{\xi}_i, \max_k \sqrt{\frac{\sum_{i=1}^{N-k} (\alpha_{i+k} - \alpha_i)^2}{k \Delta s (N-k)}} \right).$$

This function has a tremendous advantage: in the case of a fixed deposition position, for $N \to \infty$, it delivers the parameters directly, since $\mathscr{F}(\mathbf{D}_{surro}(\mathbf{P})) = \mathbf{P}$ and, thus, $\hat{\mathbf{P}} = \mathscr{F}(\mathbf{D}_{rod})$. In the case of a non-trivial reference curve, we apply a quasi Newton method to solve the minimization problem using the approximated Jacobi matrix $\mathbf{I}$ from the trivial case and the estimated initial solution $\mathscr{F}(\mathbf{D}_{rod})$. Figure 7 shows the results of such a parameter identification.

### 4.4.2 Model Extensions, Ergodicity, and Asymptotic Limits

At the Fraunhofer ITWM, the standard model for filament deposition (17) was developed as an alternative for the very time-consuming Cosserat rod simulations—that is, for highly pragmatic reasons. The model stands out by virtue of its significant potential for generalization (smooth filament curves, 3D microstructures) and, as a result of the degenerated diffusion, raises interesting analytical questions concerning long-term behavior (existence and ergodicity). For this reason, it is of great interest to both modelers and analysts. Important papers have been produced, some independently [46, 60, 62, 66, 73], and some in cooperation with the authors [5, 6, 11–13] from L.L. Bonilla (Madrid), J. Dolbeault (Paris), T. Götz (Koblenz), M. Herty (Aachen), M. Kolb (Warwick), S. Martin (London), S. Motsch (Toulouse), C. Mouhot (Cambridge), M. Savov (Oxford), C. Schmeiser (Vienna), A. Wübker (Osnabrück), as well as from M. Grothaus, A. Klar, J. Maringer, and P. Stilgenbauer (all from Kaiserslautern).

The filament curve of the standard model is continuous but non-differentiable. Replacing the Brownian motion with an Ornstein–Uhlenbeck process results in a more realistic smooth model for the curve, angle, and curvature variables. Moreover, extensions to 3D models for the direct generation of the fleece microstructure are of particular interest. The most elegant point of entry into this class of models is the formulation of geometric Langevin systems on sub-manifolds in the Stratonovich calculus [62]. As a two dimensional special case, these lead to the standard model (17). Here as well, smooth models can be designed analogously. The introduction of an anisotropy parameter takes into account the fact that filament tangents tend to lie parallel to the conveyor belt. The corresponding 3D model leads in borderline cases to a perfectly isotropic model or to the 2D model [12, 13]. For an industrial application and comparison with computer tomography data, see [8].

The aspect of degenerated diffusion increases the challenge of mathematically analyzing this class of models and calls for systematic new developments and extensions. The ergodicity of the 2D model was initially investigated with Dirichlet forms and semi-group operator techniques and delivered the first estimates of the convergence rate for a fixed deposition position [60]. Using hypocoercivity strategies [46, 61] that extend the work of [44, 47], along with probability methods [73], it was possible to deliver satisfactory results (including existence results) for the simplest straight reference curve. For the complete model class, the asymptotic relationships (3D/2D, anisotropic/isotropic, smooth/standard) have been clarified and the borderline cases of low noise (stochastic averaging techniques [6]) and high noise (analogous to the Chapman–Enskog expansion [5, 38, 39]) have been investigated.

## 5 Simulation Tools

At the Fraunhofer ITWM, various tools have been developed to numerically simulate the models presented here for filament dynamics and fleece deposition. These are used for contract research, continuously extended as elements of projects in applied basic research, and also licensed to customers. The FIDYST Suite provides software with a high level of compatibility. It includes the software tool FIDYST (*Fiber Dynamics Simulation Tool*), for simulating elastic Kirchhoff rods in turbulent flows, and the software tool SURRO (*Surrogate Models*), which is coupled to FIDYST via modules for parameter identification and used to virtually generate complete fleece microstructures. Moreover, VISFID (*Viscous Fiber Dynamics*) is a MATLAB-FLUENT toolbox that can be used to treat steady-state, aerodynamically-driven spinning processes with full coupling of filament and flow dynamics. For details about the commercial software products MATLAB and FLUENT, please refer to the supplier web pages www.mathworks.com and www.ansys.com, respectively.

## 5.1 FIDYST—Elastic Filaments in Turbulent Flows

FIDYST, the core building-block of the FIDYST Suite, is a C++ based simulator for the dynamics of Cosserat rods, with inertia-free geometry model, Kirchhoff constraint, and Bernoulli–Euler's moment-curvature relation as material law. This software tool covers a large class of possible applications for filament dynamics. With its link to SURRO (Sect. 5.2), however, it is aimed strongly towards the swirling and deposition phases in fleece production processes and has proven itself in numerous industrial applications.

FIDYST is based on the generalized string formulation (7), with $M = 0$, since a stress-free and moment-free filament end is always assumed in $s_b = 0$. Along with the elimination of the torsion term, this results in the following definition of the boundary conditions for this filament end: $\partial_{ss}\mathbf{r} = \mathbf{0}$, $\partial_s((EI)\partial_{ss}\mathbf{r}) = \mathbf{0}$, $N = 0$. In analyzing it as a first order system with the constraint $\|\partial_s\mathbf{r}\| = 1$, this corresponds to the definition of five of the ten total available degrees of freedom. The user has a great deal of flexibility in assigning the remaining five degrees of freedom for the boundary conditions on the other end of the filament (inlet). A freely definable function $s_a(t) < 0$ can be used with $\mathbf{r}(s_a(t), t) = \mathbf{r}_0(t)$ and $\partial_s\mathbf{r}(s_a(t), t) = \boldsymbol{\tau}_0(t)$ to specify time-dependent inflow speed ($v_0 = |\partial_t s_a|$), position, and direction. The filament diameter and, thus, the bending stiffness and line density, can by varied at the inlet. Along with gravitation, deterministic aerodynamic forces (10) are taken into account as external forces. Turbulence can be included as either an uncorrelated (15) or a correlated stochastic force (14) by the user. Moreover, geometry contacts and the deposition on a conveyor belt are accounted for by means of contact and friction algorithms. FIDYST can process flow data of various types, as well as geometry information in the EnSight format, and therefore works ideally in combination with FLUENT as a CFD tool. In both cases (flow and geometry), interpolations are made between discrete time points for transient information, so that, for example, the above-mentioned contact algorithm also works with movable machine parts. In practical application, for example when impact elements are used for filament diversion and distribution, this plays an important role.

FIDYST has a user-friendly GUI for initiating simulations, as well as for the accompanying visual simulation guidance and control (Fig. 8). One hallmark is the 3D viewer integrated into the GUI, which has diverse depiction options for geometry, flow data, and filament curves. All model and algorithmic parameters are accessible using the GUI. The software tool uses the EnSight format for the output as well, which allows deposition images to be fed into SURRO for further processing or simulation results to be transmitted

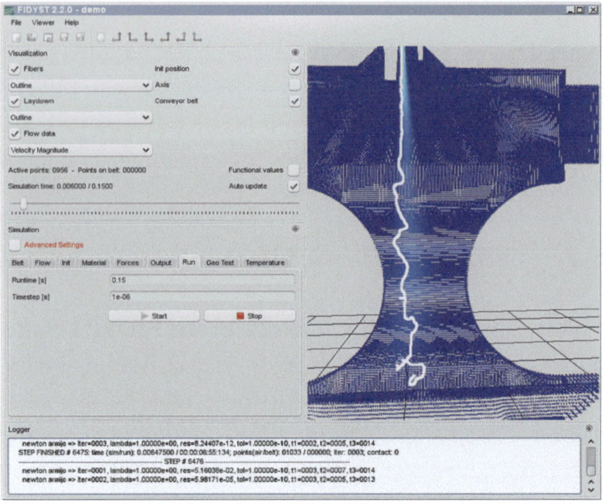


**Fig. 8** FIDYST GUI with 3D viewer: simulation run (Graphic: Simone Gramsch, Fraunhofer ITWM)

to powerful post processors, such as ParaView (see www.paraview.org). To round off this overview, we will discuss the fundamentals of the discretization and the algorithmic basis of the contact model.

**Discretization** We formulate (7) for $M = 0$ as a first order system in the variables $\mathbf{r}$, $\mathbf{v}$, and $N$ and use a semi-discretization based on finite volumes in the arc-length parameter $s_i = (i-1)\Delta s$, $i = 1, \ldots, K$, with constant grid size $\Delta s$. We characterize integral averages between $s_i - \Delta s/2$ and $s_i + \Delta s/2$ with an index $i$ and function values at the point $s_i - \Delta s/2$ with an index $i - 1/2$, and obtain the following for $i = 3, \ldots, K-2$:

$$\partial_t \mathbf{r}_i = \mathbf{v}_i, \qquad (\rho A)\partial_t \mathbf{v}_i = \mathbf{flux}_{i+1/2} - \mathbf{flux}_{i-1/2} + \mathbf{k}_i, \qquad \|\partial_s \mathbf{r}\|_{i-1/2} = 1$$
$$\mathbf{flux}_{i+1/2} = N_{i+1/2}(\partial_s \mathbf{r})_{i+1/2} - \big(\partial_s\big((EI)\partial_{ss}\mathbf{r}\big)_{i+1/2}, \quad i = 2, \ldots, K-2.$$

The resulting first and third derivatives of $\mathbf{r}$ are approximated by first order finite differences. Here, we see the merits of the staggered grid, in which $\mathbf{r}$ and $\mathbf{v}$ are assigned to the nodes, but $N$, and consequently also the constraint, are assigned to the edges. Due to the constraint, the rod in the discretization thus becomes a polygon line with a fixed geometrical spacing for the spatial points associated with the nodes (filament points). The number of nodes $K(t)$ is defined indirectly by means of the requirements $s_a(t) \in [s_2, s_3)$ and $s_b = s_K - (3/2)\Delta s$. The boundary conditions can be approximated using the ghost points $s_1, s_2, s_{K-1}, s_K$. We approximate the external forces by $\mathbf{k}_i \simeq (\mathbf{k}_{i-1/2} + \mathbf{k}_{i+1/2})/2$. For the aerodynamic forces, this has the advantage that the resulting tangents are, once again, only needed on the edges; the filament velocities, however, must be averaged across the neighboring nodes. The necessary flow data is interpolated at the associated positions. The case of an uncorrelated stochastic aerodynamic force is also handled accordingly, and what remains after integration in $s$ is a Wiener process in $t$. All in all, the semi-discretization thus leads to a DAE or a stochastic DAE system, which we discretize temporally with an implicit Euler or Euler–Maruyama method. Although the aerodynamic forces in the core (in the filament tangent and velocity) are also implicitly included, the flow data that appears in them is queried with the filament positions of the old time step, so that we can solve the resulting large nonlinear equation system using a Newton method with analytical Jacobi matrix and Armijo step-size control. The resulting linear systems are treated with a band solver. The method is so well optimized with regard to assembling the Jacobian that the main effort per time step is due to the equation solver itself.

**Geometry Contacts** FIDYST treats the geometry contacts of the filaments as non-holonomic constraints. Based on the assumption that the existing geometry has been largely triangulated, a smooth, generalized, signed distance function $H(\cdot, t) \in \mathscr{C}^2$ is generated, so that $H > 0$ represents an approximation of the admissible space for filament motion. Where needed, extrapolation is used to ensure that complete flow data are also available for this space. To establish the contact, a further Lagrange parameter $\lambda$ is introduced into the momentum balance:

$$(\rho A)\partial_{tt}\mathbf{r} = \cdots + \lambda \frac{\nabla H}{\|\nabla H\|}, \qquad (\lambda = 0 \wedge H > 0) \vee (\lambda > 0 \wedge H = 0).$$

In the semi-discretized variant, a Lagrange parameter $\lambda_i$ and a Boolean variable $\delta_i \in \{0, 1\}$ are assigned algorithmically to each node (filament point) and used to characterize the movement type as either non-contacting (free) ($\delta_i = 0$) or contacting ($\delta_i = 1$). For the time-step $t^n$ to $t^{n+1}$, the equations of motion are now solved in dependence on $\delta_i$:

$$(\rho A)\partial_{tt}\mathbf{r}_i = \cdots + \delta_i \lambda_i \frac{\nabla H}{\|\nabla H\|} \quad \text{and} \quad \begin{cases} \lambda_i = 0 & : \quad \delta_i = 0 \\ H = 0 & : \quad \delta_i = 1. \end{cases}$$

Note here that the Lagrange parameters $\lambda_i$ are distributions. For a finite Euler step, however, this creates no problems. If, at the end of the time step, the condition $H(\mathbf{r}_i, t^{n+1}) > 0$ for free points ($\delta_i = 0$) or $\lambda_i > 0$ for contacting points ($\delta_i = 1$) is violated, the Boolean variable is switched to the other value and the entire time-step is repeated for all points. This procedure is iterated until all points move consistently, that is, until it is no longer necessary to switch the Boolean variables. For purposes of modeling the deposition, this contact model is combined in FIDYST with a Coulomb friction model (kinetic and dynamic), in which the Lagrange parameters $\lambda_i$ act as normal forces according to their physical significance.

## 5.2 SURRO—Virtual Fleece Production

In the FIDYST Suite, SURRO is the congenial partner of FIDYST for simulating virtual fleece production and, thus, for analyzing fleece deposition processes. As a C++ tool, SURRO simulates the surrogate model introduced in (17) for the deposition process and

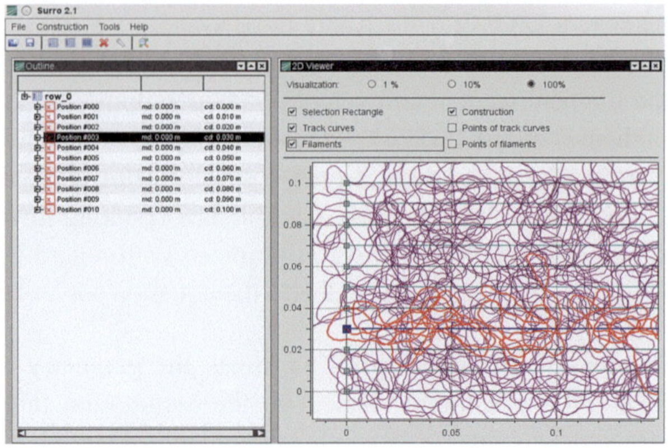


**Fig. 9** SURRO GUI: deposition with 11 spinning positions (Graphic: Simone Gramsch, Fraunhofer ITWM)

makes possible the virtual representation of large production facilities having thousands of filament spinning positions. By means of an intuitive GUI, the user defines the spinning positions and assigns to them the parameters of the surrogate model (Fig. 9). In carrying out this assignment, all spinning positions, groups of spinning positions (rows, blocks, etc.), or even single specific positions can be easily selected. If desired, the reference curve can be specified as an analytical function.

To assign the remaining parameters in SURRO, one implements the previously described identification procedure (20), which is fed with filament deposition images from FIDYST. In practical applications, many or all of the spinning positions can often be viewed as having equivalent flow and filament dynamics. In these cases, only one representative FIDYST simulation must be executed for each class of equivalent positions. Even such single simulations can be very time-consuming, however. Once the parameters have been identified, SURRO can use them as a basis for simulating even large fleece production processes in a few seconds. SURRO provides a series of post-processing functionalities for analyzing quality characteristics. For example, fluctuations in weight-per-area can be visualized and quantitatively evaluated using freely selectable scales. Practitioners are frequently interested in the integrated width and length distributions. The same holds true for other quality criteria of the virtual fleece, such as strip appraisal.

The structure depicted in SURRO is initially two-dimensional. However, one can use the arc-length parameter to decide on the crossing arrangement at points where one filament intersects either with itself or another filament. Using this information, SURRO forms the filaments into a 3D microstructure, taking their bending stiffness into consideration. This microstructure can be exported in EnSight format. Using the instruments of microstructure analysis, its flow or strength characteristics can then be further investigated and evaluated by means of other tools. Due to the short computation times, numerous stochastic realizations (samples) of the microstructure can be generated for Monte Carlo simulations without all too much effort.

The interplay of the two FIDYST Suite partners makes it possible to calculate the influence of production parameters on the primary quality features (e.g., distribution of weight-per-area) of the virtual fleece and to generate the associated microstructures for purposes

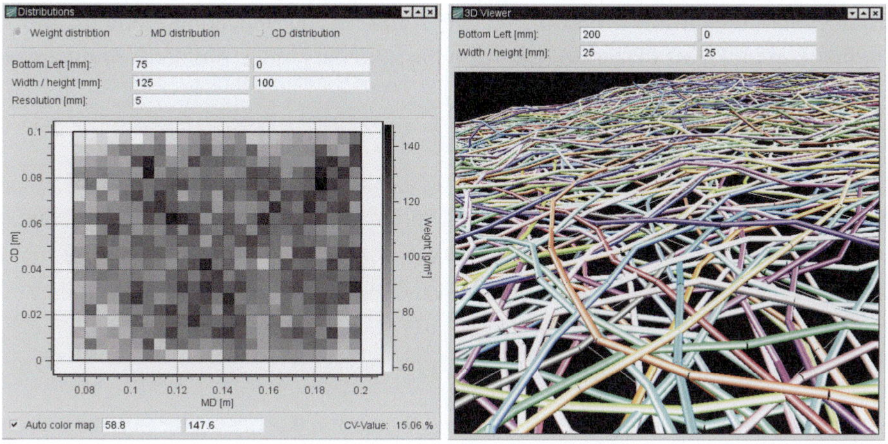


**Fig. 10** SURRO post-processing: analysis of weight-per-area distribution and generation of a 3D microstructure (Graphics: Simone Gramsch, Fraunhofer ITWM)

of appraising secondary quality features (flow and strength characteristics) (Fig. 10). This has been impossible up to now, and microstructure simulations, for example, have always had to rely on measured structures or on those generated without a direct link to the production process. This gap has been closed by the FIDYST Suite, so that the design and improvement of production facilities can now be oriented on material characteristics and quality.

## 5.3 VISFID—Coupled Fluid-Filament Simulations

VISFID, in contrast to FIDYST and SURRO, is not a finished software tool, but a MATLAB-FLUENT toolbox containing a number of core building-blocks that can be combined, modified, and/or extended, depending upon the application context. VISFID focuses on the simulation of spinning processes involving high filament densities. The toolbox treats these as steady-state processes and, pursuant to the homogenization strategies presented in Sect. 4.3.3, accomplishes full coupling with the surrounding airflow. The steady-state restriction for the filament continuum, and thus for its representatives, leads to boundary value problems for all the Cosserat rod models considered (see Sect. 7.2.1 for an application example). These problems can be solved robustly by means of a continuation-collocation method realized in MATLAB (see details in [1]). For the spinning processes we consider, the viscous rods from (8) in the Eulerian description ($e = 1$), along with the models for aerodynamic forces and heat exchange (10), represent the bases of the model. Elastic or viscoelastic filaments can also be treated in a corresponding fashion, however.

**Continuation-Collocation Method** Runge–Kutta-based collocation methods [64] represent the state of the art for numerically solving boundary value problems of the form

$$\frac{\mathrm{d}}{\mathrm{d}s}\mathbf{y} = \mathbf{f}(\mathbf{y}), \qquad \mathbf{g}\big(\mathbf{y}(0), \mathbf{y}(1)\big) = \mathbf{0}.$$

One such method of fourth order is realized in the MATLAB solver `bvp4c`. Here, for the collocation points $0 = s_0 < s_1 < \cdots < s_N = 1$, with $h_i = s_i - s_{i-1}$ and $\mathbf{y}_i = \mathbf{y}(s_i)$, a nonlinear system with $N + 1$ equations for $(\mathbf{y}_i)_{i=0,\ldots,N}$ is assembled:

$$\begin{aligned} \mathbf{g}(\mathbf{y}_0, \mathbf{y}_N) &= \mathbf{0} \\ \mathbf{y}_{i+1} - \mathbf{y}_i - \frac{h_{i+1}}{6}(\mathbf{f}(\mathbf{y}_i) + 4\mathbf{f}(\mathbf{y}_{i+1/2}) + \mathbf{f}(\mathbf{y}_{i+1})) &= \mathbf{0} \\ \text{with } \mathbf{y}_{i+1/2} = \frac{1}{2}(\mathbf{y}_{i+1} + \mathbf{y}_i) - \frac{h_{i+1}}{8}\big(\mathbf{f}(\mathbf{y}_{i+1}) - \mathbf{f}(\mathbf{y}_i)\big). \end{aligned}$$

This is solved using a simplified Newton method, so that the usability and convergence of the method depend crucially and sensitively on the initial Newton method estimator. For use with the desired coupling in MATLAB-FLUENT, we need a robust and completely automated method. This can be achieved using a continuation approach (homotopy). Here, we consider a generalization of the boundary value problem

$$\frac{\mathrm{d}}{\mathrm{d}s}\mathbf{y} = c\mathbf{f}(\mathbf{y}) + (1-c)\mathbf{f}_0(\mathbf{y}), \qquad \mathbf{g}\big(\mathbf{y}(0), \mathbf{y}(1)\big) + (1-c)\mathbf{g}_0\big(\mathbf{y}(0), \mathbf{y}(1)\big) = \mathbf{0}$$

with the continuation parameter $c \in [0, 1]$. The new functions $\mathbf{f}_0$ and $\mathbf{g}_0$ in the system and boundary conditions are selected so that a solution can be defined for $c = 0$—whenever possible, analytically. Through the choice of a suitable step width $\Delta c$, the parameter $c$ is then displaced from its starting value $c = 0$ to $c = 1$, by using the solution of each previous step as the estimator for each new step. In this manner, for $c = 1$, the original system is ultimately solved. In variations of this method, multiple continuation parameters are used to first remove and then smoothly reinsert specific terms of the ODE system. The art of performing a robust continuation lies in choosing and controlling the step width and, with multiple continuation parameters, in navigating through the sometimes highly-dimensional parameter space. The step width control we have developed calculates two half-steps $\Delta c/2$ for each full step $\Delta c$ and compares the solutions using criteria such as computation time and the number of collocation points needed. The call-ups of `bvp4c` executed here are cloaked in a try-catch routine and are crash-proof. The step width is adjusted according to the result. To navigate in higher-dimensional parameter spaces, we use a recursively programed reverse tree search on a grid across the parameter space. This method can be used to reliably and quickly solve even complex examples, such as the coiling problem of viscous rods described in [93].

**Algorithm 1** Iterative coupling of flow and filaments

1: Compute $\Psi_a^{(0)} = \mathscr{S}_{air}(\cdot)$, unloaded, i.e., without filaments
2: Set $k = 0$
3: **repeat**
4: Compute $\Psi^{(k)} = \mathscr{S}_{rod}(\Psi_a^{(k)})$ with line sources $(\mathbf{k}_{air}, q_{air})(\Psi^{(k)}, \Psi_a^{(k)})$
5: Average the filament data in the cells of the flow domain
6: Compute filament length in each cell of the flow domain
7: Compute $\Psi_a^{(k+1)} = \mathscr{S}_{air}(\Psi^{(k)})$ with volume sources $(\mathbf{k}_{rod}, q_{rod})(\Psi^{(k)}, \Psi_a^{(k+1)})$
8: Increment $k$
9: **until** $\|\Psi^{(k)} - \Psi^{(k-1)}\| < tol$

**Coupling Algorithm** The core idea of the iterative coupling algorithm (Algorithm 1) is as follows: instead of explicitly coupling, that is, simply updating the source terms step by step in the iteration between flow and filament dynamics—which raises stability concerns—we proceed implicitly, that is, we take into account the current flow and/or filament fields for the flow and/or filament calculation in the source terms. We accomplish this in FLUENT by means of appropriate user-defined functions (UDF). The coupling algorithm for the filament fields $\Psi$ and flow fields $\Psi_a$ can then be outlined in the following way. The coupling algorithm is run with FLUENT as master tool. In preprocessing, we first mesh the flow domain and make this information accessible to MATLAB and FLUENT. After each flow simulation $\mathscr{S}_{air}$, FLUENT starts the MATLAB main program, which in turn starts a MATLAB executable for each filament, in order to parallelize the filament simulation $\mathscr{S}_{rod}$. The MATLAB main program gathers the information from these computations and averages them on the grid of the flow domain. At this point, the filament length density from (16) is also computed. Here, for a rod parameterized using equidistant arc-lengths, one simply counts the filament points in each cell. As previously mentioned, FLUENT uses this data in the new flow simulation by means of a UDF for the source terms.

## 6 Production of Fleeces—The Spunbond Process

For some fifty years, fleeces have been an uninterrupted success story, one driven forward primarily by high-efficiency production processes. In contrast to textiles and similar, well-structured fabrics, fleeces are characterized by a non-ordered and tangled structure that is the result of the production process itself. The related term 'non-woven fabric,' sometimes used to describe fleeces, also makes this clear. In typical fleece processes, the melting of a polymer, the simultaneous filament formation from thousands of capillaries, the swirling in a turbulent open air jet, the deposition of these filaments on a conveyor belt, and their strengthening by means of thermal or mechanical measures are all combined into a single process (Fig. 11). Two significant process classes fall under the rubrics 'spunbond'

**Fig. 11** Spun-fleece process installation with three spinning beams: spunbond–meltblown–spunbond (Photo: Oerlikon Neumag)

and 'meltblown' [84, 90, 115], and both have been investigated in detail at the Fraunhofer ITWM. The meltblown process is characterized by an assault of transonic airflows directly at the capillary exit and, thus, by a direct connection between spinning and turbulent deposition. In the spunbond process, however, these process steps are spatially separated, and the propulsion is effected via a rapid airflow, with an interposed cooling zone characterized by low air velocities. After the filaments exit this airflow at the so-called slot, they are swirled in a turbulent open air jet and deposited on a conveyor belt. Spunbond generates filament diameters on the order of 10 μm, whereas meltblown filament diameters range from 1 μm downward. In some applications, both processes are combined in one line, so that the inner layer of the fleece contains the finer fibers and the outer layer, the somewhat coarser.

In the following discussion, we will take a closer look at the spunbond process of our industrial partner Oerlikon Neumag. This spunbond process is based on the ASON technology, which Neumag acquired a decade ago and whose further development since then has proceeded through several evolutionary stages. This development has been accompanied by the Fraunhofer ITWM in its entirety. In the beginning, our model-supported and simulation-supported perspective served primarily to deepen our insight into the processes involved and to establish simulation as an instrument in the process engineering toolkit. Subsequently, however, the focus shifted to supporting process development and installation design. Along this path, simulation studies were used to help develop and optimize two substantial evolutionary steps of the Neumag spunbond installation in the swirling and deposition zone. Below, we offer a detailed description of the process and its quality criteria, outline the simulation approaches used, and, finally, take a look at the simulation-based development of the process machinery.

## 6.1 Process Description and Quality Criteria

Along with the previously mentioned characteristics typical of spunbond processes, one distinguishing feature of the Neumag variant is a hydraulic platform that allows one to adjust the distance between capillaries and slot and between slot and deposition belt (Fig. 12). In its cross-section, the slot consists of a narrow conduit into which the filaments are introduced from above and discharged from below. In the spinning element above the slot, the filaments that are forming have already cooled significantly and thinned to nearly their final diameter. In the slot entrance, compressed air is actively blown in via two lateral feed channels, which establishes an airflow directed from top to bottom. This airflow creates a traction force along the length of the filaments, which results in their elongation in the spinning-cooling zone located above. At the slot exit, a smooth open air jet forms, which continuously weakens as it moves toward the deposition belt. The turbulent fluctuations arising in the jet initiate stochastic movements in the filaments, which, in turn, cause the filaments to decelerate, swirl, and lay down overlapping one other.

The quality of the resulting fleece is judged primarily by the homogeneity of its weight-per-area and by its strength. The homogeneity of weight-per-area is quantified by means of the $C_v$-value, as the relative standard deviation of stamped pieces of a defined size. The motion of the filament curtain exhibits fluctuations in the production direction that are typical for the spunbond process. Such fluctuations are ultimately a significant cause of the resulting cloud-like nature of the finished fleece. The strength of the fleece is determined in tensile tests. Due to the conveyor belt motion and the structure of the airflow, there are marked differences in strength between the machine direction (MD) and the cross direction (CD); a strength ratio of greater than 1.5 (MD/CD) is not unusual.

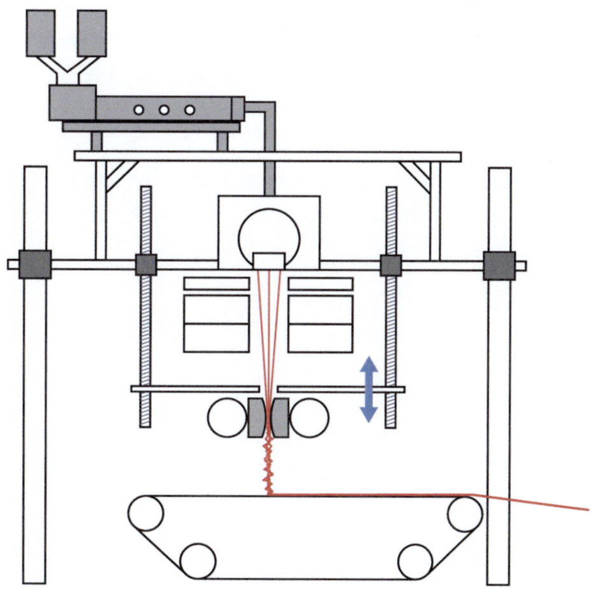

**Fig. 12** Schematic drawing of Oerlikon Neumag's spunbond process (Graphic: Oerlikon Neumag)

From an economic perspective, fleece production is characterized by high raw material costs and comparatively low processing costs. With regard to further processing and application, the required material amounts are frequently determined by the strength that must be achieved. A trivial, although costly, consequence is that greater strength can be achieved by using more material. The goal of innovations in process design is to save on materials while maintaining equally high quality.

## 6.2 Spunbond from a Simulation Perspective

In the spunbond process, the spinning-and-cooling phase is described using viscous rods (8) in aerodynamic flows, along with corresponding models for temperature-dependent viscosity. The swirling of the hardened filaments is based on elastic Kirchhoff rods (7) in a flow whose turbulence enters into the models either reconstructed or in the asymptotic limit, as white noise. For the deposition phase—and thus for the generation of the virtual fleece structure—we use the stochastic surrogate model (17). With VISFID, FIDYST, and SURRO (Sect. 5), we have software tools at our disposal for simulating all these aspects. They allow us to continuously evaluate the influence of the process parameters on the material characteristics. Here, representative single-filament dynamics from VISFID and FIDYST are used to parameterize SURRO. In the final step of this simulation chain, the generated microstructure can be analyzed in SURRO. With regard to the homogeneity of weight-per-area, this is a simple post-processing step involving the virtually-determined $C_v$-value. With regard to strength, we have already made some early progress in reconstructing tensile strength tests on the virtual microstructure. However, a more practicable way was chosen on the basis of the extensive studies conducted for Oerlikon Neumag: starting with the reasonable assumption that the strength is characterized significantly by the degree of overlapping of the filaments, the size of the deposition area serves as a substitute criterion for strength. The deposition area is defined directly by the FIDYST simulation, by correcting the deposited filament curve $\boldsymbol{\eta}$ for the conveyor belt movement $\boldsymbol{\gamma}$. In terms of the stochastic reduced model, the deposition corresponds exactly—for a fixed deposition position—or approximately—for low belt velocities—to the steady-state distribution of the withdrawn process $\boldsymbol{\xi} = \boldsymbol{\eta} - \boldsymbol{\gamma}$. In the FIDYST simulations, we observe the distributions

$$p_S(\boldsymbol{\xi}) = c \exp\left(-\frac{1}{2}\boldsymbol{\xi} \cdot \mathbf{C}^{-1} \cdot \boldsymbol{\xi}\right)$$

(see also (18) and (19)), in which the main directions of $\mathbf{C}$ correspond approximately to MD and CD. The roots of the eigenvalues (throwing ranges) describe the deposition area. The MD to CD ratios in the measured strengths are qualitatively well-reflected in the simulated MD to CD ratios of the throwing ranges. This, along with other validating measurements, confirm that the deposition area (as substitute criterion) and the $C_v$-value are suitable optimization parameters.

## 6.3 Improving Fleece Deposition in the Spunbond Process

Two significant evolutionary steps in the Neumag spunbond process were developed and optimized using the simulation tools and methods described here for the deposition zone. The primary driving force was the desired increase in strength, so as to ultimately achieve the same functional characteristics using less material. For the first evolutionary step, we carried out a comprehensive simulation study with diverse geometric variants (e.g., changes in physical dimensions, addition of components for routing air). On the basis of airflow computations from FLUENT, simulations of representative individual filaments were performed with FIDYST, which made possible a comparative evaluation of the deposition areas. The expected $C_v$-values for weight-per-area were determined using the associated SURRO simulations.

The results of this study were assessed jointly by Oerlikon Neumag and the Fraunhofer ITWM. Essentially, an increase in deposition height (distance from slot to conveyor belt) leads to an increase in the deposition area, coupled with a deterioration in $C_v$-values. In the end, as the best compromise, we chose a variant that leads to a relatively large deposition area, while still maintaining acceptable weight homogeneity. In this variant, the self-forming open air jet is laterally restricted in the lower zone and quasi entrapped by two driven rollers overlying the band. During the implementation of this principle (Fig. 13) on a 7-meter and, thus, extremely wide installation, the rotating rollers sagged too much and were ultimately replaced by appropriately-shaped, rigid sheet metal. Figure 14 shows the FIDYST simulation of the implemented variant; Fig. 15 shows the accompanying SURRO simulation and an analysis of the weight-per-area distribution. The new machinery design proved successful, thus confirming the choice of a simulation-based developmental approach. Comparative measurements of the achieved improvements verified increases in strength in both directions of about 10 % (MD) and 15 % (CD), with no increase in material usage.

Because the simulation-supported development yielded such positive results, simulation analyses were also conducted prior to a further developmental step in the spunbond process. Although the improvement described above produced overall strength increases, the MD to CD ratio remained essentially the same. Since, in general, this anisotropy is not desired, but is an artifact of the production method, the lower CD strength becomes the critical quality parameter; thus, an increase in the CD value at the expense of the MD

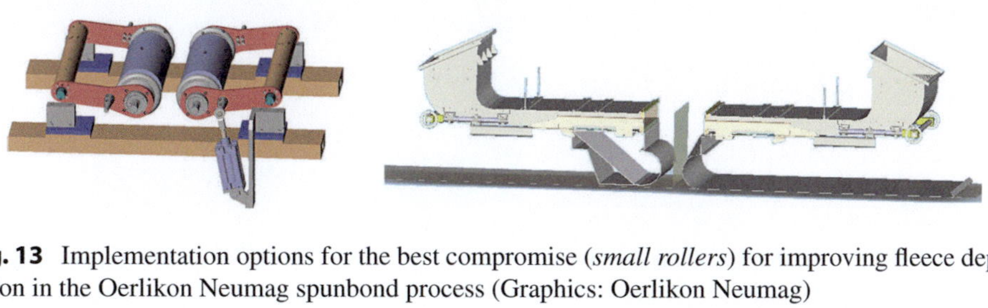

**Fig. 13** Implementation options for the best compromise (*small rollers*) for improving fleece deposition in the Oerlikon Neumag spunbond process (Graphics: Oerlikon Neumag)

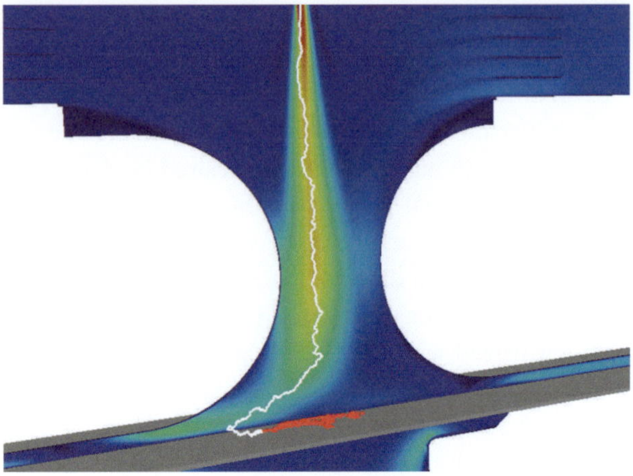

**Fig. 14** FIDYST simulation of the Neumag spunbond process: filament dynamics and airflow (Simulation: Simone Gramsch, Fraunhofer ITWM)

value is desirable. In our further cooperative efforts, our wish to equalize the directions and further improve homogeneity suggested trying a significantly altered flow routing, one which, in contrast to the previous design, leads to a completely three-dimensional flow pattern. This causes the axes of the deposition area to tilt, which significantly reduces strength differences between the production and cross directions. To summarize the cooperative work conducted thus far and to comment on the simulation approach implemented by the Fraunhofer ITWM, we cite Matthias Schemken, Vice-president and Head of Development of Oerlikon Neumag (May 2013): *"Simulations have contributed significantly to the development … of the forming zone of our spunbond process."*

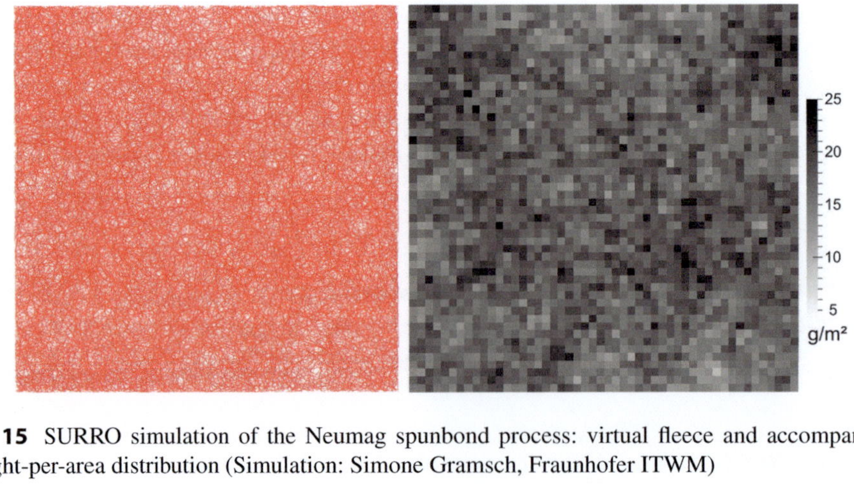


**Fig. 15** SURRO simulation of the Neumag spunbond process: virtual fleece and accompanying weight-per-area distribution (Simulation: Simone Gramsch, Fraunhofer ITWM)

# 7 Production of Glass Wool—Rotational Spinning

Producing mineral fibers out of glass or stone involves melting and fraying processes that are run at temperatures above 1000 °C. This cannot be accomplished by means of the methods typically used with polymers. The technological solution for producing glass wool consists of replacing high material pressures in a closed spinning head with high centrifugal forces in an open spinning disc. The molten glass emerging from tens of thousands of holes in the disc wall is then subsequently stretched to fibers in a hot gas stream and, finally, collected on a conveyor belt (Fig. 16). A typical production set-up for manufacturing glass wool insulation consists of four to seven such serially arranged heads.

The following sections examine in detail the installation of our industrial partner Woltz. As is perhaps already clear from the above introductory sentences, the vocabulary used to describe the processed material is branch specific: here, one speaks of fibers, not filaments. This has no impact on the models we use, however. As the industrial example selected here for illustration clearly shows, the generic model and simulation toolkit covers many, but by no means all, aspects of a real production process. After describing the process, we therefore devote ourselves to the application-specific simulation model and demonstrate its capabilities using a process management example.

**Fig. 16** Glass wool manufacturing machinery (Photo: Woltz GmbH)

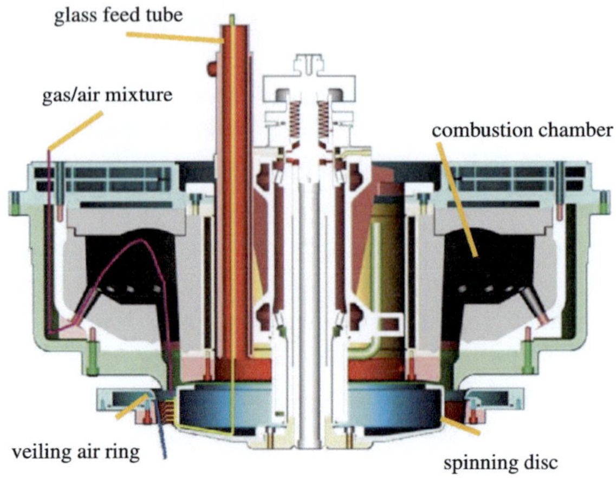


**Fig. 17** Principle behind the rotational spinning of glass wool (Graphic: Woltz GmbH)

## 7.1 Description of Rotational Spinning Process

Rotational spinning is a well-established process for manufacturing glass wool and has a significant share of the insulation material market. The process engineering principle is sketched out in Fig. 17. The raw material, containing up to 70 % recycled glass, is continuously liquefied over a large oven and discharged as a thread into the spinning disc. The centrifugal forces cause it to initially distribute itself radially on the bottom, where it migrates towards the outer perimeter and emerges through tens of thousands of holes arranged in rows in the shell surface. The amount of material that flows through any single capillary depends on the local temperature and viscosity; the thickness of the glass film and the resulting effective force; and the length and diameter of the hole. Upon discharge from the spinning disc, the fiber threads are guided through a hot gas stream and, further outwards, through a cold air jet generated in a so-called veiling air ring. The entire process is highly integrated and very sensitive, although an equilibrium is established within the spinning disc when the glass feed rate remains constant. This depends in particular on the glass volume and temperature, the rotation speed, the hot air volume and temperature, and the cold air volume.

## 7.2 Process-Specific Simulation Model

The MATLAB-FLUENT toolbox VISFID introduced in Sect. 5.3 forms the simulation core for representing the rotational spinning process. However, by itself, it only leads to systemic insights into the real production process [3], not to a comprehensive representation. This is because neither the starting conditions for the jet simulation in the hot gas stream nor the temperature boundary conditions on the spinning disc for the flow calculation can be adequately estimated. A simulator developed jointly by several groups at the Fraunhofer ITWM therefore links *Spinning in the hot gas stream* on the basis of the fiber flow simulation VISFID with *Melting phase and disc mechanics*, a COMSOL-based

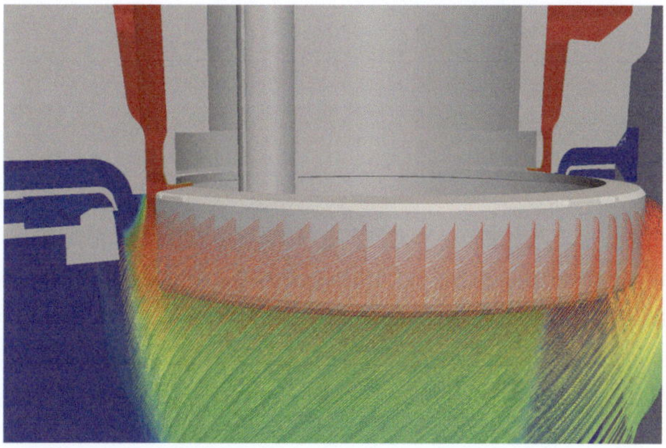

**Fig. 18** Spinning disc with selected rows of the fiber curtain (Simulation: Johannes Schnebele, Fraunhofer ITWM)

simulation of the thermo-mechanics of the spinning disc and glass reservoir. This, in turn, requires a series of analytically-based surrogate models for the melt flow in the interior of the drum [15]. For details into the commercial FEM software COMSOL, please refer to the supplier website www.comsol.com.

### 7.2.1 Spinning in the Hot Gas Stream

The perforation array in the spinning disc consists of a moderate number of rows, each with 770 equidistant holes on the perimeter. The molten glass, driven by centrifugal forces through the holes and then bent by the hot gas stream, forms of a thick curtain of glass jets (Fig. 18), which we treat as a continuum, pursuant to the homogenization strategy introduced in Sect. 4.3.3. The design of the machinery therefore suggests a rotation-invariant description of flow and fiber dynamics. For the fiber continuum, this means in particular that the totality of information pertaining to one row can be analytically captured by means of a single representative.

For the numerical treatment of the vertical direction, we choose as representatives one real spinning position per row, since this provides us with an adequate resolution of the continuum for the flow grid being used. Because we are interested in the stretching phase of the fibers, we ignore the installation equipment, restrict our investigation to the area near the nozzles, and examine fibers with given length $L$ and stress-free ends in the Eulerian description. This approach yields a steady-state for the fiber and flow dynamics in a rotating reference system with angular velocity $\Omega$ (1/s). Assuming viscous rods (8) with energy balance (9), where $e = 1$ in each case, and introducing the constant mass flow $Q = \rho A u$ (kg/s), the steady-state model equations [3] become

$$
\begin{aligned}
&\mathsf{D} \cdot \partial_s \bar{\mathsf{r}} = \mathsf{e}_3 \\
&\quad \partial_s \mathsf{D} = -\kappa \times \mathsf{D} \\
&\quad \partial_s \kappa = -\frac{\rho}{3Q}\frac{\kappa n_3}{\mu} + \frac{4\pi\rho^2}{3Q^2}\frac{u}{\mu}\mathsf{P}_{3/2} \cdot \mathsf{m} \\
&\quad \partial_s u = \frac{\rho}{3Q}\frac{u n_3}{\mu}
\end{aligned}
$$

$$\partial_s \mathsf{n} = -\kappa \times \mathsf{n} + Qu\kappa \times \mathsf{e}_3 + \frac{\rho}{3}\frac{un_3}{\mu}\mathsf{e}_3 + \mathsf{k}$$

$$\partial_s \mathsf{m} = -\kappa \times \mathsf{m} + \mathsf{n} \times \mathsf{e}_3 + \frac{\rho}{3}\frac{u}{\mu}\mathsf{P}_3 \cdot \mathsf{m} - \frac{Q}{12\pi}\frac{n_3}{\mu}\mathsf{P}_2 \cdot \kappa + \mathsf{l}$$

$$\partial_s T = \frac{1}{c_p Q} q_{air}$$

with external forces (aerodynamic forces, gravitation, and inertial forces caused by rotation) and rotation-induced torques

$$\mathsf{k} = -\mathsf{D} \cdot \bar{\mathsf{k}}_{air} + Qg\frac{1}{u}\mathsf{D} \cdot \mathsf{e}_1 + 2Q\Omega(\mathsf{D} \cdot \mathsf{e}_1) \times \mathsf{e}_3 + Q\Omega^2\frac{1}{u}\mathsf{D} \cdot \left(\mathsf{e}_1 \times (\mathsf{e}_1 \times \bar{\mathsf{r}})\right)$$

$$\mathsf{l} = -\frac{Q\Omega}{12\pi}\frac{n_3}{\mu u}\mathsf{P}_2 \cdot (\mathsf{D} \cdot \mathsf{e}_1) - \frac{Q^2\Omega}{4\pi\rho}\frac{1}{u}\mathsf{P}_2 \cdot (\kappa \times \mathsf{D} \cdot \mathsf{e}_1)$$
$$- \frac{Q^2}{4\pi\rho}\frac{1}{u^2}\mathsf{P}_2 \cdot (u\kappa + \Omega\mathsf{D} \cdot \mathsf{e}_1) \times (u\kappa + \Omega\mathsf{D} \cdot \mathsf{e}_1).$$

The boundary conditions take into account the spinning position by means of the height $H$ and radial distance from the axis $R$, as well as the exit speed $U$ and exit temperature $\theta$ at the spinning disc:

$$\bar{\mathsf{r}}(0) = (H, R, 0), \qquad \mathsf{D}(0) = \mathsf{e}_1 \otimes \mathsf{e}_1 - \mathsf{e}_2 \otimes \mathsf{e}_3 + \mathsf{e}_3 \otimes \mathsf{e}_2$$
$$\kappa(0) = 0, \qquad u(0) = U, \qquad T(0) = \theta \qquad \mathsf{n}(L) = 0, \qquad \mathsf{m}(L) = 0.$$

The airflow modeling is based on the rotationally symmetric, steady-state Navier–Stokes equations with source terms $\mathbf{k}_{rod}$ and $q_{rod}$, which are projected from the three-dimensional fiber dynamics (see [3]). The coupled complete system $\mathscr{S}_{rod}$–$\mathscr{S}_{air}$ of fiber and flow dynamics can be solved using the toolbox VISFID (Sect. 5.3), but it requires the nozzle conditions $\Psi_{nozzle} = (H, R, U, \theta)$ for the jets and the disc temperature and geometry information for the flow dynamics. This data is delivered by the simulation model $\mathscr{S}_{disk}$ for *Melting phase and disc mechanics*, which is described next (see Fig. 21).

### 7.2.2 Melting Phase and Disc Mechanics

The thermo-mechanics of the disc and molten glass reservoir are at the core of the model for *Melting phase and disc mechanics*. The model is based on linear elasticity theory for large deformations and accounts for thermal expansion, convection, thermal conduction, and thermal radiation. The numerical solution $\mathscr{S}_{disk}$ of the steady-state, rotationally symmetrical problem is produced by the software tool COMSOL (Fig. 19), for which boundary conditions, flows, and heat sources are derived from four analytical surrogate models that describe the molten glass distribution (Fig. 20): A—A viscous, uniaxial string model of the glass cord supplying material in the drum interior delivers the disc temperature at the point of contact; B—The ensuing thin film approximation describes the molten glass movement toward the inner wall under the effect of centrifugal forces as creeping flow and delivers

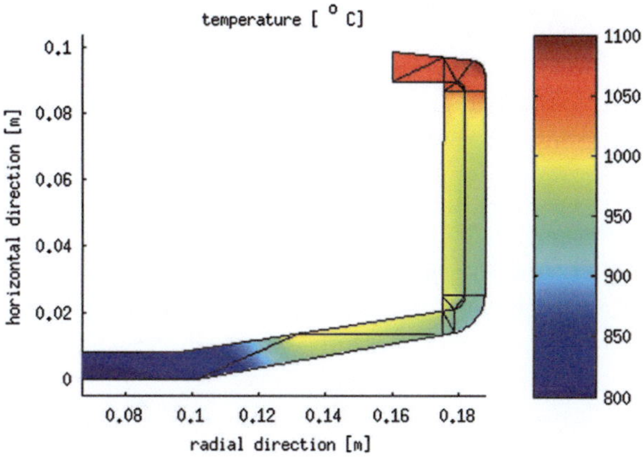


**Fig. 19** COMSOL simulation: temperature in spinning disc and glass reservoir (Simulation: Jan Mohring, Fraunhofer ITWM)

the thermal flux through the glass film into the disc; C—The free boundary value problem of the reservoir is treated in a Stokes approximation and ultimately delivers an ordinary differential equation for the thickness, as well as thickness-dependent analytical expressions for the pressure and convective speed at the perforated wall. The thickness is used as geometry information for the entire COMSOL model of the wall and the reservoir; D—A pressure-driven pipe flow for the flow through the capillaries delivers the thermal flux into the disc, along with the temperature and velocity boundary conditions of the fibers for the simulation model *Spinning in the hot gas stream*. These models are described in more detail in [15]. The COMSOL model requires the heat fluxes into the spinning disc from the flow description of the simulation model *Spinning in the hot gas stream* (see Fig. 21).

### 7.2.3 Iterative Coupling

In order to numerically solve the complete problem arising from the two simulation models *Spinning in the hot gas stream* and *Melting phase and disc mechanics*, we once again use an iterative coupling strategy (Fig. 21). This can be viewed as an extension of the VISFID coupling from Sect. 5.3, with $\Psi_{rod} = \Psi$ and $\Psi_{air} = \Psi_a$, and can be outlined algorithmically as follows (also see Algorithm 2). Because both the thermo-mechanics

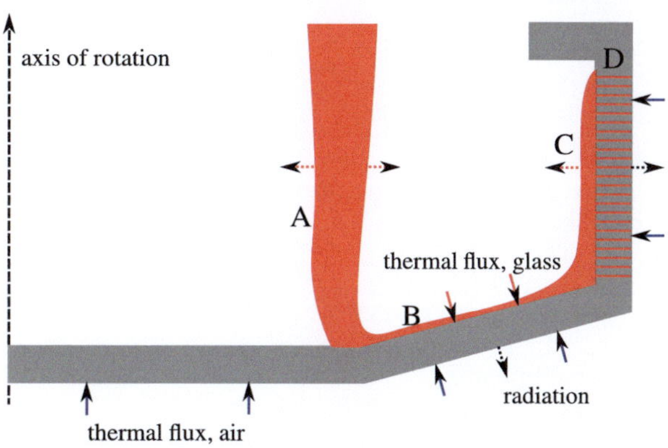


**Fig. 20** Overview of the analytical surrogate models for molten flow distribution

$\mathscr{S}_{disk}$ and the dependent fiber simulation $\mathscr{S}_{rod}$ require data from the flow equations, we perform this calculation twice in each iteration step for reasons of stability. Note that the thermo-mechanics $\mathscr{S}_{disk}$ delivers an altered geometry that must be re-meshed for the flow equations. Therefore, for clarity's sake, we have assigned an independent role to this step $\mathscr{S}_{geom}$.

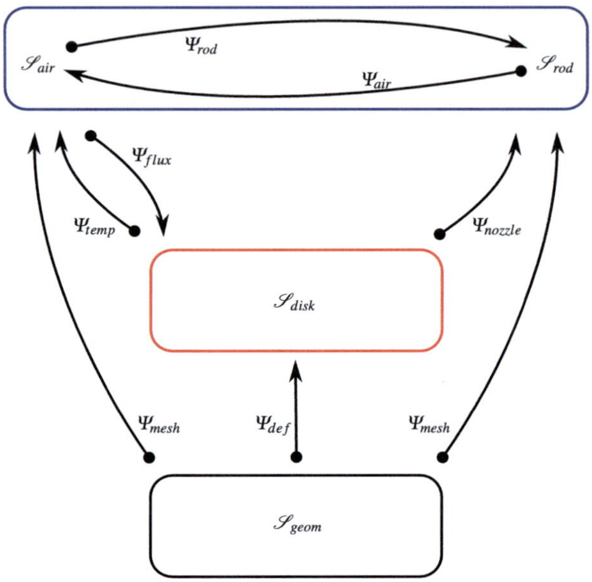


**Fig. 21** Coupling structure of the various routines

**Algorithm 2** Coupling of the simulation models

Initialize the heat fluxes with a suitable estimation and compute all initial fields (unloaded flow; i.e., without fibers)

Set $k = 0$

**repeat**

Calculate

$$\begin{aligned} \textit{Flow:}\ & \left(\cdot, \Psi_{flux}^{k+1}\right) = \mathscr{S}_{air}\left(\Psi_{rod}^{(k)}, \Psi_{temp}^{(k)}, \Psi_{mesh}^{(k)}\right) \\ \textit{Melt and disc:}\ & \left(\Psi_{nozzle}^{(k+1)}, \Psi_{temp}^{(k+1)}, \Psi_{def}^{(k+1)}\right) = \mathscr{S}_{disk}\left(\Psi_{flux}^{(k+1)}\right) \\ \textit{Mesh:}\ & \Psi_{mesh}^{(k+1)} = \mathscr{S}_{geom}\left(\Psi_{def}^{(k+1)}\right) \\ \textit{Flow:}\ & \left(\Psi_{air}^{(k+1)}, \cdot\right) = \mathscr{S}_{air}\left(\Psi_{rod}^{(k)}, \Psi_{temp}^{(k+1)}, \Psi_{mesh}^{(k+1)}\right) \\ \textit{Fibers:}\ & \Psi_{rod}^{(k+1)} = \mathscr{S}_{rod}\left(\Psi_{air}^{(k+1)}, \Psi_{nozzle}^{(k+1)}, \Psi_{mesh}^{(k+1)}\right). \end{aligned}$$

Increment $k$

**until** $\|\Psi_{rod}^{(k)} - \Psi_{rod}^{(k-1)}\| < tol$

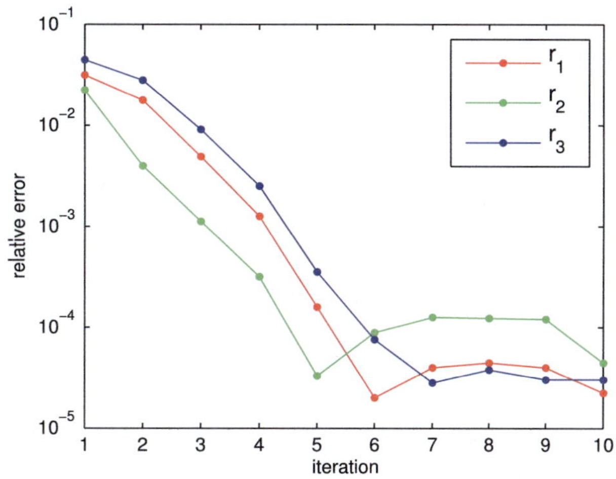


**Fig. 22** Convergence of the iterative procedure: relative $\mathscr{L}^2$-error of the fiber positions on a logarithmic scale (Simulation: Johannes Schnebele, Fraunhofer ITWM)

The fiber positions' relative $\mathscr{L}^2$-error is shown on a logarithmic scale in Fig. 22 to demonstrate the convergence of the iterative procedure. In parameter studies, four to five iterations generally suffice for a definitive simulation result; these can be carried out in a parallelized fiber simulation with a few hours of computation time.

The necessity of a completely coupled consideration of all aspects is verified in an especially impressive manner by the results of the fiber-loaded airflow. Figure 23 shows the axial velocity, the rotational velocity (swirl), and the temperature across several iterations. The influence of the fibers is made particularly clear by the presence of a relevant rotational velocity (drag effect) and in the warming of the air (a concomitant of the cooling of the fibers).

## 7.3 Simulation-Based Process Design and Management

The simulation framework introduced here allows for a comprehensive understanding of rotational spinning and forms the foundation for simulation-based process design and management. Among the parameters taken into consideration are the furnace temperature, glass throughput, cold air quantity, geometry of the cold air ring, and hole diameter of the different rows. Various aspects of the process were investigated in cooperation with our industrial partner Woltz.

One optimization goal that pervades almost all investigations is the production of fibers whose diameters are as uniform as possible, or that have at least a controlled diameter distribution. Here, one must remember that the hole rows are subjected to different pressures by the glass reservoir and pass through different flow conditions and temperatures during the spinning process. Figure 24 shows the temperature and velocity of the fibers for all hole rows. As a consequence of the steady-state continuity equation, the simple relationship $d = D\sqrt{u/U}$ applies for diameter $d$ and speed $u$, where $D$ and $U$ denote the quantities at the hole exits. Thus, the variations in speed from row to row can be dealt

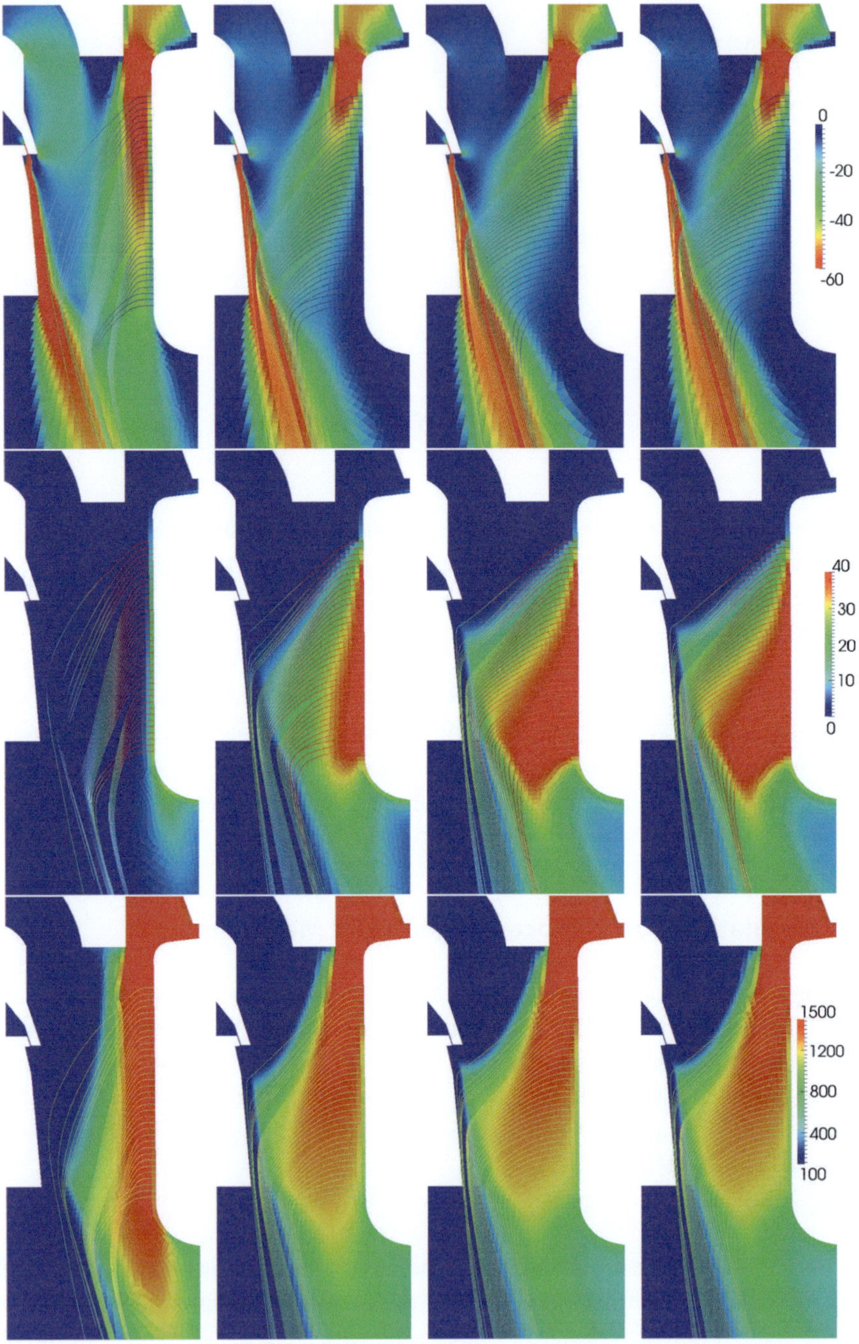


**Fig. 23** *From top to bottom*: axial velocity (m/s), rotational velocity (m/s), and temperature (°C) of the fiber-loaded airflow. *From left to right*: results of the iterative steps 0, 1, 5, and 10 (Simulation: Johannes Schnebele, Fraunhofer ITWM)

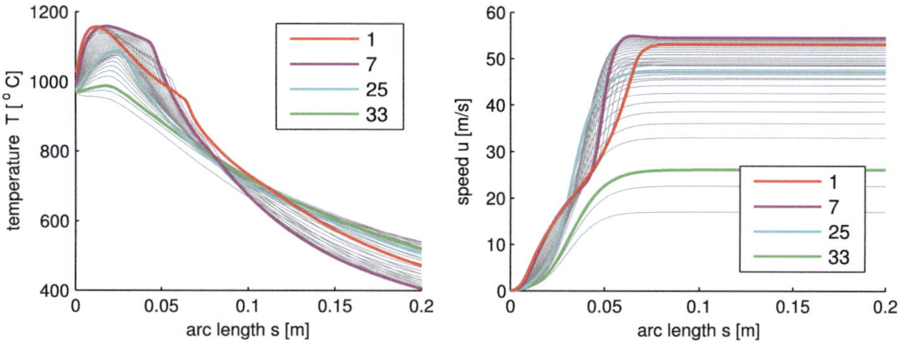


**Fig. 24** Temperature and velocity along the fiber. The numbering represents the hole rows *from top to bottom* (Simulation: Johannes Schnebele, Fraunhofer ITWM)

with by adapting the disc design with regard to the diameter $D$, in order to make the final diameter distribution uniform. Due to the coupling described above, the consequences of such measures are extremely complex, but can nonetheless be assessed quite well on the basis of simulation. The Fraunhofer ITWM has used parameter studies to generate various design proposals.

As an example of simulation-based process management, we offer our investigations into the hot-glass-induced abrasion that leads to capillary expansion in the course of operation. Figure 25 shows the hole throughput, final fiber diameter, maximal fiber speed, and exit temperature for all rows, first for the starting conditions (new disc) and then for capillaries with a 5 % expansion. The abrasion has the dramatic effect that the upper hole rows are no longer supplied with glass. Thus, unless the process parameters are tracked and reset, the disc must be replaced. The machine adjustment procedure developed at the Fraunhofer ITWM, with lowered glass melting temperature and disc rotation speed, can completely compensate for this effect and lead to throughput and fiber diameter distributions that are equivalent to the use of a new disc. This process management step significantly extends the lifetime of the disc, thus greatly reducing costs.

## 8 Summary and Outlook

In the last few years, the virtual production of filaments and fleeces has become a reality at the Fraunhofer ITWM. The models, algorithms, and software tools developed here allow us to depict highly complex production processes, so that simulation-based process design and management is now possible. This brings a new quality to the associated machine engineering work and opens up a multitude of new possibilities. In several specific modeling areas, the Fraunhofer ITWM has achieved a unique status. This is especially true for treating filament dynamics in turbulent flows, modeling fleece deposition with efficient stochastic surrogate models, and considering filament-flow cou-

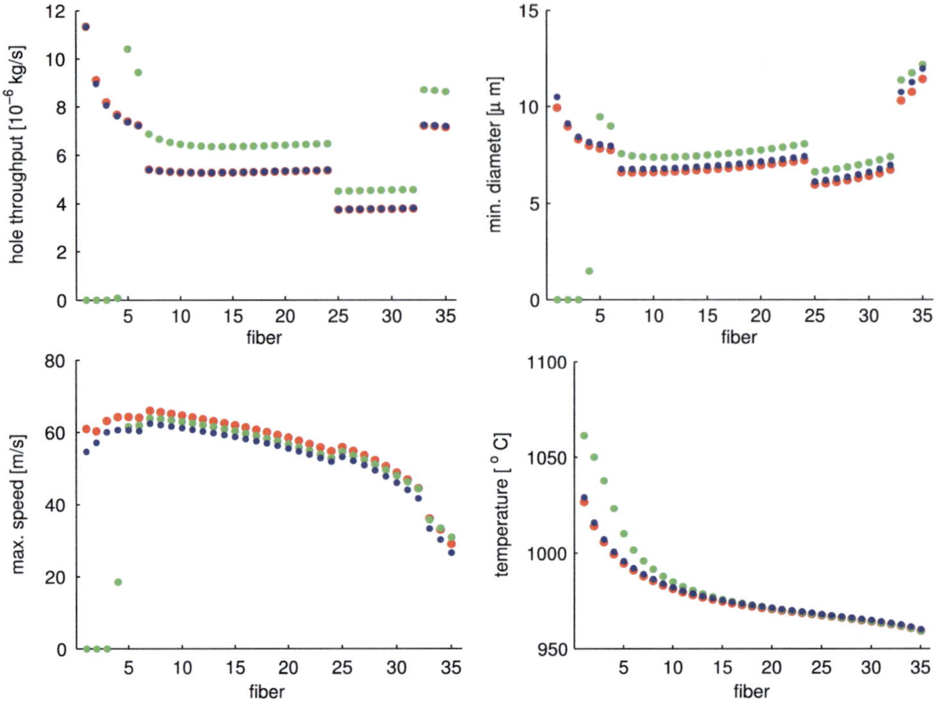


**Fig. 25** Process management: new disc (*red*), with 5 % increase in hole diameter (*green*), and with adjustment proposal for compensating abrasion (*blue*). *From left to right, above*: hole throughput and minimum fiber diameter; *below*: maximum speed and temperature as a function of hole row (Simulation: Johannes Schnebele, Fraunhofer ITWM)

pling in spinning processes with dense filament curtains. Despite such progress, this field of work remains lively: including viscoelastic effects in filament dynamics, generalizing the turbulence impact on LES simulations, handling fiber-fiber contact, extending simulation-based microstructure generation, and dealing with the feedback effects of the filaments on flow in non-steady-state and chaotic situations are only a few of the future problems and topics. As we approach these issues, we will allow ourselves to be led, in accustomed fashion, by the practical problems of our industrial partners. We look forward eagerly to the new challenges they will surely bring to us.

**Acknowledgement** The simulation results presented in this chapter are based on the work of the members of the Fraunhofer ITWM Transport Processes Department mentioned in Sect. 3. We are particularly indebted to them. Important works of the authors have been supported by the German Research Society (WE 2003/3-1, WE 2003/4-1, and MA 4526/2-1), as well as by the Federal Ministry for Education and Research (Consortium project ProFil, 05M10WEA, 05M10AMB, Consortium project OPAL, 05M13WEA, and 05M13AMD).

## References

### Publications of the Authors

1. Arne, W., Marheineke, N., Meister, A., Wegener, R.: Numerical analysis of Cosserat rod and string models for viscous jets in rotational spinning processes. Math. Models Methods Appl. Sci. **20**(10), 1941–1965 (2010)
2. Arne, W., Marheineke, N., Meister, A., Schiessl, S., Wegener, R.: Finite volume approach for the instationary Cosserat rod model describing the spinning of viscous jets. J. Comp. Phys. **294**, 20–37 (2015)
3. Arne, W., Marheineke, N., Schnebele, J., Wegener, R.: Fluid-fiber-interactions in rotational spinning process of glass wool manufacturing. J. Math. Ind. **1**, 2 (2011)
4. Arne, W., Marheineke, N., Wegener, R.: Asymptotic transition of Cosserat rod to string models for curved viscous inertial jets. Math. Models Methods Appl. Sci. **21**(10), 1987–2018 (2011)
5. Bonilla, L.L., Götz, T., Klar, A., Marheineke, N., Wegener, R.: Hydrodynamic limit for the Fokker–Planck equation describing fiber lay-down models. SIAM J. Appl. Math. **68**(3), 648–665 (2007)
6. Götz, T., Klar, A., Marheineke, N., Wegener, R.: A stochastic model and associated Fokker–Planck equation for the fiber lay-down process in nonwoven production processes. SIAM J. Appl. Math. **67**(6), 1704–1717 (2007)
7. Götz, T., Klar, A., Unterreiter, A., Wegener, R.: Numerical evidence for the non existence of solutions to the equations describing rotational fiber spinning. Math. Models Methods Appl. Sci. **18**(10), 1829–1844 (2008)
8. Grothaus, M., Klar, A., Maringer, J., Stilgenbauer, P., Wegener, R.: Application of a three-dimensional fiber lay-down model to non-woven production processes. J. Math. Ind. **4**, 4 (2014)
9. Hietel, D., Wegener, R.: Simulation of spinning and laydown processes. Tech. Text. **3**, 145–148 (2005)
10. Hübsch, F., Marheineke, N., Ritter, K., Wegener, R.: Random field sampling for a simplified model of melt-blowing considering turbulent velocity fluctuations. J. Stat. Phys. **150**(6), 1115–1137 (2013)
11. Klar, A., Marheineke, N., Wegener, R.: Hierarchy of mathematical models for production processes of technical textiles. Z. Angew. Math. Mech. **89**, 941–961 (2009)
12. Klar, A., Maringer, J., Wegener, R.: A 3d model for fiber lay-down in nonwoven production processes. Math. Models Methods Appl. Sci. **22**(9), 1250020 (2012)
13. Klar, A., Maringer, J., Wegener, R.: A smooth 3d model for fiber lay-down in nonwoven production processes. Kinet. Relat. Models **5**(1), 57–112 (2012)
14. Lorenz, M., Marheineke, N., Wegener, R.: On simulations of spinning processes with a stationary one-dimensional upper convected Maxwell model. J. Math. Ind. **4**, 2 (2014)
15. Marheineke, N., Liljo, J., Mohring, J., Schnebele, J., Wegener, R.: Multiphysics and multimethods problem of rotational glass fiber melt-spinning. Int. J. Numer. Anal. Model. B **3**(3), 330–344 (2012)
16. Marheineke, N., Wegener, R.: Fiber dynamics in turbulent flows: General modeling framework. SIAM J. Appl. Math. **66**(5), 1703–1726 (2006)
17. Marheineke, N., Wegener, R.: Fiber dynamics in turbulent flows: Specific Taylor drag. SIAM J. Appl. Math. **68**(1), 1–23 (2007)
18. Marheineke, N., Wegener, R.: Asymptotic model for the dynamics of curved viscous fibers with surface tension. J. Fluid Mech. **622**, 345–369 (2009)
19. Marheineke, N., Wegener, R.: Modeling and application of a stochastic drag for fiber dynamics in turbulent flows. Int. J. Multiph. Flow **37**, 136–148 (2011)

20. Panda, S., Marheineke, N., Wegener, R.: Systematic derivation of an asymptotic model for the dynamics of curved viscous fibers. Math. Methods Appl. Sci. **31**, 1153–1173 (2008)
21. Tiwari, S., Antonov, S., Hietel, D., Kuhnert, J., Olawsky, F., Wegener, R.: A meshfree method for simulations of interactions between fluids and flexible structures. In: Griebel, M., Schweitzer, M.A. (eds.) Meshfree Methods for Partial Differential Equations III. Lecture Notes in Computational Science and Engineering, vol. 57, pp. 249–264. Springer, Berlin (2006)

## Dissertations on This Topic at the Fraunhofer ITWM

22. Arne, W.: Viskose Jets in rotatorischen Spinnprozessen. Ph.D. thesis, Universität Kassel (2012)
23. Cibis, T.M.: Homogenisierungsstrategien für Filament–Strömung–Wechselwirkungen. Ph.D. thesis, FAU Erlangen-Nürnberg (2015)
24. Dhadwal, R.: Fibre spinning: Model analysis. Ph.D. thesis, Technische Universität Kaiserslautern (2005)
25. Leithäuser, C.: Controllability of shape-dependent operators and constrained shape optimization for polymer distributors. Ph.D. thesis, Technische Universität Kaiserslautern (2013)
26. Lorenz, M.: On a viscoelastic fibre model—Asymptotics and numerics. Ph.D. thesis, Technische Universität Kaiserslautern (2013)
27. Marheineke, N.: Turbulent fibers—On the motion of long, flexible fibers in turbulent flows. Ph.D. thesis, Technische Universität Kaiserslautern (2005)
28. Maringer, J.: Stochastic and deterministic models for fiber lay-down. Ph.D. thesis, Technische Universität Kaiserslautern (2013)
29. Panda, S.: The dynamics of viscous fibers. Ph.D. thesis, Technische Universität Kaiserslautern (2006)
30. Repke, S.: Adjoint-based optimization approaches for stationary free surface flows. Ph.D. thesis, Technische Universität Kaiserslautern (2011)
31. Schröder, S.: Stochastic methods for fiber-droplet collisions in flow processes. Ph.D. thesis, Technische Universität Kaiserslautern (2013)

## Further Literature

32. Antman, S.S.: Nonlinear Problems of Elasticity. Springer, New York (2006)
33. Audoly, B., Clauvelin, N., Brun, P.T., Bergou, M., Grinspun, E., Wardetzky, M.: A discrete geometric approach for simulating the dynamics of thin viscous threads. J. Comp. Phys. **253**, 18–49 (2013)
34. Audoly, B., Pomeau, Y.: Elasticity and Geometry. Oxford University Press, Oxford (2010)
35. Barrett, J.W., Knezevic, D.J., Süli, E.: Kinetic Models of Dilute Polymers: Analysis, Approximation and Computation. Nećas Center for Mathematical Modeling, Prague (2009)
36. Batchelor, G.K.: Slender-body theory for particles of arbitrary cross-section in Stokes flow. J. Fluid Mech. **44**(3), 419–440 (1970)
37. Bechtel, S.E., Forest, M.G., Holm, D.D., Lin, K.J.: One-dimensional closure models for three-dimensional incompressible viscoelastic free jets: von Karman flow geometry and elliptical cross-section. J. Fluid Mech. **196**, 241–262 (1988)
38. Bonilla, L.L., Klar, A., Martin, S.: Higher order averaging of linear Fokker–Planck equations with periodic forcing. SIAM J. Appl. Math. **72**(4), 1315–1342 (2012)

39. Bonilla, L.L., Klar, A., Martin, S.: Higher order averaging of Fokker–Planck equations for nonlinear fiber lay-down processes. SIAM J. Appl. Math. **74**(2), 366–391 (2014)
40. Chiu-Webster, S., Lister, J.R.: The fall of a viscous thread onto a moving surface: a 'fluid-mechanical sewing machine'. J. Fluid Mech. **569**, 89–111 (2006)
41. Cosserat, E., Cosserat, F.: Théorie des corps déformables. Hermann, Paris (1909)
42. Cox, R.G.: The motion of long slender bodies in a viscous fluid. Part 1. General theory. J. Fluid Mech. **44**(4), 791–810 (1970)
43. Decent, S.P., King, A.C., Simmons, M.J.H., Parau, E.I., Wallwork, I.M., Gurney, C.J., Uddin, J.: The trajectory and stability of a spiralling liquid jet: Viscous theory. Appl. Math. Model. **33**(12), 4283–4302 (2009)
44. Desvilettes, L., Villani, C.: On the trend to global equilibrium for spatially inhomogeneous entropy-dissipating systems: The linear Fokker–Planck equation. Commun. Pure Appl. Math. **54**, 1–42 (2001)
45. Dewynne, J.N., Ockendon, J.R., Wilmott, P.: A systematic derivation of the leading-order equations for extensional flows in slender geometries. J. Fluid Mech. **244**, 323–338 (1992)
46. Doulbeault, J., Klar, A., Mouhot, C., Schmeiser, C.: Exponential rate of convergence to equilibrium for a model describing fiber lay-down processes. Appl. Math. Res. Express **2013**, 165–175 (2013)
47. Doulbeault, J., Mouhot, C., Schmeiser, C.: Hypocoercivity for linear kinetic equations conserving mass. arXiv:1005.1495 (2010)
48. Eggers, J.: Nonlinear dynamics and breakup of free-surface flow. Rev. Mod. Phys. **69**, 865–929 (1997)
49. Eggers, J., Dupont, T.: Drop formation in a one-dimensional approximation of the Navier–Stokes equation. J. Fluid Mech. **262**, 205–221 (2001)
50. Elliott, F., Majda, A.J.: A new algorithm with plane waves and wavelets for random velocity fields with many spatial scales. J. Comp. Phys. **117**, 146–162 (1995)
51. Entov, V.M., Yarin, A.L.: The dynamics of thin liquid jets in air. J. Fluid Mech. **140**, 91–111 (1984)
52. Ferziger, J.H., Perić, M.: Computational Methods for Fluid Dynamics, 3rd edn. Springer, Berlin (2002)
53. Forest, M.G., Wang, Q.: Dynamics of slender viscoelastic free jets. SIAM J. Appl. Math. **54**(4), 996–1032 (1994)
54. Forest, M.G., Wang, Q., Bechtel, S.E.: 1d models for thin filaments of liquid crystalline polymers: Coupling of orientation and flow in the stability of simple solutions. Physics D **99**(4), 527–554 (2000)
55. Frisch, U.: Turbulence. The Legacy of A.N. Kolmogorov. Cambridge University Press, Cambridge (1995)
56. Geyling, F.T., Homsey, G.M.: Extensional instabilities of the glass fiber drawing process. Glass Technol. **21**, 95–102 (1980)
57. Gidaspow, D.: Multiphase Flow and Fluidization: Continuum and Kinetic Theory Descriptions. Academic Press, San Diego (1994)
58. Glowinski, R., Pan, T.W., Hesla, T.I., Joseph, D.D., Périaux, J.: A fictitious domain approach to the direct numerical simulation of incompressible viscous flow past moving rigid bodies: Application to particulate flow. J. Comp. Phys. **169**, 363–426 (2001)
59. Gospodinov, P., Roussinov, V.: Nonlinear instability during the isothermal drawing of optical fibers. Int. J. Multiph. Flow **19**, 1153–1158 (1993)
60. Grothaus, M., Klar, A.: Ergodicity and rate of convergence for a non-sectorial fiber lay-down process. SIAM J. Math. Anal. **40**(3), 968–983 (2008)

61. Grothaus, M., Klar, A., Maringer, J., Stilgenbauer, P.: Geometry, mixing properties and hypocoercivity of a degenerate diffusion arising in technical textile industry. arXiv:1203.4502 (2012)
62. Grothaus, M., Stilgenbauer, P.: Geometric Langevin equations on submanifolds and applications to the stochastic melt-spinning process of nonwovens and biology. Stoch. Dyn. **13**(4), 1350001 (2013)
63. Hagen, T.C.: On viscoelastic fluids in elongation. Adv. Math. Res. **1**, 187–205 (2002)
64. Hairer, E., Nørsett, S.P., Wanner, G.: Solving Ordinary Differential Equations I, Nonstiff Problems, 2nd edn. Springer, Berlin (2009)
65. Hartmann, S., Meister, A., Schäfer, M., Turek, S. (eds.): Fluid-Structure Interaction—Theory, Numerics and Application. Kassel University Press, Kassel (2009)
66. Herty, M., Klar, A., Motsch, S., Olawsky, F.: A smooth model for fiber lay-down processes and its diffusion approximations. Kinet. Relat. Models **2**(3), 489–502 (2009)
67. Hlod, A., Aarts, A.C.T., van de Ven, A.A.F., Peletier, M.A.: Three flow regimes of viscous jet falling onto a moving surface. IMA J. Appl. Math. **77**(2), 196–219 (2012)
68. Hoerner, S.F.: Fluid-Dynamic Drag. Practical Information on Aerodynamic Drag and Hydrodynamic Resistance. (1965) Published by the author. Obtainable from ISVA
69. Howell, P.D., Siegel, M.: The evolution of a slender non-axisymmetric drop in an extensional flow. J. Fluid Mech. **521**, 155–180 (2004)
70. Jung, P., Leyendecker, S., Linn, J., Ortiz, M.: A discrete mechanics approach to Cosserat rod theory—Part I: Static equilibria. Int. J. Numer. Methods Eng. **85**, 31–60 (2010)
71. Keller, J.B., Rubinow, S.I.: Slender-body theory for slow viscous flow. J. Fluid Mech. **75**(4), 705–714 (1976)
72. Kirchhoff, G.: Über das Gleichgewicht und die Bewegung eines unendlich dünnen elastischen Stabes. J. Reine Angew. Math. **56**, 285–316 (1859)
73. Kolb, M., Savov, M., Wübker, A.: (Non-)ergodicity of a degenerate diffusion modeling the fiber lay down process. SIAM J. Math. Anal. **45**(1), 1–13 (2013)
74. Kurbanmuradov, O., Sabelfeld, K.: Stochastic spectral and Fourier-wavelet methods for vector Gaussian random fields. Monte Carlo Methods Appl. **12**(5–6), 395–445 (2006)
75. Kutoyants, Y.: Statistical Inference for Ergodic Diffusion Processes. Springer, London (2004)
76. Lamb, H.: On the uniform motion of a sphere through a viscous fluid. Philos. Mag. **6**(21), 113–121 (1911)
77. Launder, B.E., Spalding, B.I.: Mathematical Models of Turbulence. Academic Press, London (1972)
78. Love, A.E.H.: A Treatise on the Mathematical Theory of Elasticity, 4th edn. Cambridge University Press, Cambridge (1927)
79. Lu, Q.Q.: An approach to modeling particle motion in turbulent flows—I. Homogeneous isotropic turbulence. Atmos. Environ. **29**(3), 423–436 (1995)
80. Maddocks, J.H.: Stability of nonlinearly elastic rods. Arch. Ration. Mech. Anal. **85**(4), 311–354 (1984)
81. Maddocks, J.H., Dichmann, D.J.: Conservation laws in the dynamics of rods. J. Elast. **34**, 83–96 (1994)
82. Mahadevan, L., Keller, J.B.: Coiling of flexible ropes. Proc. Roy. Soc. Lond. A **452**, 1679–1694 (1996)
83. Majda, A.J.: Random shearing direction models for isotropic turbulent diffusion. J. Stat. Phys. **75**(5–6), 1153–1165 (1994)
84. Malkan, S.R.: An overview of spunbonding and meltblowing technologies. Tappi J. **78**(6), 185–190 (1995)
85. Matovich, M.A., Pearson, J.R.A.: Spinning a molten threadline. Steady-state isothermal viscous flows. Ind. Eng. Chem. Fundam. **8**(3), 512–520 (1969)

86. Monaghan, J.J.: Smoothed particle hydrodynamics. Rep. Prog. Phys. **68**, 1703–1759 (2005)
87. Pearson, J.R.A.: Mechanics of Polymer Processing. Elsevier, New York (1985)
88. Pearson, J.R.A., Matovich, M.A.: Spinning a molten threadline. Stability. Ind. Eng. Chem. Fundam. **8**(3), 605–609 (1969)
89. Peskin, C.S.: The immersed boundary method. Acta Numer. **11**, 479–517 (2002)
90. Pinchuk, L.S., Goldade, V.A., Makarevich, A.V., Kestelman, V.N.: Melt Blowing: Equipment, Technology and Polymer Fibrous Materials. Springer Series in Materials Processing. Springer, Berlin (2002)
91. Pismen, L.M., Nir, A.: On the motion of suspended particles in stationary homogeneous turbulence. J. Fluid Mech. **84**, 193–206 (1978)
92. Renardy, M.: Mathematical analysis of viscoelastic flows. Annu. Rev. Fluid Mech. **21**, 21–36 (1989)
93. Ribe, N.M.: Coiling of viscous jets. Proc. Roy. Soc. Lond. A **2051**, 3223–3239 (2004)
94. Ribe, N.M., Habibi, M., Bonn, D.: Stability of liquid rope coiling. Phys. Fluids **18**, 084102 (2006)
95. Ribe, N.M., Lister, J.R., Chiu-Webster, S.: Stability of a dragged viscous thread: Onset of 'stitching' in a fluid-mechanical 'sewing machine'. Phys. Fluids **18**, 124105 (2006)
96. Rubin, M.B.: Cosserat Theories. Kluwer, Dordrecht (2000)
97. Schewe, G.: On the force fluctuations acting on a circular cylinder in cross-flow from subcritical up to transcritical Reynolds numbers. J. Fluid Mech. **133**, 265–285 (1983)
98. Schlichting, H.: Grenzschicht Theorie. Verlag G. Braun, Karlsruhe (1982)
99. Schultz, W.W., Davis, S.H.: One-dimensional liquid fibres. J. Rheol. **26**, 331–345 (1982)
100. Shah, F.T., Pearson, J.R.A.: On the stability of non-isothermal fibre spinning. Ind. Eng. Chem. Fundam. **11**, 145–149 (1972)
101. Simo, J.C., Vu-Quoc, L.: Three-dimensional finite strain rod model. Part I: Computational aspects. Comput. Methods Appl. Mech. Eng. **58**, 79–116 (1986)
102. Simo, J.C., Vu-Quoc, L.: On the dynamics in space of rods undergoing large motions—a geometrically exact approach. Comput. Methods Appl. Mech. Eng. **66**, 125–161 (1988)
103. Stokes, Y.M., Tuck, E.O.: The role of inertia in extensional fall of viscous drop. J. Fluid Mech. **498**, 205–225 (2004)
104. Sumer, B.M., Fredsoe, J.: Hydrodynamics Around Cylindrical Structures. World Scientific, New Jersey (2006)
105. Taylor, G.I.: Analysis of the swimming of long and narrow animals. Proc. Roy. Soc. Lond. A **214**, 158–183 (1952)
106. Tiwari, S., Kuhnert, J.: Finite pointset method based on the projection method for simulations of the incompressible Navier–Stokes equations. In: Griebel, M., Schweitzer, M.A. (eds.) Meshfree Methods for Partial Differential Equations. Lecture Notes in Computational Science and Engineering, vol. 26, pp. 373–387. Springer, Berlin (2003)
107. Tomotika, S., Aoi, T.: An expansion formula for the drag on a circular cylinder moving through a viscous fluid at small Reynolds number. Q. J. Mech. Appl. Math. **4**, 401–406 (1951)
108. Tomotika, S., Aoi, T., Yosinobu, H.: On the forces acting on a circular cylinder set obliquely in a uniform stream at low values of Reynolds number. Proc. Roy. Soc. Lond. A **219**(1137), 233–244 (1953)
109. VDI-Gesellschaft: VDI-Wärmeatlas, 10th edn. Springer, Berlin (2006)
110. Wallwork, I.M., Decent, S.P., King, A.C., Schulkes, R.M.S.M.: The trajectory and stability of a spiralling liquid jet. Part 1. Inviscid theory. J. Fluid Mech. **459**, 43–65 (2002)
111. Whitman, A.B., DeSilva, C.N.: An exact solution in a nonlinear theory of rods. J. Elast. **4**, 265–280 (1974)
112. Yarin, A.L.: Free Liquid Jets and Films: Hydrodynamics and Rheology. Longman, New York (1993)

113. Yarin, A.L., Gospodinov, P., Gottlieb, O., Graham, M.D.: Newtonian glass fiber drawing: Chaotic variation of the cross-sectional radius. Phys. Fluids **11**(11), 3201–3208 (1999)
114. Zdravkovich, M.M.: Flow Around Circular Cylinders. Fundamentals, vol. 1. Oxford University Press, New York (1997)
115. Ziabicki, A., Kawai, H.: High Speed Melt Spinning. Wiley, New York (1985)

# Modeling and Simulation of Filtration Processes

Oleg Iliev, Ralf Kirsch, Zahra Lakdawala, Stefan Rief, and Konrad Steiner

## 1 Industrial Challenges in Filtration

Filtration and separation processes are very important for our everyday life. Finding advanced filtration and separation solutions is often critical for the development of highly efficient and reliable products and tools, as well as for ensuring a high quality of life for the general public. It is difficult to find an industry or area of life where filters do not play an important role. In a single car, for example, one finds filters for the transmission, fuel, engine air, cabin air, coolant, and brake systems. Furthermore, the quality of our drinking water, the treatment of wastewater, the air we breathe—everything is critically dependent on filtration solutions. The filtration and separation business is expanding rapidly, with scores of large companies and thousands of SMEs competing to develop better filters. The industrial demand for innovative filtration and purification solutions is growing steadily, thus promoting the use of Computer Aided Engineering in designing filter media and filter elements. An important class of filtration processes, namely, solid-liquid separation, i.e., filtering solid particles out of liquid, is discussed in this chapter. Furthermore, the focus is mainly on dead-end filtration, where all the contaminated fluid is forced to pass through the filtering medium.

Three main criteria that determine the performance of a filter are as follows:

- the flow rate—pressure drop ratio,
- the size of the penetrating particles,
- the dirt storage capacity.

O. Iliev (✉) · R. Kirsch · Z. Lakdawala · S. Rief · K. Steiner
Fraunhofer-Institut für Techno- und Wirtschaftsmathematik, Fraunhofer-Platz 1,
67663 Kaiserslautern, Germany
e-mail: oleg.iliev@itwm.fraunhofer.de



H. Neunzert, D. Prätzel-Wolters (eds.), *Currents in Industrial Mathematics*,
DOI 10.1007/978-3-662-48258-2_7

The first criterion corresponds to the energy efficiency of the filter (i.e. the energy spent to push the fluid through the filtering medium). The second criterion defines the requirements for filtration efficiency (for example, 99.9 % of particles bigger than one micron have to be captured). The third criterion is a measure of the lifetime of a filter (e.g., the frequency with which a filter element has to be cleaned or replaced).

Obviously, the three criteria imply that manufacturers must deal with contradicting requirements. For example, higher energy efficiency can be achieved with a more porous filtering medium, but this will result in worse filtration efficiency; higher storage capacity can be achieved by using a thicker filtering medium, but this will result in increased energy demand; and so on. These contradicting requirements impose a number of challenges for designing highly efficient filters.

In general, there are many possibilities to design the filter media and filter element for increased life-time performance. Besides the detailed structuring of the pore volume within a filter media, a filter media can be designed to have a multi-layered structure made of different single porous media. Often, the active surface is increased by folding the media (i.e., star-like pleated filters). All together, a complete filter system (e.g., sewage plant, water purification plant) is usually built out of single filters combined in parallel and/or series and may have a dimension of several meters, whereas the effective filtration process happens in the pores on the microscale or nanoscale. In many cases, the effects of the filter media (including the multi-layers) and of the filter system (folding and housing) can only be analyzed and optimized separately under test conditions. However, sometimes filter media and/or a complete filter system that demonstrates high performance on the test bench does not work well under real application conditions, which is a hint that the design should simultaneously treat the microscale and macroscale filtration processes.

For several years, computer simulation methods and CAE software have been more and more actively used to provide detailed information about flow, transport, and capturing processes in order to save the time and expense of intensive functional performance testing on laboratory filtration test benches. However, the general purpose CFD software used by many companies in conjunction with the simulation of filtration processes does not account for specific modeling and numerical treatments of filtration effects and does not address the intrinsic multi-scale and multiphysics behavior of the filtration processes. Commercial CAE software tools are only able to simulate single effects, such as fluid flow through the filter medium or in the filter housing. The highly coupled interaction between the flow regimes in the filter medium and housing, including the correct interface conditions and the coupling of different length scales, often can not be efficiently solved with existing commercial software tools. Industry needs customized software tools that demonstrate high efficiency when solving filtration problems.

## 1.1 Filter Media Design

In this chapter, the discussion is restricted mainly to nonwoven filtering media (see Fig. 7), keeping in mind, however, that many other filtration materials, such as paper, foam, sand, ceramic and polymer membranes, sieves, etc., face the same challenges. Crucial factors influencing the performance of non-woven filtering media include:

- the microstructure of the media;
- grading;
- the material used to manufacture the fibers;
- the fibers' surface treatment or charging;
- the deformations in the media under operating conditions;
- customizing filtering media for particular types of particles.

*Microstructure* of the filtering media is influenced by the shape of the fibers, the fiber diameter distribution, the fiber anisotropy, etc. Today, fibers can be manufactured with a large range of dimensions, with diameters varying from nanometers to micrometers, and with a variety of shapes ranging from traditional cylindrical shapes to trilobal and ellipsoidal shapes. It is a challenging task for the manufacturers to select the right combination of fibers for the filtering medium to achieve the desired performance for a specified class of dust particles.

*Graded* and/or multilayered filtering media are used often nowadays. It is an extremely demanding task to design graded media so as to improve energy efficiency and dirt storage capacity without reducing the filtration efficiency at the same time.

*Materials* used to manufacture the fibers significantly influence the filter media performance. The chemical industry continuously supplies new materials and selecting the best material for a specific filtration application remains a difficult task.

*Charging* the filtering media is known to be a reasonable approach for increasing the filtration efficiency without reducing the energy efficiency (i.e., flow rate—pressure drop ratio), especially in the case of air filtration. At the same time, understanding the interplay between microstructure and charging and the role of the captured particles in shielding the electrostatic field, etc., would allow one to further improve the performance of the filtering media.

*Deformations* often occur under operating conditions, both at the fiber scale and at the scale of the filtering medium. These can significantly change the microstructure of the filtering medium and, thus, significantly influence the performance of the filter. It is a complex task to understand and predict the qualitative and quantitative changes in the filter's performance resulting from deformations.

*Customizing* filtering media for particular types of particles, or even just selecting filter media that provide better performance for particular real particles, is a very difficult task. The International Standardization Organization (ISO) tests are usually performed with a specific dust (e.g., Arizona fine, or coarse) having a fixed distribution of particle diameters

**Fig. 1** *Left*: A pleated filtering medium with corrugation. *Center*: A pleated cartridge for fuel filtration. *Right*: Disassembled housing of a automatic transmission oil filter. The filtering medium is perforated and the plastic covering has a supporting rib structure optimized for low flow resistivity

with a relatively simple particle shape. In reality, the dust may differ considerably from the one used in the laboratories, which means that filter performance under real operating conditions may differ from the performance measured in the laboratories.

## 1.2 Filter Element Design

Some designs of typical filter elements are shown in Fig. 1. The main factors influencing the performance of a filter element are:

- the selection of the filtering medium;
- the sizing of the filtering medium (e.g. pleating);
- the stabilization of the filtering medium (e.g. the design of the supporting mesh or supporting ribs);
- the sizing of the filter element.

*The selection of the filtering medium* can be a challenge for filter element and filter system manufacturers. Average characteristics provided by filter media manufacturers, such as grammage and porosity, may be too rough to evaluate the performance of the filter element for a particular dust. Standard ISO tests, such as single pass and/or multipass tests, Transmission Filter Effectiveness Method (TFEM), are performed to test the filter elements using flat pieces of the filtering medium. Such tests provide useful information, but also need careful interpretation in order to evaluate the performance of a filter element with a filter media and/or housing having a complex shape.

Designing a filter cartridge with the *optimal pleat count* is a difficult task for many filtration applications. Negative factors, such as pleat deflection and/or pleat crowding, can dramatically change the performance of the filter element. Even in the case of rigid filter media, determining the optimal pleat count is not trivial. Usually the pleat count is selected in a way that balances the pressure loss due to the filtering medium and the pressure loss due to the channels (narrow space) between the pleats. However, this is done usually for clean filter media. Because the resistance of the media changes as it becomes loaded, the optimum pleat counts for clean, partially loaded, and heavily loaded media may differ significantly.

*Stabilization* is often used with both flat and pleated media. In some cases, the supporting mesh or ribs may block ten percent or more of the surface of the filtering media. Furthermore, some meshes, and all ribs, create additional resistance for the flow, thus reducing the energy efficiency of the filter element. However, if the support is not properly sized, the deflection of the filtering media may cause even larger reductions in efficiency. Optimizing the support is an urgent task.

The *sizing of a filter element* is a non-trivial task. In some cases (e.g., transmission filters), the shape and size of the filter element may be limited by the free space allocated in the engine design. In other cases, e.g., round, pleated liquid filters, the height of the filter element has to be properly chosen, so that there will be enough pressure to push the liquid up to the bottom of the filter element and then through the filtering medium, when the flow inlet is at the top of the filter element.

The partially *changing operation conditions* of filter elements are also a big challenge for product development. For example, the dynamics of the operating conditions in automotive applications are influenced by rapidly turning pumps and start-stop fuel saving systems. This leads to an immediate change of flow conditions in the filters and sometimes even a release of captured dust. Similarly, during the typical industrial cleaning process of back-flushing, due to heterogeneity, the so-called channeling effect in the filter media can occur, which may significantly reduce filtration efficiency and lifetime.

To summarize, the design of efficient filter media and filter elements is a challenging task. For a long time, industrial design has relied mainly on lab experiments, despite the fact that manufacturing prototypes and performing lab measurements are expensive and time consuming procedures. In the last decade, mathematical modeling and computer simulation have been more and more widely used in supporting the design process. Computer Aided Engineering, CAE, is a part of the everyday work for many filter media and filter element manufacturers. Virtual material design and virtual design of filter elements have proven to be extremely effective, since they significantly reduce the number of prototypes, shorten the design time, and reduce total costs. Industrial mathematics is a driving force and a key component of these approaches. The next section describes the challenges confronting industrial mathematics, particularly in the field of modeling and simulation of filtration processes.

## 2 Mathematical Challenges in Modeling and Simulation of Filtration Processes

Many different interesting mathematical tasks have to be addressed to model and simulate filtration processes.

The principal aims of the mathematical modeling of filtration processes are to describe

- the fluid flow through the filtering (porous) medium and within the filter housing;
- the interaction between the fluid and the filtering medium;
- the transport of the dirt particles;
- the filtration process itself, i.e., the capturing and deposition of the dirt particles;
- the interaction of dissolved particles with each other, especially in the case of highly contaminated fluid.

A major challenge is posed by the fact that all these processes are coupled. Clearly, the flow influences the transport of the dissolved particles and their deposition in the filtering medium. The deposition behavior, in turn, changes the geometry of the pore spaces in the medium with corresponding effects on the velocity field and the pressure distribution.

Analytical solutions for the above models are rarely available, and computer simulation must be used to find solutions. The principal tasks of computer simulations are to

- develop proper numerical algorithms;
- implement these algorithms in the proper software tools;
- define the computational domain and generate a grid;
- perform simulations;
- analyse the obtained results.

In this section, we will briefly discuss the challenges faced by industrial mathematics in conjunction with modeling and simulating filtration processes. The subsections below are devoted to discussing: (1) modeling approaches at the pore scale, (2) modeling approaches at the filter element scale, (3) modeling of deformable filtering media, (4) modeling of multiscale filtration processes, (5) numerical algorithms at the pore scale, and (6) numerical algorithms at the filter element scale.

### 2.1 Specific Challenges at Microscale

Some of the main challenges in modeling filtration processes at the pore scale are:

*(i) Modeling the pore scale geometry for random microstructures.* Filtration media are often nonwoven materials, foams, membranes, or other materials having a *stochastic geometry at the pore scale*. There is no standard approach for modeling such media or for modeling other random geometries in general. Special models must be developed for each

class of geometry and a basic way to do this is to use stochastic geometry approaches. Using computerized tomography or images of material samples from a scanning electron microscope (SEM) with subsequent image processing, one can characterize the particular class of pore geometries (see e.g. [27]), and use this information as input for the stochastic geometry models. The latter can be used to create (generate) different realizations of virtual porous materials (see [37]) to be considered as computational domains in filtration simulations. Open questions in this area are developing models for the random geometries for many classes of filtration materials (e.g., various membranes, papers); achieving stationarity of the stochastic processes in generating microgeometries, especially in the case of multiparameter models; and improving models for graded filtering media.

*(ii) Modeling particle transport and the interaction of particles with the solid skeleton.* Deriving mathematical models for the *transport and capturing of particles at the pore scale* is a difficult task, and models are available mainly for simple cases. In the case of laminar flow and spherical particles, a known model consists of a coupled system of equations, including Stokes or Navier–Stokes equations describing the flow, a Langevin stochastic ordinary differential equation (see Sect. 4.1 for details) describing the transport of the particles, equipped with various adhesion mechanisms, e.g., direct interception, inertial impact, diffusional deposition, size sieving, and clogging. The Langevin equation, which accounts for the Brownian motion of the small particles, has been widely studied in the literature for no boundaries or adsorbing boundaries. There are some studies of the Langevin equation with reflection boundary conditions, but the case which is of most interest for filtration, namely, boundary conditions describing various adhesion mechanisms, has hardly been investigated mathematically. Furthermore, during their motion, the particles are treated as material points, and the volume is accounted for only in calculating the resistance. At the same time, the deposited particles have a volume, and thus the deposition of the particles changes the microgeometry. The latter leads to a change in the flow, and so on. In certain cases (e.g., some of the regimes for air filtration), this system should be enriched by an equation describing the electrostatic field [24]. Even if the charge for a filtering medium is known in advance, the deposition of particles changes the electrostatic field and it has to be recomputed. Open questions in this area include developing models for non-spherical particles and deformable particles. Furthermore, modeling efforts are needed to better understand flow at the pore scale for gases or non-Newtonian fluids. In gas filtration with media composed of nanofibers, one can reach Knudsen number regimes, for which (Navier–)Stokes equations no longer represents a proper choice and one has to consider kinetic models. Treating charged particles or macromolecules (having a chain structure) is another very big challenge.

*(iii) Modeling particle-particle interaction.* There are models for the interaction of particles in pure fluid regions [86] that include breakage and agglomeration of particles. However, adaptation of these models to the flow in the porous space of a filter medium, when the particles interact with each other and also with the solid walls of the pores, is still far from complete.

## 2.2 Problems Appearing at Macroscale

In certain cases, such as periodic or stochastically homogeneous microstructure of the filtering medium and slow flows, mathematical models of filtration processes at the macroscale can be derived from microscale models via asymptotic homogenization [70, 76] or via volume averaging [83]. Alternatively, they can be postulated directly at the macroscale based on conservation laws, and equipped with constitutive relations, if needed.

Even for the flow of clean fluid, the flow modeling is far from trivial.

*Slow flows in porous media* are usually modeled using the Darcy equation [66] or the Brinkman equation [62]. *Open questions* include defining apparent permeability for highly heterogeneous media (no REV); determining the viscosity in the porous media for the Brinkman equation; properly modeling the stochasticity for macroscale heterogeneity of the filtering media.

*Fast flows in porous media* are more difficult to model. Most often they are modeled using the quadratic Forchheimer equation [73, 83] (sometimes called Ergun equation), which in addition to the permeability contains a coefficient in front of a quadratic velocity term, which still has to be determined from experiments or parameter identification. A theoretical paper [56] based on asymptotic homogenization states that fast flows in porous media have to be described by a cubic (with respect to velocity) equation. Furthermore, some researchers claim that Navier–Stokes–Brinkman equations suffice to describe fast flows in porous media,. Finally, in an interesting paper [57], an equation having rational terms with respect to the velocity is introduced. *Open questions* in this area include determining the area of applicability for each of the models and developing reliable models in the case of turbulence. In addition to the non-linear pressure drop behavior, more understanding is needed about the turbulence in the free fluid and its interaction with the flow in the porous medium.

One more topic that is very important in modeling flows within a filter element is the topic of the *interface conditions between the plain media (unconfined fluid) and the porous media*. For flows parallel to porous media, the famous Beavers–Josef [59] condition is usually used in conjunction with flat porous media, the Stokes model for the free fluid, and the Darcy model for the porous media. This interface condition is experimentally determined, and later rigorously derived in [80]. In the case of a Stokes–Brinkman system, interface conditions are derived by Ochoa-Tapia and Whitaker [87] based on volume averaging. These interface conditions, similar to the case of Beavers–Josef, work well for flat media and parallel flow [42]. The interface conditions for non-flat media, inclined flow, or deforming filtering media are an open question and a subject of research.

Since the *macroscale modeling of filtration efficiency* must consider many different aspects, there exist different models that rely on additional assumptions, e.g., constant velocity, constant permeability, and porosity. By *filtration efficiency*, we mean the percentage of captured particles. There is extensive literature on developing macroscale models of filtration efficiency, considering only the filtering medium and ignoring the influence of the filter housing. A representative collection of models is discussed in Sect. 4.2.1. Furthermore, very little has been done on modeling the more complicated case, when the influence of the filter element housing has to be accounted for. An approach based on developing *lookup tables* and combining them with parameter identification and the above-mentioned macroscale models of filtration, is discussed e.g., in [41]. This approach reflects a basic requirement for modeling industrial processes: integrating the best research results for different stages of the studied process into a monolithic approach that can produce quantitative results and support industry in finding innovative solutions.

*Open questions* in the area of macroscopic models of filtration processes also include: developing macroscale models of filtration processes for more complicated situations, including particle diameter distribution; combined filtration effects (e.g., sieving and deposition); graded and/or multilayered filtering medium; time dependent inflow velocity and inflow concentration; robust and reliable parameter identification procedures; and analysis of stability and sensitivity of the lookup tables approach.

## 2.3 Mathematical Modeling of Deformable Filtering Media

To this point, the filtering medium was regarded as a "rigid" structure. However, in more and more fields of application in filtration, the interaction between the flow and the filtering medium can no longer be neglected.

This leads to another coupling effect: The shape of the filtering medium is influenced by the pressure distribution and, in turn, the flow field depends on the shape of the porous medium. The deformation of filtering media can have a tremendous effect on the performance of a filter element. Well-known examples are the crowding (grouping) of filter pleats and the collapse of filter pleats, i.e., the closing (some) of the inter-pleat channels under the flow-induced pressure. Of great interest is the influence of the deformations on the permeability and the filtration efficiency of the medium.

The interaction of fluids with solid structures (FSI) is a widely studied and active field in physics, engineering, and applied mathematics. However, very little is known at present about the deformation of porous media, in general, and Fluid-Porous-Structure Interaction (FPSI), in particular. It is obvious that the behavior of a deformable porous medium may be very different than a solid structure. The fluid can enter the medium, for example, which shows the importance of a proper modeling of the effects at the fluid-porous interface. Moreover, the influence of phenomena inside the filtering medium, such as the pore pressure, must be accounted for.

Classical theories, such as the pioneering works by Biot (see [60, 61]), were motivated by the settlement (consolidations) of soils. A first open question here is to what extent such models can be applied to filtering media. Another issue is that corresponding measurements are far from being trivial, and it can become difficult to perform experimental validations for any model derived. As is the case for other aspects in filtration, the modeling has to be done at both the pore space level and the macroscopic scale.

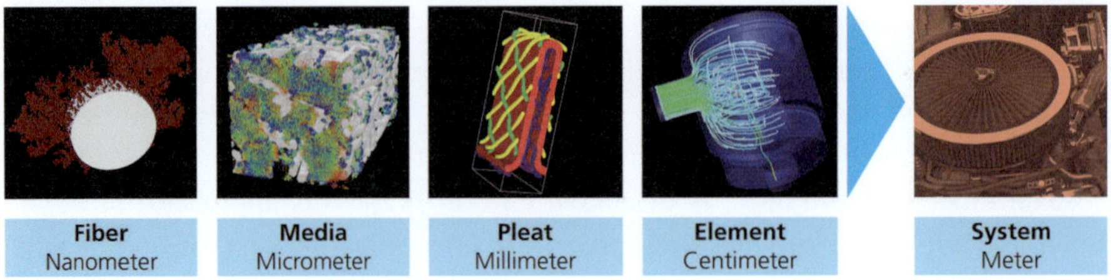


**Fig. 2** An overview of the scale magnitudes important in filtration (Grafik: S. Grützner, Fraunhofer ITWM, Simulationen: Fraunhofer ITWM, Abteilung SMS, Foto: iStockphoto)

First promising steps towards a better understanding of poroelasticity and FPSI have been taken. We will briefly discuss them in the next section.

## 2.4 Multiscale Modeling and Simulation of Filtration Processes

Filtration in general, and the dead-end depth filtration of solid particles out of fluid in particular, is intrinsically a multiscale problem (see, e.g., Fig. 2 for illustration). The deposition (capturing of particles) essentially depends on the local velocity, the microgeometry (pore scale geometry) of the filtering medium, and the diameter distribution of the particles. The deposited (captured) particles change the microstructure of the porous media, which leads to a change in the permeability. The changed permeability directly influences the velocity field and pressure distribution inside the filter element. To close the loop, we mention that the velocity influences the transport and deposition of particles. In certain cases, one can evaluate the filtration efficiency by considering only microscale or only macroscale models. In general, however, an accurate prediction of the filtration efficiency requires multiscale models and algorithms.

Filtration studies share some mathematical multiscale problems with the studies of other industrial and environmental processes. At the same time, there are some multiscale mathematical challenges that are specific to filtration problems. For example, (i) rigorously deriving macroscale (averaged) equations assuming that Stokes or Navier–Stokes are valid for flow at the pore scale and (ii) finding permeability as a function of pore scale geometry are both mathematical problems that concern any flow in porous media. On the other hand, investigating the interplay between the microscopic nature of particle-capturing and the macroscopic velocity within a filter element is a task that is specific to filtration processes.

Let us briefly discuss some of the mathematical challenges related to the multiscale modeling and simulation of filtration processes.

Mathematical modeling using asymptotic homogenization is a powerful approach for studying multiscale problems. It has several prominent features: (i) the rigorous derivations give additional confidence in experimentally discovered models; (ii) the rigorous derivations offer direct relationships between the particular microstructure and the effective (upscaled) property; and (iii) the rigorous derivations can be used to derive models for cases where no measurements are done. A typical example is the Darcy law describing slow, single-phase incompressible flow through a rigid porous medium [70, 76]. It was derived experimentally by Darcy in 1856 for isotropic porous media. The derivation via homogenization, for example, offers a reliable extension to the case of anisotropic media and also yields the algorithm for computing the permeability tensor in this case. Despite this example, there are other examples where the microscale and macroscale can be decoupled (so-called scale separation). In such situations, solving multiscale problem reduces to a two-stage procedure: (a) solve the microscale "cell-problem" and use its solution to upscale the effective properties of the multiscale media; (b) solve the upscaled (macroscale) equations with effective coefficients. It is important to note that homogenization theory provides all the components needed for solving a multiscale problem: interscale connection operators, the type of coarse scale equations (which may be different from the type of equations at the fine scale), estimates for the difference between the fine and coarse scale solutions, etc. The challenges here are related to the fact that rigorous derivations are done only for periodic and statistically homogeneous media and for slow incompressible flow. Rigorous derivation of macroscopic equations for flows in porous media for a broad class of fast flows and compressible flows is still an area of active research.

For complex filtration processes, the cell problem [70, 76] (see Sect. 4.3 for details) has to be solved numerically in order to obtain the effective properties of a filter media. This might be a challenge in and of itself. Some of the difficulties arising in this case are discussed in Sect. 4.1.

The separation of scales in filtration is not always possible, even just for the flow problem. Numerical upscaling approaches can be used here, such as the multiscale finite element method (MsFEM), [68], the multiscale finite volume method (MSFV), [81], the heterogeneous multiscale method (HMM), [101], the variational multiscale method (VMS), [55], etc. These approaches allow one to attack the multiscale problems related to flow in porous media, but they are still computationally very expensive. In fact, their adaptation to filtration problems is a nontrivial task needing further active research. Some recent developments will be mentioned in the next section. The above mentioned numerical upscaling approaches are computationally still rather expensive, and further efforts in the area of model order reduction (MOR) and reduced basis (RB) approaches are needed in order to handle practical filtration problems.

When the transport and capturing of particles are considered along with the flow problem, the situation becomes even more challenging. The Langevin stochastic differential equation describing particle transport mentioned earlier (see Sect. 4.1 for further details) can be upscaled to a concentration equation at the macroscale, but focused mathematical research is needed for upscaling the filtration mechanisms (such as interseption, sieving, etc.). There is almost nothing in the literature concerning multiscale modeling and simulation of filtration. One approach for solving such problems will be discussed in the next section and described briefly in Sect. 4.3.

To summarize, modeling and simulation of filtration processes are challenging mathematical tasks. The available literature deals mainly with some components of these processes but rarely suggests a complete solution for an industrial filtration problem. The existing commercial software tools, in general, are not adapted to the simulation of filtration processes. To fill the gap between the incomplete mathematical studies in the area of filtration processes and the needs of industry for systematic studies and complete solutions, the SMS Department of the Fraunhofer ITWM has developed a number of algorithms and software tools in the last decade that are dedicated to the simulation of filtration processes.

## 3 ITWM's Developments in the Modeling and Simulation of Filtration Processes

For more than a decade, the Fraunhofer ITWM has been involved in developing models, algorithms, and software for modeling and simulating industrial filtration processes [29]. Many specific mathematical problems have been solved, although the emphasis has been on providing complete solutions for industrial filtration applications. The latter requires integrating the developed algorithms into customized software tools [30]. A short overview of the achievements can be found in the following subsections.

### 3.1 Virtual Filter Media Design

A major difficulty for computer simulation on the pore scale of filter media is given by the complexity of the flow domain due to the random features of the geometry. Most of the existing commercial and academic 3D grid generation software tools fail at grid generation in the complicated pore structure of the filtering media. To overcome this bottleneck, keeping in mind that 3D CT images are defined on voxels anyway, the use of voxel grids is proposed and successfully exploited in [35]. Existing computer power allows filtration processes to be simulated only in a small piece of the filtering medium when the pore scale geometry is fully resolved, and this implies that special attention has to be paid to the efficiency of the developed algorithms. An idea of the microscale simulation of filtration processes can be gained, e.g., from [2, 25, 30, 92].

For more than a decade, *the virtual material laboratory GeoDict* has been under constant development at the Fraunhofer ITWM. The modular software toolbox provides a huge variety of algorithms to generate virtual porous media, in particular, filtering media. Among others, there are modules for the generation of

- nonwoven structures originating e.g. from textile applications,
- woven textiles and metal wire meshes,

- sintered ceramics used in diesel particate filters, for instance,
- paper, i.e., cellulose materials, including fillers and fines, and
- foams.

Moreover, GeoDict provides software interfaces for importing CAD data and micro-CT data sets, along with image processing tools for cutting, rotating, and filtering the images. The second set of modules addresses the computation of fully pore-scale-resolved porous media in order to determine the effective (macroscopic) properties of the porous media. These modules compute

- the porosity and pore size distributions of the media,
- permeabilities and flow resistivities based on solving the Navier–Stokes equations,
- two-phase properties, such as relative permeabilities, and
- filter efficiency and pressure drop evolution in single-pass and multi-pass setups.

The interest in GeoDict has continued to grow over the years and, in 2011, the Math2Market GmbH was founded as a spin-off company of the Fraunhofer ITWM. Math2Market focuses on the development, marketing, and dissemination of the GeoDict software suite.

Furthermore, for performing structural mechanics simulations at the microscale, the Fraunhofer ITWM developed an elasticity solver for composite and porous materials (FeelMath). The fundamental approach is the formulation of the elasticity problem as a Lippmann–Schwinger-type equation, which can be solved very efficiently using the Fast Fourier Transform (FFT) (see [33] and the references therein). Using this tool, one can study the local stretching and/or compression of the medium that leads to a change in pore geometry and compute the effective elasticity properties of the material. The corresponding permeabilities for the non-deformed and deformed states can be computed using the GeoDict software (see Fig. 3).

There is ongoing research in close collaboration with filter manufactures on this matter (see e.g. [22]), and there are also research activities concerning the macroscopic level (see Sect. 3.2).

## 3.2 Computer-Aided Design of Filter Elements

Fraunhofer ITWM is also active in the field of mathematical modeling, numerics, and software development for the filter element scale. Here, we will restrict the presentation to a selection of achievements related to the challenges described in Sect. 2.2.

For industrial applications, the selection of the appropriate model for the fluid flow and particle movement is crucial. In most cases, the (incompressible) Navier–Stokes–Brinkman equation is the basis for computing the fluid flow within filter elements, and the convection–diffusion–reaction equation is used to describe the transport and capturing

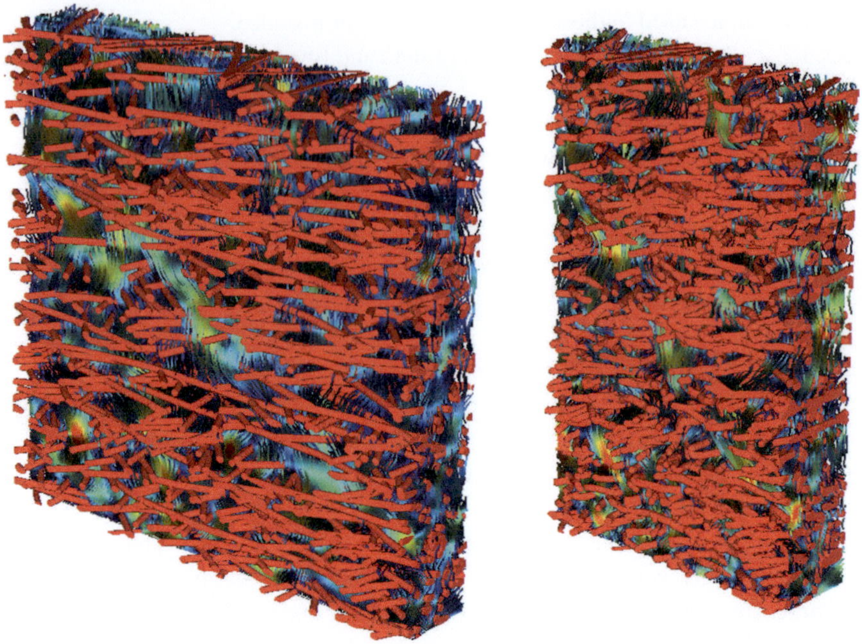

**Fig. 3** Numerical study of the influence of the deformation of filtering media on the permeability. Streamline representation of the velocity for a non-deformed sample (*left*) and the corresponding compressed structure (*right*)

of the particles. Arguments as to why the Brinkman equation is a more suitable model for filtration processes, as opposed to using the Darcy model in the porous media and coupling it to Navier–Stokes equations in the free fluid, can be found in [17, 42]. The model for filtration efficiency simulation is described in detail in [41]. Models for fast flows were the subject of a recent paper [11], in which mathematical models together with numerical and experimental results in this area were collected and discussed. Further work on determining the area of applicability for each of the models is needed. A discussion of different types of interface conditions and their applicability to filtration processes can be found in [17, 42]. A representative collection of macroscopic models for filtration efficiency can be found in [10], see also Sect. 4.2. An approach based on developing *look-up tables* and combining them with parameter identification and the above-mentioned macroscale models of filtration is discussed in [10, 41].

On the macroscopic scale, the geometry of the computational domain is mostly given in the form of CAD data (Computer Aided Design). Since the shapes of the housing, the media, etc. can be quite sophisticated, grid generation can be a non-trivial task. Robust methods for the generation of uniform Cartesian grids (voxel grids) and the efficient numerical solution of the Navier–Stokes–Brinkman equations in the context of filtration were, amongst others, the subject of the works [41, 42] and [5, 17, 34]. For a robust and accurate numerical method, special attention needs to be paid to the proper treatment of

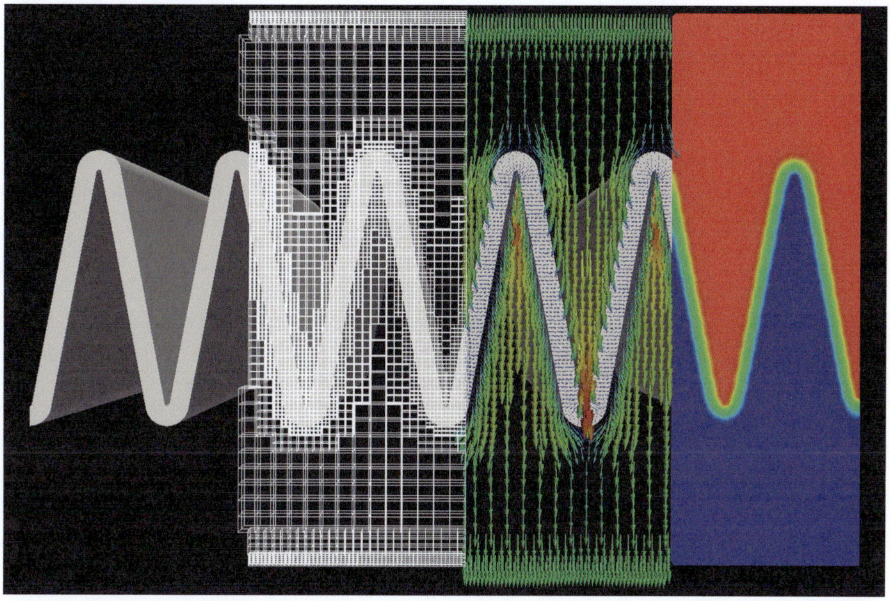

**Fig. 4** Overview of the different stages of macroscopic simulation of the flow through a pleated panel. *From left to right*: CAD geometry (surface), locally refined 3D grid for numerical simulation, resulting velocity field, and pressure distribution

the interface between the fluid and the porous filter medium. This is true for both the transient case, when using discretization schemes of the Chorin type (see e.g. [72]), and the stationary problem, when using SIMPLE (see e.g. [71]) and its variants. A problem-adapted discretization for the fluid-porous interface region was proposed in [5] that allows for recovery of a linear pressure drop profile (in accordance with Darcy's law), even if the medium is represented in the grid by only a single voxel layer.

Part of the dissertation [41] was devoted to including the transport and deposition of particles in the macroscopic simulation of filters. In particular, a method to create and use look-up tables for the handling of dynamical effects during the filtration process was developed in that work.

These numerical approaches are the basis for two software tools developed at the Fraunhofer ITWM: The *Suction Filter Simulation (SuFiS®)* (cf. e.g. [14]) and the *Filter Element Simulation Toolbox* (FiltEST).

SuFiS® is a tailor-made simulation solution for the optimal design of oil filters in the automotive sector. The software has been under development for more than a decade in close collaboration with the company IBS Filtran (see also Sect. 5.1).

The **Filter Element Simulation Toolbox (FiltEST)** is a collection of software modules for the analytical study and numerical simulation of the performance of filter elements used in solid-liquid and solid-gas separation. The core of this software family consists of the modules to perform the numerical simulation of

- the flow through the filter housing and medium,
- the particle transport with the flow, and
- the particle deposition in the filter medium.

These will be described in more detail in Sect. 4.2.

The knowledge obtained about the velocity field, the pressure distribution, the concentrations of particles, and the deposition allow for the evaluation of a filter design's main performance properties, which leads to a significant reduction in the number of manufactured real-world prototypes and accelerates the development process.

Among others, there are modules for

- the import of CAD geometries (see Fig. 4, left) and their conversion into appropriate computational grids,
- robust estimation of filtration model parameters from experimental data (see Sect. 4.2.8), and
- exporting the computed results to file formats that allow for effective visualization (see Fig. 4, right) and further processing using worksheets, etc.

Advances in computer hardware enable the users of simulation software to deal with more and more challenging problems, particularly in terms of memory requirements. This involves an increase in the computational cost of the simulation. Therefore, there is a need to allocate computer resources so that the focus is on relevant simulation setups and the most promising designs. FiltEST addresses this issue in two ways: For the important special case of pleated filters, a tool based on analytical methods can estimate the optimal pleat count (in terms of pressure drop) for both the clean medium and the loading stage. The analytical models are based on certain simplifying assumptions, so that the computations can be done within seconds and the relevant parameter range can be narrowed down quite quickly. The second technique is based on post-processing: Simulation results and/or measured data for a series of design parameters are collected in a data base and examined by a data mining software. Once the data miner is sufficiently "trained", it can predict the key quantities for designs that were not part of the collected data. Both approaches help to avoid wasting precious computer resources (and producing prototypes) on sub-optimal designs.

Thanks to its modular structure, FiltEST can be extended by customized solutions for specific application needs. Recent developments include non-linear pressure drop effects (see [11] and Sect. 4.2.1).

The Fraunhofer ITWM is very active in the mathematical and numerical treatment of deformable porous media [44] and of *Fluid-Porous-Structure Interaction (FPSI)* on the macroscopic level. In internal and international research projects, the numerical solvers for flow through porous media and elasticity were adapted to the specific needs of poroelasticity and combined for the coupled simulation of FPSI (see Fig. 5). The results obtained were very promising and received very positive feedback in the filtration commu-

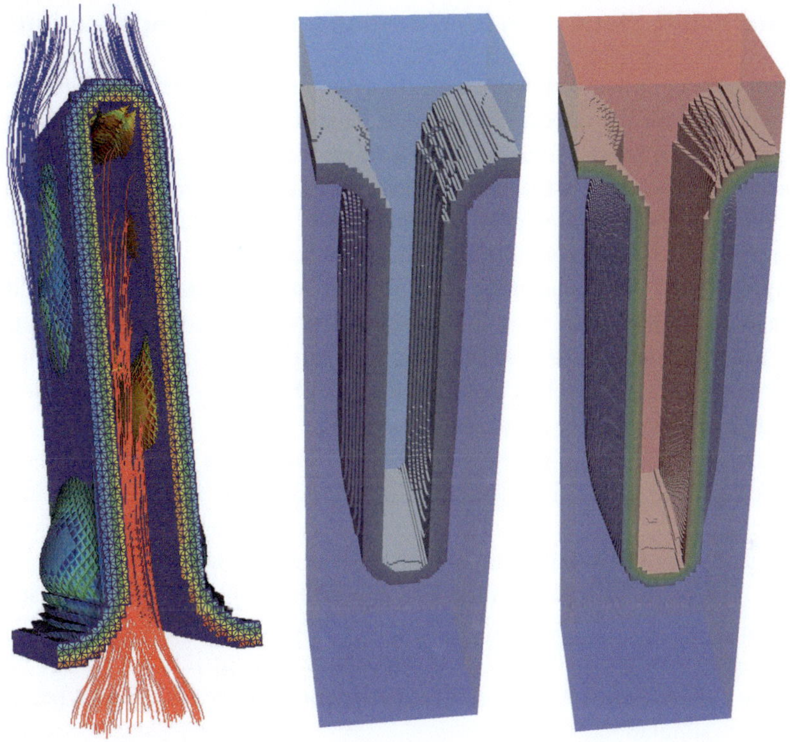

**Fig. 5** Fluid-Porous-Structure Interaction: Simulation results for the deformation of a clean filter pleat (*left*) and a partially loaded pleat with lower permeability (*right*)

nity (see [1, 13]). Some recent numerical results, as well as comparisons with lab measurements, can be found in [3].

Analogously to 3D solid mechanics models, dimension-reduced, plate models are also available for poroelasticity (see [95]). This was one of the motivations to start with the scientific investigation of FPSI on a broader basis. In the French–German Fraunhofer–Carnot project FPSI_Filt, the phenomenon has been studied analytically, numerically, and experimentally by combining the expertise of the Fraunhofer ITWM, the Department of Mathematics at the University Lyon, and the Laboratory for Mechanics of Fluids and Acoustics (LMFA), in Lyon. In this framework, new poroelastic plate models were rigorously derived (see [84]) and implemented in a numerical software. Validation against known exact solutions and experiments shows that the approach is very effective (see [8, 9]). Ongoing work is being done on the derivation of poroelastic shell models, further improvement of the coupling algorithms, and the theoretical and experimental study of turbulent flows near and through porous media.

Recently, algorithms and software for simulating cross flow filtration and reverse and forward osmosis processes were developed. A short description of these algorithms and software, as well as first numerical results, can be found in [4].

### 3.3 Multiscale Simulation for Filtration and Similar Applications

As stated at the beginning of this section, the phenomena at the microscopic level have a strong influence on the effects observed at the macroscale. However, the converse is also true: In many industrial filter element designs, the filtering medium is supported by wired meshes or rib structures that influence the flow field. Consequently, even if the clean medium can be regarded as a macroscopic homogeneous continuum, this will not remain true when the medium is non-uniformly loaded with particles.

A straightforward, direct numerical simulation of an entire filter element would be very costly, since the whole range from nanometers (fibers, dirt particles) up to centimeters (housing, inlet and outlet pipes) would have to be treated on a common computational grid. A more efficient approach is to do modeling and simulation for each scale and then to couple them in a multiscale simulation approach.

Numerical upscaling and coupled micro–macro-simulations are powerful tools for obtaining a full picture of the filtration process. The former is a well-established field in applied mathematics, with a long list and history of applications. In the filtration context, a first application is the microscale simulation of the flow through the porous filtering medium. A representative volume of the filtering medium with resolved micro-structure is selected. This subdomain is chosen for the numerical solution of the so-called *cell problem* in order to get the *effective* permeability of the volume (see e.g. [20, 29] and the references therein). The obtained effective value of the permeability serves as an input parameter for the macroscale simulation.

An approach that has asymptotically the same complexity but allows one to recover more details of the fine scale solution (or even the complete fine scale solution) is proposed in [18]. A general framework for multiscale problems, based on the variational multiscale method, and utilizing iterations between scales, is developed there. It deals with upscaling a Stokes–Brinkman problem to a Stokes–Brinkman problem and includes the concept of recalculating the "permeability" of coarse blocks. However, it does not deal with particles and changing geometries. One-way coupling from microscale (Navier–Stokes–Brinkman) to macroscale (Navier–Stokes–Brinkman) was considered in [16], but it deals only with the flow and does not consider back-coupling from macroscale to microscale. The above methods are still too expensive to be applied to the filtration problem considered here, and in general, they do not consider Navier–Stokes equations and particle tracking at the microscale. A truly multiscale model is considered in the recent paper [15], see also Sect. 4.3 for details. In fact, heterogeneous multiscale methods (HMM) can also be considered just as a general approach for solving multiscale problems. In this sense, our developments could be classified as HMM for filtration problems. More details on multiscale simulation and examples of their application in filtration will be given in Sects. 4.2.9 and 5.1.

Based on preliminary results in [41], recent progress has been made in the field of multiscale simulations of filtration processes (see [15]): Beginning with a macroscopic computation of the flow through the entire filter element, the macroscopic flow field is used as input data for microscale simulations that compute the flow and filtration efficiency in properly selected subdomains. The latter are located at "critical" points of the

**Fig. 6** *Left*: Press section of a paper machine (Foto: Voith GmbH). *Right*: Microstructure simulation of a press felt

filter element. The results obtained at these special subdomains are interpolated across the whole domain occupied by the filtering medium, respectively, and used for another flow simulation on the macroscopic level.

Another topic is the design of the pressing section of a paper machine is a specific industrial application in which multiscale models are used for the virtual design of press felts. During mechanical dewatering of the wet paper in the press nip, the water is squeezed into a porous felt. At the Fraunhofer ITWM, a complete multi-scale simulation for the press nip, including micro-structure models of press felts, has been developed [28]. Today, these simulation tools are used in the paper machine industry to develop new press felts and to virtually test the virtually developed felts in a multiscale press nip simulation for different paper machine configurations. Essential research questions about multi-scale-modeling and the numerical simulation of fast flows in thin layered porous media in press nips of a paper machine application are answered in [19, 46, 47].

During the last years, multiscale models and efficient simulation methods have been developed at the Fraunhofer ITWM within the framework of PhD theses on filtration and separation processes [41–43, 46–48, 51] as well as on other similar industrial applications [39, 40, 45, 49, 50, 52].

Summing up, it can be stated that a variety of models, algorithms, and software tools have been developed at the Fraunhofer ITWM to close the gap between the (rather) incomplete mathematical research on filtration and the industry's need for systematic studies and problem-adapted solutions. For more than ten years, the work done on this matter at the Department of Flow and Material Simulation has had a significant impact on mathematical research and the development of methods and simulation tools.

# 4 Modeling and Simulation of Filtration Processes on Different Length Scales

In this section, we will give an overview of some mathematical models and numerical methods for simulating the flow through filter media and the filtration of dissolved particles. We will exclusively consider the so-called *dead end filtration*, for which one distinguishes between the following two cases/phases:

- *Depth filtration:* The particles penetrate the filter medium, where some of them are deposited.
- *Cake filtration:* The particles are captured at the upstream interface of the fluid and the medium. The deposit forms a so-called *filter cake* on the surface of the medium. In general, this cake contributes to the filtration process.

We start with the treatment of filtration phenomena on the level of the particles and pores in the medium. After the presentation of corresponding models for the flow and deposition, some examples for the computer-aided investigation of the filtration efficiency and pressure drop will be discussed. The second subsection is devoted to the macroscopic level of the entire element, especially to depth filtration and the corresponding change in the permeability of the medium. After a quite short discussion of the numerical approximation, a robust method for the estimation of the filtration model parameters will be presented. The third subsection treats multi-scale methods for filtration.

## 4.1 Modeling and Simulation of Filtration Processes at the Pore Scale

In this section, we present the mathematical modeling and computer simulation of filtration processes at the pore scale. Since the main idea of microstructure modeling is the calculation of the filtration processes on the real pore structure of the filtering media, the inputs for the geometry are highly resolved 3D images. The images (CT, FIB-SEM) are usually represented on a structured tensorial gird; in most cases, on a regular voxel grid. Since the modeling and simulation should be able to work with large 3D images, the modeling of virtual filtration processes and the numerical simulation techniques are closely related and make essential use of the regular data structure.

Therefore, this section combines modeling and simulation aspects and is organised as follows: First, the basic principles of virtual geometry generation are introduced. Then, we explain the approach for the flow field and particle filtration model. Finally, we explain our iterative Euler–Lagrangian approach to solve for the flow, particle transport, and deposition.

### 4.1.1 Modeling and Simulation of Virtual Filter Media Geometries

The starting point of any filter simulation is a realistic three-dimensional computer model of the geometry. With regard to virtual material design, the possibility of relying on purely computer generated structures is essential. In Figs. 7 and 8, four virtual structures are

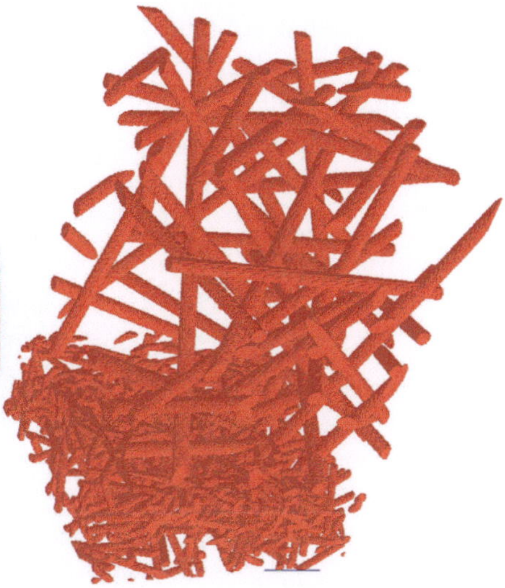

**Fig. 7** Virtual structures: woven structure (*left*) and two layered nonwoven (*right*)

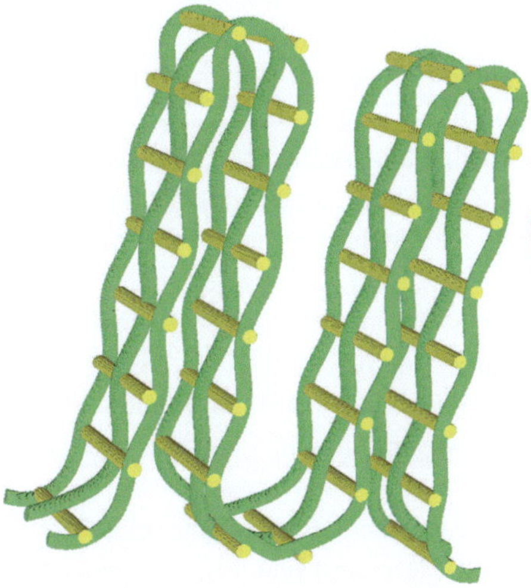

**Fig. 8** Virtual structures: wire mesh (*left*) and sinter ceramic (*right*)

shown that can be used in a filtration application. The geometries are generated by the Fraunhofer software GeoDict [35].

A crucial point when simulating filter media is that one must resolve the entire thickness, since the structure is possibly graded initially, but will become inhomogeneous due to particle loading. The modeling of virtual structures is based on the theory of deterministic and stochastic modeling of 3D images [88]. Input information for the virtual modeling may be obtained by the geometric analysis of real material images, as described in [26]. In general, all geometries are modeled on regular cubical meshes. This approach requires a huge number of cells, which are called voxels. On the other hand, extremely efficient algorithms exist that exploit the highly structured mesh. Moreover, it ensures high flexibility, so that the simulation chain can also be fed by tomographic data sets.

Subsequently, the methods for generating these structures are described.

**Virtual Nonwoven** The generation of virtual nonwoven is based on a stochastic Poisson point process [26, 88]. The generator is controlled by a set of parameters that is readily interpreted geometrically. For instance, the porosity or fiber volume fraction is selectable. Additionally, several fiber properties, such as densities, lengths, cross sections, and orientations can be prescribed. Finally, the resolutions of the underlying voxel mesh and the overall dimension have to be entered. After specification of these parameters, the fiber generation starts by randomly computing fiber axes and entering the fibers into the voxel mesh. The generation is stopped when the specified fiber volume fraction is reached. If the volume fraction could not be reached, the initialization of its random number generators is changed and the process starts again. The algorithms do not work completely randomly, but are designed to guarantee prescribed properties within a selectable tolerance. On the other hand, by adjusting the initialization of the random number generator, all geometries are reproducible.

In view of the simulations, the dimensions should be chosen sufficiently large to give representative results. Representative means that the results do not change when the dimensions are further enlarged. To give an idea of what this means, the geometries in Fig. 7 possess this property with respect to flow simulations. For filter simulations, to be representative frequently implies to entirely resolve the medium in flow direction. The size of the required mesh may reach several million voxels: Let us consider a medium with a thickness of 1.5 mm and smallest fiber diameter of 20 μm. To ensure reasonable results in a flow computation, the smallest fiber diameter should be resolved by at least 4 voxels. Hence, the edge length of a voxel is 5 μm, and we need 300 voxels in the flow direction. Having approximately the same lateral dimension, we end up with 27 million voxels.

**Virtual Woven** Virtual woven structures require precise deterministic rules following the weaving pattern of their real counterparts. On the left hand side, Fig. 7 shows a virtual woven structure possessing a basket weave pattern. Moreover, the yarns consist of many thin fibers. These fibers do not follow a strict deterministic rule, but have some built-in randomness reflecting certain irregularities also present in the real woven material.

**Virtual Sinter Structures** The generation of sinter structures comprises two steps: First, a stochastic point process [26] is used to create packings of spheres and cylinders. To achieve satisfactory results, the shape and size distribution of the real sinter grains are compared and matched with the virtual distributions as well as possible. During the second step, morphological operations [88] are applied to generate the sinter necks. Iteratively, using the operations dilatation and erosion, one creates exactly the intended connectivity.

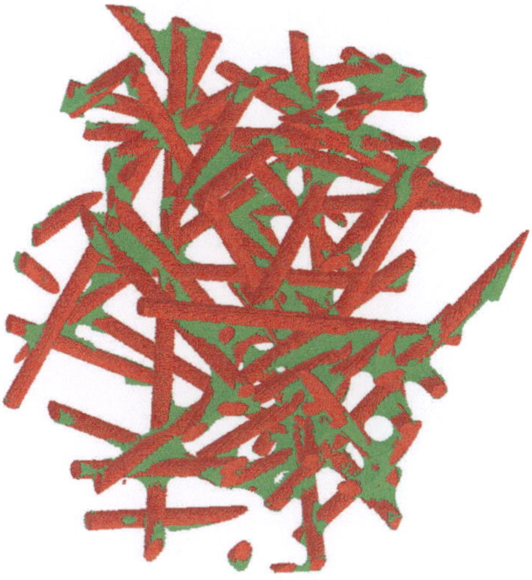

**Fig. 9** Virtual nonwoven with binder

**Complex Geometries** The methods presented in the previous sections can be considered as elementary building blocks for more complex geometries. The voxel mesh approach naturally allows for combining layers of elementary structures. Thus, media having gradients with respect to some property are easily created (Fig. 7, right). Another interesting example is the nonwoven with binder in Fig. 9. The binder is added into the nonwoven in complete analogy to the sinter necks in the previous section.

Regarding virtual material design, we want to finally mention an interesting opportunity for tomographic data sets. If one is interested in redesigning a certain layer of an existing medium, one can substitute this layer by a virtual structure. The effect of the replacement can then be studied by simulation.

**Comparison and Validation of Real and Virtual Structures** For virtual structures that are intended to reproduce existing media, quality measures are needed. Quite often, 2-D SEM (= scanning electron microscope) images or even 3-D tomography data of a real sample are available. This information can be used to compute and compare various geometrical properties, such as porosity, cord length distribution, pore size distribution, specific surface area, fiber orientation, etc. In industry, permeability tests of porous samples are standard. Moreover, flow simulations in typical porous media regimes, i.e., slow flow regimes, are very reliable. Hence, measured and computed flow properties can be compared quite confidently and provide meaningful results.

### 4.1.2 Microstructure Modeling of Filtration Processes

Depending on the filtration application in question, the simulation efforts to compute certain filter properties may differ significantly. In some situations, a single flow simulation is sufficient to determine the requested effective permeabilities or flow resistivities. Computing filter efficiencies also requires the solution of particle transport through the medium. Certainly, the most demanding application is the simulation of an entire filter lifetime.

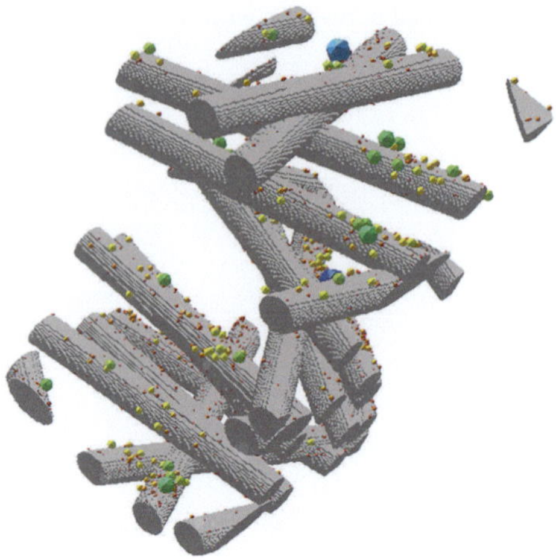

**Fig. 10** Simulation of a fibrous medium with deposited dust particles

In general, we can assume that the time scale of the flow-field changes much more slowly than the time scale of the corresponding particle transport, due to the small size of the particles. Hence, the initial stationary flow field for the clean medium is computed, a certain number of particles are tracked and, in case of collisions, deposited. After a while, the influence of the deposited dust can no longer be neglected and it is time to recompute the flow field. This iterative algorithm is repeated until a certain pressure drop is reached, for instance. At the end of Sect. 4.1.3 subsection, a simulation of a realistic diesel particulate filter medium is shown.

**Modeling of Filtration Processes** Slow flow regimes are typical for most filtration processes. Hence, flow solvers for the solution of the Stokes equations are well-suited for the simulation. The Stokes equations describe incompressible viscous flow for low velocities, i.e., when inertia is negligible:

$$-\mu\Delta\mathbf{u} + \nabla p = \mathbf{f} \quad (\textit{conservation of momentum}) \tag{1}$$

$$\nabla\cdot\mathbf{u} = 0 \quad (\textit{conservation of mass}) \tag{2}$$

$$+\ \textit{boundary conditions}. \tag{3}$$

In (1) and (2), $\mathbf{u}$, $p$, $\mathbf{f}$ denote the velocity vector, the pressure and the external body force, respectively. To solve the system, boundary conditions have to be prescribed, e.g., velocity profiles at the inlet and outlet of the computational domain.

For high velocity flows, the incompressible Navier–Stokes equations should be used. This system is quite similar to (1), (2), but contains an additional convective term accounting for inertia effects:

$$-\mu\Delta\mathbf{u} + (\rho\mathbf{u}\cdot\nabla)\mathbf{u} + \nabla p = \mathbf{f} \quad (\textit{conservation of momentum}) \tag{4}$$

$$\nabla\cdot\mathbf{u} = 0 \quad (\textit{conservation of mass}) \tag{5}$$

$$+\ \textit{boundary conditions}. \tag{6}$$

In Eq. (4), $\rho$ denotes the fluid density. Combining free and porous flows enables the modeling of diesel particulate filters (see Sect. 4.1.3), where deposited soot particles are not resolved by voxels, but modeled as porous media. For this type of application, we employ the Navier–Stokes–Brinkman equations [42]:

$$-\mu\Delta\mathbf{u} + (\rho\mathbf{u}\cdot\nabla)\mathbf{u} + \mu K^{-1}\mathbf{u} + \nabla p = \mathbf{f} \quad \textit{(conservation of momentum)} \tag{7}$$

$$\nabla\cdot\mathbf{u} = 0 \quad \textit{(conservation of mass)} \tag{8}$$

$$+ \textit{ boundary conditions.} \tag{9}$$

$K^{-1}$ is the reciprocal of the permeability of a porous medium. In the free flow domain, the permeability is infinite, simplifying (7) into (4).

In the porous medium, where $K^{-1}$ is quite large, $\mathbf{u}$ is small. In such a case, the first two terms in (7) are negligible, and we obtain Darcy's law:

$$\mathbf{u} = -\frac{K}{\mu}(\nabla p - \mathbf{f}). \tag{10}$$

Darcy's law was found experimentally in 1856 [66]. It expresses the linear relation between velocity and pressure drop for slow flows in porous media.

**Modeling of Particle Transport and Deposition** The first step in computing initial filter efficiencies is the computation of the fluid flow in the virtual geometry (see Sect. 4.1.2). The second step consists of particle tracking. Here, we make certain assumptions: The particles are spherical, there is no particle-particle interaction (= low particle concentration), and the particles do not influence the flow field. After specifying the particle size distribution and a few additional parameters, particle motion is modeled by Newton's Second Law:

$$\mathbf{F} = m\mathbf{a}, \tag{11}$$

where $\mathbf{F}$ denotes the force exerted on the particle, $m$ is the particle mass, and $\mathbf{a}$ is the particle acceleration. The particle moves due to its inertia, due to fluid friction, and due to Brownian motion. Additionally, an electrostatic force may influence the particle trajectory. For brevity, we refer to [24, 31] for further details on electrostatics. Besides inertia, which is inherent to (11), all effects are modeled as a superposition of forces. We finally solve the following system of stochastic differential equations:

$$d\mathbf{v} = \gamma\big(\mathbf{v}(\mathbf{x}(\mathbf{t})) - \mathbf{u}(\mathbf{x}(\mathbf{t}))\big)dt + \sigma d\mathbf{W}(t) + \frac{q\mathbf{E}(\mathbf{x}(t))}{m}dt, \tag{12}$$

$$\frac{d\mathbf{x}}{dt} = \mathbf{v}. \tag{13}$$

In (12) and (13), $t$, $\mathbf{x}$, $\mathbf{v}$ and denote time, position, and velocity of the particle, respectively. The first term on the right hand side of (12) describes the force due to friction. It is proportional to the difference of the particle velocity and the fluid velocity. The coefficient is given for slow flow and spherical particles by:

$$\gamma = \frac{6\pi\mu R}{m} \quad (\textit{Stokes friction}) \tag{14}$$

Here, $R$ denotes the particle radius.

The second term on the right hand side of (12) models Brownian motion by a three-dimensional Wiener-process $\mathbf{W}$. Let $T$ be the temperature and $k$ be the Boltzmann constant. Then, we have, by the fluctuation-dissipation theorem:

$$\sigma^2 = \frac{2k_B\gamma T}{m}. \tag{15}$$

For further details of the model, we refer to [23, 38]. The last term models the influence of an electric field $\mathbf{E}$ on particles with charge $q$.

### 4.1.3 Simulation of Filter Media

As mentioned previously, only numerical methods that make essential use of the regular voxel structure are used for industrial applications, due to the complex geometries of the real or virtual geometric structure. The natural approach for the numerical solution of systems (1) and (2) or, in general, (7) and (8) is the Lattice–Boltzmann method [7]. Lattice–Boltzmann methods make use of the relation between the Boltzmann equation and the (Navier–)Stokes equation in a discrete way on a regular voxel grid. The primary quantity is a discrete distribution function and the velocity and pressure are moments of the distribution.

Therefore, the method can be directly applied to the voxel grid and it uses an explicit update rule to converge to the stationary solution. Since at least the discrete distribution function must be stored on each voxel, other matrix free methods are alternatively considered.

Actual industrial simulation techniques make use of a finite volume or finite difference discretization on the voxel grid and solve the resulting system with the fast Fourier transform [36]. Concerning the implementation of all solvers, we want to remark that two ingredients are of predominant importance:

- exploiting the regularity of the voxel meshes and avoiding any kind of overhead to restrict the storage and
- scalable parallelization of the algorithms.

Keeping both ingredients in mind, one can achieve computation times ranging from minutes up to a few hours on modern multi-core workstations to solve the CFD problems.

The particle transport equations (12), (13) can be independently solved for each single particle by an implicit Euler method. In addition to the computation of the particle motion, we have to check for collisions of the particles with the geometry in each time step. If a collision is detected, the particle stops and it is marked as deposited. After the simulation

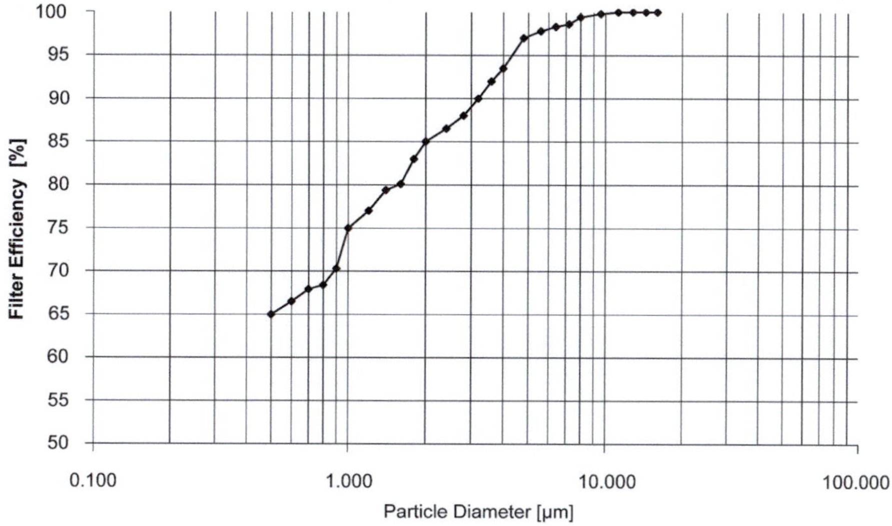


**Fig. 11** Simulation of the initial filter efficiency of an air filter medium

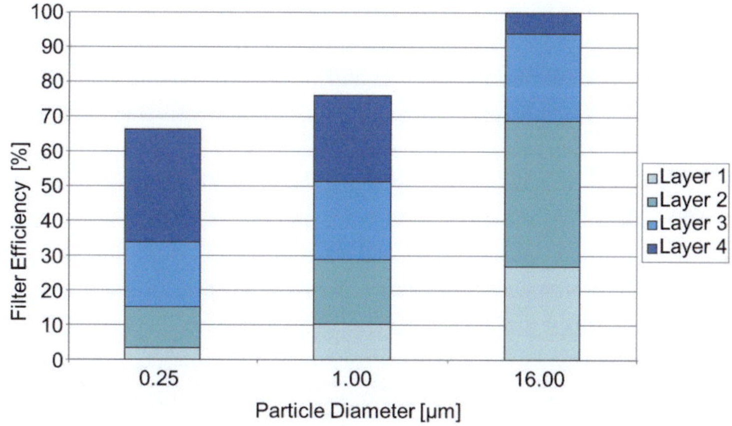


**Fig. 12** Cumulative initial filter efficiency of a four-layered air filter medium

run, we calculate for each particle size the fraction of deposited to total number of particles (see Fig. 11). Since the simulation provides complete information about each individual particle, it is possible to create diagrams like Fig. 12. To date, this detailed analysis is unavailable experimentally for many filtration applications.

**Simulation of Filter Lifetimes—Solid and Porous Deposition Mode** Filter lifetime simulations iteratively compute the flow field and particle transport as introduced in Sect. 4.1.2. An essential third simulation component comes into play, since we need to *modify* the geometry due to deposited particles. Therefore, we keep track of the volume fraction filled by deposited particles for every voxel of the geometry. Basically, there exist two modes how volume fractions influence forthcoming simulations. The first mode is called *solid deposition mode*. It is intended to be used whenever particle diameters are greater than or equal to the voxel length. Hence, particles are resolvable by the voxel mesh. When the volume fraction of a voxel reaches 1, it is marked as *solid*. The flow solver treats this voxel as an obstacle, and the particle tracking treats it as a collision voxel, where a particle may deposit. The second mode is called *porous deposition mode*. It is used when particles are much smaller than a voxel and, hence, build up porous substructures. Depending on its volume fraction, a permeability value is assigned to the voxel. Consequently, the flow computation is based on the Navier–Stokes–Brinkman approach (7), (8). Particle deposition in a porous voxel is possible as long as a prescribed maximum volume fraction is not reached. When a voxel exceeds the maximum fill-level, it becomes a collision voxel. Obviously, filter lifetime simulations using the porous deposition model depend on the prescribed permeability and maximum volume fraction. In Sect. 4.1.3, we show how the parameters can be obtained by highly resolved single fiber simulations.

**Example: Simulation of a Diesel Particulate Filter** The aim of this section is to simulate the evolution of the pressure drop of the diesel particulate filter medium in Fig. 13. We will briefly summarize the essential steps and refer to [32] for further details.

The clean medium in bright grey consists of a ceramic substrate with an additional fiber layer. Both geometries are purely virtual and are created by applying the methods described in Sect. 4.1.1. The resolution is 1 μm, and the dimension of the geometry is $150 \times 150 \times 650$ voxels. With respect to porosity and cord length distribution, the virtual structure has quite similar properties to its real counterpart. The comparison is done using SEM images. Moreover, the simulated initial pressure drop is in good accordance with the measured pressure drop. Since the particle diameters vary between 20 and 300 nm, the filter lifetime simulation is run in the porous deposition mode. To determine the parameters of the subgrid model, i.e., the maximum volume fraction and permeability of the porous medium, highly resolved single fiber simulations in the solid deposition mode are performed (see Fig. 14). The voxel length is set to 10 nm.

We determine the permeability and the maximum volume fraction by investigating the porous layer in the upstream direction of the fiber. We obtain 15 % as the maximum volume fraction and a corresponding permeability of $10^{-3}$ Darcy. Both parameters are then used in the filter lifetime simulation of the filter medium. Figure 15 shows the characteristic S—profile when filtration switches from depth to surface filtration. The results are in good qualitative agreement with measurements. Repeating the same simulation with the ceramic substrate only, i.e., without the fiber layer, we observe a slightly reduced initial pressure drop, but a much faster and unwanted transition to surface filtration. Thus, the simulation qualifies the design with the fiber layer as the better medium. The same result is achieved experimentally.

**Fig. 13** Simulation of soot deposition in a diesel particulate filter medium

**Fig. 14** Simulation of soot deposition on a single fiber

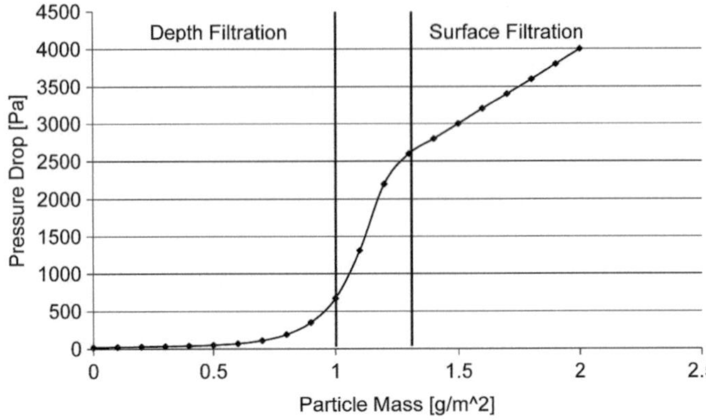


**Fig. 15** Typical pressure drop evolution of a diesel particulate filter

## 4.2 Modeling and Numerics for the Macroscopic Scale

The principal tasks of mathematical modeling on the macroscopic level are the same as for the micro-scale: Based on an appropriate description of the domain, one has to identify suitable models for the flow, for particle transport, and for particle deposition in the filtering media. Compared to the microscale problem, the main differences in the macroscopic approach are the following:

- A filtering medium is described in the spirit of Darcy's law (10) as a homogenized porous material with a permeability $K$, i.e., in contrast to the microscopic description, there is no distinction between pore spaces and the solid regions of the fibers.
- Regarding transport and deposition of dirt from the perspective of simulation, it would be very costly to deal with the contamination of the fluid in terms of individual particles. Moreover, such detailed information is usually not required on this scale. Therefore, the distribution of dirt is described via the concentration of particles.
- The previous two aspects require a different approach to account for the change in permeability of the filtering media due to the loading with dirt: If the clean medium is already treated as a homogenized continuum with no distinction between pore spaces and solid skeleton, then there is no way to "add" the captured dirt to the solid part in the domain. In other words, the constriction of the pore spaces due to the deposition requires a different modeling approach on the macroscopic level.

We will discuss the following aspects of modeling and simulation of macroscopic filtration:

- models for slow and fast flows through the filter element,
- various models for the deposition of dirt in the filtering medium with the focus on depth filtration processes,

- an approach to estimate the deposition model parameters from given measurement data,
- models for the time evolution of the permeability of the loaded media,
- numerical algorithms for the simulation of flow and particle transport and capturing at the scale of filter element, and
- validation and calibration of simulation tools together with their benefits for the filter element design.

For the sake of self-consistency in this section, we will repeat some formulas already given in the corresponding section on the microscopic models.

### 4.2.1 Macroscopic Modeling of Filtration Processes

On the level of filter elements, the description of the domain under consideration is somewhat different from the microscopic case. On the macroscopic scale, one has to take into account the geometric features of the filter element's housing and the filtering media. Moreover, the inflow and outflow regions can have a non-trivial shape. Therefore, it seems obvious to decompose the filter element's domain $\Omega^{\text{macro}} \subset \mathbb{R}^d$ into (at least) three parts:

- the subdomain $\Omega_f^{\text{macro}}$, occupied by the "free" fluid outside of the filtering media,
- the solid part $\Omega_s^{\text{macro}}$, occupied by the filter housing, ribs, supporting meshes, etc., and
- the porous subdomain $\Omega_p^{\text{macro}}$, occupied by the filtering media.

Obviously, it holds that

$$\Omega^{\text{macro}} = \Omega_f^{\text{macro}} \cup \Omega_s^{\text{macro}} \cup \Omega_p^{\text{macro}},$$

$$\Omega_f^{\text{macro}} \cap \Omega_s^{\text{macro}} \cap \Omega_p^{\text{macro}} = \emptyset.$$

Examples of filter element geometries are shown in Fig. 16. It is worth noting that depending on the field of application, a filter element can contain several filtering media of different types. So, in general, one has

$$\Omega_p^{\text{macro}} = \bigcup_{i=1}^{N_m} \Omega_{p,i}^{\text{macro}}, \qquad \bigcap_{i=1}^{N_m} \Omega_{p,i}^{\text{macro}} = \emptyset,$$

where $N_m$ denotes the number of filtering media involved and $\Omega_{p,i}^{\text{macro}}$ is the subdomain occupied by the $i$-th filtering medium.

The boundary $\Gamma = \partial\Omega^{\text{macro}}$ also decomposes into several parts:

$$\Gamma = \Gamma_i \cup \Gamma_o \cup \Gamma_s, \tag{16}$$

where $\Gamma_i$ is the inlet (or inflow) boundary, $\Gamma_o$ is the outlet (or outflow) boundary, and $\Gamma_s$ is the solid (or wall) boundary part. In some cases, there are several inlets and/or outlets and, depending on the application, pressure-controlled bypass valves may also be part of the filter housing.

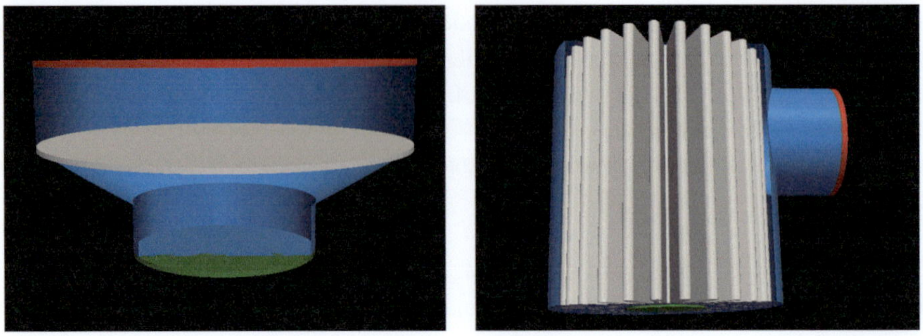

**Fig. 16** Simple geometries of filter elements. *Left*: A conical filter element with a flat disc-shaped filtering medium. *Right*: A pleated cartridge element. The fluid part $\Omega_f^{\text{macro}}$ of the domain is colored *blue*, the porous part $\Omega_p^{\text{macro}}$ is shown in *white*, the inlet is marked in *red*, and the outlet in *green*

### 4.2.2 Macroscopic Flow Models

Assuming that the fluid under consideration is incompressible and that the flow is laminar, there are several ways to describe the flow inside the filter element:

- The decomposition of the domain into a fluid region $\Omega_f^{\text{macro}}$ and the porous subdomain $\Omega_p^{\text{macro}}$ suggests a coupled system of Navier–Stokes and Darcy equations.
- Alternatively, one can use the Navier–Stokes–Brinkman equations as a flow medium.

Usually, (clean) filtering media are highly porous, such that the usage of Brinkman models is justified. Since we will focus on industrial applications, the coupled Navier–Stokes and Darcy system will not be discussed here. A more detailed argumentation and comparison of the two formulations can be found, e.g., in [42]. With the notations used so far, the unsteady Navier–Stokes–Brinkman equations read

$$\nabla \cdot \mathbf{u} = 0,$$
$$\rho\left(\frac{\partial \mathbf{u}}{\partial t} + (\mathbf{u} \cdot \nabla)\mathbf{u}\right) - \nabla \cdot (\tilde{\mu}\nabla \mathbf{u}) + \mu K^{-1}\mathbf{u} = -\nabla p + \mathbf{f}. \tag{17}$$

Note that the momentum equation contains the *effective viscosity* $\tilde{\mu}$, which might differ from $\mu$ in $\Omega_p^{\text{macro}}$. Outside of the porous region $\Omega_p^{\text{macro}}$, it holds that $\tilde{\mu} = \mu$ and clearly $K^{-1} = 0$, such that in $\Omega_f^{\text{macro}}$, the equation is identical with the usual Navier–Stokes system.

A typical set for the boundary conditions is the following:

$$\mathbf{u}(x) = \mathbf{u}_{\text{in}}(x), \quad x \in \Gamma_i; \qquad p(x) = 0, \quad x \in \Gamma_o; \qquad \mathbf{u}(x) = 0, \quad x \in \Gamma_s. \tag{18}$$

The boundary conditions can be naturally extended to the case of multiple inlets and/or outlets.

As already mentioned, the choice of the interface conditions between the free fluid subdomain and the porous subdomain is of special importance when modeling the flow. In the case of depth filtration, when the flow is (essentially) perpendicular to the porous media, requiring the continuity of the velocity and of the normal component of the stress tensor are a natural choice (see, e.g., [42, 54]).

We refer to ([42]) for other variants and further details on this matter.

For moderate volumetric flow rates and corresponding flow speed through the filtering medium, the Navier–Stokes–Brinkman system has proven to be a suitable mathematical model. However, it has been observed that the proportionality between the pressure gradient and fluid velocity according to Darcy's law does not hold for fast flows and there are relevant application fields (e.g., air filtration) where this has to be taken into account.

There is a good deal of literature (see, e.g., [58, 82, 83] and references therein) on different models for fast flows through porous media. We shall refer to these models as *non-Darcy* models. In order to include the nonlinearity in the Navier–Stokes–Brinkman equations (17), the permeability $K$ is replaced by a so-called *apparent* permeability term $K_{\text{app}}$. We list some examples for this.

If the Darcy law is valid, then the Navier–Stokes–Brinkman equations do not need any modification. Therefore,

$$\mu K_{\text{app}}^{-1} = \mu K^{-1}. \tag{19}$$

For the classical Forchheimer model (see [73]), one has

$$\mu K_{\text{app}}^{-1} = \mu K^{-1} + \frac{\rho}{2} F|\mathbf{u}|, \tag{20}$$

with the Forchheimer coefficient $F$. The Ergun-type models (see [69, 100]) also incorporate a quadratic term. They read

$$\mu K_{\text{app}}^{-1} = \mu K^{-1} + \frac{\rho}{\sqrt{K}} E|\mathbf{u}|, \tag{21}$$

with the Ergun coefficient $E$. These two models are widely used in the engineering literature and commercial software packages for the simulation of the flow. A more recent example for non-linear pressure drop models was given by Barree and Conway (see [57]), which is based on the pore space Reynolds number

$$Re = \frac{\rho u \xi}{\mu},$$

where $u$ is the Darcy speed through the porous medium and $\xi$ is the characteristic length, related to the (average) pore size of the medium. The apparent permeability depends on the flow as follows:

$$\mu K_{\text{app}}^{-1} = \mu \left( K_{\text{min}} + \frac{K - K_{\text{min}}}{(1 + Re^F)^E} \right)^{-1}. \tag{22}$$

This model provides a single formula to cover the entire range from low to quite high volumetric flow rates. At low flow rates, this model coincides with Darcy's law and it agrees well with Forchheimer's law for intermediate flow rates. Furthermore, the model features asymptotic behavior at rather high flow rates, which indicates there is a constant permeability (or minimum permeability $K_{\min}$), which is consistent with laboratory measurements.

Based on the method of asymptotic homogenization, the authors of the paper [56] conclude that the nonlinear term cannot be quadratic, but should be cubic instead. In this context, it should be noted that some experimental papers report a linear dependence of the Forchheimer coefficient on the velocity, which means that there is a cubic term with respect to the velocity.

It is worth noting that the concept of an apparent permeability is not standard when modeling the flow through porous media. However, this approach allows for a simple and compact description of a class of models and the corresponding numerical algorithms that are able to handle fast flow scenarios. Furthermore, although fast flows are discussed here, all the considered models describe laminar flows. Possible turbulent effects are outside the scope of our considerations.

### 4.2.3 Macroscopic Models for Particle Transport and Deposition

As already mentioned at the beginning of this section, the concentration of dissolved particles is a suitable quantity for describing the distribution of dirt on the macroscopic level. For a given particle type, this can be the count per unit volume or, equivalently, the particle mass per unit volume. In the following, we will denote by $C$ the concentration of dissolved particles and by $M$ the concentration of deposited dirt. We will restrict ourselves to the mathematical treatment of depth filtration processes. The modeling of cake filtration on macroscale is not considered here, although it is a subject of very active research (see, e.g. [89, 96] and [53] as well as the references therein).

Assuming that the particles travel with the velocity field and taking into account the diffusion and deposition in the porous medium, the evolution of the concentration of dissolved particles is described as follows:

$$\frac{\partial C}{\partial t} + \mathbf{u} \cdot \nabla C - D \Delta C = \begin{cases} 0, & x \in \Omega_f, \\ -\frac{\partial M}{\partial t}, & x \in \Omega_p. \end{cases} \tag{23}$$

Here, $D$ denotes the diffusion constant. Thus, the evolution of the concentration $C(x,t)$ of dissolved particles is modeled by a convection–diffusion–reaction equation. The reactive term is non-zero in the porous regions only (depth filtration) and is given by the *absorption rate*, i.e., the time derivative of the deposited number (or mass) of particles $M$.

The following boundary conditions are usually imposed for the concentration:

$$\forall x \in \Gamma_i : C(x,t) = C_{\text{in}}(t), \tag{24}$$

where $C_{\text{in}}$ denotes the prescribed inlet concentration and

$$\forall x \in \Gamma_s \cup \Gamma_o : \frac{\partial C}{\partial n}(x) = 0. \tag{25}$$

### 4.2.4 Macroscopic Models of Depth Filtration Processes

On the macroscopic level, the filtration process is modeled by defining the absorption rate appearing on the right hand side of (23). Here, we point out that dusts consist of particle mixtures with different particle sizes and materials, so that under real operating conditions, each species can exhibit a different filtration behavior. This leads to the necessity of modeling individual filtration behavior for each of these components.

It is certainly beyond the scope of this short section to provide a complete survey of depth filtration deposition models. The discussion here is restricted to presenting a few examples that have proven useful in industrial filtration processes. A survey of classical models can be found e.g. in [77] and [97].

A very simple (yet useful) model is obtained by assuming that the absorption rate is proportional to the concentration of dissolved particles, i.e.,

$$\frac{\partial M}{\partial t} = \alpha C, \tag{26}$$

where $\alpha > 0$ is called the *absorption coefficient*. This model is valid in most cases for the initial stage of filtration, when the filtering medium is still (quite) clean. Similarly to the filter cake on the surface, the deposit in the medium can also have an influence on the capture rate. Therefore, most macroscopic depth filtration models acquire the following form:

$$\frac{\partial M}{\partial t} = \alpha \Phi(M) C - \Psi(M). \tag{27}$$

The function $\Phi$ describes the influence of the deposit on the capturing of dissolved particles, whereas $\Psi$ serves as a "loss term," modeling the desorption (washing out) of captured particles. Denoting once again by $\alpha$ the absorption coefficient of the clean filtering medium, we have $\Phi(0) = 1$ and $\Psi(0) = 0$. If the absorption rate is increased by the deposit, the upstream layers of the filtering medium will collect more dust than the downstream layers. This leads to the so-called *clogging* of the medium on the upstream side. A *linear clogging* model was introduced in [78] for water filtration and later studied in [63] for aerosols. It reads:

$$\frac{\partial M}{\partial t} = \alpha \left(1 + \frac{M}{M_0}\right) C, \tag{28}$$

i.e., we have $\Psi(M) \equiv 0$ and $\Phi(M) = 1 + M/M_0$. The absorption rate depends linearly on the deposit already present in the medium, as described by the parameter $M_0$.

As already mentioned in Sect. 4.1, many filtration processes begin with an initial depth filtration stage, which is followed by cake filtration phase (see also Fig. 15). During the transition from depth to cake filtration, one usually observes a sudden increase of the absorption rate. In order to model this transition, a *two-stage* model was introduced in [90]:

$$\frac{\partial M}{\partial t}=\begin{cases}\alpha(1+\frac{M}{M_0})C, & M\le M_1\\ \alpha(1+\frac{M}{M_0}+a\frac{M-M_1}{M_0})C, & M>M_1.\end{cases} \tag{29}$$

Once the deposit exceeds a certain value $M_1$, a second accelerated filtration stage begins. The "acceleration" is described by the parameter $a>0$. Such a behavior is well-known in aerosol filtration but, depending on the operating conditions of the filter element, it can also be observed for liquids.

Especially in liquid filtration applications, one can observe that a part of the deposit is released into the fluid. In order to describe such phenomena, the following model was introduced in [85]:

$$\frac{\partial M}{\partial t}=\alpha C-\gamma M. \tag{30}$$

Here, $\gamma>0$ denotes the *release coefficient* (re-entrainment parameter). The models discussed above describe depth filtration, which was the main topic for the Fraunhofer ITWM in studying macroscopic filtration processes. Recently, research was initiated on the macroscopic modeling and simulation of cake filtration, and more importantly, on combined cake filtration (for large particles) and depth filtration (for small particles). First results can be found in [12].

### 4.2.5 Models for Permeability

From our considerations of the microscopic level, we know that captured particles lead to a constriction of the pore spaces in the medium, which results in increasing flow resistance.

Therefore, the permeability has to be treated as a quantity depending on both space $x$ and time $t$. The question arises as to how to transfer information about the deposits to the changed permeability. Experimental and theoretical studies have shown that the (local) permeability strongly depends on the (local) *porosity* of the filtering medium

$$\phi=\frac{V_{\text{pores}}}{V}, \tag{31}$$

which can be extended to a porosity distribution $\phi(x)$ by considering representative elementary volumes in the medium. The majority of macroscopic permeability models are of the form

$$K(x)=r^2F\big(\phi(x)\big),$$

where $r$ denotes the (average) radius of the fibers in fibrous media or the particles in granular media, respectively.

A well-known example for such a permeability model is the one derived in [79] for fibrous porous media:

$$K_{JJ}(\phi)=-r_{\text{fib}}^2\frac{3}{20}\frac{\ln(1-\phi)+0.931}{1-\phi}. \tag{32}$$

For granular porous media, the Kozeny–Carman model is widely used (see e.g. [83]):

$$K_{KC}(\phi) = r_{\mathrm{par}}^2 c_{\mathrm{par}} \frac{\phi^3}{(1-\phi)^2}, \tag{33}$$

where the parameter $c_{\mathrm{par}}$ is related to the sphericity of the particles.

When a fibrous medium is loaded with (more or less) spherical particles, the contribution of the deposit to the total flow resistivity will in general differ significantly from the resistance of the medium, such that the use of a permeability law for fibrous materials or for purely granular materials cannot be expected to produce good results. It therefore seems both natural and promising to use models combining the above two permeability models,

$$K(t) = \left( \frac{1}{K_{\mathrm{clean}}} + \frac{1}{K_{\mathrm{load}}(\overline{\phi_+}(t))} \right)^{-1}, \tag{34}$$

i.e., the permeability component due to the deposit is assumed to depend on the increment of solid volume fraction $\bar{\phi} = 1 - \phi$ at time $t$:

$$\overline{\phi_+}(t) = \bar{\phi}(t) - \bar{\phi}(0).$$

An example for such a combined permeability model was introduced in [1]. The impact of the use of such combined permeability models on the quality of numerical simulation results will be shown in Sect. 4.2.9.

Note that, in general, the permeability depends not only on the porosity, but also on the microscopic structure of the filtering media and the deposited particles. This suggests the use of homogenization methods for the computation of the permeability and in fact, these methods have proven to produce accurate results. Examples of these multiscale models and corresponding algorithms for filtration processes are discussed in Sect. 4.3.

### 4.2.6 Numerical Algorithms for the Simulation of Filtration Processes on the Macroscopic Scale

In most cases, the time scale for the consideration of filtration phenomena on the pore scale is very different from the time scale used for the modeling of filtration on the macroscopic level: The deposition of individual particles (which is typical for the pore level modeling) does not immediately lead to a significant change in the permeability of the homogenized porous medium (which is characteristic for modeling on the filter element scale). Therefore, the following quasi-stationary approach has proven to be very effective for computer simulation on the macroscopic scale (see Fig. 17):

1. Compute the flow field by solving the Navier–Stokes–Brinkman equations (17).
2. Solve the Transport-Diffusion-Reaction equation (23) and compute the captured mass $M(x,t)$ in the filtering medium.
3. Update the porosity $\phi(x,t)$ and, using a suitable model, the permeability $K(x,t)$ of the filtering medium.

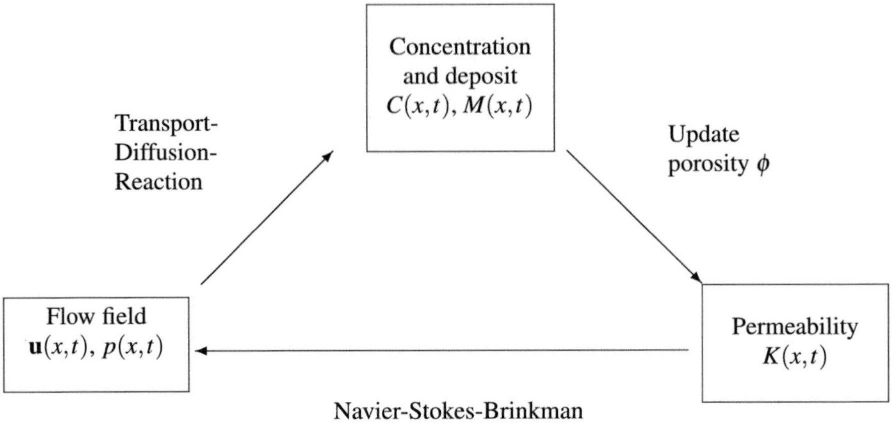


**Fig. 17** The algorithmic principle for filtration simulations on the macroscale

4. Repeat steps 2. and 3. until the (average) permeability of the filter medium has changed to a certain degree since the last computation of the flow. Once this is the case, the flow field cannot be regarded as valid any longer and the algorithm continues starting from step 1.

Specifying a threshold for the permeability change is already a first step towards computational efficiency, since the computation of the flow field can be very costly. On the other hand, an up-to-date knowledge of the flow field is crucial to a correct prediction of the particle transport to and through the medium.

A comprehensive discussion of numerical algorithms for Navier–Stokes–Brinkman equations and for convection–diffusion–reaction equations is certainly beyond the scope of this text. We will restrict ourselves to algorithms which have proven effective in the context of filtration simulation.

### 4.2.7 Space and Time Discretization

As usual in Computational Fluid Dynamics (CFD), the following steps are involved in the numerical simulation of filter elements:

- Generation of a computational grid (in most practical cases, based on CAD data),
- Time stepping and spatial discretization of the Navier–Stokes–Brinkman equations (17) on this grid,
- Time stepping and spatial discretization of the concentration equation(s) (23) on this grid, and
- solution of the (usually) large-scale linear systems obtained by these discretizations.

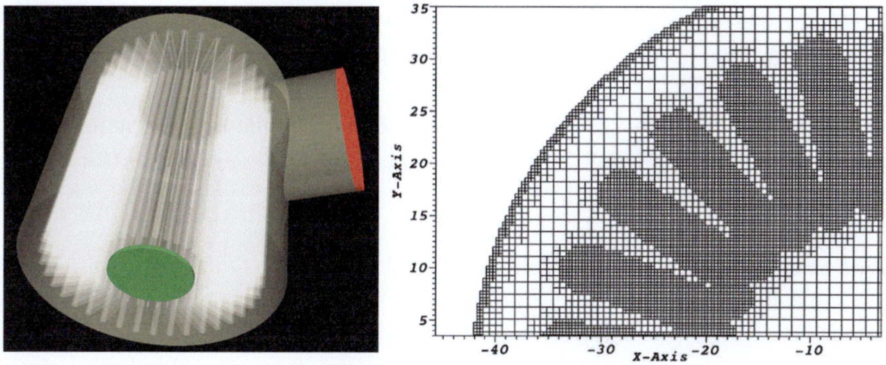


**Fig. 18** *Left*: A CAD geometry of a pleated cartridge filter element with housing, filtering medium (*white*), inlet boundary (*red*) and outlet (*green*). *Right*: Three-level grid for the simulation of the pleated cartridge (magnified part of the cross-section of the 3D grid)

**Grid Generation** Various structured and unstructured grids are used in CFD, (see e.g. [71, 72, 102] and references therein). As mentioned in Sect. 3.2, the grid generation for complicated filter element geometries can be a quite challenging task. Here, we restrict our considerations to simple grids which allow for both easy and relatively robust generation algorithms, i.e., we present the numerical discretization on uniform Cartesian grids based on voxel cells and grids obtained from these grid by local refining or local coarsening (see Fig. 18).

The number of grid cells is directly related to the number of unknowns and therefore, the computational cost. Therefore, the *grid size* should be minimized. On the other hand, a sufficient resolution of geometrical features is crucial to the accuracy of simulation results. A coarse grid can accelerate the computations, but the quality of the results can be rather poor. A reasonable trade-off is the decomposition of the computational domain into subregions requiring a fine grid resolution and other parts allowing for a relatively coarse grid.

When applying this local grid resolution strategy to filtration simulation, material interfaces and boundaries have proven to be suitable criteria for the selection of the subregions:

- Near the boundary $\Gamma$ of the domain, the finest resolution is used for the grid, in order to account for boundary effects.
- In the porous region $\Omega_p^{\mathrm{macro}}$ of the filter medium, the finest resolution is used, so that the pressure drop across the medium and the local changes in the particles' concentration and the distribution of the captured mass are resolved with good accuracy.
- The fluid-porous interface is also resolved on the finest level in order to account for effects on the pressure and concentration at the material interfaces.
- The remaining parts of the fluid subregion $\Omega_f^{\mathrm{macro}}$ can be resolved by a coarser grid.

Using this method recursively, several resolution levels can be achieved in one grid, leading to a further reduction of the grid size. An example for such a multi-resolution grid applied to a pleated cartridge filter is shown in Fig. 18.

In combination with state-of-the-art solution methods for the linear systems, this grid generation technique can result in a five-fold speed-up of the simulation for the same geometry. Further details about the voxel grids considered here can be found e.g. in [41].

**Time Discretization of the Flow Problem** The discretization of incompressible Navier–Stokes and Navier–Stokes–Brinkman equations is non-trivial, because there is no explicit equation for the pressure, but only an implicit constraint given by the continuity equation.

There is a good deal of literature on how to overcome this difficulty, such as splitting algorithms and/or preconditioning for the Navier–Stokes equations (see, e.g. [72, 74, 94, 98]).

A straight-forward application of these algorithms to the discretization of the Navier–Stokes–Brinkman equations is not recommendable, since the Darcy term in this equation requires special consideration. In the following discussion, possible ways to do this will be presented.

As a preliminary remark, note that the apparent permeability in the non-Darcy cases is linearized using Picard's method. This is not the most robust and efficient approach in general, but it allows for a unified formulation of the algorithms.

We will use the following notation: The discretized operators for the convection and diffusions terms are denoted by $\mathbf{C}(\mathbf{u})\mathbf{u}$ and $\mathbf{D}\mathbf{u}$, respectively. $\mathbf{G}$ is the discrete gradient, $\mathbf{G}^\top$, the discrete divergence operator and $\mathbf{B}\mathbf{u}$, the discretized Darcy term in the momentum equations.

The actual form of these operators depends on the kind of spatial discretization, which will be the subject of the subsequent paragraph. The superscripts $\cdot^{k+1}$ and $\cdot^{k}$ denote the new and old time step, respectively, and $\tau$ is the time step length, i.e., $\tau = t^{k+1} - t^k$.

A splitting scheme for the time discretization of the Navier–Stokes–Brinkman equations (17) can be formulated as follows:

$$\left(\rho\mathbf{u}^* - \rho\mathbf{u}^k\right) + \tau\left(\mathbf{C}\left(\mathbf{u}^k\right) - \mathbf{D} + \mathbf{B}\right)\mathbf{u}^* = \tau\mathbf{G}p^k \tag{35}$$

$$\left(\rho\mathbf{u}^{k+1} - \rho\mathbf{u}^*\right) + \tau\left(\mathbf{B}\mathbf{u}^{k+1} - \mathbf{B}\mathbf{u}^*\right) = \tau\left(\mathbf{G}p^{k+1} - \mathbf{G}p^k\right) \tag{36}$$

$$\mathbf{G}^\top\rho\mathbf{u}^{k+1} = 0. \tag{37}$$

Operator splitting or projection methods are different names related to very similar concepts (cf. [65, 71, 72, 74]). All these methods first solve the momentum equations and then a pressure correction equation. This is why the algorithm above can be regarded as a Brinkman variant of the well-known Chorin method for the Navier–Stokes equations.

For the solution of the momentum equations (35), the pressure value of the previous time step is used on the right-hand side. Thus, one obtains a prediction $\mathbf{u}^*$ for the velocity (*intermediate* velocity).

The equation for the pressure correction

$$q = p^{k+1} - p^k$$

is obtained by applying the discrete divergence operator to (36):

$$\tau \mathbf{G}^\top \mathbf{G} q = \mathbf{G}^\top \big(\rho \mathbf{u}^{k+1} - \rho \mathbf{u}^*\big) + \tau \mathbf{G}^\top \big(\mathbf{B}\mathbf{u}^{k+1} - \mathbf{B}\mathbf{u}^*\big).$$

Together with the continuity equation (37) and the assumption that the pressure profile in the flow direction is essentially linear (i.e. $\mathbf{G}^\top \mathbf{B}\mathbf{u}^{k+1} \approx 0$), this simplifies to

$$\tau \mathbf{G}^\top \mathbf{G} q = -\mathbf{G}^\top \big(\rho \mathbf{u}^* + \tau \mathbf{B}\mathbf{u}^*\big). \tag{38}$$

This is nothing but a Poisson equation for the pressure correction with constant coefficients.

The drawback of this approach for the pressure correction is the lack of information about the porous medium in the Poisson operator and the fact that the continuity equation is satisfied only approximately in the filter medium region $\Omega_p^{\mathrm{macro}}$. This variant is therefore not advisable for the numerical simulation of the flow through porous media with non-constant permeability.

In order to improve the pressure correction, consider the following reformulation of (36):

$$\left(\mathbf{I} + \frac{\tau}{\rho}\mathbf{B}\right)\rho \mathbf{u}^{k+1} - \left(\mathbf{I} + \frac{\tau}{\rho}\mathbf{B}\right)\rho \mathbf{u}^* = \tau \mathbf{G} q,$$

where $\mathbf{I}$ denotes the identity operator. Since the Brinkman operator is positive definite, $\mathbf{I} + \frac{\tau}{\rho}$ is invertible, such that

$$\rho \mathbf{u}^{k+1} - \rho \mathbf{u}^* = \tau \left(\mathbf{I} + \frac{\tau}{\rho}\mathbf{B}\right)^{-1} \mathbf{G} q. \tag{39}$$

As before, we apply the divergence on both sides and use the continuity equation, giving the following variant of the pressure correction equation:

$$\tau \mathbf{G}^\top \left(\mathbf{I} + \frac{\tau}{\rho}\mathbf{B}\right)^{-1} \mathbf{G} q = -\mathbf{G}^\top \rho \mathbf{u}^*. \tag{40}$$

In the fluid region $\Omega_f^{\mathrm{macro}}$, this coincides with the previous pressure correction equation (38), because there, we have $\mathbf{B} \equiv 0$. In the porous subdomain $\Omega_p^{\mathrm{macro}}$ however, the Brinkman term is dominant. There, this variant is an approximation to the Poisson problem that would be obtained by applying the divergence to Darcy's law. We see that the pressure correction (40) is equally suitable for both the fluid and the porous regions, in contrast to (38).

After solving the pressure correction equation, the pressure is updated,

$$p^{k+1} = p^k + q, \tag{41}$$

and the new velocity value $\mathbf{u}^{k+1}$ is computed according to (39):

$$\rho \mathbf{u}^{k+1} = \rho \mathbf{u}^* + \tau \left( \mathbf{I} + \frac{\tau}{\rho} \mathbf{B} \right)^{-1} \mathbf{G} q. \tag{42}$$

**Time Discretization of the Concentration Equation** Using the general form (27) for the deposition model, a straight-forward time-implicit discretization of the concentration equation (23) reads:

$$\frac{C^{n+1} - C^n}{\Delta t} + \mathbf{u} \cdot \left( \mathbf{G} C^{n+1} \right) - \mathbf{D} C^{n+1}$$
$$= \begin{cases} 0 & \text{in } \Omega_f^{\text{macro}} \\ -(\alpha \Phi(M^n) C^{n+1} - \Psi(M^n)) & \text{in } \Omega_p^{\text{macro}}. \end{cases}$$

The discrete times $t^n$ $(n = 0, 1, \dots)$ and corresponding time steps $\Delta t = t^{n+1} - t^n$ belong to the time scale on which the loading of the medium occurs and therefore, they are not to be confused with the ones used for the flow problem above. This discretization approach is obviously one of the simplest choices, but it has proved to be very efficient for quite a lot of scenarios.

At this stage of the numerical simulation, the choice of suitable deposition models and the knowledge of proper values of the corresponding model parameters are crucial to the reliability of the simulation results. We will address the question of how this can be done in the next subsection.

After the problem for the concentration is solved, the solution $C^{n+1}$ is used to update the local captured mass $M^{n+1}$. This in turn is used to update the local porosity and permeability, and a re-computation of the flow is performed, if necessary.

Finally, relevant macroscopic key performance indicators, such as the total pressure drop across the filter element, its filtration efficiency, and the total captured mass, can be deduced from the numerical results produced by the flow and concentration simulation.

**Spatial Discretization** For the numerical solution of flow problems involving porous media, the following discretization methods can be used:

- Finite volume methods (FVM),
- Finite difference methods (FDM),
- Finite element methods (FEM),
- Lattice Boltzmann Method (LBM), particularly for pore-scale problems, and
- Meshfree methods.

In order to make the solution procedure as robust as possible and to ensure the local conservation of mass, momentum, and particle concentration, the finite volume method is a good choice. There is much literature devoted to the FVM discretization of Navier–Stokes equations (cf. e.g. [71, 99, 102] and the references therein). Therefore, we will limit our considerations here to those aspects that need special attention when the FVM discretization is applied to the Navier–Stokes–Brinkman equations.

If a *cell-centered* scheme is used, i.e., if both pressure and velocity values are located in the centers of the grid cells, there is a risk of oscillations occurring in the numerical results. The pressure is especially sensitive to these oscillations. This effect can be avoided or at least dampened by using the Rhie-Chow interpolation (see [91]). Another way to overcome this problem is the use of *staggered grids*, i.e., the velocity components are located on grids that are shifted by half a cell length in the corresponding coordinate direction (cf. e.g. [99]).

We already mentioned that, in many cases, the filter media are very thin compared to the typical length of the geometry (housing dimension). Even if a multi-resolution grid is used, the finest grid in the medium may have a relatively coarse resolution compared to the thickness of the medium. In order to ensure good numerical results in these cases, special interpolation techniques are required at the fluid-porous interfaces.

For further details on these matters, we refer the interested reader to [41, 42] and [5].

### 4.2.8 Parameter Estimation for Depth Filtration Models

We now turn to the question of how to identify proper values for the parameters found in the filtration models. The choice of the appropriate model will in general depend on the combination of the following influencing factors:

- the fluid,
- the types of filtering media (material, porosity, etc.),
- the dirt/dust to be removed, in terms of material(s) involved and particle size (distribution), and
- the operating conditions of the filter element (temperature/viscosity of the fluid, volumetric flow rate, etc.).

In most cases, it is not clear a priori which model is most suitable for a given setup, so measurements have to be done to find the appropriate model and the corresponding parameter values.

This should be done for several experimental setups in order to compensate for effects that are not taken into account by the (sometimes quite) simplified deposition models. As already stated in Sect. 2.1, on the pore scale, the filtration efficiency depends on many factors and most of these depend, in turn, more or less directly on the flow velocity in the medium. Consequently, the parameter values should be obtained for the range of flow velocities at which the filter element will operate. Once this is accomplished, look-up tables can be created from which the parameter values for the actual velocity can be determined during the simulation.

For the sake of simplicity, let us assume that the test dust consists of only one particle type (monodisperse dust), such that it is sufficient to consider one particle concentration. A simple experimental concept is the so-called *single pass* experiment (see Fig. 19). Here,

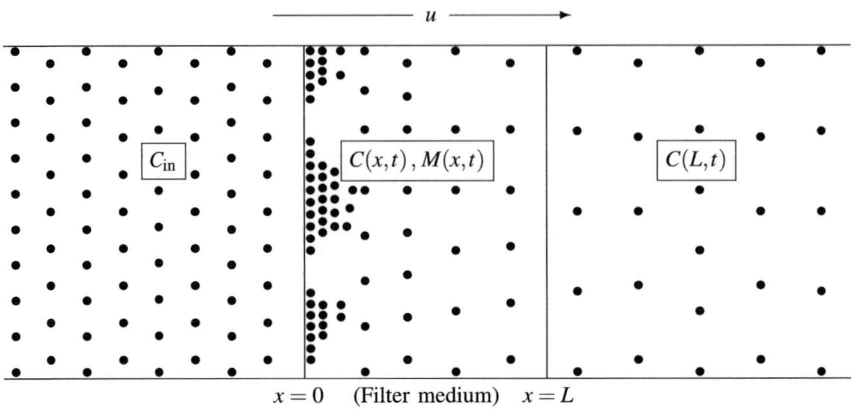


**Fig. 19** Single-Pass experiment with constant upstream concentration

a flat probe of the filtering medium is installed in a sufficiently wide flow channel, such that boundary effects can be neglected. The contaminated fluid flows through the porous medium at a constant flow rate. Using particle counters, the concentrations upstream and downstream of the medium are measured at certain times $t_i$ $(i = 0, \ldots, N)$. A common quantity is the so-called *beta ratio*, which is defined as the ratio of the concentrations upstream and downstream of the filtering medium. For a medium of thickness $L$ and a measurement performed at time $t$, this reads

$$\beta(t) = \frac{C(0,t)}{C(L,t)}. \tag{43}$$

Since the upstream concentration is kept constant in the single-pass setup and local variations of the concentrations are negligible in the channel upstream of the porous medium, we can assume

$$C(0,t) \equiv C_{\text{in}},$$

where $C_{\text{in}}$ denotes the inlet concentration.

In addition, let us suppose that for a given combination of fluid, dirt, and filtering medium, a suitable filtration model is identified. The goal is then to determine the values of the filtration model parameters from a series of measured beta ratios

$$\beta_i = \beta_{\text{exp}}(t_i), \quad i = 0, 1, \ldots, N.$$

Due to the inevitable noise in the experimental data, the estimation method should be as robust as possible.

(Semi-)discretized methods based on finite differences are flexible with regard to the experimental setup, but the approximations will in general create an additional sensitivity to the noise in the data. A more robust approach is based on exact solutions, which can be obtained for a quite large class of depth filtration models, provided that the experimental data are obtained in a single-pass setting. In this case, one can assume that the change in local concentration in the filtering medium is caused by transport and filtration only. Diffusion effects can either be neglected before-hand (e.g. in oil filtration) or accounted for in the simulation later by providing suitable look-up tables obtained for different experimental conditions. Under these conditions, the concentration equation (23) simplifies significantly to a one-dimensional stationary problem:

$$\forall t \geq 0, \quad x \in [0, L] : u\frac{\partial C}{\partial x} = -\frac{\partial M}{\partial t}, \tag{44}$$

and boundary conditions

$$\forall t \geq 0 : C(0, t) \equiv C_{\mathrm{in}}.$$

Let the filtering medium be clean at initial time $t_0 = 0$, i.e.,

$$\forall x \in [0, L] : M(x, 0) \equiv 0.$$

With these assumptions, one obtains the following exact solutions for the deposition models we have considered so far: The concentration for the simple filtration model (26) reads

$$C(x, t) = C_{\mathrm{in}} \mathrm{e}^{-\frac{\alpha}{u}x}, \tag{45}$$

and for the linear clogging model (28), one has

$$C(x, t) = \frac{C_{\mathrm{in}}}{1 + \mathrm{e}^{\frac{\alpha C_{\mathrm{in}}}{M_0}t}(\mathrm{e}^{\frac{\alpha}{u}x} - 1)}. \tag{46}$$

For the two-stage clogging model (29), one has to distinguish several cases: Denote by

$$t_c = \frac{M_0}{\alpha C_{\mathrm{in}}} \ln\left(1 + \frac{M_1}{M_0}\right),$$

the time at which the second, accelerated stage of the filtration process begins, i.e., for $t \leq t_c$, the model is identical to the previous one and therefore, the expression for the concentration is the same. For $t > t_c$, a certain upstream part of the filter depth is described by the second accelerated stage and the remaining part is still described by the initial linear clogging model. For $t > t_c$, the interface of these two zones is located at the position

$$x_c(t) = \frac{u}{\alpha} \frac{M_0}{M_0 - aM_1} \ln\left(1 + \frac{M_0 - aM_1}{M_1(1+a)}\left(1 - \mathrm{e}^{-\alpha \frac{1+a}{M_0} C_{\mathrm{in}}(t - t_c)}\right)\right).$$

For $t > t_c$ and $x \leq x_c(t)$, the concentration reads

$$C(x, t) = \frac{C_{\mathrm{in}}}{1 + \frac{M_0 + M_1}{M_0 - aM_1} \mathrm{e}^{\alpha \frac{1+a}{M_0} C_{\mathrm{in}}(t - t_c)} (\mathrm{e}^{\frac{M_0 - aM_1}{M_0} \frac{\alpha}{U} x} - 1)}, \tag{47}$$

whereas for $t > t_c$ and $x > x_c(t)$, we have

$$C(x,t) = C_{\text{in}}\left(\frac{1}{1+a}\left(\left(a\frac{M_1}{M_0} - 1\right) + \left(1 + \frac{M_1}{M_0}\right)e^{\alpha\frac{1+a}{M_0}C_{\text{in}}(t-t_c)}\right)\right.$$
$$\left.\times\left(\left(1 + \frac{M_0}{M_1}\right)e^{\frac{\alpha}{u}(x-x_c(t))} - 1\right)\right)^{-1}. \tag{48}$$

Finally, the analytical expression for the concentration in the case of the linear release model (30) is

$$C(x,t) = C_{\text{in}}e^{-\frac{\alpha}{u}x}\left(e^{-\gamma t}I_0\left(2\sqrt{\frac{\alpha\gamma}{u}xt}\right) + \gamma\int_0^t e^{-\gamma\tau}I_0\left(2\sqrt{\frac{\alpha\gamma}{u}x\tau}\right)d\tau\right), \tag{49}$$

where $I_0$ denotes the modified Bessel function of the first kind of order zero. The analytical beta ratios for a given model are obtained from the corresponding exact solution by setting $x = L$ and $t = t_i$ for $i = 0, 1, \ldots, N$:

$$\beta(t_i) = \frac{C_{\text{in}}}{C(L,t_i)}.$$

For the estimation of the corresponding filtration model parameters, a least-squares approach reads

$$\sum_{i=0}^{N}\left(\beta(t_i) - \beta_i\right)^2 \to \min.$$

The solution of these non-linear equations can be computed using a quasi-Newton algorithm. In order to test the robustness of this approach in a reproducible way, a set of exact parameters was chosen for each of the models. Synthetic measurement data were created by evaluating exact beta ratios evaluated at times $t_i$ and adding noise. This set of noisy beta ratios served as input for the parameter estimation. Finally, the reconstructed curves were compared to the original, exact ones. Qualitative test results are shown in Figs. 20, 21 and 22.

Exact solutions not only serve as a means to obtain deposition model parameters in a robust way. They can also be used for a first, rigorous validation of the simulation software by running numerical tests for single-pass experiments in simple geometries.

### 4.2.9 Simulation of Filter Elements: Examples, Validation, and Benefits

In this section, we briefly discuss how a filtration simulation software is validated and calibrated by applying it to a simple reference problem. Once this step is accomplished, simulations can be done for more sophisticated real-world designs.

**Conical Housing with Flat Sheet Filtering Medium** In order to validate and calibrate the simulation software, a test setup should be chosen that is equally accessible to both measurements in the lab and simulations in a reproducible way. The geometrical shapes of

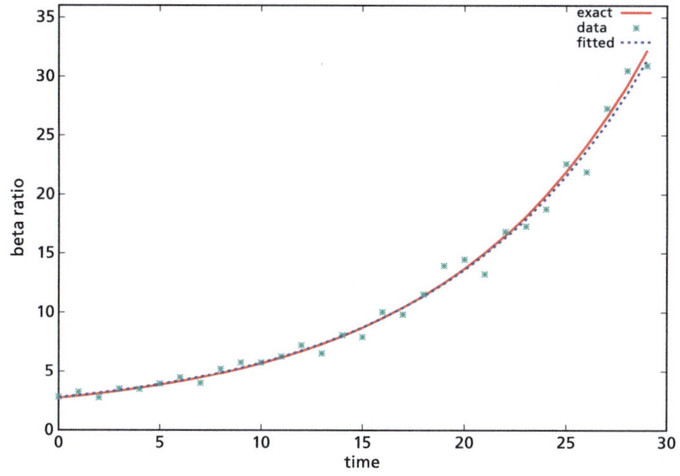


**Fig. 20** Parameter identification for the linear clogging model (28)

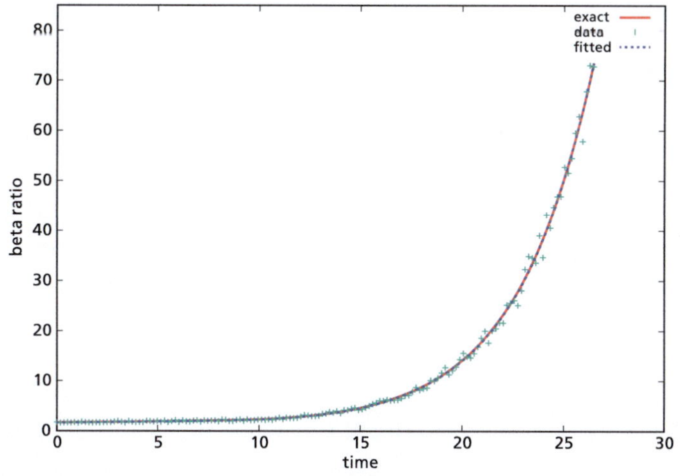


**Fig. 21** Parameter estimation for the two-stage model (29)

the housing and the filter medium should be as simple as possible, so that their influence on the results are minimal. This aspect is important in order to validate the models (and identify the corresponding parameters) for filtration and permeability change.

As an example, we consider a conical filter housing with a flat sheet filter medium, as shown in Fig. 16, on the left.

As for the parameter estimation, a constant volumetric flow rate is prescribed, together with the specification of a standardized test dust, the type of fluid used, and other relevant experimental parameters, such as the temperature, etc. Assuming that the flow model given by the Navier–Stokes–Brinkman equation (17) is appropriate for the test fluid, what

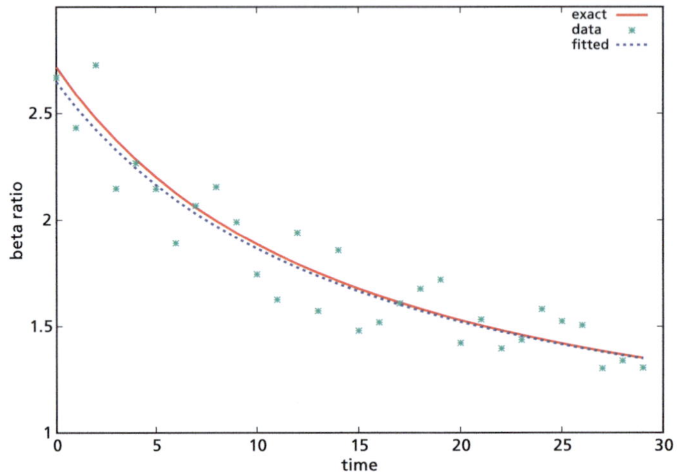


**Fig. 22** Parameter estimation for the release model (30)

remains is to validate the concentration/deposition models and the model chosen for the permeability.

With this well-defined experimental setup, the validation proceeds as follows:

- First, the clean permeability model is determined by both measuring and simulating the total pressure drop across the conical filter element. With these measurements, one can check whether Darcy's law is valid for the entire range of flow rates or which non-linear pressure drop model has to be chosen. In the latter case, the model parameters can be deduced from the experimental data.
- After having chosen a series of measurement times $t_i$ $(i = 0, 1, \dots, N)$, the medium is loaded with the test dust and the series of beta ratios are measured. In the simulation, these beta ratios are easily derived from the solution of the concentration equation (see Fig. 23, left). Note that this procedure is not necessarily restricted to single-pass tests; other test types (e.g. multi-pass) can also be treated, provided they are implemented in the simulation code. If good agreement of measured and simulated beta ratios is observed, the deposition models and all parameters involved are validated. In particular, it follows that the captured mass $M$ computed in the simulation (see Fig. 23, right) agrees well with the corresponding mass deposited in the real medium in the experiment. Using gravimetric analysis, the captured mass can also be measured in the experiment.
- Once it is clear that the choice of deposition model is suitable, the permeability model used for the simulation must be validated. This can be done by direct comparison of the measured total pressure drop and the corresponding values obtained in the simulation at the times $t_i$.

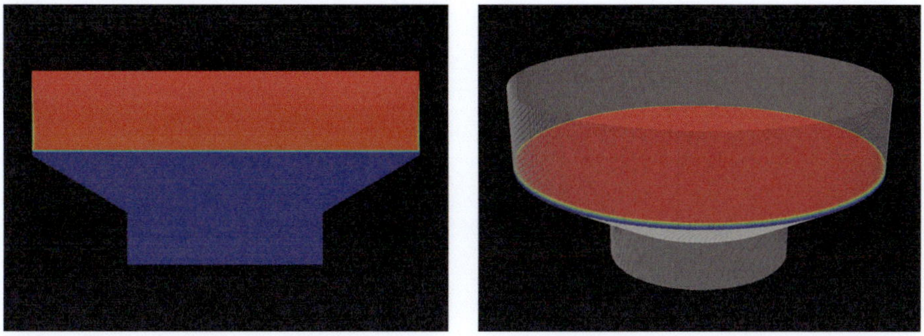

**Fig. 23** Simulation results for the test filter element with conical housing. *Left*: Concentration of dissolved particles (central section, *red*: high concentration, *blue*: low concentration). *Right*: Distribution of captured mass in the filter medium

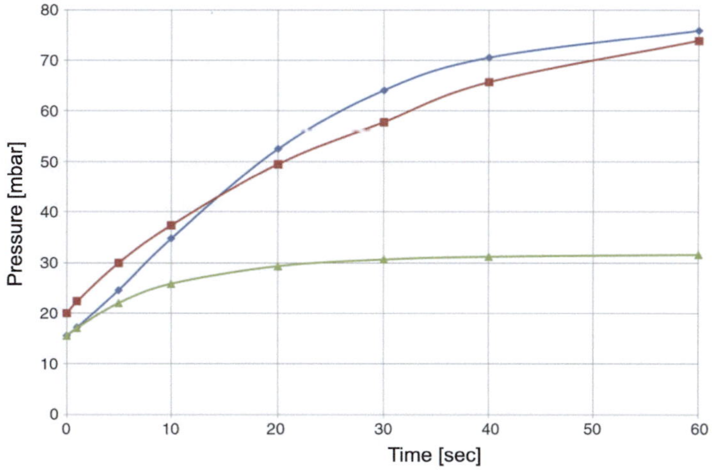


**Fig. 24** The increase in pressure drop as a consequence of the loading of the filter medium with dust. Comparison of the measured pressure values (*red*) with the simulation results when using the Jackson–James model (*green*) and the combined permeability model (*blue*) (Graphics: IBS Filtran)

In general, this procedure is done for several different conditions (viscosities, flow rates, etc.), to ensure that the models and parameters are able to cover a certain range of operating conditions.

Figure 24 shows the time evolution of the total pressure drop in a test filter element for a flat sheet of a highly efficient medium that is loaded with test dust. Depicted are the pressure curves that were measured in the laboratory and two simulated curves obtained by the software SuFiS®. As one can see, the proper choice of the permeability model is crucial to a reliable prediction of the pressure drop by the computer simulation. In the study shown here, one can see that the combined permeability model (34) produces much better results than the Jackson–James model (32) alone.

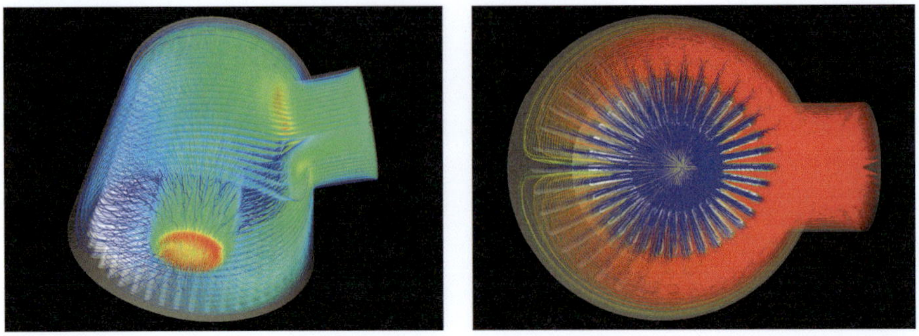

**Fig. 25** Simulation results for the pleated cartridge filter element. *Left*: Streamline representation of the simulated velocity field of the flow through the element colored by magnitude (*red*: high flow speed, *blue*: low flow speed). *Right*: Streamline representation of the concentration of dissolved particles in the filter element (*red*: high concentration, *blue*: low concentration)

**Application to Real Filter Element Designs** After successfully validating the models and identifying parameters, one can proceed with the simulation of more sophisticated filter element designs for the market. Figure 18 (left) shows a CAD geometry of a pleated cartridge filter element. This a classical filter design, since a large area of filtering medium can be enclosed by a relatively small housing. The geometry of the element is given as an STL surface (*Surface Tesselation Language*), which is a common output format for CAD tools. From this geometrical description, a computational grid for the Finite Volume Method is generated as depicted in Fig. 18 (right).

Figure 25 (left) shows the computed velocity field of the fluid flow inside the pleated cartridge filter element. It can be seen that the highest flow speeds are close to the inlet and outlet of the filter. With simulations, the velocities and pressure values at different cross sections of the filter element can be visualized and analyzed. Among other things, this gives an idea about the optimal pleat count and shaping. Figure 25 (right) shows the particle concentrations in a selected cross-section of the filter element.

By providing such detailed information, the simulation gives valuable insight into the distribution of "filtering activity" regions in the element for different times during the operation of the device. It would require substantial efforts to retrieve a comparable level of information experimentally, if it were possible at all. The product developer will benefit from the simulation results by receiving suggestions on the further optimization of a design without the need to construct prototypes.

## 4.3 Multiscale Modeling and Simulation of Filtration Processes

Filtration processes, as specified earlier, are intrinsically multiscale. The mathematical challenges in multiscale modeling and simulation of filtration processes were discussed briefly in Sect. 2. In this section, we will present some details on:

- Modeling the permeability of filtering media;
- Subgrid algorithms for the simulation of flow within filter elements; and
- Multiscale modeling and simulation of filtration processes.

The first topic is rather general and concerns not only filtration processes, but also any flow in porous media. While the last two topics can also be considered from a general viewpoint, here we discuss developments done exactly for the need of filtration simulation.

### 4.3.1 Modeling the Permeability of Filtering Media

Permeability plays a key role in simulation of the filtration processes, and we will discuss it in some detail. Rough models for permeability were discussed in Sect. 4.2.1. These are the Kozeny–Carman (33) and Jackson and James (32) formulae. They link permeability to the porosity for granular media or to the porosity and fiber diameter for nonwoven filtering media. Although very useful, these formulae have a limited area of applicability (e.g., only scalar permeability, only monodiameter distribution for (32), etc.). A more accurate modeling of permeability can be done on the basis of asymptotic homogenization. Let us recall some known results, which are important for understanding the approaches discussed later on. Consider for a moment slow, incompressible laminar flow (described by the steady state Stokes equation) in a periodic microstructure, with $Y_F$ being the fluid part of the periodicity cell and $Y_S$ being the solid part of the periodicity cell. As known (see, e.g., [70, 76]), the permeability in this case is given by

$$K_{ij} = \langle \nabla_y \omega^i : \nabla_y \omega^j \rangle_Y, \tag{50}$$

where $\omega^i$, $\nabla\pi^i$ are solutions of the following cell problem:

$$\begin{aligned} -\Delta\omega^i + \nabla\pi^i &= e^i && \text{in } Y_F \\ \nabla\cdot\omega^i &= 0 && \text{in } Y_F \\ \omega^i &= 0 && \text{on } \partial Y_S \\ \omega^i,\ \pi^i & && Y\text{-periodic.} \end{aligned} \tag{51}$$

The governing equations at the macroscale (effective porous media) are the well-known Darcy equation and mass conservation equation:

$$\begin{cases} \mathbf{u} = \frac{1}{\mu} K(f - \nabla p) & \text{in } \Omega \\ \nabla\cdot\mathbf{u} = 0 & \text{in } \Omega \\ \mathbf{u}\cdot\nu = 0 & \text{on } \partial\Omega \end{cases} \tag{52}$$

In the engineering literature, another approach for calculating the permeability is usually used. The Darcy equation at the macroscale is not derived, as in the case of homogenization, but is assumed to be valid, and is used to compute the permeability. More precisely, the microscale Stokes problem is formulated in some REV (Representative Elementary Volume)

$$\begin{aligned}\nabla p_\epsilon^i - \mu \Delta \mathbf{u}_\epsilon^i &= f^i && \text{in } Y_F^\epsilon\\ \nabla \cdot \mathbf{u}_\epsilon^i &= 0 && \text{in } Y_F^\epsilon\\ \mathbf{u}_\epsilon^i &= 0 && \text{on } \partial Y_S^\epsilon\\ \mathbf{u}_\epsilon^i,\ \pi_\epsilon^i & && Y^\epsilon\text{-periodic,}\end{aligned} \tag{53}$$

where $\mathbf{u}_\epsilon^i$, $p_\epsilon^i$ are microscale velocity and pressure, respectively. Assuming that the Darcy equation with a permeability tensor is valid,

$$\langle u_{\epsilon,j}^i \rangle_{Y_\epsilon} = \frac{1}{\mu} \sum_{k=1}^{3} K_{jk} \left( f_k^i - \left\langle \frac{\partial p_\epsilon^i}{\partial x_k} \right\rangle_{Y_\epsilon} \right), \tag{54}$$

then the permeability is computed as follows:

$$K_{ij} = K_{ji} = \langle u_{\epsilon,j}^i \rangle_{Y_\epsilon}. \tag{55}$$

It is shown in [75] that, in the case of Stokes flow in periodic microstructure, both approaches are equivalent. However, there are no formal requirements for periodicity of the media for the volume averaging approach; the cell problem can be easily reformulated, omitting the periodicity requirement, and be formally applied to Navier–Stokes equations as well. Although this approach can not be rigorously justified, we will use it in connection with upscaling algorithms for the filtration problems presented in the next two subsections.

### 4.3.2 Subgrid Algorithm for Simulation of Flow Within the Filter Element

The motivation for developing the subgrid algorithm is the fact that filter element housings can have very complicated shapes, e.g., featuring very thin media and comparably thick walls and supporting ribs of different sizes and shapes. On a single uniform grid, such differences in size can be accounted for in the simulation only by using a sufficiently fine grid. However, such fine grids often make the simulation very costly or even impossible. On the other hand, simulations on coarse grids do not provide enough accuracy for the pressure drop in the filter element. Local grid refinement or coarsening is a possible remedy, as already seen in Sect. 4.2, but even this can become computationally challenging in cases in which the length scales involved differ too greatly. For this class of problems, the subgrid algorithm has proven to be a suitable solution strategy. In this approach, one solves the problem on a coarser grid, but accounts for the unresolved geometrical features by solving local auxiliary problems on an underlying finer grid in some properly selected coarse cells.

Subgrid methods have been used in connection with other applications, namely, in solving the Darcy problem for the fine and for the coarse resolution (cf. [55, 64]) and for the Navier–Stokes–Brinkman equations in [51] and [18]. The upscaling approach presented here is similar to the one proposed and justified in [75].

For a given computational domain of the filter element, we consider a fine and a coarse grid, with the property that the interior of each coarse cell is the union of the fine grid cell volumes contained within it. Each fine cell is assigned a material (value), depending on whether it belongs to the fluid part, the porous medium, or the solid region of the domain. Accordingly, the resolution of the fine grid is chosen so that the material distribution in the domain is represented sufficiently well. The (matching) resolution of the coarse grid is selected in order to ensure an efficient numerical solution of the Navier–Stokes–Brinkman equation

$$\rho\frac{\partial \mathbf{u}_0}{\partial t} - \nabla\cdot(\tilde{\mu}\nabla\mathbf{u}_0) + \mu\tilde{K}_{\text{eff}}^{-1}\mathbf{u}_0 + \nabla p_0 = \mathbf{f}_0, \tag{56}$$

where the subscript 0 indicates that the corresponding quantity is considered on the coarse grid.

The major purpose of the local fine grid simulations is the suitable upscaling of the permeability of the filter media, thus providing the effective permeability $\tilde{K}_{\text{eff}}$ for the coarse grid computations. To this end, a pre-processing step identifies the so-called *quasi porous* cells, i.e., the coarse cells that contain more than one material in terms of the material distribution on the fine grid. The auxiliary problems for the upscaling are solved in the quasi porous cells on the fine grid, using the *actual* permeability (distribution) of the medium in the porous cells. The effective permeability is then determined using volume averaging. Either one can compute $\tilde{K}_{\text{eff}}$ for each quasi porous cell or for a union of several quasi porous cells. In the latter case, the resulting effective permeability is assigned to all coarse cells that were involved in the computation. This variant is recommended e.g. if the coarse resolution is too fine in the sense that the grid cells cannot be regarded as representative elementary volumes.

The subgrid approach can be summarized as follows:

1. Choose/identify the quasi-porous cells on the coarse grid.
2. Solve the local cell problems on the fine grid. In some cases, it is sufficient to solve the auxiliary problems only in some pre-selected quasi-porous subdomains.
3. Compute the effective permeabilities in these quasi porous cells.
4. Solve the macroscopic flow problem (56)) using the coarse grid and the upscaled permeability $\tilde{K}_{\text{eff}}$.

Remark:

(a) Both the macroscopic and the local cell simulation use the same numerical method. However, the criteria for convergence etc. will, in general, be different.

(b) If the number of quasi-porous cells is relatively large and/or the local grids contain a relatively large number of fine cells, the computation of the effective permeability may represent a major part of the total numerical costs. On the other hand, the auxiliary problems are completely independent of each other and, therefore, they allow for effective parallelization.

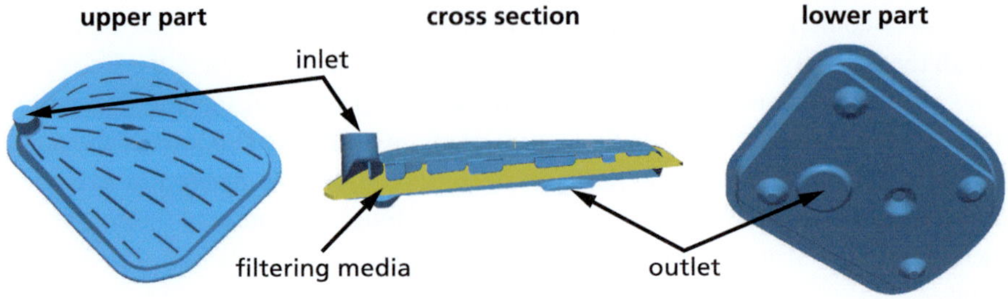


**Fig. 26** Filter element housing developed by IBS Filtran (*blue*), with flat filter medium (*yellow*)

A more detailed presentation of the method can be found in [16]. A practical application of the subgrid algorithm will be subject of Sect. 5.

### 4.3.3 Multiscale Modeling and Simulation of Flow and Filtration Efficiency Processes

The above subgrid algorithm for solving flow problems in filter elements aims at solving local problems at a finer scale and, via upscaling, bringing the information to the macroscale. The algorithm is suitable for flow problems when there is no back coupling from the macroscale to the finer scales. This general concept must be further developed when filtration efficiency has to be modeled and simulated. The processes at the microscale and macroscale are not independent of each other in this case. The microscale geometry changes due to the deposited particles, which leads to changes in the permeability used in the macroscale equations. Conversely, the macroscopic velocity influences the transport and capturing of the particles.

A sketch of one time step of the coupled microscale—macroscale simulation algorithm for the simulation of filtration processes in an industrial filter (e.g., such as the IBS Filtran-designed filter shown in Fig. 26) is as follows.

1. At the selected locations, as shown in Fig. 27 of the filtering porous media, Navier–Stokes–Brinkman problems are solved;
2. The computed velocities are input for the Langevin equation, which together with the prescribed deposition mechanism, is used to simulate particle transport and deposition in these selected locations using the approach described in Sect. 4.1;
3. Based on a consecutive upscaling procedure, these results are used to update permeability in the selected locations (see Fig. 27);
4. A proper interpolation procedure is used to calculate the permeability in the full porous medium (see Fig. 28);
5. The updated permeability is used to compute the macroscopic velocity and pressure in the whole filter element (see Fig. 29) using the approach described in Sect. 4.2; and
6. The computed velocities and the concentration of the particles are used as input for the micro scale computations at the next time step.

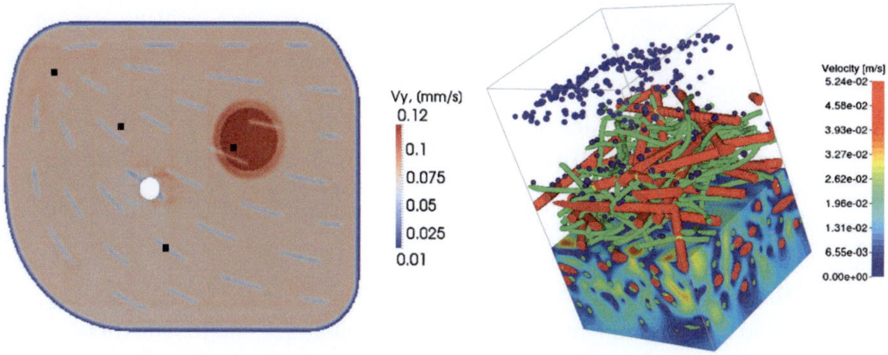


**Fig. 27** *Left*: velocity profile in a cross-section of the filter element. The points mark the positions of the "observer cells" for the filtration simulation on the pore level. *Right*: result of the microsimulation in an observer cell

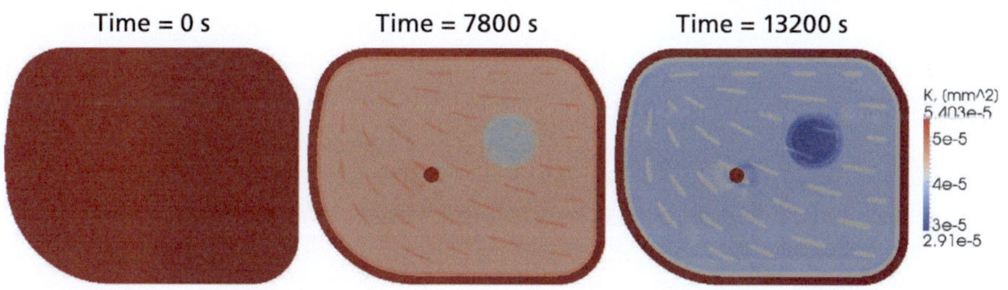


**Fig. 28** Permeability distribution in filter medium calculated by interpolation. The locally varying flow leads to significant inhomogeneities

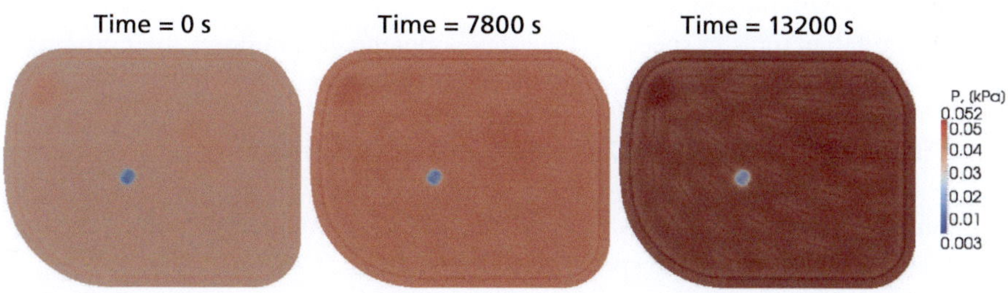


**Fig. 29** Pressure distribution calculated from the macroscale simulation

These steps are repeated until some prescribed stopping criterion is satisfied (e.g., a certain duration of the filtration process or pressure limit). The selection of the locations (windows) in which the microscale problem is solved can be tricky, but in some cases, simple criteria can be applied based on experience, e.g., selecting locations near the inlet and at the inlet, near the ribs, in the middle of a channel, etc. Of course, one needs to have enough information about the microstructure of the filtering media and the particle size distribution of the dust. From an algorithmic view point, the fact that microscale and macroscale filtration processes occur at different time scales can be used to select different time steps for the microscale and macroscale simulations.

The computational complexity of the coupled multiscale simulations depends on the number of selected locations where the microstructure is resolved. Here, however, as in many other algorithms for multiscale problems, the simulations in the selected windows can be performed in parallel.

Selected results for a practically relevant application (see Fig. 26) are presented in Figs. 27 to 29. One can observe that the deposition process in the initially heterogeneous locations leads to non-uniform changes in their permeabilities and to redistribution of the macroscopic flow.

We have seen how mathematical research can help to handle complex filtration processes both qualitatively and quantitatively by providing suitable models and numerical algorithms. The calibration and validation is done by using well-defined and reproducible experiments, which have to be done anyway, in most cases. Using simulations does not mean at all that measurements in the lab won't be needed any longer. In fact, the computer-aided study of filtration processes supports the optimization of experimental efforts in the sense that the lab resources can be used in a more effective way for the innovation and improvement of products.

## 5 Successful Industrial Applications

In this section, we will present two cases in which filter manufacturers successfully applied numerical simulation tools to speed up their developments and improve their products.

### 5.1 Accelerating Product Development at IBS Filtran

IBS Filtran[1] develops and produces parts for the automotive industry, such as suction filters, pressure filters, oil pans, and specialized filter media.

In order to assist the company with the design of filter elements, the Fraunhofer ITWM has developed the tailor-made software SuFiS® (Suction Filter Simulation), in close collaboration with IBS Filtran (cf. e.g. [6]). The main focus of the collaboration has been the development of a dedicated simulation tool specialized for the computer-aided design and optimization of automatic transmission fluid filters. The purity of the transmission fluid is crucial to the efficient operation of automatic transmissions in vehicles, so that the quality and performance of the filter elements is of special importance. The development of these

[1] IBS-Filtran GmbH, Morsbach, Germany, www.ibs-filtran.com.

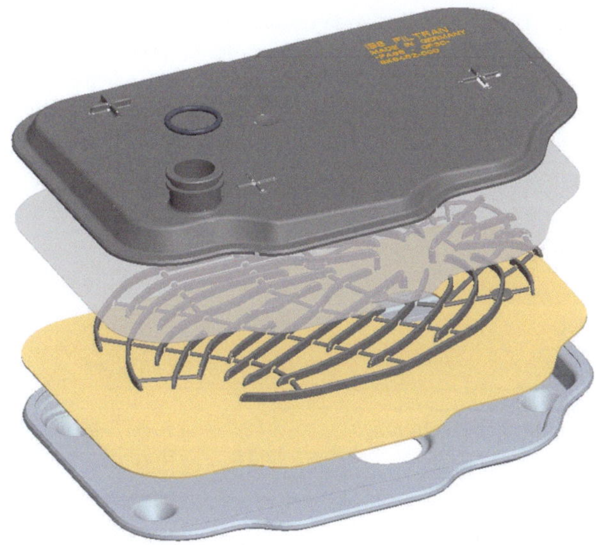

**Fig. 30** A *combi filter* element design by IBS Filtran. *From top to bottom*: filter upper cover (plastics), screen mesh, rib tray spacer, non-woven filter medium, plate bottom pan (plate) (Graphics: IBS Filtran)

products is very challenging, requiring continuous adaptation of the filter element designs to changes in the operating conditions and the available installation space.

An example of such a transmission filter element is shown in Fig. 30. Any optimization of such a filter element has to be done under certain constraints. As already mentioned, the installation space is very limited in the engine compartment, so that there is not much freedom to vary the geometry of the filter housing. In order to minimize the pressure drop while providing a high efficiency and dirt holding capacity, the following two parts of the filter element were major optimization targets:

- The geometrical structure of the rib tray should provide mechanical stability while the flow resistivity should be as small as possible.
- A highly efficient non-woven filter medium was selected for the purification of the transmission oil. Without modification, the pressure drop across such a medium would increase much too rapidly during the loading, especially when the transmission oil is cold (high viscosity). The idea by IBS Filtran was to introduce a perforation of the medium which should decrease the differential pressure, while still providing good filtration efficiency. The identification of an optimal combination of number, size and distribution of the holes in the medium should be based on numerical simulations.

The length scale of the filter element is measured in centimeters, while the perforation holes are on the millimeter scale or even smaller. Doing a numerical simulation on a single grid would require a very fine resolution for the holes, causing very high computational costs for the entire element. On the other hand, when using a coarse resolution, the perfo-

**Table 1** SuFiS® simulation results for perforated filter media. Top: Computational cost and pressure drop (dP) when using a high resolution grid. Bottom: Computational cost, relevant component of the permeability $K_{22}$ and pressure drop (dP) when using the subgrid method. Here, $T_{SG}$ denotes the CPU time for the solution of the auxiliary problems and $l$ is the number of corresponding additional fluid layers

| Grid resolution [mm] | Number of CVs | Memory [MB] | $T_{\mathrm{CPU}}$ [s] | dP [kPa] |
|---|---|---|---|---|
| 2 | 5445 | 4,42 | 63,54 | 16,762 |
| 1 | 37400 | 26,49 | 465,02 | 11,627 |
| 0,5 | 299200 | 182,70 | 28271,47 | 1,579 |
| 0,25 | 2393600 | 1338,37 | 518425,0 | 1,762 |

| Grid/subgridresolution [mm] | $l$ [1] | $K_{22}$ [mm$^2$] | $T_{SG}$ [s] | $T_{\mathrm{CPU}}$ [s] | dP [kPa] |
|---|---|---|---|---|---|
| 1/0,25 mm | 4 | 0,00972 | 1126,3 | 1829,4 | 1,7984 |

rations would not be "visible" and therefore their influence on the pressure drop could not be studied.

Therefore, the task was a perfect application case for the subgrid method presented in Sect. 4.3: Here, the quasi-porous cells are those coarse grid cells containing the filter medium with the perforation holes. In addition, a certain number of grid layers in the fluid upstream and downstream of the medium were added to the quasi-porous cells.

In Table 1 (top), the results of typical single grid simulations are presented for different grid resolutions. One can see that for the coarse grids, the holes are not (properly) resolved, leading to an overestimation of the pressure drop. As one would expect, the results improve with finer grid resolutions. But this goes hand in hand with a corresponding increase in computational time and memory requirements.

The bottom part of Table 1 presents the corresponding data when using the subgrid method: One can see that the computational efficiency is much higher, while at the same time providing comparable accuracy in the results.

Using the SuFiS® software, it is possible to simulate standardized efficiency tests for filter elements (e.g. ISO 16889, TFEM). The development engineers at IBS Filtran use these features to predict the evolution of the efficiency and/or the pressure loss under test conditions. For the perforated filter medium, the computer simulations enabled the designers to predict the influence of the perforation pattern on the filter element's efficiency and dirt holding capacity (see Fig. 31) without producing corresponding prototypes that would have required lengthy measurements.

The computer-aided identification of possible weaknesses in a design has led to a substantial speed-up in the company's developmental process (see Fig. 32). For the final design of the perforated medium, IBS Filtran now holds a patent (cf. [67]).

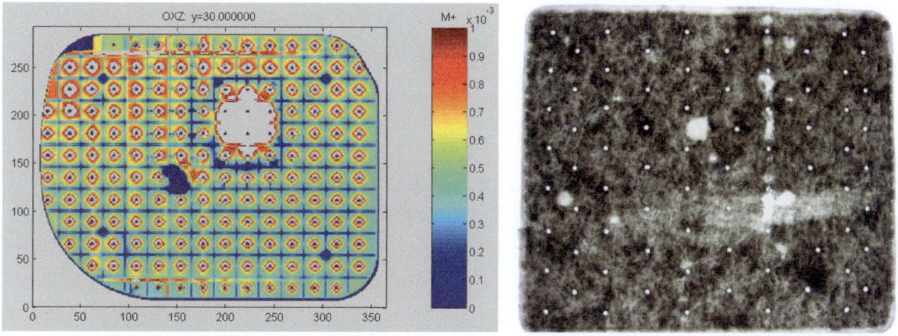


**Fig. 31** *Left*: Particle deposition in the perforated filter medium, computed by SuFiS®. *Right*: The deposition in the medium as observed in the lab (Photo: IBS Filtran)

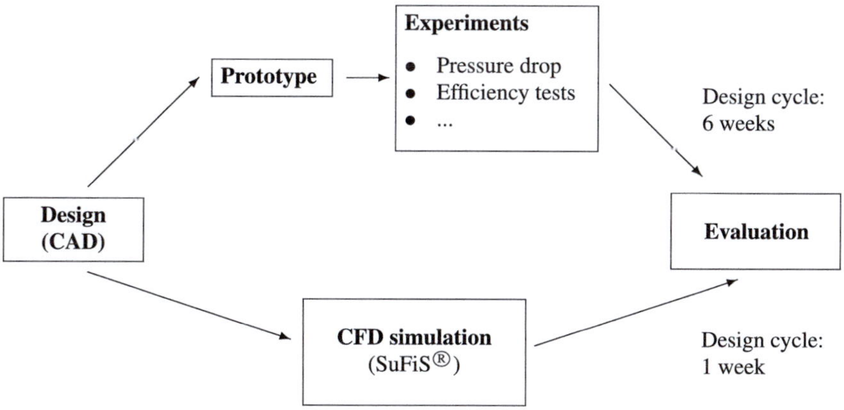


**Fig. 32** Acceleration of the product development due to the use of SuFiS® at IBS Filtran

## 5.2 Optimization of Pleat Support in Hydraulic Filters at ARGO-HYTOS

The company ARGO-HYTOS[2] is a leading manufacturer of systems and components in the field of hydraulic machine engineering. A major part of the business is the development and production of filter elements for mobile hydraulic systems. Due to their favorable ratio of filtration area to housing volume, these elements often use pleated filtering media.

The challenge in the design of filter elements for hydraulic applications is the huge pressure at which the hydraulic fluid is flowing through the system. As we already mentioned in the introductory sections, the interaction of the fluid with the filtering medium can easily lead to pleat crowding and/or pleat collapse. A common way to prevent this is to support the pleated medium by a mesh made of some robust material (e.g. steel). Obviously, there are numerous ways to design the supporting mesh (e.g., thickness of the mesh

[2]ARGO-HYTOS GmbH, Kraichtal-Menzingen, Germany, www.argo-hytos.com.

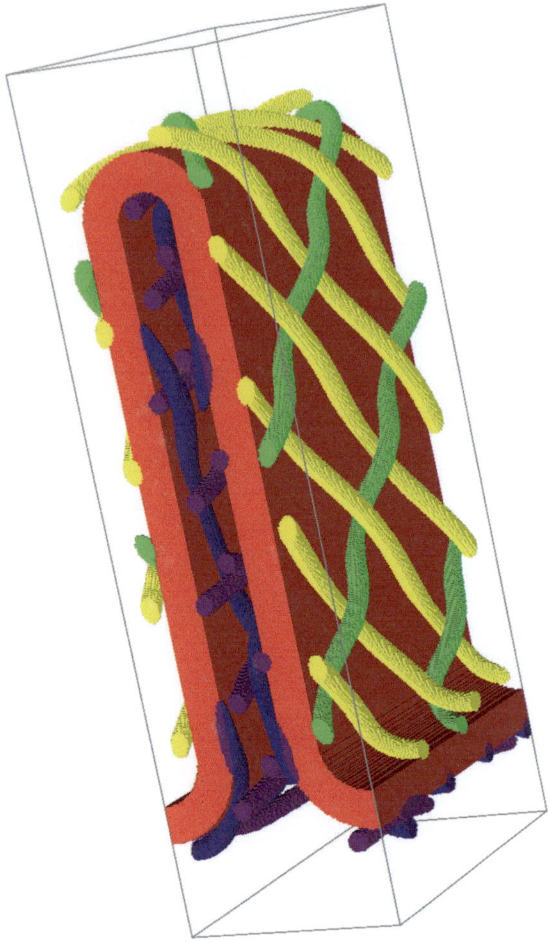

**Fig. 33** Computer model of a filtering pleat with supporting mesh (zoomed in on the *right*)

wires, shape and dimension of mesh openings, etc.). We state here only the main aspects that have to be taken into consideration for an optimal design:

- The supporting mesh should not block too much of the filtration area.
- The supporting mesh should not produce too much additional pressure drop.
- Even for the highest pressures, the wires of the mesh should be arranged such that the fluid can still flow through the pleat channel, i.e., this should be ensured even if the wires of adjacent pleats touch each other.

The latter requirement clearly indicates that one has to employ a 3D model to address this issue. In order to solve the problem, the GeoDict module PleatGeo was used to create a computer model of the pleat together with the supporting mesh structure (see Fig. 33). The corresponding computational grid was created so that the wires were resolved sufficiently

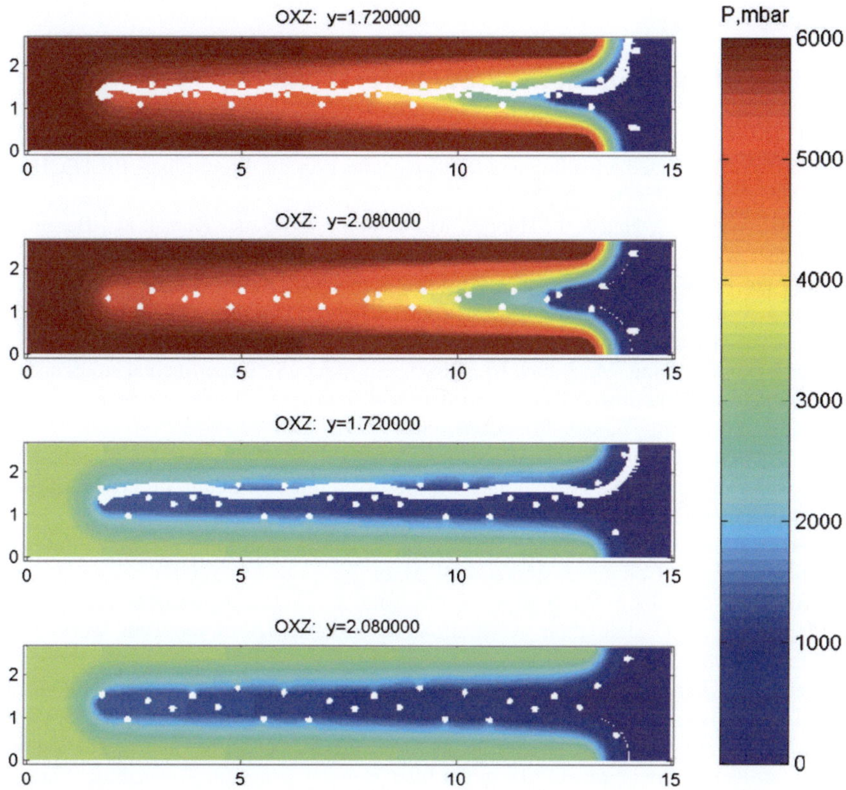


**Fig. 34** Simulated pressure drop for two different designs of the supporting mesh (at two different cross-sections). The *white areas* indicate the wires

well. Finally, the pressure drop was computed using the Navier–Stokes–Brinkman FVM solver (see [21]). As one can see in Fig. 34, the arrangement of the wires in the supporting mesh has a significant effect on the pressure drop.

Using the computer simulations, an optimal design of the support mesh could be identified without producing prototypes for each conceivable variant. With the optimized mesh, the total pressure drop could be reduced up to 35 %. This significant improvement eventually led to a patent for ARGO-HYTOS (see [93]).

The two examples presented here show that the combination of mathematical, numerical, and engineering expertise significantly helps to improve product development in filtration. In situations in which purely empirical approaches are unreliable, costly, or practically impossible, computer simulations provide a powerful tool for both innovation and optimization of filtration products. For markets with tough competition, this is certainly a strong asset for the R&D departments in the industry.

## References

### Publications of the Topic at the Fraunhofer ITWM


1. Andrä, H., Iliev, O., Kabel, M., Lakdawala, Z., Kirsch, R., Starikovičius, V.: Modelling and simulation of filter media loading and of pleats deflection. In: Proceedings FILTECH Conference 2011, vol. I, pp. 480–486 (2011)
2. Beier, H., Vogel, C., Haase, J., Hunger, M., Schmalz, E., Sauer-Kunze, M., Bergmann, L., Lieberenz, K., Fuchs, H., Frijlink, J.J., Schmidt, G., Wiesmann, A., Durst, M., Best, W., Burmeister, A., Wiegmann, A., Latz, A., Rief, S., Steiner, K.: Vliesstoffe für technische anwendungen. In: Fuchs, H., Albrecht, W. (eds.) Vliesstoffe: Rohstoffe, Herstellung, Anwendung, Eigenschaften, Prüfung, pp. 539–637. Wiley-VCH, Weinheim (2012)
3. Bernards, D., Dedering, M., Kabel, M., Kirsch, R., Staub, S.: Experimental study and numerical simulation of the flow-induced deformation of filtering media in automotive transmission filters. In: Proceedings Filtech 2015 Conference (2015). L15-02-P076
4. Calo, V., Nicolò, E., Iliev, O., Lakdawala, Z., Leonard, K., Printsypar, G.: Simulation of osmotic and reactive effects in membranes with resolved microstructure. In: Proceedings Filtech 2015 Conference (2015). M07-03-P101
5. Ciegis, R., Iliev, O., Lakdawala, Z.: On parallel numerical algorithms for simulating industrial filtration problems. Comput. Methods Appl. Math. **7**(2), 118–134 (2007)
6. Dedering, M., Stausberg, W., Iliev, O., Lakdawala, Z., Ciegis, R., Starikovičius, V.: On new challenges for CFD simulation in filtration. In: Proceedings of the 10th World Filtration Congress (2008)
7. Ginzburg, I., Klein, P., Lojewski, C., Reinel-Bitzer, D., Steiner, K.: Parallele Partikelcodes für industrielle Anwendungen (2001)
8. Iliev, D., Iliev, O., Kirsch, R., Dedering, M., Mikelić, A.: Modelling and simulation of fluid-porous structure interaction (FPSI) on the filter element scale. In: Proceedings Filtech 2013 Conference (2013). G16-03-138
9. Iliev, D., Iliev, O., Kirsch, R., Mikelic, A., Printsypar, G., Calo, V.: Efficient simulations of poroelastic deformations in pleated filters. In: Proceedings Filtech 2015 Conference (2015). F05-01-P097
10. Iliev, O., Kirsch, R., Lakdawala, Z.: On some macroscopic models for depth filtration: analytical solutions and parameter identification. In: Proceedings Filtech Conference 2011 (2011)
11. Iliev, O., Kirsch, R., Lakdawala, Z., Starikovičius, V.: Numerical simulation of non-Darcy flow using filter element simulation toolbox (FiltEST). In: Proceedings Filtech 2013 Conference (2013). G18-02-148
12. Iliev, O., Kirsch, R., Osterroth, S.: Cake filtration simulation for poly-dispersed spherical particles. In: Proceedings Filtech 2015 Conference (2015). L10-03-P112
13. Iliev, O., Lakdawala, Z., Andrä, H., Kabel, M., Steiner, K., Starikovičius, V.: Interaction of fluid with porous structure in filteration processes: modelling and simulation of pleats deflection. In: Proc. Filtech Europa, vol. 2, pp. 27–31 (2009)
14. Iliev, O., Lakdawala, Z., Kirsch, R., Steiner, K., Toroshchin, E., Dedering, M., Starikovičius, V.: CFD simulations for better filter element design. In: Proceedings Filtech Conference 2011. Filtech Congress 2011 (2011)
15. Iliev, O., Lakdawala, Z., Printsypar, G.: On a multiscale approach for filter efficiency simulations. Comput. Math. Appl. **67**(12), 2171–2184 (2014)
16. Iliev, O., Lakdawala, Z., Starikovičius, V.: On a numerical subgrid upscaling algorithm for Stokes–Brinkman equations. Comput. Math. Appl. **65**(3), 435–448 (2013)
17. Iliev, O., Laptev, V.: On numerical simulation of flow trough oil filters. Comput. Vis. Sci. **6**, 139–146 (2004)

18. Iliev, O., Lazarov, R., Willems, J.: Variational multiscale finite element method for flows in highly porous media. Multiscale Model. Simul. **9**(4), 1350–1372 (2011)
19. Iliev, O., Printsypar, G., Rief, S.: A two-dimensional model of the pressing section of a paper machine including dynamic capillary effects. J. Eng. Math. **01/2013**, 81–107 (2013)
20. Iliev, O., Rybak, I.: On numerical upscaling for flows in heterogeneous porous media. Comput. Methods Appl. Math. **8**, 60–76 (2008)
21. Iliev, O., Schindelin, A., Wiegmann, A.: Computer aided development of hydraulic filter elements—from theory to patent and products. In: Berns, K., Schnindler, C., Dreßler, K., Jörg, B., Kalmar, R., Hirth, J. (eds.) Proceedings of the 1st Commercial Vehicle Technology (CVT 2010). Commercial Vehicle Technology 2010, pp. 68–75 (2010)
22. Kabel, M., Andrä, H., Hahn, F., Lehmann, M.: Simulating the compression of filter materials. In: Proceedings Filtech 2013 Conference (2013). G16-01-057
23. Latz, A.: Partikel- und Wärmetransport durch Strömung in Mikrostrukturen. Tech. Rep. 2. Report, Stiftung Rheinland-Pfalz für Innovation (2003)
24. Latz, A., Rief, S., Wiegmann, A.: Research note: computer simulation of air filtration including electric surface charges in 3-dimensional fibrous microstructures. Filtration **6**(2), 169–172 (2006)
25. Latz, A., Wiegmann, A.: Filtermaterialdesign per software. Laboratory IT User Service **1/04** (2004)
26. Ohser, J., Schladitz, K.: 3d images of material structures—processing and analysis. Imaging Microsc. **11**(4), 21 (2009)
27. Prill, T., Schladitz, K., Jeulin, D., Wieser, C.: Morphological segmentation of FIB-SEM data of highly porous media. J. Microsc. **250**(2), 77–87 (2013)
28. Probst-Schendzielorz, S., Rief, S., Wiegmann, A., Andrä, H., Schmitt, M.: Simulation of press dewatering. In: Progress in Paper Physics Seminar, pp. 301–302 (2011)
29. Proceedings of 2nd Annual meeting of Bulgarian Section of SIAM: Modelling and Simulation of Multiscale Problems in Industrial Filtration Processes (2007)
30. Rief, S., Iliev, O., Kehrwald, D., Latz, A., Steiner, K., Wiegmann, A.: Simulation und virtuelles Design von Filtermedien und Filterelementen. In: Durst, G.M., und Klein, M. (eds.) Filtration in Fahrzeugen. Expert-Verlag, Renningen (2006)
31. Rief, S., Wiegmann, A., Latz, A.: Computer simulation of air filtration including electric surface charges in three-dimensional fibrous micro structures. Math. Model. Anal. **10**(3), 287–304 (2005)
32. Schmidt, K., Rief, S., Wiegmann, A.: Simulation of dpf media, soot deposition and pressure drop evolution. In: Proc. Filtech Europa, vol. 2, pp. 74–80 (2009)
33. Spahn, J., Andrä, H., Kabel, M., Müller, R.: A multiscale approach for modeling progressive damage of composite materials using fast Fourier transforms. Comput. Methods Appl. Mech. Eng. **268**, 871–883 (2014). ISSN 0045-7825
34. Starikovičius, V., Ciegis, R., Iliev, O., Lakdawala, Z.: A parallel solver for the 3d simulation of flows through oil filters. In: Parallel Scientific Computing and Optimization. Springer Optimization and Its Applications, vol. 27, pp. 181–191 (2008)
35. Wiegmann, A.: GeoDict. Fraunhofer ITWM Kaiserslautern. www.geodict.com
36. Wiegmann, A.: A Fast Fictitious Force 3D Stokes Solver. Fraunhofer ITWM Kaiserslautern (2007)
37. Wiegmann, A., Becker, J.: Virtual characterization of the pore structure of nonwoven. In: Proceedings of the International Nonwoven Technical Conference (2007)
38. Wiegmann, A., Zemitis, A.: Electrostatic Fields for Filtration Simulations in Fibrous Air Filter Media. Fraunhofer ITWM Kaiserslautern (2006)

## Dissertations on This Topic at Fraunhofer ITWM


39. Buck, M.: Overlapping domain decomposition preconditioners for multi-phase elastic composites. Ph.D. thesis, Technical University Kaiserslautern (2013)
40. Kronsbein, C.: On selected efficient numerical methods for multiscale problems with stochastic coefficients. Ph.D. thesis, Technical University Kaiserslautern (2013)
41. Lakdawala, Z.: On efficient algorithms for filtration related multiscale problems. Ph.D. thesis, Technical University Kaiserslautern (2010)
42. Laptev, V.: Numerical solution of coupled flow in plain and porous media. Ph.D. thesis, Technical University Kaiserslautern (2004)
43. Nagapetyan, T.: Multilevel Monte Carlo method for distribution function approximation with application to asymmetric flow field flow fractionation. Ph.D. thesis, Technical University Kaiserslautern (2014)
44. Naumovich, A.: Efficient numerical methods for the Biot poroelasticity system in multilayered domains. Ph.D. thesis, Technical University Kaiserslautern (2007)
45. Niedziela, D.: On numerical simulations of viscoelastic fluids. Ph.D. thesis, Technical University Kaiserslautern (2006)
46. Printsypar, G.: Mathematical modeling and simulation of two-phase flow in porous media with application to the pressing section of a paper machine. Ph.D. thesis, Technical University Kaiserslautern (2012)
47. Rief, S.: Nonlinear flow in porous media—numerical solution of the Navier–Stokes system with two pressures and application to paper making. Ph.D. thesis, Technical University Kaiserslautern (2005)
48. Schmidt, K.: Dreidimensionale Modellierung von Filtermedien und Simulation der Partikelabscheidung auf der Mikroskala. Ph.D. thesis, Technical University Kaiserslautern (2011)
49. Schmidt, S.: On numerical simulation of granular flow. Ph.D. thesis, Technical University Kaiserslautern (2009)
50. Strautins, U.: Flow-driven orientation dynamics in two classes of fibre suspensions. Ph.D. thesis, Technical University Kaiserslautern (2008)
51. Willems, J.: Numerical upscaling for multiscale flow problems. Ph.D. thesis, Technical University Kaiserslautern (2009)
52. Zemerli, C.: Continuum mechanical modelling of granular systems. Ph.D. thesis, Technical University Kaiserslautern (2013)


## Further Literature


53. Abboud, N.M., Corapcioglu, M.Y.: Numerical solution and sensitivity analysis of filter cake permeability and resistance to model parameters. Transp. Porous Media **10**(3), 235–255 (1993)
54. Angot, P.: A fictitious domain model for the Stokes–Brinkman problem with jump embedded boundary conditions. C. R. Math. **348**(11–12), 697–702 (2010)
55. Arbogast, T., Bryant, S.: A two-scale numerical subgrid technique for waterflood simulations. SPE J. **7**(4), 446–457 (2002)
56. Balhoff, M., Mikelić, A., Wheeler, M.: Polynomial filtration laws for low Reynolds number flows through porous media. Transp. Porous Media **81**(1), 35–60 (2010)
57. Barree, R., Conway, M.: Beyond beta factors: a complete model for Darcy, Forchheimer and trans-Forchheimer flow in porous media. In: 2004 Annual Technical Conference and Exhibition, Paper SPE 89325 (2004)
58. Bear, J., Bachmat, Y.: Introduction to Modeling of Transport Phenomena in Porous Media. Kluwer Academic, Dordrecht (1990)

59. Beavers, G., Joseph, D.: Boundary conditions at a naturally permeable wall. J. Fluid Mech. **30**, 197–207 (1967)
60. Biot, M.: General theory of threedimensional consolidation. J. Appl. Phys. **12**(2), 155–164 (1941)
61. Biot, M.: Theory of elasticity and consolidation for a porous anisotropic solid. J. Appl. Phys. **26**(2), 182–185 (1955)
62. Brinkman, H.: A calculation of the viscous force exerted by a flowing fluid on a dense swarm of particles. Appl. Sci. Res. A **1**, 27–34 (1949)
63. Brown, R., Wake, D.: Loading filters with monodisperse aerosols: macroscopic treatment. J. Aerosol Sci. **30**(2), 227–234 (1999)
64. Chen, M., Durlofsky, L., Gerristen, M., Wen, X.: A coupled local-global upscaling approach for simulating flow in highly heterogeneous formations. Adv. Water Resour. **26**(10), 1041–1060 (2003)
65. Chorin, A.: Numerical solution of Navier–Stokes equation. In: Mathematics of Computation, vol. 22, pp. 745–760 (1968)
66. Darcy, H.: Les fontaines publiques de la ville de Dijon. Victor Dalmont, Paris (1856)
67. Dedering, M.: Filter medium for an oil filter. Patent DE102008027663 (2009)
68. Efendiev, Y., Hou, T.: Multiscale Finite Element Methods. Springer, Berlin (2009)
69. Ergun, S.: Fluid flow through packed columns. Chem. Eng. Prog. **48**, 89 (1952)
70. Espedal, M., Fasano, A., Mikelić, A.: Filtration in Porous Media and Industrial Application. Springer, Berlin (1998)
71. Ferziger Peric, M.: Computational Methods for Fluid Dynamics. Springer, Berlin (1999)
72. Fletcher, C.: Computational Techniques for Fluid Dynamics. Springer, Berlin (1991)
73. Forchheimer, P.: Wasserbewegung durch Boden. Z. Ver. Deutsch. Ing. **45**, 1782 (1901)
74. Gresho, P., Sani, R.: Incompressible Flow and the Finite Element Method. Advection–Diffusion and Isothermal Laminar Flow, vol. 1. Wiley, Chichester (1998). In collaboration with M.S. Engelman
75. Griebel, M., Klitz, M.: Homogenisation and numerical simulation of flow in geometries with textile microstructures. Multiscale Model. Simul. **8**(4), 1439–1460 (2010)
76. Hornung, U.: Homogenization and Porous Media. Springer, New York (1996)
77. Ives, K.: Rapid filtration. Water Res. **4**, 201–223 (1970)
78. Iwasaki, T.: Some notes on sand filtration. J. Am. Wat. Wks. Ass. **29**(10), 1591–1602 (1937)
79. Jackson, G.W., James, D.F.: The permeability of fibrous porous media. Can. J. Chem. Eng. **64**(3), 364–374 (1986)
80. Jäger, W., Mikelić, A.: On the boundary conditions at the contact interface between a porous medium and a free fluid. Ann. Sc. Norm. Super. Pisa, Cl. Sci. **23**(3), 403–465 (1996)
81. Jenny, P., Lunati, I.: Multi-scale finite-volume method for elliptic problems with heterogeneous coefficients and source terms. PAMM **6**(1), 485–486 (2006)
82. Johnson, R.W. (ed.): The Handbook of Fluid Dynamics. CRC Press, Boca Raton (1998)
83. Kaviany, M.: Principles of Heat Transfer in Porous Media. Mechanical Engineering Series. Springer, Berlin (1995)
84. Marciniak-Czochra, A., Mikelić, A.: A rigorous derivation of the equations for the clamped Biot-Kirchhoff-Love poroelastic plate. Arch. Ration. Mech. Anal. **215**(3), 1035–1062 (2015). doi:10.1007/s00205-014-0805-2
85. Mints, D.M.: Kinetics of the filtration of low concentration water suspensions. Dokl. Akad. Nauk SSSR **78**(2), 315–318 (1951)
86. Morris, J.: An Overview of the Method of Smoothed Particle Hydrodynamics (1995)
87. Ochoa-Tapia, J., Whitaker, S.: Momentum-transfer at the boundary between a porous-medium and a homogeneous fluid. Int. J. Heat Mass Transf. **38**, 2635–2655 (1995)

88. Ohser, J., Mücklich, F.: Statistical Analysis of Microstructures in Materials Science. Wiley, New York (2000)
89. Olivier, J., Vaxelaire, J., Vorobiev, E.: Modelling of cake filtration: an overview. Sep. Sci. Technol. **42**(8), 1667–1700 (2007)
90. Podgórski, A.: Macroscopic model of two-stage aerosol filtration in a fibrous filter without reemission of deposits. J. Aerosol Sci. **29**, S929–S930 (1998)
91. Rhie, C., Chow, W.: Numerical study of the turbulent flow past an airfoil with trailing edge separation. AIAA J. **21**(11), 1525–1532 (1983)
92. Saleh, A., Hosseini, S., Tafreshi, H.V., Pourdeyhimi, B.: 3-d microscale simulation of dust-loading in thin flat-sheet filters: a comparison with 1-d macroscale simulations. Chem. Eng. Sci. **99**, 284–291 (2013)
93. Schindelin, A., Schadt, W.: Gewelltes oder gefaltetes Flachmaterial. Patent DE102007040892 A1 (2009)
94. Silvester, D., Elman, H., Kay, D., Wathen, A.: Efficient preconditioning of the linearized Navier–Stokes equation. J. Comput. Appl. Math. **128**, 261–279 (2001)
95. Taber, L.A.: A theory for transversal deflection of poroelastic plates. J. Appl. Mech. **59**, 628–634 (1992)
96. Tien, C.: Introduction to Cake Filtration: Analyses, Experiments and Applications. Elsevier, Amsterdam (2006)
97. Tien, C., Payatakes, A.C.: Advances in deep bed filtration. AIChE J. **25**(5), 737–759 (1979)
98. Turek, S.: Efficient Solvers for Incompressible Flow Problems. An Algorithmic and Computational Approach. Lecture Notes in Computational Science and Engineering, vol. 6. Springer, Berlin (1999)
99. Versteeg, H., Malalasekera, W.: An Introduction to Computational Fluid Dynamics: The Finite Volume Method. Pearson Education, Upper Saddle River (2007)
100. Ward, J.: Turbulent flow in porous media. J. Hyd. Div. ASCE **90**, 1–12 (1964)
101. Weinan, E., Engquist, B.: The heterognous multiscale methods. Commun. Math. Sci. **1**(1), 87–132 (2003)
102. Wesseling, P.: Principles of Computational Fluid Dynamics. Springer, Berlin (2001)

# Maximal Material Yield in Gemstone Cutting

Karl-Heinz Küfer, Volker Maag, and Jan Schwientek

## 1 Optimum Material Usage—A Must with Expensive Resources

The quest for optimum material cutting is one of the basic principles of industrial production, since the sales price of a manufactured good is not only a function of the production costs, but often depends predominantly on the necessary raw material usage. Hence, the range of problems involving maximizing material usage is large.

A tradesman papering walls, for example, will seek to minimize the number of rolls of wallpaper he uses. In so doing, he will try to manage his use of remnants so that the final waste pieces are as small as possible. A carpenter cutting molding to size deals with the same challenge, as does a metalworker using ready-made metal profiles. This one-dimensional problem—only the length of the pieces matters here—is known in the mathematical literature as the *Cutting Stock Problem* (see [18, 38], for example). Even in its simple form, it proves to be NP-hard, which is the same as saying that there can be no efficient algorithm for minimizing waste.

Cutting shapes from standard wooden panels, pieces of clothing from fabric rolls, or shoe elements from leather hides represents an even more difficult material usage optimization problem; here, in addition to the geometry of the cut-outs, one must also consider their orientation—as with a fiber's running direction in a fabric—or cut around flaws in the material—as with knots in a wooden board or injuries to the animal supplying the hide.

Analogous problems also exist in three dimensions: a dispatcher, for example, when picking and packing goods will search for the smallest package that will hold all the pieces, in order to minimize shipping costs. A diamond or colored-gemstone producer will also

K.-H. Küfer · V. Maag (✉) · J. Schwientek
Fraunhofer-Institut für Techno- und Wirtschaftsmathematik ITWM, Fraunhofer-Platz 1, 67663 Kaiserslautern, Germany
e-mail: volker.maag@itwm.fraunhofer.de



H. Neunzert, D. Prätzel-Wolters (eds.), *Currents in Industrial Mathematics*,
DOI 10.1007/978-3-662-48258-2_8

**Fig. 1** Exploiting gemstones: raw stones and a selection of cut jewels from Paul Wild oHG

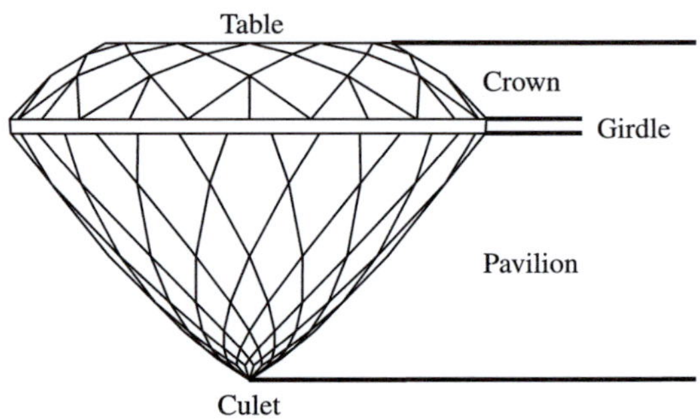


**Fig. 2** The elements of a faceted stone

strive to cut the largest and thus most valuable jewels possible from the raw material he receives from the mine, taking into consideration the preferred orientation and such flaws as cracks and inclusions (see Fig. 1). In the literature, the optimization task in the 2D or 3D situation is often referred to as a *Nesting Problem* (see [19], for example).

## 1.1 Gemstone Production—An Ancient Craft Using Scarce Raw Materials

This chapter deals with the optimal cutting of gemstones, although most of the methods developed here can be applied in an analogous manner to the other examples mentioned earlier. To promote a better understanding of the practical questions, we have compiled some background information about gemstone cutting.

For more than 500 years, the most common form of jewel has been the *faceted stone*. This is a cut and polished gemstone whose surface consists of small, planar areas known as *facets*. The gemstone is divided into three elements: the *crown*, the *girdle*, and the *pavilion* (see Fig. 2).

The crown and pavilion are polyhedral. The girdle is bordered by planar or curved surfaces and determines the base form of the faceted stone. There are many *faceted stone shapes*, the best-known of which are shown in Fig. 3.

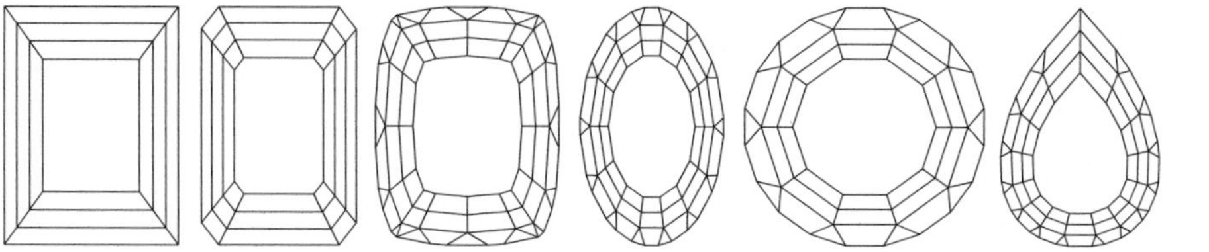

**Fig. 3** The best-known faceted stone shapes, *from left to right*: baguette, emerald, antique, oval, round, and pear

**Fig. 4** The round shape in a brilliant cut and a step cut; Fig. 2 depicts the Portuguese cut.

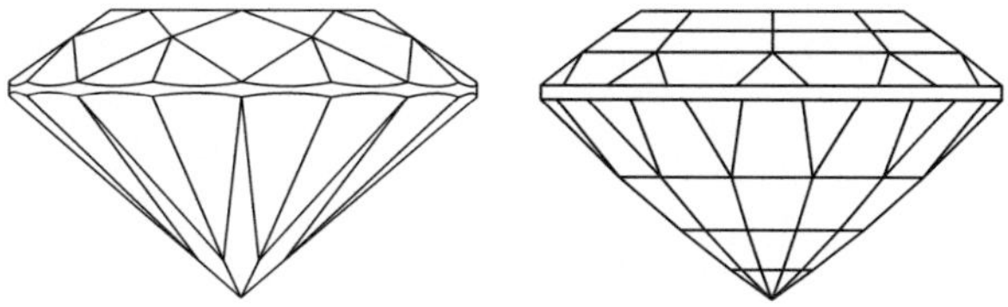

Along with the base form, there are various basic types of crown and pavilion cuts, which we will subsequently refer to as *facetings* (see Fig. 4, [25]). Some are possible for every base form; others are not. Moreover, with some cuts, the number of facets is predefined, whereas for others, the number of facets depends on the size of the finished stone.

Besides its base form and cut, a faceted stone is also characterized by a variety of size parameters, such as the height, length, and width of the crown, girdle, and pavilion. For optical and esthetic reasons, there are upper and lower limits on certain ratios between these parameters, which we will refer to subsequently as *proportions*. With diamonds, for example, the transparency of the material and the laws of optics dictate that faceting patterns and proportions be held within very narrow limits, in order to promote the most favorable light transmission paths. Here, it is typical that standard faceted stone shapes are merely scaled to fit the raw material and rotated in order to maximize yield. With colored gemstones, the rules for proportions and faceting are significantly less stringent. This has a favorable impact on the optimization tolerances, but it also makes the resulting mathematical problem considerably harder to solve. For this reason, we consider the more general problem of colored gemstone cutting in the following discussion.

In the past, size-dependent cuts and weak constraints on proportions led to the facets not being cut directly into the raw material. The process chain for producing a faceted stone contains four steps:

(1) *Sectioning*: First, the raw material is sectioned into "clean" pieces containing no flaws or cracks, which we will refer to as *rough stones*. In the end, each rough stone delivers one faceted stone.
(2) *Pre-forming*: Here, the rough stones are coarsely pre-cut, or ebauched. This defines the base form and the approximate proportions of the subsequent faceted stone.
(3) *Grinding*: Next, the facets of the preferred cut are applied to these pre-cut forms.
(4) *Polishing*: Finally, the facets are polished to a high gloss finish.

A faceted stone is appraised according to four criteria, the so-called *Four C's*: *Carat*, *Clarity*, *Color*, and *Cut*. The carat is a measure of weight equaling 0.2 grams. The value of a faceted stone is directly proportional to its weight. The clarity indicates the absence of inclusions, cracks, and surface flaws. The greater the clarity, the more valuable the faceted stone. The natural color of a gemstone and/or the enhanced effect created during its processing also have a substantial impact on the value of a faceted stone. Because this factor can hardly by influenced, however, it will not be discussed further. The cut of a faceted stone has a decisive influence on its ability to reflect and refract light. An increase in a faceted stone's reflective and refractive characteristics increases its value. Moreover, the faceting contributes significantly to a stone's overall esthetic qualities.

The value of a faceted stone is thus appraised according to its weight and its esthetic qualities.

Today, gemstones and diamonds are still manufactured largely by hand. Although industrial saws and modern grinding machines are used here, all geometric determinations rely solely on the practiced eye and skilled craftsmanship of the jewel makers. Because the processes involved are complex and expensive, and because there are not enough apprentices learning the trade in the old industrialized nations, most production has long since shifted to the countries of South Asia.

In the first processing step, the sectioning of larger stones into rough stones so as to avoid flaws in the material, about half of the raw material is lost. In converting the rough stones from step (1) into faceted stones in steps (2)–(4), approximately two-thirds more of the precious material is lost. Thus, the loss of weight from the original raw material to the finished product is about five-sixths of the total.

## 1.2 Automation as a Chance for Better Material Utilization

Given the losses described above, it is natural to ask if mathematical modeling and algorithmic concepts that optimize the sectioning of raw material and the embedding of a faceted stone in a rough stone might not be able to significantly increase the yield above that achieved by the skill of the craftsman. In order to answer this question, however, a number of challenges must be met, the most important of which are mentioned here:

- *Data acquisition*: The first step toward using mathematical models is collecting input data. Here, the geometry of the rough stones must be depicted for the entire process (steps 1–4) by means of 3D imaging. This can be accomplished using CT technology, for example. However, due to the limited resolution of the available technology, it is very difficult to represent hairline cracks and very small air inclusions in the material. If one assumes only clean individual stones (steps 2–4), then the digitalization can be limited to depicting the stones' surfaces, which can be accomplished with stripe projection or laser scanning technology.

How does one prepare the large data sets so that they are suited for the subsequent optimization problems?

- *Mathematical model*: Two questions must be answered when dealing with optimization problems: What is feasible? and What is good? Neither of these questions can be easily answered for colored gemstone production. Weak constraints on the proportion rules and the large variety of base forms and faceting patterns make it hard to completely describe the alternative sets mathematically. Even harder is bringing the wish for maximum weight—which is directly proportional to volume—into harmony with minimum esthetic demands, which depend on individual taste and cultural background.

How does one mathematically formulate esthetic requirements?

- *Exploitation algorithms*: From a mathematical perspective, the resulting optimization problems are extremely complex. This is due less to the above-mentioned large data sets arising from the digitalization of rough stones than to the geometric principles, which, although actually quite simple, are laborious to mathematize. These principles demand that the resulting faceted stones must be completely contained within the rough stone and may not overlap each other. A second issue is the simultaneous existence of continuous variables, such as size and proportion, and discrete variables, such as the number of facets.

How does one mathematically model the containment and non-overlapping conditions? Is it conceivable to de-couple the combinatorics of the faceting from the optimal sizing of the proportions?

- *Fully-automated production process*: If one wants to use mathematical models and algorithmic concepts to optimize the cutting of rough gemstones, it becomes necessary to automate production; one cannot simply present a craftsman with a good plan and then wish him luck with it. Simple studies show that even the smallest deviations from the optimal positioning of the faceted stones in the rough stone can lead to marked deteriorations in yield. Thus, there is no way around the implementation of an industrial production process involving the use of CNC technology.

How can one clamp the individual work pieces during processing? How can the geometry be transferred from one process step to the next with the required precision? Which handling technology should be used? Which saws, grinders, and polishers are appropriate? Can one continue to use the techniques of manual production, or will it be necessary to develop new ways and means?

## 2 Optimum Volume Yield—Is This a Mathematically Challenging Problem?

A person less trained in mathematics might think: a problem that is so easy to put into words and so easy to understand cannot be so difficult to solve. After all, it's just a matter of packing a few faceted stones into a rough stone in an economically favorable manner; what's so hard about that? Unfortunately, this first impression is deceptive, and a look in the mathematical literature or a search of the Internet under the keywords Cutting Stock or Nesting Problem brings a rude awakening. Only the simplest variants, such as rectangular or ball packaging, are well understood mathematically—and even these have only been partially solved. More generalized problem statements and solution approaches are extremely rare. Thus, in 2003, as the ITWM began work on this problem, the first task was to find a model that suited the problem.

### 2.1 Mathematically Modeling the Optimization Problem—Or, what Is an Acceptable Design for a Jewel?

The central question for modeling the problem is how to mathematically describe a faceted stone. The initial idea of describing the most common convex base forms as polyhedrons failed, since the girdle that separates the crown from the pavilion is, in many cases, a smooth, curved surface, whereas the crown and pavilion have a polyhedral structure. Another question is even more complicated: what is the class of acceptable facet patterns belonging to a given base form? The craftsmen have rules-of-thumb for the number of facets on the girdle, and these depend on the size of the stone; they know the approximate number of facet rows or steps on the crown and pavilion; they know the size of the limiting angles between the facets and the girdle. Facets should decrease in height as one moves away from the girdle; they should be kite-shaped on Portuguese cut stones and the half-axes should divide the kites approximately into golden cuts; and much more. Regarding the proportions, the following guidelines apply: the crown contributes about one-third of the total height, the pavilion, about 50–55 %, and the girdle makes up the rest. The pavilion should not be too "bellied," but not too slender either—otherwise, too much volume is lost, etc. And the most important point of all is this: at the end of the day, the stone must be beautiful; rules and guidelines alone are not enough.

The above discussion indicates the all too typical dilemma of putting mathematical optimizations into practice: the mathematician needs clear-cut rules to do his work. The alternative set—in this case, the feasible faceted stones—from which favorable solutions should ultimately be selected, must be described exactly, according to fixed rules. There is no room for vagueness. Moreover, to optimize, one also needs a target quantity to help in comparing the quality of two possible solutions. At first glance, this would seem to be simple for gemstone cutting: the stones should be as large as possible. This increases the number of carats, i.e., the weight, thus raising their value. At second glance, however, there is a problem here as well.

If the stone is merely large, but not beautiful, no one will buy it. Therefore, we need a definition of "beautiful" that can be incorporated into the description of the alternative set. Or, at a minimum, we need measurement quantities that correlate well with "beautiful," so that we can then optimize them as objectives in balance with solutions that are "large" or "heavy."

The geometric problem that seems at first so easy to formulate now proves to be mathematically challenging indeed. Gemstone cutting seems somehow to be an art or perhaps a craft—in any event, not a science. Peering over the shoulder of the practitioner might provide us with some clues. How does a cutter answer the above questions? Does he simply start cutting away, or does he use rules-of-thumb containing mathematical principles that we can imitate with our models?

Observations of the craftsman at work are quite revealing: after sectioning the raw material, he then closely inspects the shape of a resulting rough stone to see which base form the final faceted stone might have and how this base form is oriented inside the rough stone. Then he starts by cutting the base form's girdle. The crown and pavilion are coarsely pre-formed; as of this point, there are no facets. This pre-forming process determines the proportions of the stone, the height ratio and degree of belliedness as well as the base angles to the girdle. After pre-forming, the facet rows and counts are assigned and the crown and pavilion are faceted. Figure 5 shows the pre-cut form and intended proportions for the faceted stone depicted in Fig. 2.

The manual production process is thus divided into two parts: pre-forming and faceting. This inspired us, in our mathematical modeling, to de-couple the continuous variables, such as the height and proportions of the faceted stone, from the discrete variables, such as the number of rows and facets in a given facet pattern.

The approach of de-coupling continuous and discrete variables simplifies the structure of the optimization problem significantly and allows the esthetic boundary conditions to be more easily described in the reduced variable sets. But what is the best way to implement this approach?

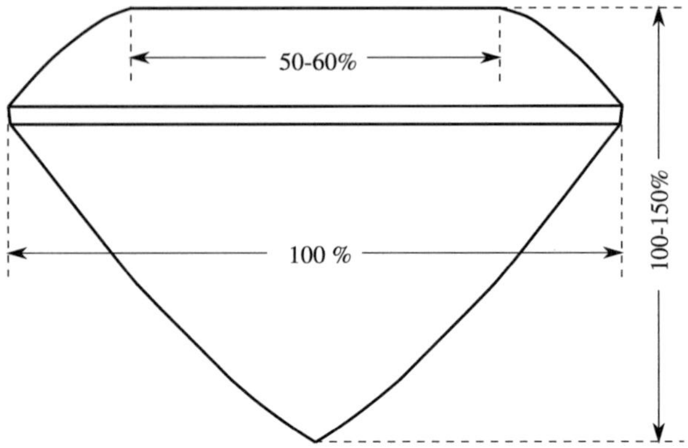


**Fig. 5** Pre-cut form and proportions for a round-cut stone: table diameter and total height in relation to girdle diameter

The implementation involves introducing a parameterized equivalent to the smooth pre-cut form, which we refer to as the *calibration body*. This is then optimized toward the end of maximizing material yield. Considerations regarding the appearance of a suitable faceting are relegated to a second step, which is discussed in detail in Sect. 5.1

Let us now turn to the optimization problem of the parameterized calibration body. For a single stone, this is closely related to the design-centering problem known in the literature (see [30]), when one describes the quantities relevant for the proportions, such as height, width, and degree of belliedness, as calibration body parameters (i.e., *design* parameters) and takes position and scaling as further degrees of freedom for the optimization. If one now places limits on the proportion parameters so as to ensure a more-or-less satisfactory esthetic result, then one is left with the question of how to achieve the largest possible volume of a parameterized gem design.

In the following discussion, the requirement that the faceted stone be completely contained within the rough stone is called the *containment condition*. This is simple and easy to understand, but how can it be mathematically implemented?

Putting it another way, the containment condition requires that each point of the design, that is, the calibration body, must also be a point of the container, that is, the rough stone. We have here, then, an *infinite* number of constraints for a finite number of parameters, which must be fulfilled for a feasible calibration body. Problems of this sort are referred to as *semi-infinite optimization problems*. Further challenges revolve around the questions of whether one can mathematically describe in a similar manner the localization of flaws in the resulting jewel or the non-overlapping of two faceted stones in cases where more than one jewel is embedded in a single rough stone. This *non-overlapping condition* is closely allied to the containment condition. The approach to dealing with both of these questions is discussed in Sect. 4.

A generalization results when one also requires minimum separation distances. Thus, when sectioning raw material into blanks or embedding multiple stones in one rough stone, it is important to arrange the blanks or stones so as to maintain the minimum separation distances required for the production process. Moreover, the production process may also demand adherence to other arrangement principles. For example, if circular saw technology is being used, one must ensure that the arrangement allows for consecutively executed through-cuts, also known as *guillotine cuts* (for more, see Sect. 6.2).

## 2.2 The Algorithms—How to Find Optimal Solutions

If one keeps to the above modeling approach, the algorithmic challenge in gaining an optimal calibration body then becomes developing numerical solution concepts for *semi-infinite optimization problems* that robustly solve high-dimensional, non-convex problems in an acceptable computation time.

To do so, one must first work on reducing the problem size. Here, the goal is to depict the rough stone—discretized via volume or surface data—using the most economical representation possible. Ideally, this is accomplished in a model-friendly form that allows for reduction to a finite problem (see Sect. 5.1.3). To depict the rough stone, one enlists the smallest possible number of simple, smooth parametrical functions that permits numerically non-problematical evaluation.

What remains is a global optimization problem, which commonly has numerous local extreme solutions. If one can characterize the local extremes in the general case using a first-degree optimality condition—such as the Karush–Kuhn–Tucker condition (KKT condition)—then the challenge is to select a suitable strategy for finding an approximately globally optimal solution. Here, there is no generic approach. A hybrid strategy must be found for enumerating favorable local extremes and/or excluding unfavorable ones.

When one has found good calibration bodies for approximating feasible faceted stones, then one can turn to the second optimization task: finding a favorable faceting; that is, one that both follows the standard rules of the gemstone cutter's art and minimizes volume reduction of the calculated calibration body. At first, it seems obvious that using enough small facets should guarantee such an approximation. However, upon closer inspection, it becomes clear that the standard facet patterns used in the gemstone industry do not allow every calibration body to be approximated adequately. Thus, a certain coupling of faceting and base form once again sneaks in through the back door, so to speak. For fixed facet patterns, the problem of faceting can also be modeled as a non-linear global optimization problem. Here, the question arises as to how one can suitably integrate into the optimization problem the number of facets and facet rows as free optimization variables.

# 3 ITWM Projects Dealing with This Topic

## 3.1 Projects with the Gemstone Industry

The idea of increasing material yield during gemstone production by using mathematical optimization methods and automation was prompted by Paul Wild oHG (oHG = general partnership). This family-managed, mid-sized firm located in Kirschweiler, Rheinland-Pfalz, near Idar-Oberstein, is one of Europe's leading producers of precious colored gemstones. The Company has its own mines in Africa, South America, and Asia, which ensure its supply of raw materials. Production of jewelry stones takes place predominantly in Asia, whereas administration and sales are headquartered in Kirschweiler.

As is typical for the industry, Wild's jewelry stone production was carried out exclusively by hand until 2003. Up to that point, there had been no significant attempts to industrialize or automate production processes. Some experiments in improving yields in the 1990's using a semi-automatic installation from Israel gave managing director Markus P. Wild the idea that it ought to indeed be possible to produce colored gemstones in a fully-automated industrial process, one optimized for each individual rough stone. Since 2003, Markus P. Wild has been pursuing this vision, in collaboration with the Fraunhofer-Gesellschaft and other partners from the machine engineering sector.

### 3.1.1 First Steps—Preliminary Feasibility and Profitability Studies

The Spring of 2003 marked the first contact between Markus P. Wild and the Fraunhofer-Gesellschaft. As a result, the Fraunhofer Institute for Industrial Mathematics ITWM, in Kaiserslautern, the Fraunhofer Institute for Applied Optics and Precision Engineering IOF, in Jena, and the Fraunhofer Institute for Manufacturing Technology and Advanced Materials IFAM, in Bremen, were commissioned in the Fall of 2003 and 2004 to conduct a series of preliminary studies toward the end of preparing a concept for the automatic production of colored jewelry stones:

- A study into 3D measurement of raw gemstones by means of the stripe projection method (Fraunhofer IOF, Jena)
- A study into calculating optimal cutting volumes of colored raw gemstones (Fraunhofer ITWM, Kaiserslautern)
- A study into bonding colored gemstones to metallic processing pins by means of UV-hardened or hot-melt adhesives (Fraunhofer IFAM, Bremen)

In the course of these preliminary studies, the basic feasibility of colored gemstone production with regard to pre-forming, grinding, and polishing in an industrial process was adequately verified. Thus, the development of an automatic cutting process in the context of an industrial research project could be started with acceptable prospects for success. This project was funded from 2005 to 2007 by the mid-sized company promotion foundation of Rheinland-Pfalz via the Investitions- und Strukturbank (ISB). An experimental setup was developed that was able to demonstrate, with scientific rigor, the feasibility of fully-automated colored gemstone processing.

### 3.1.2 Pioneer Work—The First Industrial Automation of Pre-forming, Grinding, and Polishing

The preliminary results were promising, and considerably higher volume yields could be achieved while still retaining excellent quality for the automatically processed jewelry stones. Thus, as a follow-up to the ISB-sponsored R&D endeavor, Wild oHG commissioned the construction of a fully-automated CNC-controlled production line. Although the most significant technological risks had been dealt with in the context of the ISB project, there were still some hurdles to overcome before a practicable industrial process could be implemented on the new production equipment. These were indeed overcome and, since 2008, the world's first fully automated production line for colored gemstones has been in operation at Wild oHG.

The operation of the production line quickly showed that, for efficient utilization, an integrated, multi-criteria decision-making process would be needed that considers all of the four C's–carat, color, clarity, and cut. In cooperation with the Fraunhofer ITWM in Kaiserslautern, in the course of a project sponsored by the German Economics Ministry from 2009 to 2011, a novel decision-support system was developed that facilitates the different types of production decisions: Proposals resulting from the cutting optimization are visualized within the rough stones before production starts; interactive 3D representation permits comparisons of the variants of proportion and faceting; production supervisors can check the quality of the variants before cutting begins; and the marketing department can integrate customers into the decision-making process via the Internet.

The research work in the Fraunhofer ITWM-Wild consortium was praised in the press and described as trailblazing. More than 70 articles appeared in such newspapers and journals as Die Zeit, FAZ (Frankfurter Allgemeine Zeitung), Handelsblatt, and Bild der Wissenschaft. Moreover, the accomplishments of the research consortium were honored in 2009 with the Joseph-von-Fraunhofer prize in a ceremony attended by the German Chancellor Angela Merkel.

The decision was finally made at the end of 2009 to guide the gemstone production machine to series maturity and bring it to market. In 2010, a modular pilot machine was built at the Fraunhofer Center in Kaiserslautern and, starting in the same year, control software was developed (see Fig. 6). The machine has been ready for marketing since the autumn of 2013, and is now being shown to potential buyers. The statements of interest that have already been received from more than 70 companies and technology brokers around the world are indeed very promising. Property rights that protect the machine concept have been granted. To this point, demonstrations at trade fairs have been avoided, so as not to aid potential product counterfeiters located in areas outside the patent protection zone.

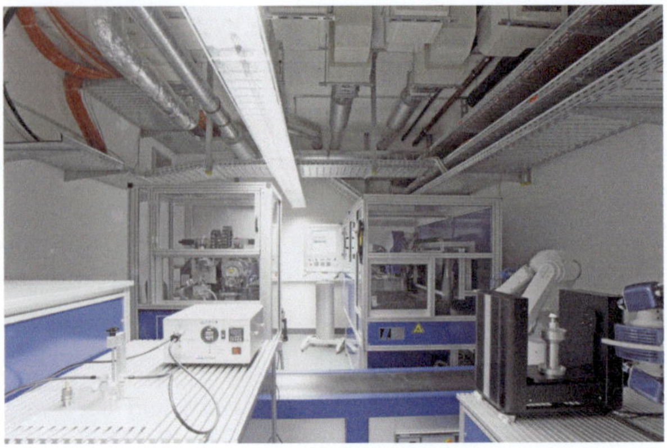

**Fig. 6** Pilot-production prototype developed at the Fraunhofer ITWM (Photo: G. Ermel, Fraunhofer ITWM)

### 3.1.3 The New Horizon—Automating the Sectioning Process

The earlier projects, dating from the years up to 2008, revolved primarily around the question of how to garner a single faceted stone from a rough stone. Beginning in 2009, however, the question of how to automate the sectioning process moved into the sights of the project group gathered around Wild oHG. Although one can produce individual stones from clean raw material by merely collecting data about the stone surface, one must collect volume data for the sectioning process, in order to distinguish between exploitable material and impurities, inclusions, and cracks. The method of choice for gaining such 3D data is high-resolution computer tomography (CT). Thus, Wild oHG commissioned testing of CT devices for their suitability for collecting volume data about colored raw gemstones. In 2010, a suitable system based on a two-frequency measurement process was located in the industry. The system was not yet being produced serially, however.

In addition to collecting volumetric data, automating the sectioning process also required a comprehensive study into which cutting technology would be appropriate for such automation. As with the cutting of individual stones, imitation of the manual production process seemed to be the safest path. To this point in time, raw material had always been sectioned by the most experienced craftsmen with the aid of diamond-studded circular saws. In 2009, Wild oHG and the Fraunhofer ITWM initiated the project "Development of a fully-automated sectioning process for colored gemstones," which was sponsored by the ISB Rheinland-Pfalz and concluded in late June, 2011. The results confirmed that one can indeed use a circular saw to section a colored gemstone in a fully automated process. A prototype of a sectioning machine was then built in the manufacturing center in Kirschweiler. During the actual operation of this machine, however, several obstacles became apparent that made its practical use uneconomical. Thus, some other technologies were also taken into consideration. In 2013, Wild oHG eventually bought a high-pressure waterjet cutting machine. An extension of the ISB-sponsored sectioning project, conducted in cooperation with the Fraunhofer ITWM, is now aiming for a fully-automated sectioning process based on the use of CT and waterjet cutting technologies. A detailed discussion of the sectioning process can be found in Sect. 6.

### 3.2 Relevant Competences of the ITWM Optimization Department and Related Projects

Since the beginning of its cooperation with Wild oHG, the Fraunhofer ITWM's Optimization Department has been systematically expanding its competences in modeling and solving industrial problems with semi-infinite optimization. Alongside the main project of gemstone cutting, questions stemming from other domains having comparable structures are also being treated with the help of these techniques.

In the area of nonlinear optimization, the Department has been utilizing its own algorithms from its inception. But it has also drawn upon commercial methods stemming mainly from the academic world, which are each adapted individually to the problem being treated. Here, a broad field of work is the hierarchic decomposition of problems into simpler sub-problems, or complexity reduction by means of adaptive discretization, or model reduction in optimization problems through the use of simplified/surrogate models.

In addition to those of the gemstone project, the following problems have been modeled and solved with the aid of semi-infinite optimization methods:

- Optimizing cooling systems of injection molds and pressure casting dies
- Optimizing the applicator position for radio frequency ablation

Both of these optimization problems deal with how to optimally distribute heat in a geometrically complex environment. With injection and pressure casting, a cavity must be cooled as homogeneously as possible; with radio frequency ablation, tumor tissue must be heated as homogeneously as possible. In each case, a suitable, enveloping isotherm must be established around the cooling or heating zone. If one models the heat distribution at equilibrium, then the requirement that the cooling or heating zone lie within the suitable isotherm is analogous to the containment condition of a faceted stone within a rough stone. Moreover, as with the gemstone problem, one can describe the non-overlapping of cooling channels and mold cavities or the non-puncturing of blood vessels by the applicator using semi-infinite constraints, which permits usage of the algorithm from the gemstone application.

Along with the above-mentioned semi-infinite modeling examples, the Fraunhofer ITWM's Optimization Department also considers numerous other decomposition problems from various industrial branches. Due to their character, however, these are solved using discrete enumeration techniques:

- Optimal arrangement of electronic components and switches for system-in-package applications
- Optimal cross-sections for cutting conifer woods in large sawmills
- Optimal cutting patterns for pants in the textile industry
- Optimal layouts for photovoltaic installations

### 3.3 Scientific Studies and Collaborations Involving Optimal Volume Yield

A whole series of scientific inquiries from the aforementioned domains led to graduate theses and publications. In a seminal degree thesis, semi-infinite optimization methods were applied for the first time to the problem of optimizing the material yield of gemstones. More specifically, [11] deals with the approximation of the rough stone using planes and quadrics and the volume optimization of a faceted stone using generalized semi-infinite optimization on the basis of a simple calibration-body model. The ideas originating here were then further developed and supplemented in a dissertation [16]. The topics of this work are volume optimization using realistic calibration-body models, as well as modeling multi-body embedding problems as a generalized semi-infinite optimization problem and developing a feasible method for generalized semi-infinite optimization problems. The most significant results were published in [2, 10, 12].

Other sub-problems were treated in three degree theses. In [6], the authors calculated the faceting for a given calibration body using methods of 3D-body reconstruction from two-dimensional drawings. The goal in [3] was to improve the rough stone approximation using splines. The topic in [7] was generating better starting points by comparing the rough stone geometries.

An alternative to the semi-infinite modeling approach for volume optimization of a faceted stone is described in [4]. Here, the idea was to apply methods of collision detection from algorithmic geometry to triangulations of the rough and faceted stones.

The more complex problems of sectioning and embedding multiple designs in one container are probed in the dissertation [14]. This study involved volume optimizing multiple calibration bodies using generalized semi-infinite optimization; extending the modeling of multi-body embedding problems as a generalized semi-infinite optimization problem; and developing two methods for generalized semi-infinite optimization problems.

One method used in this context to solve the semi-infinite optimization problems is to reformulate them as usual nonlinear problems (see Sect. 4.5.1). These are ill-posed, however, in the sense that the usual regularity requirements are not all fulfilled. As a consequence, the customary solution methods don't work directly; first, a regularization is required, that is, a softening of the original problem to a similar one having better characteristics. In [5], this idea of softening was transferred to the surface-minimized packing of rectangles, formulated as a nonlinear optimization problem to prevent the optimization from getting stuck in local optima.

The related thematic areas of cooling systems and radio frequency ablation mentioned in the previous section each yielded a dissertation [13, 15], and the latter also resulted in a publication [1].

Our studies into gemstone cutting also resonated strongly in the mathematical community. Along with a cover story in the SIAM news on gemstone cutting, the work was reported on in the American Mathematical Society's *Mathematical Moments* and a podcast was created.

In addition to the already mentioned Joseph von Fraunhofer Prize, awarded for the gemstone project, the two first-mentioned dissertations were also honored with a prize by the Kreissparkassen Foundation of the University City of Kaiserslautern for the best dissertations of the year in the field of mathematics.

## 4 Modeling and Solving Maximum Material Yield Problems

From a mathematical perspective, volume optimization in gemstone cutting represents a cutting and packing problem, more precisely, a *maximum material yield problem* (MaxMY).

In maximum material yield problems, the goal is to work out from a large body—the so-called *container*—a set of smaller bodies—the so-called *designs*–so that as little of the container material as possible is left over as scrap. If the container has flaws in it, the designs must also avoid these.

When modeling such problems, two different types must be distinguished:

(1) If the designs are fixed in size, then one searches within the set of all designs that can be generated from the container for the subset that best exploits it.
(2) If the designs are variable in size and possibly also in shape, then one searches for the variant of the designs that fits in the container and possesses the largest total volume.

**Notation Conventions** Let $\mathbb{N}$ be the set of natural numbers $\{1, 2, \ldots\}$, $\mathbb{N}_0 := \mathbb{N} \cup \{0\}$, $\mathbb{R}_+$, the set of non-negative real numbers, and $\mathbb{R}_{++}$, the set of positive real numbers.

We denote the set of real $m$-dimensional vectors as $\mathbb{R}^m$. The denotations $\mathbb{R}^m_+$ and $\mathbb{R}^m_{++}$ transfer accordingly. Vectors are essentially column vectors and printed in lower-case, bold type: $\mathbf{a}$. We denote the null vector with $\mathbf{0}$.

We denote the set of real $m \times n$ matrices with $\mathbb{R}^{m\times n}$. Matrices are printed in upper-case, bold type: $\mathbf{A}$. The matrix $\text{diag}(\mathbf{a})$ is the diagonal matrix, which possesses the components of the vector $\mathbf{a}$ as diagonal elements.

Sets (of scalars, vectors, etc.) are printed in upper-case, normal type: $A$. We denote the cardinality with $|A|$, the interior with $\text{int}(A)$, and the power set of a set $A$ with $2^A$.

We denote the gradients of a differentiable function $f : \mathbb{R}^m \to \mathbb{R}$ at the point $\bar{\mathbf{x}}$ with $\nabla f(\bar{\mathbf{x}})$. If the function depends on two (or more) vectors, that is, $f : \mathbb{R}^m \times \mathbb{R}^n \to \mathbb{R}$, then $\nabla_{\mathbf{x}} f(\bar{\mathbf{x}}, \bar{\mathbf{y}})$ is the vector of the first-order derivatives of $f$ in $(\bar{\mathbf{x}}, \bar{\mathbf{y}})$ with regard to the $\mathbf{x}$ variables. Optimization problems are printed in upper-case, sans serif type: $\mathsf{P}$.

## 4.1 Set-Theoretical Models

In the following section, we formalize the verbal description and derive a set-theoretical model for both types of maximum material yield problems.

### 4.1.1 Problems with Fixed Designs

We first turn to type (1) problems, which we call maximum material yield problems with *fixed* designs (MaxMY-FD). With $C$, we denote the container, with $F_k, k \in K := \{1, \ldots, r\}$, the flaws, and with $D_l, l \in L := \{1, \ldots, s\}$, the designs. Each of these objects is represented by a non-empty, compact subset of $\mathbb{R}^n$, $n \in \mathbb{N}$ (in general $n \in \{2, 3\}$).

While the container can be given with its flaws in an arbitrary position, we assume that the designs are located in a defined position. In order to be able to verify whether a design can be arranged in the container without overlapping the other designs and the flaws, the designs must be transformed into the container. For a maximum material yield problem with fixed designs, for which design rotations are not allowed, we search for a subset $L^* \subseteq L$ of designs and *translation vectors* $\boldsymbol{\sigma}_l \in \Sigma_l \subseteq \mathbb{R}^n$, $l \in L^*$, such that the design $D_l$ translated by $\boldsymbol{\sigma}_l$ (see Fig. 7, *left*) fulfills for $l \in L^*$ all arrangement conditions (containment in the container, non-overlapping with flaws, and non-overlapping with other designs).

If design rotation is allowed, one also searches for parameters $\boldsymbol{\theta}_l \in \Theta_l$, $l \in L^*$, of a *rotation matrix* $\mathbf{R} = \mathbf{R}(\boldsymbol{\theta}_l) \in \mathbb{R}^{n \times n}$, so that the design $D_l$, which is rotated by means of $\mathbf{R}(\boldsymbol{\theta}_l)$ and translated by $\boldsymbol{\sigma}_l$ (see Fig. 7, *right*), fulfills the arrangement conditions for $l \in L^*$. In many practical applications, the ranges $\Theta_l$, $l \in L$, of the rotation parameters are severely restricted or even finite sets.

This therefore yields the following set-theoretical model for maximum material yield problems with fixed designs:

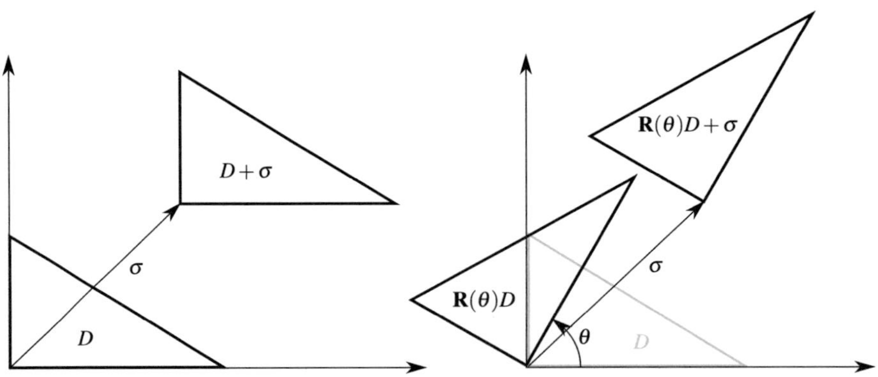


**Fig. 7** *Left*, translation, *right*, rotation and translation of a triangular design $D$

$$\text{MaxMY-FD:} \quad \max_{\substack{L^* \subseteq L \\ \sigma_l \in \Sigma_l \\ \theta_l \in \Theta_l}} \sum_{l \in L^*} \text{Vol}(D_l)$$

$$\text{s.t.} \quad \mathbf{R}(\boldsymbol{\theta}_l) D_l + \boldsymbol{\sigma}_l \subseteq C,$$

$$l \in L^*, \tag{1}$$

$$\mathbf{R}(\boldsymbol{\theta}_l) D_l + \boldsymbol{\sigma}_l \cap \text{int}(F_k) = \emptyset,$$

$$l \in L^*, \ k \in K, \tag{2}$$

$$\mathbf{R}(\boldsymbol{\theta}_{l_1}) D_{l_1} + \boldsymbol{\sigma}_{l_1} \cap \text{int}\big(\mathbf{R}(\boldsymbol{\theta}_{l_2}) D_{l_2} + \boldsymbol{\sigma}_{l_2}\big) = \emptyset,$$

$$l_1, l_2 \in L^*, \ l_1 < l_2, \tag{3}$$

where $\text{int}(A)$ refers to the interior of the set $A$, thus allowing the designs to contact one another as well as the flaws.

### 4.1.2 Problems with Variable Designs

We now consider type (2) problems, which we call maximum material yield problems with *variable* designs (MaxMY-VD). In addition to the previously introduced notation, we use $\mathbf{p}_l \in \mathbb{R}^{d_l}$ to denote the size and form parameters of the $l$-th design and $P_l$ to denote the associated set of the feasible parameter values. The simplest example of a purely size-variable design is a circle with variable radius. An example of a design that is both size and form variable is a so-called *superellipse*:

$$D^{\text{SE}}(\mathbf{p}) := \left\{ \mathbf{y} \in \mathbb{R}^2 \;\middle|\; \left(\frac{y_1^2}{p_1^2}\right)^{p_3} + \left(\frac{y_2^2}{p_2^2}\right)^{p_3} \leq 1 \right\}, \quad \mathbf{p} \in P = \mathbb{R}^3_{++}. \tag{4}$$

Variations in $p_1$ or $p_2$ yield changes in size; variations in $p_3$ yield changes in form. For $p_3 = 1/3$, $D^{\text{SE}}(\mathbf{p})$ is a generalized astroid; for $p_3 = 1/2$, a rhombus; for $p_3 = 1$, a usual ellipse; and for $p_3 \to \infty$, $D^{\text{SE}}(\mathbf{p})$ approaches a rectangle (see Fig. 8).

Because the designs are now at least size-variable, the search for an optimal subset of the set of all designs no longer makes sense, since, in principle, the designs of each subset can be arranged in the container if they are only made small enough.

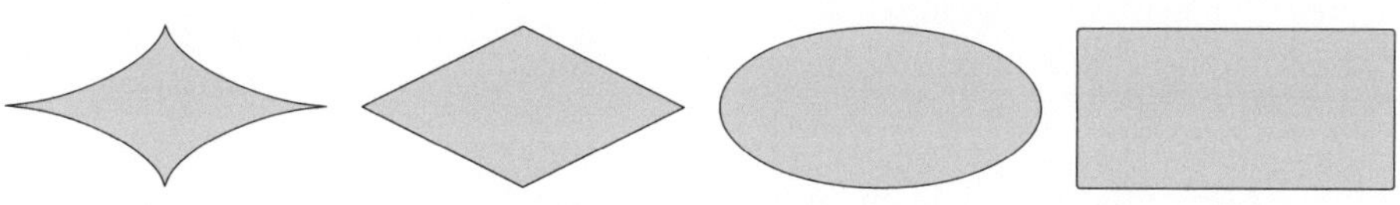

**Fig. 8** Superellipse for $p_1 = 2$ and $p_2 = 1$ and various values of $p_3$, *from left to right*: $p_3 = 1/3$, $p_3 = 1/2$, $p_3 = 1$, and $p_3 = 50$

Therefore, for maximum material yield problems with variable designs, we have the following set-theoretical model:

$$\text{MaxMY-VD:} \quad \max_{\substack{\sigma_l \in \Sigma_l \\ \theta_l \in \Theta_l \\ \mathbf{p}_l \in P_l}} \sum_{l \in L} \mathrm{Vol}\big(D_l(\mathbf{p}_l)\big)$$

$$\text{s.t.} \quad \mathbf{R}(\boldsymbol{\theta}_l) D_l(\mathbf{p}_l) + \boldsymbol{\sigma}_l \subseteq C,$$

$$l \in L, \tag{5}$$

$$\mathbf{R}(\boldsymbol{\theta}_l) D_l(\mathbf{p}_l) + \boldsymbol{\sigma}_l \cap \mathrm{int}(F_k) = \emptyset,$$

$$l \in L,\ k \in K, \tag{6}$$

$$\mathbf{R}(\boldsymbol{\theta}_{l_1}) D_{l_1}(\mathbf{p}_{l_1}) + \boldsymbol{\sigma}_{l_1} \cap \mathrm{int}\big(\mathbf{R}(\boldsymbol{\theta}_{l_2}) D_{l_2}(\mathbf{p}_{l_2}) + \boldsymbol{\sigma}_{l_2}\big) = \emptyset,$$

$$l_1, l_2 \in L,\ l_1 < l_2. \tag{7}$$

Whereas the model MaxMY-FD possesses a combinatorical component, the model MaxMY-VD does not. Nonetheless, it is also conceivable here that one might vary over subsets of the set of considered designs or various design numbers. What the two models have in common is the structure of the constraints, which we now turn to in the following discussion.

## 4.2 Handling Containment and Non-overlapping Conditions

The set-theoretical constraints (1) to (3) or (5) to (7) are of two different types. Whereas constraints (1) and (5) represent containment conditions, the other equations represent non-overlapping conditions. However, each type can be transformed into the other: A set $A \subseteq \mathbb{R}^n$ is contained in a set $B \subseteq \mathbb{R}^n$ if and only if it does not overlap with the complement $\mathbb{R}^n \setminus B$ of set $B$:

$$A \subseteq B \quad \Longleftrightarrow \quad A \cap \mathrm{int}\big(\mathbb{R}^n \setminus B\big) = \emptyset.$$

Therefore, in the following discussion, we will also use the expression "non-overlapping" as a substitute for "containment."

However, the abstract formulation of the constraints (1) to (3) or (5) to (7) isn't numerically tractable.

> In order to obtain computable problems, the set-theoretical constraints must be transformed into usual constraints of mathematical optimization.

In some cases, this is possible on the basis of geometrical considerations. For example, a circle is contained within a second circle if and only if the distance between their centers is less than or equal to the difference between the radii of the second and first circles (see

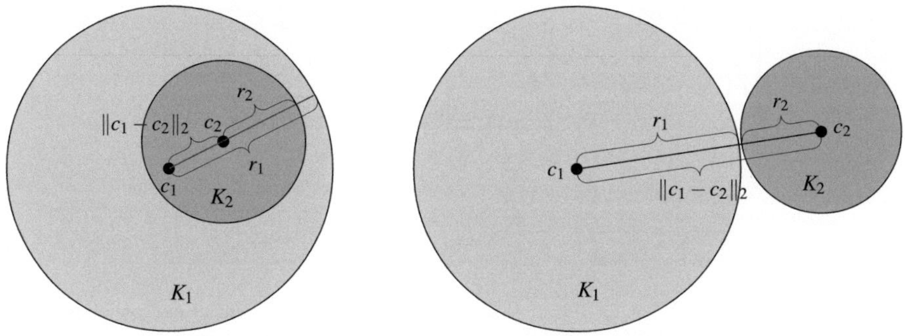


**Fig. 9** Positional relationship of two circles: *left*, containment; *right*, non-overlapping

Fig. 9, *left*). Moreover, two circles do not overlap if and only if the distance between their centers is greater than or equal to the sum of their radii (see Fig. 9, *right*).

In cases involving complicated objects, this kind of approach is usually fruitless. In the following discussion, we describe two generally valid solution approaches. The first approach uses the methods of computational geometry, more precisely, collision detection. The second approach presupposes a functional description of the objects and transforms the set-theoretical constraints into semi-infinite ones.

## 4.3 Treating the Non-overlapping Constraints Using Collision Detection Methods

In the present context, we understand the term "collision detection" (see [24], for example) to refer to methods used primarily in the fields of computer games and physical simulations to quickly establish whether two objects are overlapping or not. The methods were developed for three-dimensional space and presuppose that the objects are given explicitly as either triangulations—where an object's surface is approximated by means of triangles—or as polyhedrons. The critical feature of these methods is the efficiency with which non-overlapping can be tested. One way to make the test as efficient as possible is to pre-process the triangulations by placing a box around each triangle. The boxes are then, in turn, repeatedly pooled together in an appropriate fashion. The result is a tree of boxes, a so-called *bounding box tree (BBT)*, in which each box covers one part of the object, and the box at the root of the tree covers it entirely (see Fig. 10). If a triangulation is now given, one can use the tree to quickly determine which of its triangles might possibly be intersected by the surface of a second object. In this way, one must usually only test a relatively small number of triangles, even when the triangulation contains many of them, as is typically the case for the triangulation of rough stone, for example. For problems with fixed designs, one can directly verify non-overlapping in this fashion, since translation and rotation can be applied directly to the triangulation and the BBT. For problems with variable designs, the triangulation and associated BBT must be newly generated each

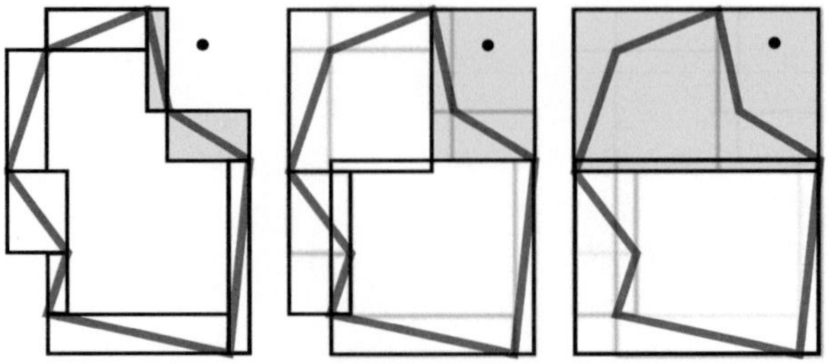

**Fig. 10** Design of a BBT in 2D: each line of the starting object is covered by a box. These are then iteratively pooled together—pair-wise, for example—and covered by another box, until only one remains. To check whether the point at the upper right is contained within the object, one need only test the shaded boxes.

time. Often, this can prove too costly. In our case, however, the complex triangulation of the rough stone remains unchanged, and the triangulation of a faceted stone and its corresponding BBT can be generated quickly. The application of this idea to gemstone cutting is described in detail in [4].

## 4.4 Transforming the Non-overlapping Conditions into Semi-Infinite Constraints

Let us turn now to the re-formulation of non-overlapping conditions as semi-infinite constraints. First, we introduce our understanding of the latter. Let $2^A$ denote the power set, i.e., the set of all subsets, of a set $A$ and let $|A|$ denote its cardinality.

**Definition 1** (Semi-infinite constraint, infinite index set) Let $m, n \in \mathbb{N}$, $g : \mathbb{R}^m \times \mathbb{R}^n \to \mathbb{R}$ be a scalar-valued function, and let $Y : \mathbb{R}^m \to 2^{\mathbb{R}^n}$ be a set-valued mapping with $|Y(\mathbf{x})| = \infty$ for all $\mathbf{x} \in \mathbb{R}^m$. Then, the condition

$$g(\mathbf{x}, \mathbf{y}) \leq 0 \quad \text{for all } \mathbf{y} \in Y(\mathbf{x}) \tag{8}$$

is called a *general semi-infinite constraint*. If $Y(\mathbf{x}) \equiv \bar{Y} \subset \mathbb{R}^n$ for all $\mathbf{x} \in \mathbb{R}^m$, then the condition (8) is called a *standard semi-infinite constraint*. In both cases, the set $Y(\mathbf{x})$ is referred to as the *infinite index set*.

If the function $g$ does not depend on $\mathbf{x}$, this does not affect the terminology.

For our subsequent analysis, we summarize the translation, rotation, and size/shape parameters for each design $D_l$, $l \in L$ in a vector $\tilde{\mathbf{p}}_l$; introduce the set of feasible parameter values $\tilde{P}_l := \Sigma_l \times \Theta_l \times P_l$; and write $D_l(\tilde{\mathbf{p}}_l)$ instead of $\mathbf{R}(\boldsymbol{\theta}_l)D_l(\mathbf{p}_l) + \boldsymbol{\sigma}_l$.

If the container can be represented as the solution set of a system of inequalities, that is, if

$$C = \left\{\mathbf{y} \in \mathbb{R}^n \;\middle|\; c_i(\mathbf{y}) \leq 0,\ i \in I_0\right\},$$

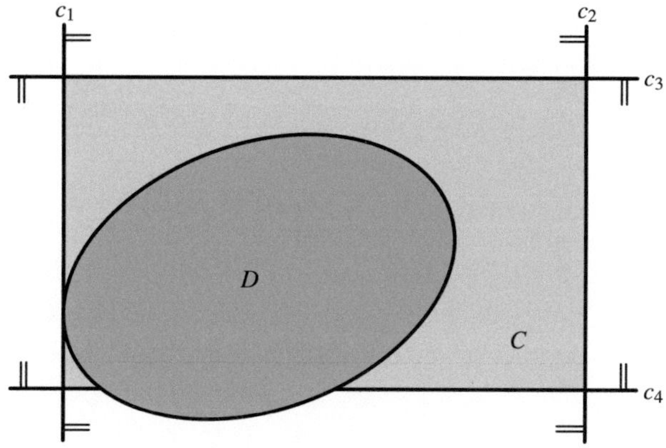


**Fig. 11** Transformation of a containment condition into a semi-infinite constraint

where $I_0$ is a finite index set and $c_i$, $i \in I_0$, are real-valued functions, then the transformation of the containment conditions (1) or (5) into semi-infinite constraints is straightforward (see Fig. 11 for a graphical illustration):

$$D_l(\tilde{\mathbf{p}}_l) \subseteq C \quad \Leftrightarrow \quad c_i(\mathbf{y}) \leq 0 \quad \text{for all } \mathbf{y} \in D_l(\tilde{\mathbf{p}}_l),\ i \in I_0.$$

For the semi-infinite reformulation of the non-overlapping conditions, two approaches were introduced and investigated in [16] and [14]: *mutual separation* and *separation by hyperplane*. Because only the second approach can be applied in cases where there are additional, relevant requirements stemming from the production technology (see Sect. 6.2) we will restrict our discussion to this approach. The foundation for this discussion consists of a so-called *separation theorem*:

**Theorem 1** (Separation theorem, see [20], for example) *Let $A, B \subset \mathbb{R}^n$ be two non-empty, convex sets, of which at least one is open. Then $A$ and $B$ are non-overlapping if and only if a vector $\boldsymbol{\eta} \in \mathbb{R}^n \setminus \{\mathbf{0}\}$ and a number $\beta \in \mathbb{R}$ exist, such that the following holds:*

$$\boldsymbol{\eta}^T \mathbf{y} \leq \beta \quad \textit{for all } \mathbf{y} \in A$$

*and*

$$\boldsymbol{\eta}^T \mathbf{z} \geq \beta \quad \textit{for all } \mathbf{z} \in B.$$

The hyperplane $H(\boldsymbol{\eta}, \beta) := \{\mathbf{y} \in \mathbb{R}^n \mid \boldsymbol{\eta}^T \mathbf{y} = \beta\}$, which separates the sets $A$ and $B$, is called a *separating hyperplane*.

If the flaws and designs are convex, the above theorem delivers a semi-infinite formulation of the non-overlapping conditions (2) and (3) or (6) and (7) (for a graphical illustration, see Fig. 25, *right*, with $\delta = 0$):

(1) $D_l(\tilde{\mathbf{p}}_l) \cap \mathrm{int}(F_k) = \emptyset$ if and only if a vector $\boldsymbol{\eta}_{l,k}^{\mathrm{DF}} \in \mathbb{R}^n \setminus \{\mathbf{0}\}$ and a number $\beta_{l,k}^{\mathrm{DF}}$ exist, such that

$$\left(\boldsymbol{\eta}_{l,k}^{\mathrm{DF}}\right)^T \mathbf{y} \leq \beta_{l,k}^{\mathrm{DF}} \quad \text{for all } \mathbf{y} \in D_l(\tilde{\mathbf{p}}_l) \tag{9}$$

and

$$\left(\boldsymbol{\eta}_{l,k}^{\mathrm{DF}}\right)^T \mathbf{z} \geq \beta_{l,k}^{\mathrm{DF}} \quad \text{for all } \mathbf{z} \in F_k. \tag{10}$$

(2) $D_{l_1}(\tilde{\mathbf{p}}_{l_1}) \cap \mathrm{int}(D_{l_2}(\tilde{\mathbf{p}}_{l_2})) = \emptyset$ if and only if a vector $\boldsymbol{\eta}_{l_1,l_2}^{\mathrm{DD}} \in \mathbb{R}^n \setminus \{\mathbf{0}\}$ and a number $\beta_{l_1,l_2}^{\mathrm{DD}}$ exist, such that

$$\left(\boldsymbol{\eta}_{l_1,l_2}^{\mathrm{DD}}\right)^T \mathbf{y} \leq \beta_{l_1,l_2}^{\mathrm{DD}} \quad \text{for all } \mathbf{y} \in D_{l_1}(\tilde{\mathbf{p}}_{l_1}) \tag{11}$$

and

$$\left(\boldsymbol{\eta}_{l_1,l_2}^{\mathrm{DD}}\right)^T \mathbf{z} \geq \beta_{l_1,l_2}^{\mathrm{DD}} \quad \text{for all } \mathbf{z} \in D_{l_2}(\tilde{\mathbf{p}}_{l_2}). \tag{12}$$

Whereas the conditions (9), (11), and (12) represent general semi-infinite constraints, condition (10) is a standard semi-infinite one. The condition $\boldsymbol{\eta} \neq \mathbf{0}$ is problematic from an optimization perspective, but can be suitably reformulated by means of normalization, for example, $\|\boldsymbol{\eta}\|_2^2 = 1$, where $\|\cdot\|_2$ is the Euclidean norm.

Let

$$\mathbf{x} := \left(\tilde{\mathbf{p}}_1, \ldots, \tilde{\mathbf{p}}_s, \boldsymbol{\eta}_{1,1}^{\mathrm{DF}}, \beta_{1,1}^{\mathrm{DF}}, \ldots, \boldsymbol{\eta}_{s,r}^{\mathrm{DF}}, \beta_{s,r}^{\mathrm{DF}}, \boldsymbol{\eta}_{1,2}^{\mathrm{DD}}, \beta_{1,2}^{\mathrm{DD}}, \ldots, \boldsymbol{\eta}_{s-1,s}^{\mathrm{DD}}, \beta_{s-1,s}^{\mathrm{DD}}\right)$$

be the vector of all parameters (design and hyperplane parameters) and let

$$X := \left\{ \mathbf{x} \;\middle|\; \begin{array}{c} \tilde{\mathbf{p}}_l \in \tilde{P}_l, \ l \in L, \\ \|\boldsymbol{\eta}_{l,k}^{\mathrm{DF}}\|_2^2 = 1, \ l \in L, \ k \in K, \\ \|\boldsymbol{\eta}_{l_1,l_2}^{\mathrm{DD}}\|_2^2 = 1, \ l_1, l_2 \in L, \ l_1 < l_2 \end{array} \right\}$$

be the set of feasible parameter values. Then, the reformulation of a maximum material yield problem with variable designs as a so-called *general semi-infinite optimization problem* using the separation by hyperplanes approach becomes:

$$\mathrm{GSIP}_{\mathrm{MaxMY\text{-}VD}}: \quad \max_{\mathbf{x} \in X} \sum_{l \in L} \mathrm{Vol}\big(D_l(\mathbf{p}_l)\big)$$

$$\text{s.t.} \quad c_i(\mathbf{y}) \leq 0 \quad \text{for all } \mathbf{y} \in D_l(\tilde{\mathbf{p}}_l),$$

$$i \in I_0, \ l \in L, \tag{13}$$

$$\left.\begin{array}{l} \left(\boldsymbol{\eta}_{l,k}^{\mathrm{DF}}\right)^T \mathbf{y} - \beta_{l,k}^{\mathrm{DF}} \leq 0 \quad \text{for all } \mathbf{y} \in D_l(\tilde{\mathbf{p}}_l), \\ \left(\boldsymbol{\eta}_{l,k}^{\mathrm{DF}}\right)^T \mathbf{z} - \beta_{l,k}^{\mathrm{DF}} \geq 0 \quad \text{for all } \mathbf{z} \in F_k, \end{array}\right\}$$

$$l \in L, \ k \in K, \tag{14}$$

$$\left.\begin{array}{l} \left(\boldsymbol{\eta}_{l_1,l_2}^{\mathrm{DD}}\right)^T \mathbf{y} - \beta_{l_1,l_2}^{\mathrm{DD}} \leq 0 \quad \text{for all } \mathbf{y} \in D_{l_1}(\tilde{\mathbf{p}}_{l_1}), \\ \left(\boldsymbol{\eta}_{l_1,l_2}^{\mathrm{DD}}\right)^T \mathbf{z} - \beta_{l_1,l_2}^{\mathrm{DD}} \geq 0 \quad \text{for all } \mathbf{z} \in D_{l_2}(\tilde{\mathbf{p}}_{l_2}), \end{array}\right\}$$

$$l_1, l_2 \in L, \ l_1 < l_2. \tag{15}$$

## 4.5 Solution Methods for General Semi-Infinite Optimization Problems

Now that we know how a maximum material yield problem can be transformed into a general semi-infinite optimization problem, the question arises as to how such problems can be solved numerically. We now want to answer this question.

Let us consider optimization problems of the following form:

$$\text{GSIP:} \quad \min_{\mathbf{x} \in X \subseteq \mathbb{R}^m} f(\mathbf{x}) \quad \text{s.t.} \quad g_i(\mathbf{x}, \mathbf{y}) \leq 0 \quad \text{for all } \mathbf{y} \in Y(\mathbf{x}),\ i \in I, \tag{16}$$

with

$$Y(\mathbf{x}) := \{\mathbf{y} \in \mathbb{R}^n \mid v_j(\mathbf{x}, \mathbf{y}) \leq 0,\ j \in J\} \quad \text{and} \quad |Y(\mathbf{x})| = \infty \quad \text{for all } \mathbf{x} \in X, \tag{17}$$

$I := \{1, \ldots, p\}$ and $J := \{1, \ldots, q\}$, as well as real-valued, sufficiently smooth functions $f$, $g_i$, $i \in I$, and $v_j$, $j \in J$. According to Definition 1, we identify such an optimization problem either as:

- a *general(ized) semi-infinite program*, if the set-valued mapping $Y$ depends on $\mathbf{x}$, or as
- a *(standard) semi-infinite program*, if the set-valued mapping $Y$ is constant.

The latter is then referred to as an SIP, rather than a GSIP.

The consideration of multiple infinite index sets, a situation that arises for maximum material yield problems, can proceeded directly. For clarity's sake, we will restrict ourselves in the following discussion to one infinite index set.

For a comprehensive introduction to semi-infinite optimization, we refer the reader to the review article [29] and the book [36] for the SIP problem class and to the review articles [28, 40] and the monographs [39, 50] for the more general GSIP problem class.

> Even if the difference between general and standard semi-infinite problems initially appears to be minimal, the former are substantially more complicated structurally and much more difficult to solve numerically.

For the remainder of this section, we make the following assumptions, which we need for our further considerations and which can be fulfilled very easily for maximum material yield problems by means of a suitable modeling approach.

**Assumption 1** For all $\mathbf{x} \in X$, the set $Y(\mathbf{x})$ is *non-empty* and *compact*.

**Assumption 2** For all $\mathbf{x} \in X$, the functions $g_i(\mathbf{x}, \cdot)$, $i \in I$, are *concave* and the set $Y(\mathbf{x})$ is *convex*.

**Assumption 3** For all $\mathbf{x} \in X$, the set $Y(\mathbf{x})$ possesses a *Slater point*, that is, a point $\hat{\mathbf{y}}(\mathbf{x})$, such that $v_j(\mathbf{x}, \hat{\mathbf{y}}(\mathbf{x})) < 0,\ j \in J$, holds.

The key to both the theoretical and the numerical treatment of semi-infinite optimization problems lies in their two-level structure. The parametric *lower-level problems* from GSIP are given by

$$\mathsf{Q}_i(\mathbf{x}): \quad \max_{\mathbf{y} \in \mathbb{R}^n} g_i(\mathbf{x}, \mathbf{y}) \quad \text{s.t.} \quad v_j(\mathbf{x}, \mathbf{y}) \le 0, \quad j \in J. \tag{18}$$

The term $\varphi_i(\mathbf{x})$ denotes the optimal value of $\mathsf{Q}_i(\mathbf{x})$. Accordingly, the function $\varphi_i$ is called the *optimal value function*. Obviously, a point $\mathbf{x} \in X$ is feasible for GSIP if and only if $\varphi_i(\mathbf{x}) \le 0$ for all $i \in I$. The main challenge for the numerical solution of semi-infinite optimization problems is that evaluating $\varphi_i(\mathbf{x}) \le 0$ requires computing a *global* solution of the problem $\mathsf{Q}_i(\mathbf{x})$. This is a very difficult task in general. Under Assumptions 2 and 3, however, the lower-level problems are convex, regular optimization problems. This makes a global solution computable. Moreover, under Assumptions 1 to 3, the optimal value functions $\varphi_i,\ i \in I$, are well defined and continuous. Thus, the feasible set of GSIP

$$\begin{aligned} M &:= \left\{\mathbf{x} \in X \mid g_i(\mathbf{x}, \mathbf{y}) \le 0 \text{ for all } \mathbf{y} \in Y(\mathbf{x}),\ i \in I\right\} \\ &= \left\{\mathbf{x} \in X \mid \varphi_i(\mathbf{x}) \le 0,\ i \in I\right\} \end{aligned}$$

is closed, and a minimum value exists.

To date, solution methods for general semi-infinite optimization problems have been developed primarily from a conceptual perspective. To the best of our knowledge, comprehensive numerical evaluations exist only for the explicit smoothing approach from [39, 42]. These evaluations can be found in [39], [16], and [12]. All in all, the methods developed so far are based on two concepts:

(1) the *generalization* of methods for standard semi-infinite optimization problems and
(2) the *transformation* of a general semi-infinite optimization problem into a standard semi-infinite optimization problem.

The methods stemming from concept (1) can be further subdivided:

(A) discretization and exchange methods (see [46, 47]),
(B) methods based on local reduction of the general semi-infinite problem (see [43–45, 48]), and
(C) methods based on the reformulation of GSIP into a related problem, so-called *lift-&-project* approaches (see [23, 42] and [10]).

We now introduce two methods that were developed at the ITWM in connection with two dissertations [14, 16] and tested by means of gemstone cutting problems.

### 4.5.1 A Feasible, Explicit Smoothing Method

The first method (see [16] and [10]) consists of a modification of the explicit smoothing approach from [39, 42]. With this modification, the solutions generated in the method for the surrogate problem are feasible for the original problem. We first introduce briefly the explicit smoothing approach and then take a closer look at the aforementioned modification.

**Explicit Smoothing Approach** Under Assumption 1, the semi-infinite constraints (16) are equivalent to the conditions

$$\max_{\mathbf{y}\in Y(\mathbf{x})} g_i(\mathbf{x},\mathbf{y}) \leq 0, \quad i \in I$$

(see [41]). Thus, GSIP can be written as a ***bi-level program***:

$$\begin{aligned} \mathsf{BLP}_{\mathsf{GSIP}}: \quad & \min_{\substack{\mathbf{x},\\ \mathbf{y}_1,\dots,\mathbf{y}_p}} f(\mathbf{x}) \\ & \text{s.t.} \quad g_i(\mathbf{x},\mathbf{y}_i) \leq 0, && (19)\\ & \qquad\quad \mathbf{y}_i \text{ solves } \mathsf{Q}_i(\mathbf{x}), \quad i \in I. && (20) \end{aligned}$$

Under Assumptions 2 and 3, each global solution $\mathbf{y}_i$ of the lower-level problem $\mathsf{Q}_i(\mathbf{x})$, $i \in I$, can be characterized by the first-order optimality conditions:

$$\begin{aligned} \nabla_{\mathbf{y}}\mathscr{L}_i(\mathbf{x},\mathbf{y}_i,\boldsymbol{\mu}_i) &= \mathbf{0},\\ \operatorname{diag}(\boldsymbol{\mu}_i)\mathbf{v}(\mathbf{x},\mathbf{y}_i) &= \mathbf{0},\\ \boldsymbol{\mu}_i &\geq \mathbf{0},\\ \mathbf{v}(\mathbf{x},\mathbf{y}_i) &\leq \mathbf{0}, \end{aligned}$$

where

$$\mathscr{L}_i(\mathbf{x},\mathbf{y},\boldsymbol{\mu}) := g_i(\mathbf{x},\mathbf{y}) - \boldsymbol{\mu}^T\mathbf{v}(\mathbf{x},\mathbf{y})$$

is the Lagrangian function of problem $\mathsf{Q}_i(\mathbf{x})$, $\boldsymbol{\mu}_i$ is the $\mathbf{y}_i$-associated vector of Lagrange multipliers, $\operatorname{diag}(\boldsymbol{\mu}_i)$ is the diagonal matrix with diagonal elements $\mu^i_j$, $j \in J$, and

$$\mathbf{v}(\mathbf{x},\mathbf{y}) := \left(v_1(\mathbf{x},\mathbf{y}),\dots,v_q(\mathbf{x},\mathbf{y})\right)^T.$$

Therefore, $\mathsf{BLP}_{\mathsf{GSIP}}$ can be written as a ***mathematical program with complementarity constraints***:

$$\begin{aligned}
\mathsf{MPCC_{GSIP}}: \quad & \min_{\substack{\mathbf{x},\\ \mathbf{y}_1,\ldots,\mathbf{y}_p,\\ \boldsymbol{\mu}_1,\ldots,\boldsymbol{\mu}_p}} \quad f(\mathbf{x}) \\
& \text{s.t.} \quad g_i(\mathbf{x}, \mathbf{y}_i) \le 0, && (21)\\
& \qquad \nabla_{\mathbf{y}} \mathscr{L}_i(\mathbf{x}, \mathbf{y}_i, \boldsymbol{\mu}_i) = \mathbf{0}, && (22)\\
& \qquad -\operatorname{diag}(\boldsymbol{\mu}_i)\mathbf{v}(\mathbf{x}, \mathbf{y}_i) = \mathbf{0}, && (23)\\
& \qquad \boldsymbol{\mu}_i \ge \mathbf{0}, && (24)\\
& \qquad -\mathbf{v}(\mathbf{x}, \mathbf{y}_i) \ge \mathbf{0}, \quad i \in I. && (25)
\end{aligned}$$

At this point, we do indeed have a reformulation of GSIP as a finite, one-level optimization problem. However, for optimization problems with complementarity constraints, classical regularity conditions such as MFCQ—which are of tremendous significance for numerical methods—are generally not fulfilled at any feasible point (see [37]). (Explicit) smoothing represents one possibility of regularization. The idea here is to replace the "malignant" conditions (23) with the conditions

$$-\operatorname{diag}(\boldsymbol{\mu}_i)\mathbf{v}(\mathbf{x}, \mathbf{y}_i) = \tau^2 \mathbf{1}, \quad i \in I, \tag{26}$$

where $\tau > 0$ is a perturbation parameter and $\mathbf{1} = (1, \ldots, 1)^T \in \mathbb{R}^q$. In this way, $\mathsf{MPCC_{GSIP}}$ is embedded into a parametric family of optimization problems

$$\begin{aligned}
\mathsf{P}_\tau: \quad & \min_{\substack{\mathbf{x},\\ \mathbf{y}_1,\ldots,\mathbf{y}_p,\\ \boldsymbol{\mu}_1,\ldots,\boldsymbol{\mu}_p}} \quad f(\mathbf{x}) \\
& \text{s.t.} \quad g_i(\mathbf{x}, \mathbf{y}_i) \le 0, \\
& \qquad \nabla_{\mathbf{y}} \mathscr{L}_i(\mathbf{x}, \mathbf{y}_i, \boldsymbol{\mu}_i) = \mathbf{0}, \\
& \qquad -\operatorname{diag}(\boldsymbol{\mu}_i)\mathbf{v}(\mathbf{x}, \mathbf{y}_i) = \tau^2 \mathbf{1}, \\
& \qquad \boldsymbol{\mu}_i \ge \mathbf{0}, \\
& \qquad -\mathbf{v}(\mathbf{x}, \mathbf{y}_i) \ge \mathbf{0}, \quad i \in I.
\end{aligned}$$

In [42], the authors show that the degenerateness of the complementarity constraints (23) is eliminated via the regularization described above, and that $P_\tau$ can be solved using standard software for nonlinear optimization problems. A solution of $\mathsf{P}_0 = \mathsf{MPCC_{GSIP}}$ can now be found by solving a sequence of problems $\mathsf{P}_{\tau_k}$, where $\{\tau_k\}_{k \in \mathbb{N}_0} \subset \mathbb{R}_{++}$ is a monotonically decreasing null sequence:

**Algorithm 1** Explicit smoothing method, [39, 42]

1: Choose a monotonically decreasing null sequence $\{\tau_k\}_{k\in\mathbb{N}_0} \subset \mathbb{R}_{++}$ and a starting point $\mathbf{x}^0 \in X \subseteq \mathbb{R}^m$.
2: Compute a starting point $(\mathbf{x}^{0,0}, \mathbf{y}_1^{0,0}, \dots, \mathbf{y}_p^{0,0}, \boldsymbol{\mu}_1^{0,0}, \dots, \boldsymbol{\mu}_p^{0,0})$ of $\mathsf{P}_{\tau_0}$.
3: Set $k := 0$.
4: **while** a termination criterion is not fulfilled, **do**
5: Compute a solution $(\mathbf{x}^{k,*}, \mathbf{y}_1^{k,*}, \dots, \mathbf{y}_p^{k,*}, \boldsymbol{\mu}_1^{k,*}, \dots, \boldsymbol{\mu}_p^{k,*})$ of $\mathsf{P}_{\tau_k}$ using $(\mathbf{x}^{k,0}, \mathbf{y}_1^{k,0}, \dots, \mathbf{y}_p^{k,0}, \boldsymbol{\mu}_1^{k,0}, \dots, \boldsymbol{\mu}_p^{k,0})$ as starting point.
6: Set $(\mathbf{x}^{k+1,0}, \mathbf{y}_1^{k+1,0}, \dots, \boldsymbol{\mu}_p^{k+1,0}) := (\mathbf{x}^{k,*}, \mathbf{y}_1^{k,*}, \dots, \boldsymbol{\mu}_p^{k,*})$.
7: Replace $k$ by $k+1$.
8: **end while**
9: **return** $\mathbf{x}^{k,0}$

Whereas problem $\mathsf{MPCC}_{\mathsf{GSIP}}$ is an equivalent formulation for GSIP, the parametric problem $\mathsf{P}_\tau$ represents for $\tau > 0$ merely an approximation.

In [39], the author shows that the explicit smoothing approach possesses an external approximation property (see Fig. 12 also):

**Theorem 2** ([39]) *Let $M_\tau$ be the projection of the feasible set of $\mathsf{P}_\tau$ in the $\mathbf{x}$-space. Then:*
(i) *For all $0 < \tau_1 < \tau_2$, $M_{\tau_1} \subset M_{\tau_2}$.*
(ii) *For all $\tau > 0$, $M \subset M_\tau$.*

A negative effect of this external approximation property is that the $\mathbf{x}$-components of the solutions of $\mathsf{P}_\tau$ can be infeasible for GSIP for all $\tau > 0$, although the infeasibility vanishs in the limiting case. This is a serious problem when the feasibility of the iterates plays a role.

**Feasibility in the Explicit Smoothing Approach** The dissertation [16] (and the article [10]) show how the drawback of the iterates' infeasibility can be redressed by a simple modification of the conditions (21). We will now outline how this works.

Whereas the conditions (23) to (25) characterize the global solutions of the lower-level problems, the conditions (24) to (26) describe for $\tau > 0$ the global solutions of the so-called *log-barrier problems* (see [39, 42]):

$$\mathsf{Q}_i^\tau(\mathbf{x}): \quad \max_{\mathbf{y}\in\mathbb{R}^n} b_i^\tau(\mathbf{x},\mathbf{y}) := g_i(\mathbf{x},\mathbf{y}) + \tau^2 \sum_{j=1}^{q} \ln\bigl(-v_j(\mathbf{x},\mathbf{y})\bigr), \quad i \in I. \tag{27}$$

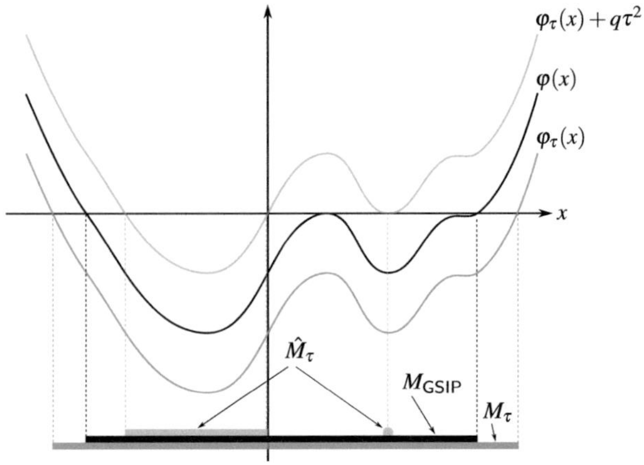


**Fig. 12** Under- and over-estimation of the optimal value function $\varphi$ by $\varphi_\tau$ and $\varphi_\tau + q\tau^2$

Using the duality theory of convex optimization, the optimal value functions $\varphi_i$, $i \in I$, can be estimated from above and thus the feasible set of GSIP can be approximated from the interior (see Fig. 12).

**Lemma 1** ([16]) *For $\tau > 0$ and $i \in I$, let $\mathbf{y}_i^\tau(\mathbf{x})$ be a global solution of $\mathsf{Q}_i^\tau(\mathbf{x})$. Then,*

$$\varphi_i(\mathbf{x}) = \max_{\mathbf{y} \in Y(\mathbf{x})} g_i(\mathbf{x}, \mathbf{y}) \leq g_i\left(\mathbf{x}, \mathbf{y}_i^\tau(\mathbf{x})\right) + q\tau^2,$$

*where $q$ is the number of functions $v_j$, $j \in J$, describing the index set $Y(\mathbf{x})$.*

Thus, the original constraints of the upper level (21) can be replaced by the conditions

$$g_i(\mathbf{x}, \mathbf{y}_i) + q\tau^2 \leq 0, \quad i \in I, \tag{28}$$

which yields the parametric optimization problem

$$\begin{aligned}
\hat{\mathsf{P}}_\tau: \quad & \min_{\substack{\mathbf{x},\\ \mathbf{y}_1,\ldots,\mathbf{y}_p,\\ \boldsymbol{\mu}_1,\ldots,\boldsymbol{\mu}_p}} \ f(\mathbf{x}) \\
& \text{s.t.} \quad g_i(\mathbf{x}, \mathbf{y}_i) + q\tau^2 \leq 0, \\
& \qquad \nabla_{\mathbf{y}} \mathscr{L}_i(\mathbf{x}, \mathbf{y}_i, \boldsymbol{\mu}_i) = \mathbf{0}, \\
& \qquad -\mathrm{diag}(\boldsymbol{\mu}_i)\mathbf{v}(\mathbf{x}, \mathbf{y}_i) = \tau^2 \mathbf{1}, \\
& \qquad \boldsymbol{\mu}_i \geq \mathbf{0}, \\
& \qquad -\mathbf{v}(\mathbf{x}, \mathbf{y}_i) \geq \mathbf{0}, \quad i \in I.
\end{aligned}$$

This modification leads to an internal approximation of the feasible set of GSIP (see Fig. 12 also):

**Theorem 3** ([16]) *Let $\hat{M}_\tau$ be the projection of the feasible set of $\hat{\mathsf{P}}_\tau$ in the $\mathbf{x}$-space. Then, for all $\tau > 0$, $\hat{M}_\tau \subset M$.*

A combination of the internal approximation property of $\hat{M}_\tau$ with the external one of $M_\tau$ leads to a "sandwiching result:"

**Corollary 1** ([16]) *Let $\{\tau_k\}_{k \in \mathbb{N}_0} \subset \mathbb{R}$ be a monotonically decreasing null sequence. Then,*

$$\bigcup_{k \in \mathbb{N}_0} \hat{M}_{\tau_k} \subseteq M \subseteq \bigcap_{k \in \mathbb{N}_0} M_{\tau_k}.$$

This result significantly improves the termination criteria, which depend on the problem structure: For a given $\tau > 0$, the objective function value for each point in $\hat{M}_\tau$ is an upper bound on the optimum value of GSIP, while the global minimum value of $f$ delivers a lower bound over $M_\tau$. Thus, in cases where the latter minimum value is numerically available, the difference between the upper and lower bounds can be used as a termination criterion.

Analogously to Algorithm 1, an optimal solution of GSIP is to be found by solving the problems $\hat{\mathsf{P}}_{\tau_k}$ for a monotonically decreasing null sequence $\{\tau_k\}_{k \in \mathbb{N}_0} \subset \mathbb{R}_+$. What is problematical here, however, is the fact that the set $\hat{M}_\tau$ can be empty for large values of $\tau$, due to the modification employed. For example, this occurs when the set defined by the tightened constraints (28)

$$G_\tau(\mathbf{x}) := \left\{\mathbf{y} \in \mathbb{R}^n \mid g_i(\mathbf{x}, \mathbf{y}) \leq -q\tau^2,\ i \in I\right\},$$

which, in the context of maximum material yield problems with only one design and no flaws, corresponds to a "*shrunken*" container,

$$C_\tau := \left\{\mathbf{y} \in \mathbb{R}^n \mid c_i(\mathbf{y}) \leq -q\tau^2,\ i \in I_0\right\}, \tag{29}$$

is empty. Therefore, in a first phase, one must find a threshold value $\bar{\tau}$ with $\hat{M}_\tau \neq \emptyset$ for all $\tau \leq \bar{\tau}$ and a $\mathbf{x} \in \hat{M}_{\bar{\tau}}$, before one then, in the second phase, proceeds as in Algorithm 1. For details, please refer to [16] and [10].

Finally, we want to graphically illustrate how the explicit smoothing method (Algorithm 1) and its feasible variant work, by means of a ***design centering problem***:

$$\mathsf{DC}: \quad \max_{\mathbf{x} \in X \subseteq \mathbb{R}^m} \ \mathrm{Vol}\big(D(\mathbf{x})\big) \quad \text{s.t. } D(\mathbf{x}) \subseteq C,$$

that is, by means of a maximum material yield problem with one variable design and no flaws. Here, an ellipse is to be embedded with maximal area in the following container (see Fig. 13):

$$C^{\mathrm{CT}} := \left\{ \mathbf{y} \in \mathbb{R}^2 \;\middle|\; \begin{array}{r} -y_1 - y_2^2 \leq 0, \\ 1/4 y_1 + y_2 - 3/4 \leq 0, \\ -y_2 - 1 \leq 0. \end{array} \right\} \tag{30}$$

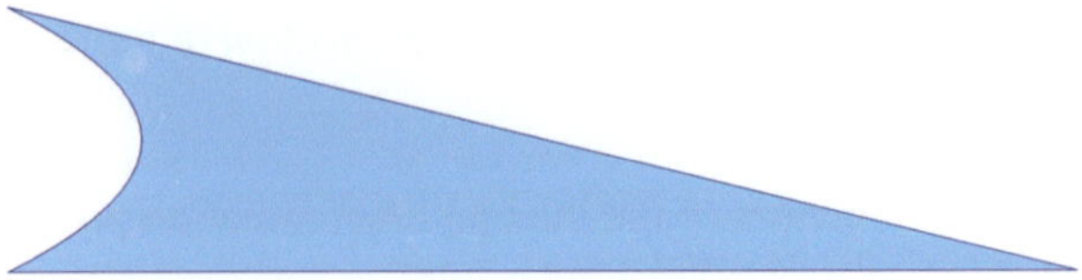

**Fig. 13** The container $C^{\mathrm{CT}}$

One possible description of an ellipse is as the affine image of the unit circle:

$$\begin{aligned} D^{\mathrm{E}}(\mathbf{x}) &:= \left\{\mathbf{A}(\mathbf{x})\mathbf{y} + \mathbf{c}(\mathbf{x}) \in \mathbb{R}^2 \,\middle|\, \|\mathbf{y}\|_2^2 \le 1\right\} \\ &= \left\{\mathbf{y} \in \mathbb{R}^2 \,\middle|\, [\mathbf{y} - \mathbf{c}(\mathbf{x})]^T \left[\mathbf{A}(\mathbf{x})\mathbf{A}(\mathbf{x})^T\right]^{-1} [\mathbf{y} - \mathbf{c}(\mathbf{x})] - 1 \le 0\right\} \end{aligned} \tag{31}$$

with

$$\mathbf{c}(\mathbf{x}) := \begin{pmatrix} x_1 \\ x_2 \end{pmatrix}, \qquad \mathbf{A}(\mathbf{x}) := \begin{pmatrix} x_3 & x_5 \\ 0 & x_4 \end{pmatrix}, \qquad \text{and} \quad \mathbf{x} \in X := \mathbb{R}^2 \times \mathbb{R}^2_{++} \times \mathbb{R}_+.$$

The area of an ellipse with this parameterization is:

$$\mathrm{Vol}_2(\mathbf{x}) = \pi x_3 x_4.$$

The formulation of the design-centering problem $\mathsf{DC}^{\mathrm{E\text{-}CT}}$ as a general semi-infinite optimization problem becomes:

$$\begin{aligned} \mathsf{GSIP}_{\mathsf{DC}^{\mathrm{E\text{-}CT}}}: \quad & -\min_{\mathbf{x} \in X} -\pi x_3 x_5 \\ \text{s.t.} \quad & -y_1 - y_2^2 \le 0 \quad \text{for all } \mathbf{y} \in D^{\mathrm{E}}(\mathbf{x}), \\ & 1/4 y_1 + y_2 - 3/4 \le 0 \quad \text{for all } \mathbf{y} \in D^{\mathrm{E}}(\mathbf{x}), \\ & -y_2 - 1 \le 0 \quad \text{for all } \mathbf{y} \in D^{\mathrm{E}}(\mathbf{x}). \end{aligned}$$

We turn first to the explicit smoothing method (Algorithm 1). We have chosen as null sequence the geometrical sequence $\{1/2^k\}_{k \in \mathbb{N}_0}$ and as starting point $\mathbf{x}^0$ the (infeasible) point $(0, 0, 1, 1, 0)$; that is, the unit circle (see Fig. 14(a) also). We have obtained an initial configuration for the solutions of the lower-level problems and the associated Lagrange multipliers by solving the log barrier problems (27). Algorithm 1 terminates when the relative error in either the solutions or the associated function values is less than or equal to $10^{-6}$ and the violation of the feasibility of the solution with regard to the underlying general semi-infinite problem is less than or equal to $10^{-6}$. Figure 14 graphically illustrates the iterative solution of the problems $\mathsf{P}_{\tau_k}$, $k \in \mathbb{N}_0$.

Using the same example, we want to now look at the feasible variant of the explicit smoothing method. To do so, we use the same null sequence and starting point. The initialization of the solutions of the lower-level problems, as well as of the associated Lagrange multipliers, takes place as above. For termination, we now only have to consider the relative error in the solutions and in the "optimum values," since a feasible solution of a problem $\hat{\mathsf{P}}_{\tau_k}$ is, per construction, also feasible for the next problem $\hat{\mathsf{P}}_{\tau_{k+1}}$. Figure 15 graphically illustrates the algorithmic procedure. Both the actual container (in light blue)

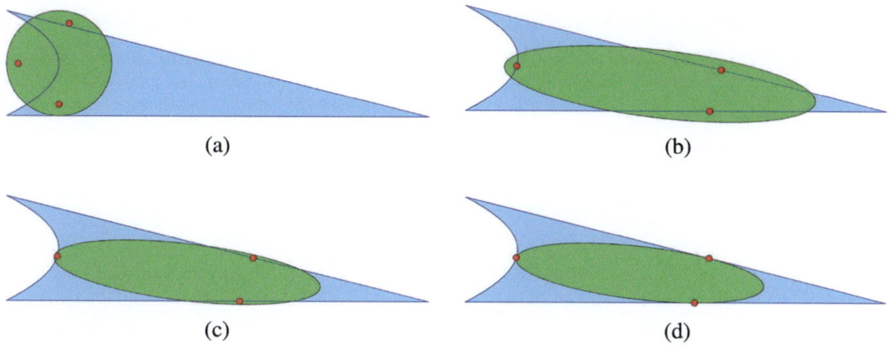


**Fig. 14** Area-maximal design-centering of an ellipse into the container $C^{CT}$ using the explicit smoothing method (Algorithm 1) [*light blue*-container, *green*-design, *red*-solutions of the log barrier problems (27)]: (**a**) initial situation ($\tau = 0.5$), (**b**) after solution of problem $\mathsf{P}_{0.5}$, (**c**) after solution of problem $\mathsf{P}_{0.25}$, and (**d**) final situation (after a total of 12 iterations, that is, for $\tau = 0.000244140625$).

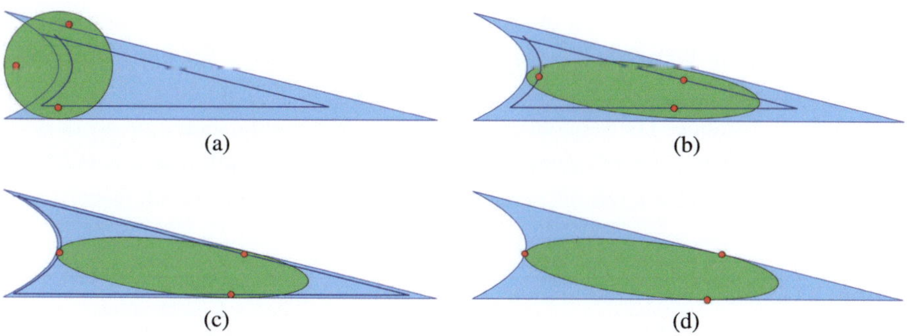


**Fig. 15** Area-maximal design-centering of an ellipse into the container $C^{CT}$ using the feasible explicit smoothing method [*light blue*-container, *dark blue*-"shrunken" container, *green*-design, *red*-solutions of the log barrier problems (27)]: (**a**) initial situation ($\tau = 0.5$), (**b**) after solution of problem $\hat{\mathsf{P}}_{0.5}$, (**c**) after solution of problem $\hat{\mathsf{P}}_{0.25}$, and (**d**) final situation (after a total of 7 iterations, that is, for $\tau = 0.0078125$)

and the "shrunken" container $C_\tau$ (in dark blue; see (29)) are depicted. With this example, it is not necessary to execute a first phase for finding a suitable threshold value $\bar{\tau}$ and feasible solution for $\mathsf{GSIP}_{\mathsf{DC}^{\mathrm{E\text{-}CT}}}$, since the "shrunken" container is not empty.

### 4.5.2 A Transformation-Based Discretization Method

We now introduce a second method developed at the ITWM for solving general semi-infinite optimization problems with convex lower-level problems. This method cleverly combines the solution approaches "discretization of infinite index sets" and "transformation into a standard semi-infinite problem," thereby circumventing the weak points of each approach. We will first discuss the two solution approaches separately.

**Discretization Methods for Standard Semi-Infinite Optimization Problems** In this section, we consider standard semi-infinite optimization problems, that is, optimization problems of the form

$$\text{SIP:} \quad \min_{\mathbf{x} \in X \subseteq \mathbb{R}^m} f(\mathbf{x}) \quad \text{s.t.} \quad g_i(\mathbf{x}, \mathbf{y}) \leq 0 \quad \text{for all } \mathbf{y} \in Y,\ i \in I,$$

where $I := \{1, \dots, p\}$, $Y$ is a non-empty, compact, infinite (index) set, and $f$, $g_i$, $i \in I$, are real-valued, sufficiently smooth functions. For $\hat{Y} \subset Y$, we introduce the optimization problem

$$\text{SIP}(\hat{Y})\text{:} \quad \min_{\mathbf{x} \in X \subseteq \mathbb{R}^m} f(\mathbf{x}) \quad \text{s.t.} \quad g_i(\mathbf{x}, \mathbf{y}) \leq 0 \quad \text{for all } \mathbf{y} \in \hat{Y},\ i \in I,$$

If the set $\hat{Y}$ is finite, SIP($\hat{Y}$) is referred to as a *discretized* SIP problem.

The basic idea of discretization methods is to successively calculate solutions of discretized SIP problems SIP($\dot{Y}^k$), $k \in \mathbb{N}_0$, using a solution method for finite optimization problems, where $\{\dot{Y}^k\}_{k \in \mathbb{N}_0}$ is a sequence of finite subsets of $Y$ that converges to the set $Y$ in the Hausdorff distance. The sequence $\{\dot{Y}^k\}_{k \in \mathbb{N}_0}$ is either established *a priori* or defined *adaptively*. In the latter case, information from the $k$-th discretization step is enlisted for defining $\dot{Y}^{k+1}$. These considerations can be algorithmically applied as follows:

**Algorithm 2** General discretization method for SIP problems, [34, 35]

Choose a sequence $\{Y^k\}_{k \in \mathbb{N}_0}$ of non-empty, compact subsets of $Y$, such that $|Y^0| < \infty$, $Y^k \subseteq Y^{k+1}$ for all $k \in \mathbb{N}_0$ and the sequence converges to $Y$ in the Hausdorff distance; a starting point $\mathbf{x}^0 \in X \subseteq \mathbb{R}^m$; and a feasibility tolerance $\varepsilon > 0$.
Set $\dot{Y}^0 := Y^0$, $\mathbf{x}^{0,0} := \mathbf{x}^0$, and $k := 0$.
**repeat**
  Compute a solution $\mathbf{x}^{k,*}$ of the discretized SIP problem SIP($\dot{Y}^k$) using $\mathbf{x}^{k,0}$ as starting point.
  Choose a set $\dot{Y}^{k+1}$ with $\dot{Y}^k \subseteq \dot{Y}^{k+1} \subseteq Y^{k+1}$.
  **for** $i = 1 \to p$ **do**
    Compute a global solution $\mathbf{y}_i^{k,*}$ of $\max_{y \in Y^{k+1}} g_i(\mathbf{x}^{k,*}, \mathbf{y})$.
    **if** $g_i(\mathbf{x}^{k,*}, \mathbf{y}_i^{k,*}) > \varepsilon$ **then**
      Set $\dot{Y}^{k+1} := \dot{Y}^{k+1} \cup \{\mathbf{y}_i^{k,*}\}$.
    **end if**
  **end for**
  Set $\mathbf{x}^{k+1,0} := \mathbf{x}^{k,*}$ and replace $k$ by $k+1$.
**until** $\max_{i=1,\dots,p} g_i(\mathbf{x}^{k-1,*}, \mathbf{y}_i^{k-1,*}) \leq \varepsilon$
**return** $\mathbf{x}^* = \mathbf{x}^{k-1,*}$.

It is not necessary that the starting point $\mathbf{x}^0$ in step 1 is feasible for SIP. In the simplest case, $Y^{k+1} := Y$ can be chosen in steps 1 and 5. In step 4, essentially any method for solving finite optimization problems can be used. The only two requirements here are that it can handle infeasible starting points and high-dimensional problems. Except for small $m$ and $|\dot{Y}^k|$, however, it is not appropriate to use a generic solution method, since such methods often solve sub-problems having the same number of constraints as the problem itself. Thus, they do not take advantage of the fact that the constraints of a discretized SIP problem stem from only a few functions. For this reason, proprietary methods have been developed to solve these special finite optimization problems (see, for example, [27, 31, 32]).

In order for the method to converge, it is crucial in step 7 to compute a global solution, or at least a good approximation.

**Transformation of a General into a Standard Semi-Infinite Problem** In order to be able to use discretization techniques for solving general semi-infinite optimization problems, the methods must either be generalized for the case of variable index sets or the general semi-infinite optimization problem must be transformed into an equivalent standard problem.

In principle, it is possible to generalize discretization and exchange methods for standard semi-infinite optimization problems to the general semi-infinite case. An additional challenge here, however, along with the rapidly growing size of the induced finite problems, is the $\mathbf{x}$-dependency of the index set $Y(\mathbf{x})$, and, thus, of its discretization. In order to guarantee that the feasible sets of the optimization problems induced by the discretizations are closed, the discretization points must be so designed that they depend at least continuously on $\mathbf{x}$, which is non-trivial (see [47]).

Using suitable assumptions, the transformation of a general into a standard semi-infinite optimization problem is, in principle, at least *locally* possible (see [45, 49]). However, such a transformation is only of practical use when it is *globally* defined. The ideal situation is as follows:

**Assumption 4** Let there be a non-empty, compact set $Z \subset \mathbb{R}^{\tilde{n}}$ and a mapping $\mathbf{t} : \mathbb{R}^m \times Z \to \mathbb{R}^n$ that is at least continuous, such that $\mathbf{t}(\mathbf{x}, Z) = Y(\mathbf{x})$ for all $\mathbf{x} \in X \subseteq \mathbb{R}^m$.

Under this assumption, the general semi-infinite constraints

$$g_i(\mathbf{x}, \mathbf{y}) \leq 0 \quad \text{for all } \mathbf{y} \in Y(\mathbf{x}),\ i \in I,$$

are clearly equivalent to the standard semi-infinite constraints

$$\tilde{g}_i(\mathbf{x}, \mathbf{z}) := g_i\big(\mathbf{x}, \mathbf{t}(\mathbf{x}, \mathbf{z})\big) \leq 0 \quad \text{for all } \mathbf{z} \in Z,\ i \in I.$$

For one-dimensional index sets $Y(\mathbf{x}) = [a(\mathbf{x}), b(\mathbf{x})]$, with $a(\cdot) \leq b(\cdot)$, such a transformation can be designed simply by means of a convex combination of the interval limits; for higher dimensional index sets, there exists such a transformation when it is star-shaped (see [45]), which is the case under Assumptions 1 to 3.

However, the transformation entails a serious disadvantage: it can destroy the convexity in the lower-level that is so important for the convergence of the discretization method (see [14], for example).

**Combination of Both Techniques** We now outline how the above-mentioned disadvantage can be circumvented, thus allowing the solution of *transformable* general semi-infinite optimization problems using discretization methods. For details, the reader is referred to [14] (along with [8] and [9]).

We begin by introducing the standard semi-infinite optimization problem induced by the transformation:

$$\widetilde{\mathsf{SIP}}: \quad \min_{\mathbf{x} \in X \subseteq \mathbb{R}^m} f(\mathbf{x})$$
$$\text{s.t.} \quad \tilde{g}_i(\mathbf{x}, \mathbf{z}) \leq 0 \quad \text{for all } \mathbf{z} \in Z,\ i \in I,$$

with $\tilde{g}_i(\mathbf{x}, \mathbf{z}) := g_i(\mathbf{x}, \mathbf{t}(\mathbf{x}, \mathbf{z}))$, $i \in I$. We denote its lower-level problems by

$$\tilde{\mathsf{Q}}_i(\mathbf{x}): \quad \max_{\mathbf{z} \in Z} \tilde{g}_i(\mathbf{x}, \mathbf{z}), \quad i \in I.$$

As already seen, the feasible sets, and thus the local and global solutions of GSIP and $\widetilde{\mathsf{SIP}}$, coincide. Consequently, a solution for the underlying general semi-infinite problem can be obtained by solving the induced standard problem. A similar result is also obtained with the global solutions of the corresponding lower-level problems.

**Theorem 4** ([14]) *Let $\mathbf{x} \in X$ and $i \in I$. Then, the point $\mathbf{z}^*$ is a global solution of $\tilde{\mathsf{Q}}_i(\mathbf{x})$ if and only if $\mathbf{y}^* = \mathbf{t}(\mathbf{x}, \mathbf{z}^*)$ is a global solution of $\mathsf{Q}_i(\mathbf{x})$.*

One can thus calculate a global solution for the non-convex problem $\tilde{\mathsf{Q}}_i(\mathbf{x})$ by finding a global solution of the convex problem $\mathsf{Q}_i(\mathbf{x})$ and transforming it via $\mathbf{t}(\mathbf{x}, \cdot)$ in $Z$. This makes it unnecessary to solve the non-convex problems $\tilde{\mathsf{Q}}_i(\mathbf{x})$, $i \in I$, using time-consuming methods of global optimization.

Using the insights from Theorem 4, we can now adapt the relevant steps in Algorithm 2 and obtain a discretization method for transformable general semi-infinite optimization problems:

**Algorithm 3** Transformation-based discretization method for GSIP problems, [14] and [8]

1: Choose a starting point $\mathbf{x}^0 \in X$ and a feasibility tolerance $\varepsilon > 0$.
2: Choose/calculate a starting discretization $\dot{Y}^0(\mathbf{x}^0) \subset Y(\mathbf{x}^0)$ and determine $\dot{Z}^0$ such that $\mathbf{t}(\mathbf{x}^0, \dot{Z}^0) = \dot{Y}^0(\mathbf{x}^0)$.
3: Set $\mathbf{x}^{0,0} := \mathbf{x}^0$ and $k := 0$.
4: **repeat**
5: Compute a solution $\mathbf{x}^{k,*}$ of $\widetilde{\mathsf{SIP}}(\dot{Z}^k)$ using $\mathbf{x}^{k,0}$ as starting point.
6: **for** $i = 1 \to p$ **do**
7: Compute a (global) solution $\mathbf{y}_i^{k,*}$ of $\mathsf{Q}_i(\mathbf{x}^{k,*})$.
8: **if** $g_i(\mathbf{x}^{k,*}, \mathbf{y}_i^{k,*}) > \varepsilon$ **then**
9: Determine $\mathbf{z}_i^{k,*}$ such that $\mathbf{t}(\mathbf{x}^{k,*}, \mathbf{z}_i^{k,*}) = \mathbf{y}_i^{k,*}$ and set $\dot{Z}^{k+1} := \dot{Z}^k \cup \{\mathbf{z}_i^{k,*}\}$.
10: **end if**
11: **end for**
12: Set $\mathbf{x}^{k+1,0} := \mathbf{x}^{k,*}$ and replace $k$ by $k+1$.
13: **until** $\max_{i=1,\dots,p} g_i(\mathbf{x}^{k-1,*}, \mathbf{y}_i^{k-1,*}) \le \varepsilon$
14: **return** $\mathbf{x}^* = \mathbf{x}^{k-1,*}$.

The requirements for the transformation-based discretization method are the same as those for Algorithm 2. If no starting discretization $\dot{Y}^0(\mathbf{x}^0)$ from $Y(\mathbf{x}^0)$ is available for step 2, one can be obtained by solving the lower-level problems and transforming the solutions. A feasible starting point for step 7 can be calculated from a feasible point from $Z$ via the transformation $\mathbf{t}(\mathbf{x}, \cdot)$. With regard to the curvature behavior of the involved functions, only the convexity of the lower-level problems is presupposed in the above method, and not the convexity of the objective function $f$ and the functions $g_i(\cdot, \mathbf{y})$, $i \in I$, for all $\mathbf{y}$. Therefore, the result $\mathbf{x}^*$ of Algorithm 3 is only as "optimal" as the results of the method used to solve the discretized SIP problems in step 5. Incidentally, this is also the case for the explicit smoothing method (Algorithm 1) and its feasible variant.

Finally, we want to illustrate how the transformation-based discretization method works, by means of an example. And here, we'll employ the same example used for the explicit smoothing method. Our goal, therefore, is once again the area-maximal embedding of an ellipse in the container $C^{\mathrm{CT}}$. For the transformation-based discretization method, we not only need a function to describe the ellipse and an area computation formula, we also need a description of the ellipse as an image of a compact set under a continuously differentiable mapping. As mentioned previously, we model the ellipse as a translated and distorted unit circle. Accordingly, one possible transformation is

$$\mathbf{t}: \mathbb{R}^5 \times [0,1]^2 \to \mathbb{R}^2 \quad \text{with} \quad \mathbf{t}(\mathbf{x}, \mathbf{z}) := \mathbf{A}(\mathbf{x}) \begin{pmatrix} z_1 \cos(2\pi z_2) \\ z_1 \sin(2\pi z_2) \end{pmatrix} + \mathbf{c}(\mathbf{x}),$$

where $\mathbf{A}(\mathbf{x})$ and $\mathbf{c}(\mathbf{x})$ are chosen as above.

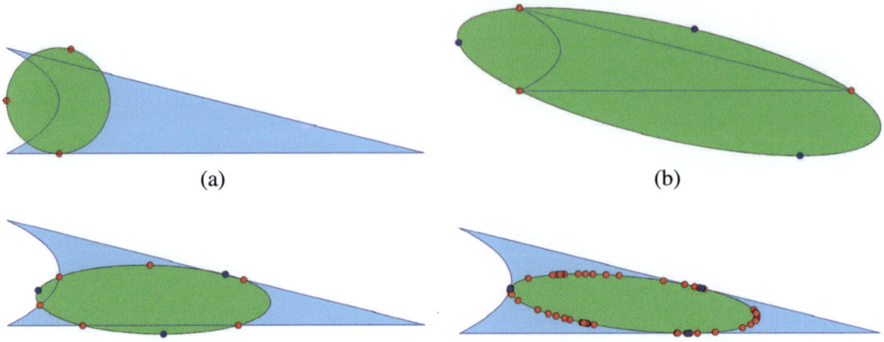


**Fig. 16** Area-maximal design-centering of an ellipse into the container $C^{\mathrm{CT}}$ using the transformation-based discretization method (Algorithm 3) [*light blue*-container, *green*-design, *red*-points of the current discretization, *dark blue*-points of the greatest violation of the container constraints, which are added to the current discretization for the next calculation]: (**a**) initial situation with calculated starting discretization, (**b**) after solving problem $\widetilde{\mathsf{SIP}}(\dot{Z}^0)$, (**c**) after solving $\widetilde{\mathsf{SIP}}(\dot{Z}^1)$, and (**d**) final situation (after a total of 30 refinements)

As starting point $\mathbf{x}^0$, we have again selected the (infeasible) point $(0, 0, 1, 1, 0)$ (see Fig. 16(a), also), and as feasibility tolerance, $\varepsilon = 10^{-6}$. The initial discretization $\dot{Y}^0(\mathbf{x}^0)$ consists of the solutions of the lower-level problems. Figure 16 illustrates graphically the successively refined discretization and the solution of the discretized SIP problems $\widetilde{\mathsf{SIP}}(\dot{Z}^k)$, $k \in \mathbb{N}_0$.

## 5 Industrial Project I—Automation of Pre-forming, Grinding, and Polishing

In Sect. 1.1, we described how gemstones are processed by hand in the traditional manufacturing setting and outlined the new automated approach developed for producing colored gemstones over the past decade—an approach derived from the traditional jewelmaker's craft. In this section, we elaborate on the resulting modeling questions and algorithmic solution approaches and discuss the implementation of the automating equipment and software.

### 5.1 Questions for Modeling an Optimization Problem—Describing Alternative Sets and Quality Measures

In order to make mathematical optimization methods of practical use, one needs an available feasibility or alternative set and well-defined target quantities, which should be characterized as favorably as possible. An easily formulated optimization goal is to maximize the material yield, that is, the sellable volume fraction of a rough stone. A simple feasibility requirement is the containment condition, that is, the requirement that the desired faceted stone is completely contained within the rough stone and that there exists, where necessary for processing reasons, an additional safety buffer between the faceted stone and the edge of the rough stone.

The esthetic requirements are markedly more difficult. For example, the cut pattern of a stone has a significant impact on the final appearance of the faceted stone. Here, a constellation of problems becomes apparent: First, beauty, as the saying goes, is in the eye of the beholder; that is, it is subjective. Second, the subjective appraisal of a person, a jeweler for example, is elusive and difficult to fix precisely. Thus, the esthetic aspect represents one of the greatest modeling challenges.

**Basic Approach** For a given rough stone, esthetically motivated conditions are placed on the proportions, and volume optimal solutions are then defined for each faceted stone base form being considered (round, oval, octagonal, etc.). The variously shaped and proportioned faceted stones thus calculated are then presented to a decision-maker via a graphic user interface. On the basis of what he considers to be the most favorable combination of material yield and esthetic considerations, the decision-maker then selects a faceted stone shape for production.

As described in Sect. 2.1, the division of the manual production process into two parts, pre-forming and faceting, motivated us to divide the modeling into two parts as well, by decoupling the continuous and discrete variables. We accomplish this by introducing the calibration body as a parameterized equivalent to the smooth pre-grinding form. We can optimize the calibration body, with an eye on the material yield and proportions, without first having to commit to a particular faceting pattern. In the following subsections, we discuss in greater detail the description of the calibration body, the faceting, and the rough stone modeling.

### 5.1.1 Faceted Stone Shapes and Calibration Body

A calibration body is characterized in part by its parameterization; parameters include, for example, position in rough stone, height, length, width, and degree of belliedness. Some are generic parameters that are independent of the faceted stone shape and others are shape specific. The parameters' feasibility domains ensure that the proportions remain within zones that result in esthetically appealing jewels. After one has defined specific values for the parameters, then the most appropriate faceting pattern can be chosen.

The calibration body is also characterized by smooth functions $\mathbf{v} : \mathbb{R}^m \times \mathbb{R}^3 \to \mathbb{R}^q$, which establish, in dependency on the parameters $\mathbf{x} \in \mathbb{R}^m$, whether a point $\mathbf{y} \in \mathbb{R}^3$ is indeed located within the calibration body ($v_j(\mathbf{x}, \mathbf{y}) \leq 0$ for all $j = 1, \ldots, q$), or whether it is not ($v_j(\mathbf{x}, \mathbf{y}) > 0$ for at least one $j \in \{1, \ldots, q\}$). The choice of functions depends of course on the shape of the faceted stone.

On the basis of a simple faceted stone shape with a circular girdle base form, we will now explain how a calibration body can be described and parameterized and how the esthetic requirements fit into the analysis.

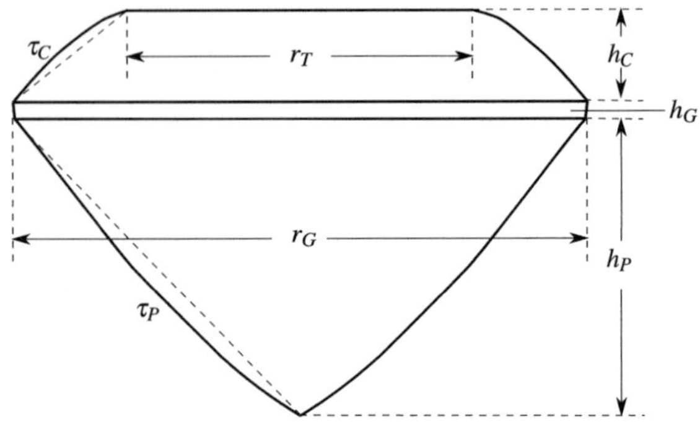


**Fig. 17** Parameterization of the round faceted stone shape using heights and radii and degree of belliedness for pavilion and crown

**Parameterizing a Calibration Body** With the help of six real-valued parameters and a suitable coordinate system, one can represent the absolute position and rotation of a calibration body within the rough stone. The description of the actual extent of the faceted stone shape depends on the base form. For a round stone it can be described using seven more parameters, as depicted in Fig. 17. For the three faceted stone elements—crown, girdle, and pavilion—the three heights $h_C$, $h_G$, and $h_P$ are further parameters; for the crown and girdle, there is a radius $r_T$ and a radius $r_G$ for the table and the girdle; and for the crown and pavilion, there is one more parameter each, $\tau_C$ and $\tau_P$, which describe the degree of belliedness or curvature. For other, more complicated faceted stone shapes, there are additional parameters, such as the ratio of the length to the height of the girdle base form.

Alternatively, one can also choose a scaling-invariant parameterization. Here, the radius of the girdle is set to 1, and all other parameters that specify a length are replaced by the ratio of that length to the girdle radius. Thus, we obtain the new parameters $\tilde{h}_C := \frac{h_C}{r_G}$, $\tilde{h}_G := \frac{h_G}{r_G}$, $\tilde{h}_P := \frac{h_P}{r_G}$, and $\tilde{r}_T := \frac{h_C}{r_G}$. The parameters $\tau_C$ and $\tau_P$ remain unchanged. Later, in the algorithmic section, we will make use of the advantages of this scaling-invariant parameterization.

**Calibration Body Proportions** The feasible value domain of the parameters is very important for the esthetic appearance of the final faceted stone. For example, parameters that specify lengths must fulfill certain proportion requirements. The ratio of the girdle radius to the crown radius, for example, is restricted by both an upper and a lower bound. The same holds true for the ratio of the total calibration body height to the girdle radius or to the individual heights of the pavilion, girdle, and crown. Likewise, there are upper and lower bounds for the belliedness parameters $\tau_P$ and $\tau_C$. It is not easy to make a good choice for the combination of these bounds, since this choice depends very strongly on the esthetic sensibilities of the decision-maker. A guideline for the mathematical model should be to not make the feasibility intervals too small, otherwise the latitude for volume optimization and the incorporation of esthetic considerations becomes too limited.

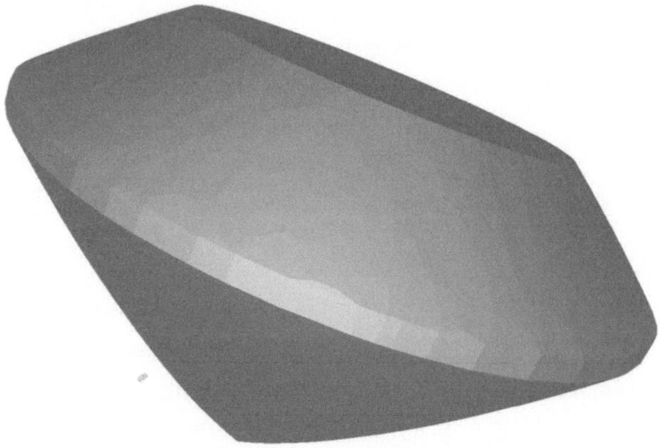

**Fig. 18** Calibration body of the round faceted stone

**Functions for Describing a Calibration Body** In order to be able to use the methods from Sect. 4.5, the calibration body must be described by convex, differentiable functions. Using the parameters described above, we specify for our example of the round faceted stone a corresponding description. Here, the calibration body is given as $D^{\text{Round}}(\mathbf{x}) := \{\mathbf{y} \in \mathbb{R}^3 \mid \mathbf{v}(\mathbf{x}, \mathbf{y}) \le 0\}$ and we define $\mathbf{v}$ as follows:

$$\mathbf{v} : \mathbb{R}^7 \times \mathbb{R}^3 \to \mathbb{R}^5 : y \mapsto \begin{cases} y_1^2 + y_2^2 - r_P(y_3) & \text{lateral boundary of the pavilion} \\ y_1^2 + y_2^2 - r_C(y_3) & \text{lateral boundary of the crown} \\ y_1^2 + y_2^2 - r_G^2 & \text{lateral boundary of the girdle} \\ y_3 - h_C - h_G & \text{boundary of the table} \\ -y_3 - h_P & \text{boundary at the pavilion apex} \end{cases} \quad (32)$$

The functions $r_P$ and $r_C$ specify the radius for a given height $y_3$ and are dependent upon $\tau_P$, $\tau_C$, $r_C$, and $r_G$. The curvatures of pavilion and crown depend upon $\tau_P$ and $\tau_C$, respectively. We refer the reader to [16] for a more detailed description of $r_P$ and $r_C$. There, and in [14] as well, one can find formulas for calculating calibration body volumes via the appropriate integrations. Figure 18 shows an approximation of the set $D^{\text{Round}}$.

### 5.1.2 Calculating the Facetings

The step from a calibration body to a beautiful faceting is more complex than one might imagine. The obvious idea of simply laying the corners of the facets on the margin of the calibration body won't work, since the resulting system of equations is over-defined, at least for the Portuguese cut.

As already hinted in Sect. 2.1, the challenges of defining suitable facetings result from numerous aspects that must be considered: The faceting determines the reflection of light and thus the stone's sparkle and inner flame. Here, the setting angles are crucial. These angles must be neither too sharp nor too shallow, in order to ensure optimal reflection. The number of rows and the number of facets per row are also important criteria and depend on the size and material characteristics of the stone. In general, the larger the stone, the

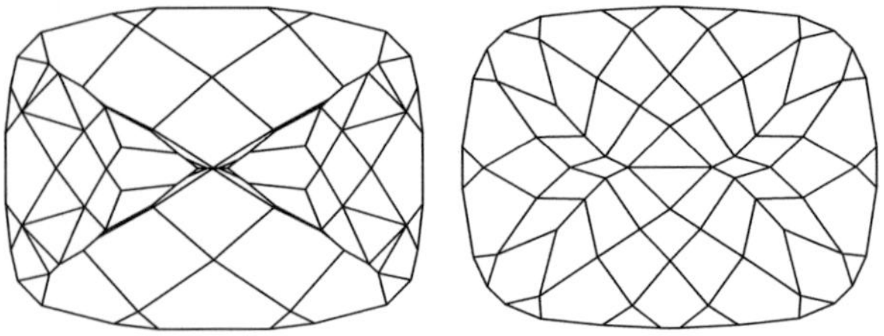

**Fig. 19** 2D projection of a pavilion faceting pattern for the antique stone shape: *left*, using the iterative approach, which delivers an unsatisfactory result; *right*, using the explicit approach

more facets it should have. The faceting must have the same axes of symmetry as the corresponding faceted stone shape. Moreover, an attractive cut pattern is one in which all the facets in a given row have approximately the same setting angle and are about the same height and width. The facets should also decrease in size as one moves away from the girdle.

Two approaches have proven successful for calculating the faceting. The first consists of iteratively adding rows of facets, starting at the girdle and working outwards towards the pavilion's apex and the crown's table. Here, a shallower setting angle is specified for each succeeding row. With this approach, however, the possibilities for influencing the faceting pattern are limited. In the second approach, the desired final facets are given at the start and parameterized via their corner points and the normals of the associated levels. If one now formulates as equations the fact that two neighboring facets share two corners or that all the facets in a given row are to have the same setting angle, then one obtains a system of equations that, although over-defined, can nevertheless be solved approximately after appropriate relaxation. This approach is explicit and allows one to exactly control the resulting faceting pattern. It also requires substantially more effort, since a different type of equation system must be set up for each faceted stone shape. Under some circumstances, however, as depicted in Fig. 19, it delivers clearly more attractive faceting than the first approach, which cannot always guarantee good results.

### 5.1.3 Describing the Rough Stone

The geometric shape of the rough stone must be captured in a suitable manner in order to make it accessible to the optimization algorithms. One possibility for digitalizing the rough stone utilizes data about its surface. Using a 3D scanner with either the stripe projection or laser scanning procedure, surface point clouds are recorded for the rough stone, which one can then convert to a surface model by means of triangulation.

In the terminology of the material cutting field, the rough stone forms the container. If we want to use the solution method from Sect. 4.5, however, then triangulation as a description of the container is hardly suited, since tremendous point clouds arise on the surface for exactness requirements of about 5–10 micrometers. Instead, the convex hull of the net is used, and larger indentations, that is, differences between the rough stone net and the convex hulls, are additionally described by means of quadrics. Because one can represent the inside of the convex hulls by a potentially large number of linear inequalities, this method yields a description of the container consisting of linear and convex quadratic functions.

## 5.2 Algorithmic Implementation

In the following, we introduce two alternative methods for achieving a maximum material yield.

The first method is based on Algorithm 1 from Sect. 4.5.1. We consider a hierarchic formulation of the problem: First, optimize the volume of the calibration body; then, approximate this calibration body using a suitable faceting pattern with as little volume loss as possible. In a post-optimization step, we can scale and/or rotate the resulting faceted stone into the rough stone in a volume-optimized manner.

The calibration body modeling described in Sect. 5.1.1 conforms to the formal design description requirements needed for the algorithm. The container is likewise described functionally, as discussed in the previous section. In principle, then, Algorithm 1 can be applied. In practice, however, the problem arises that, due to the many functions describing the container—hereafter referred to as *container functions*—the algorithm becomes very slow. This problem is dealt with by initially considering only a very small portion of the container functions. One iteratively applies the algorithm to the selected subset of container functions, calculates a solution for this relaxed problem, and checks to see whether the resulting faceted stone is feasible with regard to all container functions. If not, one expands the selected subset by including the violated container functions. Using this new, expanded selection, one then re-calculates and begins the next iteration. In practice, this procedure leads to a solution after a few iterations. This solution is feasible for the starting problem, but nonetheless, even in the final iteration, only a small number of container functions must be taken into consideration. Using the calibration body parameters found in this way, one now calculates the faceting. Because the container functions do not map the rough stone exactly and only the calibration body is considered in the optimization, the faceted stone must be adjusted by slight translation, rotation, and scaling operations in a final step, in order to ensure that it lies completely within the rough stone surface described by the triangulation.

The second approach is based on the treatment of the non-overlapping condition described in Sect. 4.3 and works directly with the triangulation of the rough stone. This allows the original surface description to be used directly. The faceting can also be understood as a triangulation if the girdle is suitably discretized. The main challenge inherent in this second approach is to quickly develop a faceting when the calibration body parameters change. Here, one advantage is that the triangulation resulting from a faceting is very small relative to the triangulation of the rough stone. A second advantage is that a change affecting the position of the faceted stone does not lead to a re-calculation of the faceting. For the algorithmic implementation of this approach, one now uses the scaling-invariant parameterization described above with scaling parameter $s$. The problem of maximum material yield can now be described as the search for a maximum scaling parameter $s^*$. Because the containment of the design within the container for a given $s$ can be quickly verified, the optimal value $s^*$ can also be quickly determined—for example using a bisection approach. In general, as shown in [4], $s^*$ depends continuously on the scaling-invariant parameters, so that common optimization methods can be used for the resulting optimization problem.

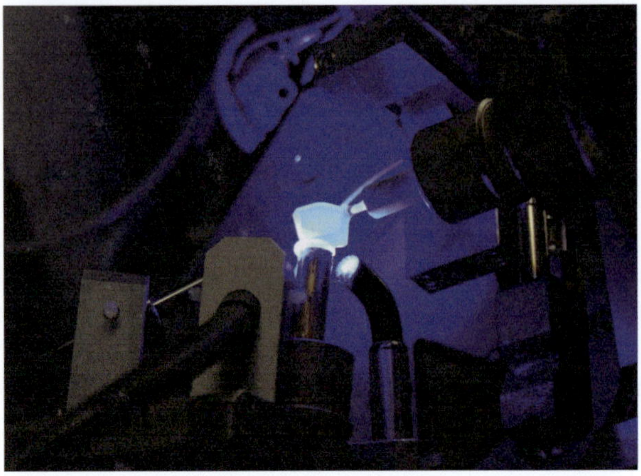

**Fig. 20** A rough stone glued to a measurement pin and the re-gluing procedure

### 5.3 Automating the Grinding and Polishing Process—The Technical Challenges

In addition to the virtualization of design and container required to make the maximum material yield problem mathematically and informationally accessible, there are also technical challenges to be mastered, such as holding and guiding the stone during the work steps and automating the grinding and polishing processes.

The starting point for successfully industrializing gemstone production is the approach used in the hand manufacturing process, which is to be suitably refined and automated. The entire process consists of the following steps:

1. The rough stone is manually glued to a measurement pin.
2. It is then measured and digitalized using a 3D scanner (see Fig. 21).
3. An optimal faceted stone is virtually embedded in the measured rough stone via mathematical optimization, as described in Sect. 5.2.
4. The corresponding difference images are converted into re-gluing, grinding, and polishing plans and transferred to the machines.
5. The rough stone is transferred from the measurement pin to a processing pin, while preserving the coordinate system (see Fig. 20).
6. Transfer to the processing station; grinding and polishing of the girdle and the front side (see Fig. 22).
7. Transfer to the pin re-positioning station; axial re-positioning on a second processing pin.
8. Transfer to the processing station; grinding and polishing of the back side.
9. The processing pin is removed by hand; the faceted stone is now finished.

The goal in designing the process was a level of accuracy in all steps such that an absolute accuracy of 5–10 micrometers could be achieved for the overall production. To describe here in detail all of the technical requirements and their mechanical engineer-

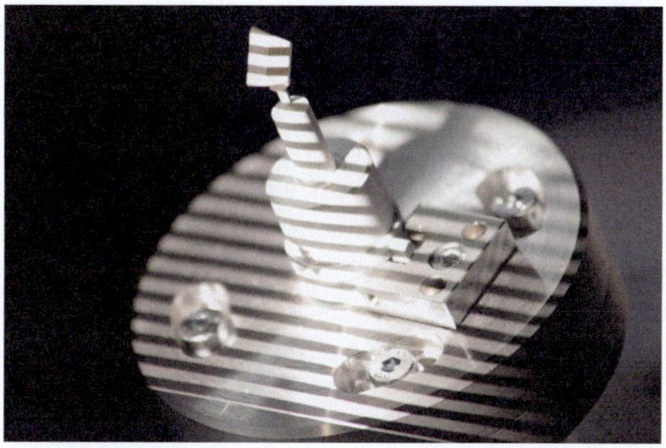

**Fig. 21** A configuration for generating a three-dimensional representation of the rough stone via the stripe projection method (Photo: G. Ermel, Fraunhofer ITWM)

**Fig. 22** Grinding a stone's first processing side

ing solutions would exceed the scope of our discussion. The pictures, however, do offer some impressions of the pre-series prototype at the Fraunhofer ITWM, which fulfilled the targeted requirements.

## 5.4 Automating the Grinding and Polishing Processes—The Software Challenges

Along with the technical challenges, there were also five software development problems to be solved:

1. Implementing the optimization algorithm: Here, the primary challenge is to efficiently implement the above-described approaches and to ensure that they are also robust in the face of very rare, pathological, numerical cases that might not arise until the process has been in operation for a length of time.

2. Scalable parallelization: Due to the high computing time requirements—about ten faceted stone shapes must be calculated for each stone—parallel execution of the optimization is necessary. Here, the calculations are distributed on multiple CPUs.
3. Centralized data-keeping is a critical element: It not only has to support the parallelization of the calculations, it must also maintain in readiness a consistent view of the data for machine controlling and the user interface. Here, the extensive functionality of modern databases is very helpful.
4. The machine control system must manage the various stations of the machines for grinding, polishing, pin re-positioning, scanning, and transporting the stone and must pick up error functions and breakdowns.
5. As the interface between operator and machine, the user display must include components for controlling and configuring the machines, for showing the virtual rough stones and calculated faceted stones with hardware-optimized 3D depictions, and for starting and configuring the optimization calculations. Here, the ease-of-use of the software and the resulting user experience—hopefully, a positive one—play an important role.

Because of the variety of functions, a professional software design is indispensable. Although the code was created originally in a dissertation according to purely scientific considerations, the current process software now has a modular, maintainable, and extendable structure, in which the individual components can be added or removed, as needed.

## 6 Industrial Project II—Gemstone Sectioning

After looking at the optimal conversion (with respect to cut and volume) of a rough stone to a faceted one in Industrial Project I, we now turn to the "gemstone sectioning" project. Here, several faceted stones are to be produced from a single rough stone, while maximizing the total volume of the final jewels and avoiding flaws. With this endeavor, we move one step closer to the goal of solving the complete gemstone cutting problem.

### 6.1 Description of the Problem

We recall that the (main) task of gemstone cutting consists of transforming a rough stone marred with surface flaws, inclusions, and cracks into faceted stones in such a way that their total value is as high as possible. Here, we consider only the volume as value-determining criterion and require that the finished faceted stones be free of flaws. The implementation of the esthetic requirements can be accomplished analogously to Sect. 5.

In order to produce several faceted stones from a single rough stone, the latter must first be sectioned into blanks, each of which yields one faceted stone. This raises the following two questions:

How many faceted stones should be produced from the rough stone, that is, into how many blanks should the rough stone be sectioned? What does such a sectioning look like?

Although, in the manual production process, "sectioning" and "grinding" are separate work steps, generally performed by different persons, the sawyer is already giving some thought to how the blanks should look in order to yield large-volume and esthetically pleasing faceted stones.

The standard tool for sectioning rough stones is the circular saw, with which only straight cuts are possible. While it is indeed possible to cut out a wedge-shaped blank using a circular saw, we want to assume that each cut is a through-cut, which is referred to in the trade as a *guillotine cut*. Moreover, since each cut consumes valuable material, one uses narrow-kerf blades and tries to keep the cuts short and few in number when sectioning the rough stone.

## 6.2 Modeling

From a mathematical perspective, the above problem is once again of the maximum material yield type. Here, however, we need to generalize the non-overlapping condition, since the faceted stones must have a specified minimum distance from one another to allow for the kerf width. As an additional requirement, they must also be present in a guillotine arrangement in order to be amenable to circular saw technology (see Fig. 23). In the following discussion, we illustrate how both requirements can be mathematically modeled in the context of Sects. 4.1 and 4.4, that is, for arbitrarily shaped containers and designs.

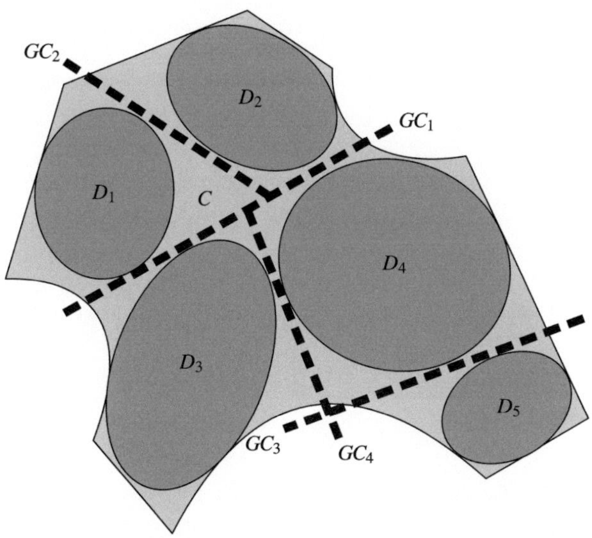


**Fig. 23** Guillotine arrangement of five elliptical designs with minimum distances in a container described by lines and quadrics

### 6.2.1 Minimum Distance Between the Designs

We first look at the requirement that the designs must have a specified minimum distance $\delta > 0$ from each other.

In some cases, geometrical considerations allow one to deduce practicable conditions. For example, two circles have at least the distance $\delta$ between them if and only if the distance between their centers is greater than or equal to the sum of their radii plus $\delta$ (see Fig. 24).

For more complicated designs, this approach is not usually expedient. However, as with the non-overlapping condition, describing the designs by means of functions also allows one here to implement this requirement using semi-infinite constraints. In [14], two modeling approaches were proposed for accomplishing this aim: via *Euclidean (norm) distance* and via *separating hyperplane*.

The first approach is intuitive. Two designs have a minimum distance $\delta$ between them if and only if each point of one design has at least a distance $\delta$ from each point of the other design (see Fig. 25, *left*). The mathematical formulation for this is

$$\|\mathbf{y} - \mathbf{z}\|_2 \geq \delta \quad \text{for all } (\mathbf{y}, \mathbf{z}) \in D_1(\tilde{\mathbf{p}}_1) \times D_2(\tilde{\mathbf{p}}_2),$$

which is clearly of semi-infinite nature.

The second approach, as the name implies, is based on the separation of the designs by means of hyperplanes. On the one hand, we know that the non-overlapping of two convex designs can be guaranteed by means of a separating hyperplane (see Sect. 4.4).

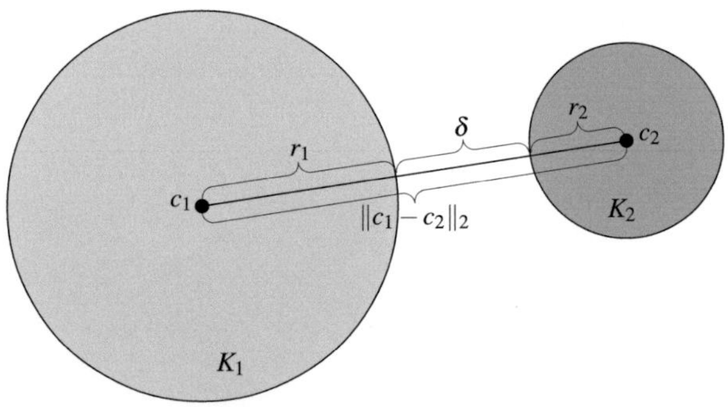


**Fig. 24** Two circles with distance $\delta$

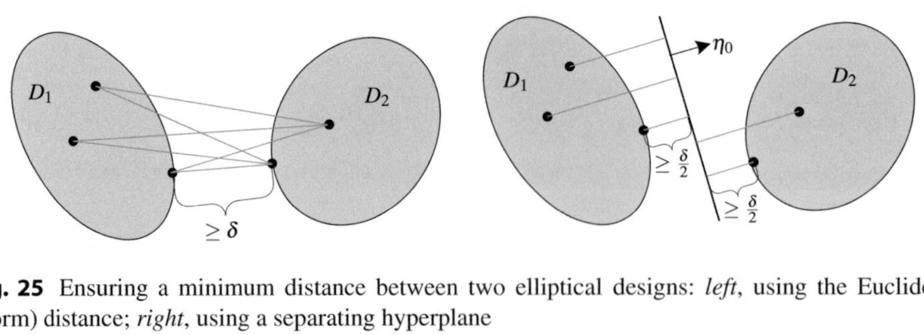


**Fig. 25** Ensuring a minimum distance between two elliptical designs: *left*, using the Euclidean (norm) distance; *right*, using a separating hyperplane

On the other hand, the distance of a point from a hyperplane can be directly calculated by inserting the point into the hyperplane equation $\boldsymbol{\eta}^T \mathbf{y} = \beta$, if the normal vector $\boldsymbol{\eta}$ of the hyperplane is normalized to 1, that is, if $\|\boldsymbol{\eta}\|_2 = 1$. If we now require that all points of one design lie on one side of the hyperplane and are at least a distance $\delta/2$ away from it, and that all points of the other design lie on the other side of the hyperplane and are at least the same distance away from it, then the designs have a minimum distance $\delta$ between them (see Fig. 25, *right*). These requirements can be formulated mathematically as the inequalities

$$\boldsymbol{\eta}_0^T \mathbf{y} - \beta \leq -\frac{\delta}{2} \quad \text{for all } \mathbf{y} \in D_1(\tilde{\mathbf{p}}_1)$$

and

$$\boldsymbol{\eta}_0^T \mathbf{z} - \beta \geq \frac{\delta}{2} \quad \text{for all } \mathbf{z} \in D_2(\tilde{\mathbf{p}}_2),$$

which are both semi-infinite in nature. Here, $\boldsymbol{\eta}_0$ denotes the normalized unit vector.

### 6.2.2 Guillotine Arrangements of Designs

We now show how the requirement of a guillotine arrangement can be implemented for maximum material yield problems.

In order to take advantage of such an arrangement using a saw, it is necessary, of course, to leave space for the saw kerf between the planned designs. However, in the following considerations, for clarity's sake, we require no minimum distance between the designs. As shown in the previous section, this requirement can be easily integrated into the model at a later stage.

Guillotine cutting problems have been investigated mathematically since the mid-1960s ([26]). The most frequently considered problem is the so-called *two-dimensional orthogonal guillotine cutting problem* (see Fig. 26):

2DOGCP: Can a given set of orthogonally rotatable rectangles be cut out of a large rectangle by a series of linear cuts that run either parallel or orthogonal to the sides of the large rectangle, i.e., by a series of *guillotine cuts*?

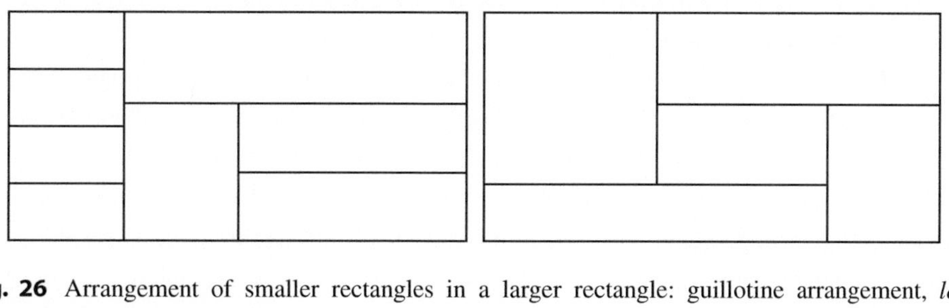

**Fig. 26** Arrangement of smaller rectangles in a larger rectangle: guillotine arrangement, *left*; non-guillotine arrangement, *right*

To date, only minor modifications of this standard problem have been investigated: guillotine arrangement of equally sized circles in a rectangle (see [21, 22]) and guillotine cuts of cuboids (see [33]), as well as hyper-cuboids (see [17]).

All familiar models and solution methods take advantage of the simple and fixed geometry of the designs and the container. The designs are translatable; rotation, in contrast, when allowed at all, is only permitted in 90° steps. The guillotine cuts run orthogonal or parallel to each other.

For the gemstone cutting problem, however, these kinds of guillotine arrangements are not suitable, due to the irregularity of the rough stone and the shape and parametrization of the faceted stones; they would lead to smaller jewels and, thus, lower yields. Instead, we want here to allow the guillotine cuts to be made in an arbitrary position, both absolutely and relative to one another, so that we can "generate" more jewel volume.

In keeping with 2DOGCP, we can, however, introduce our understanding of a guillotine arrangement of arbitrary convex designs (*general guillotine cutting problem*; see Fig. 23, also) and how one achieves it:

GGCP: Let there be an arrangement of a set of convex designs in a container. The arrangement is call a *general guillotine arrangement* when the container can be sectioned into pieces by a series of straight, through-cuts, i.e., guillotine cuts, such that each piece contains exactly one design and no design is cut into during the sectioning of the container.

A guillotine cut is therefore a hyperplane, which not only separates two designs from each other, but also one set of designs from another set of designs. In other words, in guillotine arrangements, the non-overlapping of the designs (and maintenance of a minimum distance) is always guaranteed by separating hyperplanes. Here, however, fewer hyperplanes are required than for an arbitrary arrangement. Thus, the number of optimization parameters is smaller. The number and structure of the semi-infinite constraints, however, is the same for guillotine arrangements as for arbitrary arrangements.

The procedure in GGCP generates a fully binary tree, whose nodes correspond to the container portions resulting from the successive sectioning process. Here, the inner nodes represent the guillotine cuts and the leaves represent the designs (see Fig. 27).

When the number of designs reaches four or more, then there will be at least two possible guillotine arrangements. Here, to find the best one, all the structurally different arrangements must be calculated. The number of possible guillotine arrangements increases exponentially with the number of designs.

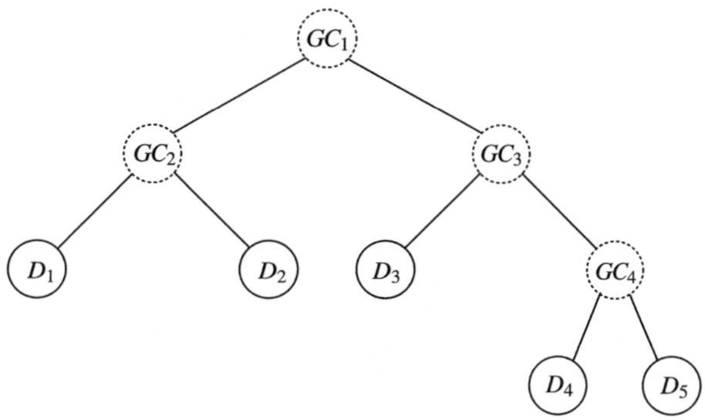


**Fig. 27** Representation of the guillotine arrangement from Fig. 23 using a fully binary tree

## 6.3 Algorithmic Implementation

To numerically solve this problem, we have chosen to use the transformation-based discretization method introduced in Sect. 4.5.2, since this works very nicely when most of the constraint functions are affine linear and the infinite index sets are very close to polyhedral. For gemstone cutting problems involving a guillotine arrangement of the faceted stones, these prerequisites are either met outright, or can be contrived (see [14]).

### 6.3.1 Modeling the Faceted Stones

For the transformation-based discretization method, one needs a functional description of the faceted stone shape and a representation of this description as the image of a compact set under a continuously differentiable mapping. Due to the complexity of the faceted stone shapes, it is impossible to represent any shape as the image of a *single* set. However, if one considers the crown, girdle, and pavilion separately, then this can indeed be done. If the shapes are very complex, further subdivisions may even be necessary.

For illustration purposes, we want to consider the girdle of a round shaped stone. For representations of the crown and pavilion—along with other shapes—as the image of one or more compact sets under continuously differentiable mapping, we refer the reader to [14]. As we know from Eq. (32), the girdle of the round shape

$$\left\{ \mathbf{y} \in \mathbb{R}^3 \;\middle|\; \begin{aligned} y_1^2 + y_2^2 - r_G^2 &\le 0 \\ 0 \le y_3 &\le h_G \end{aligned} \right\}$$

is a cylinder with radius $r_G$ and height $h_G$. Using the polar coordinate representation of a cylinder, this is, accordingly, the image of the set $[0, 1]^3$ under the mapping $\mathbf{t}((h_G, r_G), \cdot) : \mathbb{R}^3 \to \mathbb{R}^3$, with

$$\mathbf{t}(\mathbf{x}, \mathbf{z}) := \begin{pmatrix} z_2 r_G \cos(2\pi z_1) \\ z_2 r_G \sin(2\pi z_1) \\ z_3 h_G \end{pmatrix}.$$

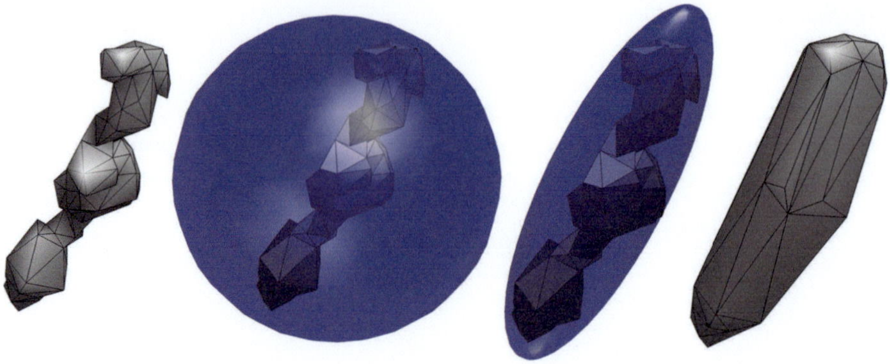

**Fig. 28** Enclosing a flaw: *from left to right*, triangulation of its surface, smallest enclosing sphere, Löwner–John ellipsoid, convex hull of the triangulation

### 6.3.2 Modeling the Rough Stone and Its Flaws

From Sect. 5.1.3, we already know how to model the rough stone surface so that the above problem can be (re-)formulated and solved as a general semi-infinite optimization problem. Therefore, in this section, we will only describe how to model the flaws. We recall that for the convexity of the lower-level problems, which result from reformulating the non-overlapping conditions between the designs and the flaws, both the designs and the flaws must be convex (see Sect. 4.4). While the former always are, the latter may not be. Thus, they must be approximated by convex sets. The simplest such approximation consists of enclosing a flaw, or enclosing the triangulation of its surface, by means of a sphere with minimal radius. A better external approximation is delivered by a so-called *Löwner–John ellipsoid*. This is an ellipsoid that encloses a set of points and has minimal volume. Finally, the convex hull of each flaw triangulation is calculated. Due to the multi-step nature of the problem solution (see Sect. 6.3.3), one approximates the flaws with various bodies, as illustrated in Fig. 28.

### 6.3.3 Determining the Starting Point

We now describe how to initialize the transformation-based discretization method (Algorithm 3, Sect. 4.5.2) for gemstone cutting problems. In this context, the starting point in step 1 of the method corresponds to assigning the initial translation, rotation, and size/shape parameters of all faceted stone designs, along with the parameters of all separating planes. The starting discretizations in step 2 correspond to an initial discretization of the faceted stone designs and flaws.

When calculating the initial faceted stone designs and separating plane parameters, we are motivated by the fact that this is a maximum material yield problem. For this reason, we proceed as follows: The problem of volume-maximal embedding of a given number of spheres in a polyhedral container with spherical surfaces and internal cavities (flaws) represents the simplest maximum material yield problem in $\mathbb{R}^3$ and can be reduced directly to a finite optimization problem, that is, a problem with a finite number of constraints. Therefore, in the first step, we solve such a problem. Except when centering a single sphere, we are dealing here with a non-convex optimization problem. Therefore, in general, a standard solution method for nonlinear problems will find no global solution. Hence, a calculated solution is repeatedly disturbed and re-optimized (a technique of global optimization known as *Monotonic Basin Hopping (MBH))*, so as to find a best possible local, perhaps even global, solution.

Unlike spheres, faceted stones have different extensions in the directions of the three main axes. For this reason, spheres are not very well suited for use as surrogate models. Therefore, in the second step, we shift to an elliptical representation of the various objects and solve the corresponding multi-body design centering problem. This, too, can still be formulated as a finite optimization problem. Due to its non-convexity, we use MBH here as well, so as to find the best possible local solution.

We ultimately obtain an initial assignment of the faceted stone design parameters by embedding their discretization in the arranged design ellipsoid while maximizing the volume of the faceted stone design. To solve the semi-infinite reformulation of the gemstone cutting problem, we shift to a polyhedral representation of the surfaces and inner cavities. We refer the reader to [14] for details regarding the entire starting point calculation (see Fig. 31, also, for a graphical illustration).

### 6.3.4 A Numerical Example

In conclusion, we want to use a numerical example to illustrate the problem dimensions of the resulting optimization problems, as well as the run-times and iteration counts of the transformation-based discretization method.

We implemented both the multi-body design centering problems and the transformation-based discretization method in MATLAB (R2012a). To solve the finite reformulations of the multi-body design centering problems and the discretized SIP problems in the context of the transformation-based discretization method, we used the SQP method of the `fmincon` routine of the `Optimization Toolbox V6.1`, with standard settings and first-order derivatives. The calculations were performed on a 32 bit Windows laptop PC with Intel Core Duo T2500 2.0 GHz processor and 2.0 GB RAM.

In this example, we do not consider the requirement of maintaining a minimum distance, since it is difficult for the observer to verify this in the two-dimensional representation of the three-dimensional situation. For the same reason, we have dispensed with the requirement that only a guillotine arrangement is allowed.

The rough stone we have selected contains three inclusions. We consider a triangulation of the rough stone surface with 576 triangles (see Fig. 29, *left*). We approximate these with 9 planes and one quadric (see Fig. 29, *right*). The convex hull of the approximated surface cavity has 24 corner points. We have enclosed the surface triangulations of each of the three inclusions in one sphere and one ellipsoid (see Fig. 29, *right*). Their convex hulls have 25, 38, and 50 corner points, respectively.

Within this rough stone, we want to embed one to five faceted stone designs of the baguette shape (see Fig. 3) with maximum total volume. We let 3 be the maximum number of non-improvements in the MBH for the design centering of both spheres and ellipsoids.

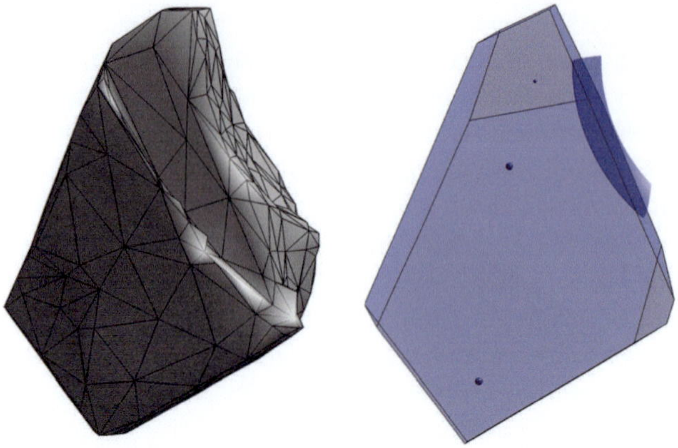

**Fig. 29** Rough stone with three inclusions: *left*, surface triangulation; *right*, approximating the rough stone surface with planes and a sphere, and enclosing the inclusions using spheres

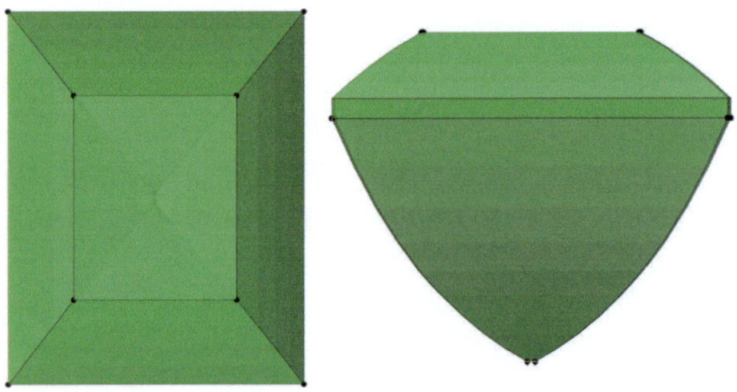

**Fig. 30** The initial discretization of a baguette-shaped faceted stone design: *left*, top view; *right*, side view

The transformation-based discretization method terminates when the maximum violation of the solution feasibility with regard to the underlying general semi-infinite problem is less than or equal to $10^{-3}$. The initial discretization of a baguette-shaped faceted stone design consists of 10 points and is shown in Fig. 30.

Table 1 shows the problem dimensions of the general semi-infinite optimization problems resulting from the maximal material yield problems. Table 2 shows the results of the calculations for the embedding of one to five baguette-shaped faceted stone designs in the rough stone approximation under consideration. The abbreviations in the tables can be deciphered as follows:

| | |
|---|---|
| # D | : Number of designs |
| D | : Designs: S = Sphere(s), E = Ellipsoid(s), FD = Faceted Stone Designs |
| # V | : Number of problem variables |

\# FC : Number of finite constraints
\# g's : Number of constraint functions of the semi-finite constraints
\# IIS : Number of infinite index sets
\# SIC : Number of semi-infinite constraints
\# I : Number of loop executions for monotonic basin hopping or transformation-based discretization method for refining the discretization
*t* : CPU-time in seconds
MBH-V : Absolute improvement of objective function value in percent as a result of MBH
Vol : Volume yield in percent

Figures 31, 32, and 33 illustrate the calculated solutions.

**Table 1** Problem dimensions of the resulting general semi-infinite optimization problems for the embedding of one to five faceted stone designs in the rough stone approximation

| # D | # V | # g's | # IIS | # SIC |
|---|---|---|---|---|
| 1 | 30 | 13 | 5 | 17 |
| 2 | 64 | 18 | 6 | 36 |
| 3 | 102 | 24 | 7 | 54 |
| 4 | 144 | 31 | 8 | 80 |
| 5 | 190 | 39 | 9 | 105 |

**Table 2** Embedding of one to five spheres, ellipsoids, and baguette-shaped faceted stone designs in the rough stone approximation: problem dimensions of the finite problems, CPU-times, and volume yields

| # D | D | # V | # FC | # I | *t* | MBH-V | Vol |
|---|---|---|---|---|---|---|---|
| 1 | S | 4 | 21 | 5 | 1.465 | <0.01 | 12.20 |
| | E | 25 | 35 | 9 | 7.815 | 3.03 | 22.40 |
| | FD | 30 182 ↗ | 472 | 7 | 14.159 | – | 35.66 |
| 2 | S | 8 | 43 | 11 | 2.476 | 8.11 | 20.04 |
| | E | 54 | 73 | 5 | 17.873 | <0.01 | 32.46 |
| | FD | 64 385 ↗ | 1605 | 6 | 67.419 | – | 56.37 |
| 3 | S | 12 | 66 | 6 | 2.169 | 4.64 | 29.39 |
| | E | 87 | 114 | 6 | 31.214 | <0.01 | 45.63 |
| | FD | 102 609 ↗ | 2274 | 6 | 93.905 | – | 62.93 |
| 4 | S | 16 | 90 | 12 | 4.191 | 10.71 | 36.64 |
| | E | 124 | 158 | 5 | 46.495 | <0.01 | 48.68 |
| | FD | 144 854 ↗ | 3554 | 7 | 335.644 | – | 69.39 |
| 5 | S | 20 | 115 | 10 | 4.658 | 1.30 | 37.49 |
| | E | 165 | 205 | 7 | 132.598 | <0.01 | 50.79 |
| | FD | 190 1120 ↗ | 5383 | 9 | 545.305 | – | 73.02 |

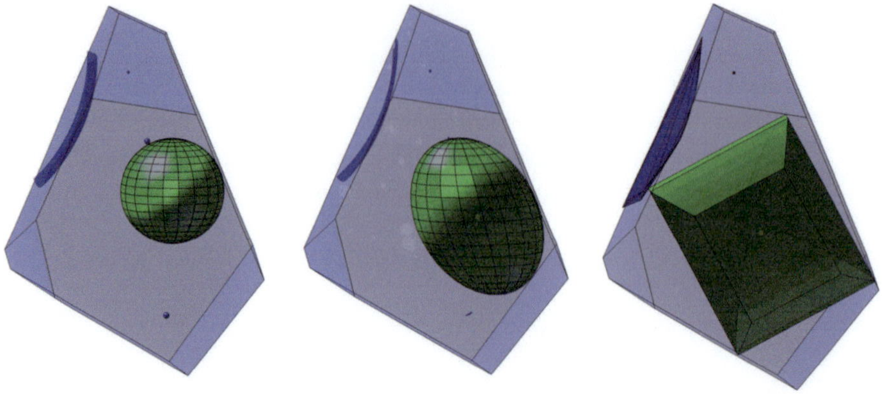

**Fig. 31** Multi-step procedure for solving gemstone cutting problems: from spherical to elliptical to polyhedral shaped objects

**Fig. 32** Calculated solution for three baguette-shaped faceted stone designs, as viewed from two perspectives

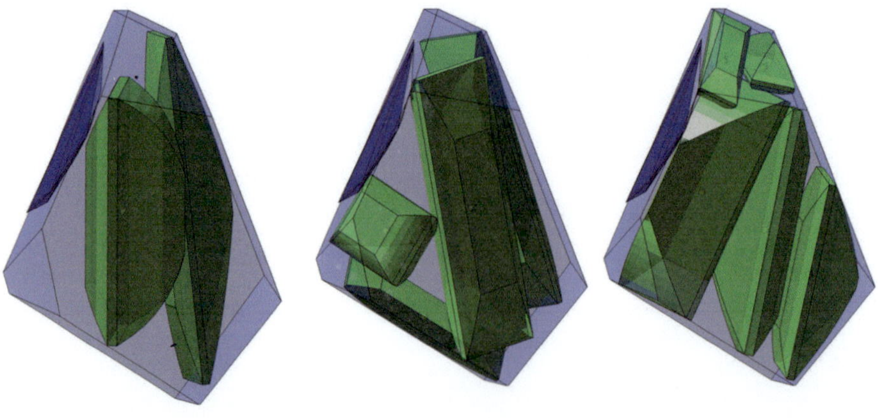

**Fig. 33** Calculated solutions for two, four, and five baguette-shaped faceted stone designs

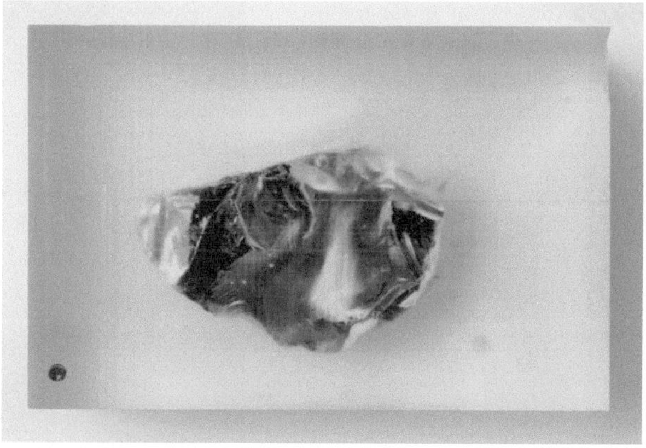

**Fig. 34** Milled-to-size resin cuboid with enclosed rough stone

## 6.4 Automating the Sectioning Process

The technical goal of this project was to automate the sectioning of raw material into blanks that can be further processed. Here, we wanted to carry over as many of the manual processing steps as possible into the automated process.

The sawing technology used in the manual process is the circular saw. Here, wooden clamps are used to hold the rough stones in place for sawing. This holding technology is not suited for automation, however, since it is impossible to predict how deeply the rough stone will sink into the wooden clamps. For an automated process, the stone must adhere precisely to the coordinate system, a problem that was ultimately solved in the following manner: The rough stone is cast in synthetic resin and the resulting block then milled down to a cuboid (see Fig. 34). The resin cuboid is glued to a cutting underlay into which the cutting disc is free to penetrate. The cutting underlay is clamped to a T-grooved plate, which is then fixed in a vice.

The circular saw is designed so that the cutting disc remains stationary, and the resin cuboid is aligned according to the cut to be executed. To allow this alignment, the vice is mounted on a rotary-swivel table that can be shifted orthogonally and parallel to the cutting disc. The cut is then made by guiding the clamping system against the cutting disc (see Fig. 35).

For this project, we needed to be able to detect flaws in the interior of the rough stone, which ruled out stripe projection as a means of data collection. Instead, we decided upon computer tomography. The rough stones, including resin cuboids, were digitalized using the ITWM's own computer tomography equipment.

Along with the optimization algorithms, we implemented two other modules that deal with the execution of the guillotine cuts: The first is a program for virtually executing the guillotine cuts and then visualizing the resulting blanks. The second is a program for calculating the machine data (angle settings of the rotation axes and position of the linear axes) for the saw prototypes in preparation for performing the actual cutting sequence.

**Fig. 35** Sawing by guiding the properly aligned resin cuboid into the cutting disc

The original plan was to have the cuts automatically executed once the resin cuboid was aligned properly. This proved to be impracticable, however, due to two problems that arose with the very first cutting trials: When initiating the cut, if the cutting disc penetrates too quickly into the resin surface or at too obtuse an angle, it can slide off and tilt. The same thing can happen when the resin work piece is withdrawn from the cutting disc. Therefore, cut initiation and work piece withdrawal must both be performed manually under the guidance of an experienced cutter.

Ultimately, the sectioning process was carried out as follows:

1. **Preparation**: encasing the rough stone in synthetic resin and milling the resulting resin cast into cuboid form
2. **Computer tomography and preparation of volume data**: photographing the resin cuboid, segmenting the resin cuboid and rough stone, and analyzing flaws
3. **Intermediate check 1**: re-adjusting the flaw classification
4. **Preparation of volume data for optimization**: generating surface data and approximating the resin cuboid, the rough stone surface, and the flaws
5. **Optimization**: calculating the optimum sectioning plan with respect to volume
6. **Intermediate check 2**: selecting a sectioning plan
7. **Generation of machine data**
8. **Preparation of the cut**: Gluing resin cuboid to an acrylic glass plate, fastening the acrylic glass plate to a aluminum T-groove plate with clamping jaws, clamping the aluminum T-groove plate in the vice of the retaining jig, aligning the retaining jig according to the calculated angles and translations
9. **Sawing the resin cuboid**: manual cut initiation, further cutting with automatic feed-in, manual withdrawal of work piece
10. **Detaching resin cuboid**
11. **Repetition of steps 7/8 to 10 until all calculated cuts have been performed**
12. **Final check**

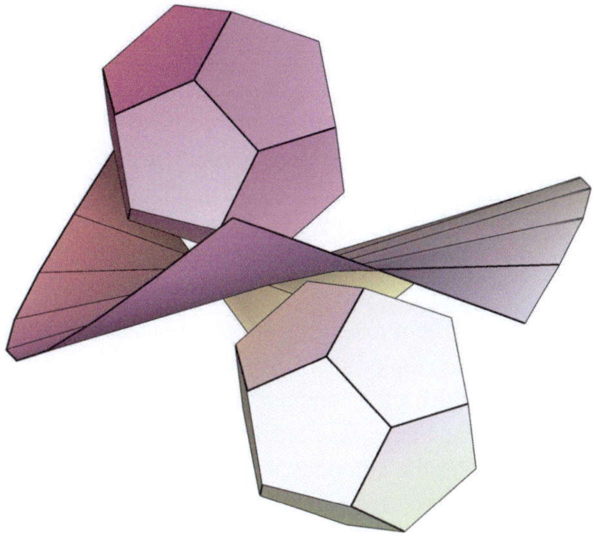

**Fig. 36** Two polyhedral designs separated by a developable surface

One problem with this process is the difficulty of guaranteeing sufficient workplace safety. Given the large forces generated by the cutting disc, manual guidance of the resin cuboid is simply too dangerous. Therefore, alternative approaches must be considered. One such alternative that appears promising is the use of high-pressure waterjet cutting for the sectioning process. This approach is discussed below.

## 6.5 Sectioning by Waterjet Cutting

In our deliberations over the best way to section the raw material, we initially rejected waterjet cutting technology, since the cutting kerfs generated by the waterjet were too wide. In 2011, however, innovations in this technology rendered its use in the gemstone industry feasible. A series of test runs commissioned by Wild oHG in Sweden and Switzerland demonstrated that a high-pressure waterjet, when combined with the correct abrasive, could indeed be used to cut gemstones without transferring significant forces into the stones, and at the same time keeping the kerf width and the depth-of-cut within acceptable bounds. Therefore, high-pressure waterjet cutting technology continues to be investigated within the framework of a current research project.

The powerful waterjet used to section the material has considerably more degrees of freedom than the guillotine cuts. As a consequence, the cut surfaces resulting from waterjet cutting are not necessarily planar, but are, in general, so-called *developable surfaces*, as depicted in Fig. 36. To model this approach for sectioning and solving the associated maximum material yield problem, the approaches used so far will have to be generalized in future research projects.

## 7 Outlook

The problem of optimum material yield in gemstone cutting is an outstanding example of using mathematical methods in the age of computer-supported, customized production. Here, one must not only master the challenges of the machine and software technology, but also the challenges presented by the mathematics involved. When Paul Wild oHG commissioned the Fraunhofer ITWM to initiate this work, there was no practicable mathematical method for calculating optimum faceted stone designs that could simply be taken off the shelf and applied to solving the problems posed in this project. It was necessary to take available algorithmic concepts for problems having containment and non-overlapping conditions and develop or alter them, so as to produce numerically robust methods that could produce results for the complex problems existing here in a reasonable amount of time.

New mathematics resulted from the ITWM projects through the development of a method of feasible solutions for general semi-infinite problems (GSIP), and a new class of methods for solving GSIPs was developed, to wit, the transformation-based discretization method. Moreover, it was demonstrated that sectioning problems and the treatment of inclusions can also be handled using the GSIP model class.

Despite these visible successes, there are still many fascinating questions waiting to be answered and problems that have not been adequately solved.

One exciting example is the sectioning of stones using waterjet technology. To date, we have only investigated guillotine cuts, as executed with circular saw technology. The new technology offers more freedom to design the cuts, which leads to the question of how this more generalized sectioning technology can be described mathematically. How can one calculate optimal sectioning processes? Which is better suited, the semi-infinite formulation or the method based on collision detection? In a new project, planned together with Wild oHG, there are exactly these questions that we will be tackling in our continuing effort to use the tools of mathematics to optimize the cutting of precious stones.

## References

### Publications on This Topic at the Fraunhofer ITWM

1. Haase, S., Süss, P., Schwientek, J., Teichert, K., Preusser, T.: Radiofrequency ablation planning: an application of semi-infinite modelling techniques. Eur. J. Oper. Res. **218**(3), 856–864 (2012). doi:10.1016/j.ejor.2011.12.014
2. Küfer, K.H., Stein, O., Winterfeld, A.: Semi-infinite optimization meets industry: a deterministic approach to gemstone cutting. SIAM News **41**(8) (2008)
3. Ludes, T.: Inverse convex approximation of irregular solids by tensor-product splines. Diplomarbeit, TU Kaiserslautern (in cooperation with Fraunhofer ITWM) (2008)
4. Maag, V.: A collision detection approach for maximizing the material utilization. Comput. Optim. Appl. **61**(3), 761–781 (2015). doi:10.1007/s10589-015-9729-5
5. Maag, V., Berger, M., Winterfeld, A., Küfer, K.H.: A novel non-linear approach to minimal area rectangular packing. Ann. Oper. Res. **179**, 243–260 (2008). doi:10.1007/s10479-008-0462-7

6. Malysheva, O.: Optimal approximation of nonlinear gemstone-models by parameterized polyhedra. Diplomarbeit, TU Kaiserslautern (in cooperation with Fraunhofer ITWM) (2008)
7. Proll, S.: Matching and alignment methods for three-dimensional objects applied to the volume optimization of gemstones. Diplomarbeit, TU Kaiserslautern (in cooperation with Fraunhofer ITWM) (2009)
8. Schwientek, J., Seidel, T., Küfer, K.H.: A transformation-based discretization method for solving general semi-infinite optimization problems. Working paper
9. Seidel, T.: Konvexitäts- und Konvergenzbetrachtungen am Beispiel des transformationsbasierten Diskretisierungsverfahrens für semi-infinite Optimierungsprobleme. Bachelorarbeit, TU Kaiserslautern (in cooperation with Fraunhofer ITWM) (2014)
10. Stein, O., Winterfeld, A.: A feasible method for generalized semi-infinite programming. J. Optim. Theory Appl. **146**(2), 419–443 (2010). doi:10.1007/s10957-010-9674-5
11. Winterfeld, A.: Maximizing volumes of lapidaries by use of hierarchical GSIP-models. Diplomarbeit, TU Kaiserslautern (in cooperation with Fraunhofer ITWM) (2004)
12. Winterfeld, A.: Application of general semi-infinite programming to lapidary cutting problems. Eur. J. Oper. Res. **191**, 838–854 (2008). doi:10.1016/j.ejor.2007.01.057


## Dissertations on This Topic at the Fraunhofer ITWM


13. Maag, V.: Multicriteria global optimization for the cooling system design of casting tools. Ph.D. thesis, TU Kaiserslautern (in cooperation with Fraunhofer ITWM) (2010). Published by Der Andere Verlag, Marburg, ISBN 978-3-89959-956-5
14. Schwientek, J.: Modellierung und Lösung parametrischer Packungsprobleme mittels semi-infiniter Optimierung–Angewandt auf die Verwertung von Edelsteinen. Ph.D. thesis, TU Kaiserslautern (in cooperation with Fraunhofer ITWM) (2013). Published by Fraunhofer-Verlag, Stuttgart, ISBN 978-3-8396-0566-0
15. Teichert, K.: A hyperboxing Pareto approximation method applied to radiofrequency ablation treatment planning. Ph.D. thesis, TU Kaiserslautern (in cooperation with Fraunhofer ITWM) (2013). Published by Fraunhofer-Verlag, Stuttgart, ISBN 978-3-8396-0783-1
16. Winterfeld, A.: Large-scale semi-infinite optimization applied to lapidary cutting. Ph.D. thesis, TU Kaiserslautern (in cooperation with Fraunhofer ITWM) (2007). Published by dissertation.de—Verlag im Internet GmbH, Berlin, ISBN 978-3-86624-301-9


## Further Literature


17. Amossen, R.R., Pisinger, D.: Multi-dimensional bin packing problems with guillotine constraints. Comput. Oper. Res. **37**(11), 1999–2006 (2010)
18. Belov, G.: Problems, models and algorithms in one- and two-dimensional cutting. Ph.D. thesis, TU Dresden (2004)
19. Bennell, J.A., Oliveira, J.F.: The geometry of nesting problems: a tutorial. Eur. J. Oper. Res. **184**(2), 397–415 (2008)
20. Boyd, S., Vandenberghe, L.: Convex Optimization, 7th edn. Cambridge University Press, Cambridge (2009)
21. Cui, Y., Chen, F., Liu, R., Liu, Y., Yan, X.: A simple algorithm for generating optimal equal circle cutting patterns with minimum sections. Adv. Eng. Softw. **41**, 401–403 (2010)
22. Cui, Y., Gu, T., Hu, W.: Simplest optimal guillotine cutting patterns for strips of identical circles. J. Comb. Optim. **15**, 357–367 (2008)

23. Diehl, M., Houska, B., Stein, O., Steuermann, S.: A lifting method for generalized semi-infinite programs based on lower level Wolfe duality. Comput. Optim. Appl. **54**(1), 189–210 (2013)
24. Ericson, C.: Real-Time Collision Detection, vol. 14. Elsevier, Amsterdam (2005)
25. Fischer, K.: Edelsteinbearbeitung, vol. 2, 3th edn. Rühle-Diebener-Verlag, Stuttgart (1996)
26. Gilmore, P.C., Gomory, R.E.: Multistage cutting-stock problems of two and more dimensions. Oper. Res. **13**, 90–120 (1965)
27. Goerner, S.: Ein Hybridverfahren zur Lösung nichtlinearer semi-infiniter Optimierungsprobleme. Ph.D. thesis, TU Berlin (1997)
28. Guerra Vázquez, F., Rückmann, J.J., Stein, O., Still, G.: Generalized semi-infinite programming: a tutorial. J. Comput. Appl. Math. **217**(2), 394–419 (2008)
29. Hettich, R., Kortanek, K.O.: Semi-infinite programming: theory, methods, and applications. SIAM Rev. **35**, 380–429 (1993)
30. Horst, R., Tuy, H.: The design centering problem. In: Global Optimization—Deterministic Approaches, pp. 572–591. Springer, Berlin (1996). Chap. C 4.1
31. Lawrence, C.T., Tits, A.L.: Feasible sequential quadratic programming for finely discretized problems from SIP. In: Reemtsen and Rückmann [36], pp. 159–193
32. Panier, E.R., Tits, A.L.: A globally convergent algorithm with adaptively refined discretization for semi-infinite optimization problems arising in engineering design. IEEE Trans. Autom. Control **34**, 903–908 (1989)
33. de Queiroz, T.A., Miyazawa, F.K., Wakabayashi, Y., Xavier, E.C.: Algorithms for 3D guillotine cutting problems: unbounded knapsack, cutting stock and strip packing. Comput. Oper. Res. **39**, 200–212 (2012)
34. Reemtsen, R.: Discretization methods for the solution of semi-infinite programming problems. J. Optim. Theory Appl. **71**(1), 85–103 (1991)
35. Reemtsen, R., Goerner, S.: Numerical methods for semi-infinite programming: A survey. In: Reemtsen and Rückmann [36], pp. 195–275
36. Reemtsen, R., Rückmann, J.J. (eds.): Semi-Infinite Programming. Kluwer Academic, Boston (1998)
37. Scheel, H., Scholtes, S.: Mathematical programs with complementarity constraints: stationarity, optimality, and sensitivity. Math. Oper. Res. **25**, 1–22 (2000)
38. Scheithauer, G.: Zuschnitt- und Packungsoptimierung: Problemstellungen, Modellierungstechniken, Lösungsmethoden. Vieweg+Teubner, Wiesbaden (2008)
39. Stein, O.: Bi-Level Strategies in Semi-Infinite Programming. Kluwer, Boston (2003)
40. Stein, O.: How to solve a semi-infinite optimization problem. Eur. J. Oper. Res. **223**(2), 312–320 (2012)
41. Stein, O., Still, G.: On generalized semi-infinite optimization and bilevel optimization. Eur. J. Oper. Res. **142**, 444–462 (2002)
42. Stein, O., Still, G.: Solving semi-infinite optimization problems with interior point techniques. SIAM J. Control Optim. **42**(3), 769–788 (2003)
43. Stein, O., Tezel, A.: The semismooth approach for semi-infinite programming under the reduction ansatz. J. Glob. Optim. **41**(2), 245–266 (2008)
44. Stein, O., Tezel, A.: The semismooth approach for semi-infinite programming without strict complementarity. SIAM J. Optim. **20**(2), 1052–1072 (2009)
45. Still, G.: Generalized semi-infinite programming: theory and methods. Eur. J. Oper. Res. **119**, 301–313 (1999)
46. Still, G.: Discretization in semi-infinite programming: the rate of convergence. Math. Program. **91**, 53–69 (2001)
47. Still, G.: Generalized semi-infinite programming: numerical aspects. Optimization **49**, 223–242 (2001)

48. Still, G.: Solving generalized semi-infinite programs by reduction to simpler problems. Optimization **53**(1), 19–38 (2004)
49. Weber, G.W.: Generalized semi-infinite optimization: on some foundations. J. Comput. Sci. Technol. **4**, 41–61 (1999)
50. Weber, G.W.: Generalized Semi-Infinite Optimization and Related Topics. Heldermann-Verlag, Lemgo (2003)

# Robust State Estimation of Complex Systems

Jan Hauth, Patrick Lang, and Andreas Wirsen

## 1 Challenges for Industry

The complexity of many technical applications and production processes is continuously increasing, due to growth in the technological possibilities of the produced goods. For biological processes, which are inherently very complex to begin with, complexity has quite different facets, which find their expression, for example, in the linkage of numerous subprocesses, in nonlinear system dynamics, and in combinations of the two. Moreover, in many cases, the descriptions of the processes and systems are plagued with significant uncertainties. In technical systems, these result from uncertainties regarding the parameters of integrated components and their time-dependent variability during operations, as well as disturbances originating in the external process environment. In biological systems, the natural fluctuations and variability typical of living systems mean that these uncertainties often play an even more important role. Therefore, when developing new medical compounds or devices, or when designing and controlling bioreactors, for example, it is imperative to take them into account.

Despite the increasing complexity of and unavoidable uncertainties in technically relevant systems, the requirements for ensuring a variety of process and system characteristics are also becoming increasingly stringent. Some of these characteristics are:

**Product Quality** The complexity and associated dynamic effects make continuous and complete monitoring of critical system parameters imperative, in order to be able to react quickly with suitable control measures to changes in system behavior and thus guarantee

J. Hauth · P. Lang (✉) · A. Wirsen
Fraunhofer-Institut für Techno- und Wirtschaftsmathematik, Fraunhofer-Platz 1,
67663 Kaiserslautern, Germany
e-mail: patrick.lang@itwm.fraunhofer.de

 
H. Neunzert, D. Prätzel-Wolters (eds.), *Currents in Industrial Mathematics*,
DOI 10.1007/978-3-662-48258-2_9

consistent product quality. The use of automated control methods also requires access to such system information.

**System Reliability** When there is a complex interplay of many components and this interplay is very dependent on the system's various operating modes, it is often impossible to give a meaningful *a priori* estimate of the lifetime of the individual components. In order to (economically) ensure continuous functioning of the total system, permanent monitoring of critical system components is therefore a sensible alternative. The replacement of components whose operating characteristics are deteriorating can then be scheduled intelligently. Such a predictive maintenance approach allows down time and maintenance costs to be minimized.

Ensuring both qualities requires access to critical information about dynamic system events. The most straightforward way to obtain the needed system and process information consists of directly measuring the crucial states and parameters using suitable sensor technology. However, directly monitoring all relevant system quantities is usually impossible, due to technical limitations in the available sensor technology and the limited number of suitable measurement sites. Moreover, due to the number of sensors that would be needed, direct measurements of all quantities would often be too expensive. Model-based state estimation offers one way around this problem. Here, system simulation on the basis of an existing system model is combined incrementally with each piece of available measurement information to derive the best possible estimate of the system's true state. The system model allows one to calculate the system quantities and parameters that are actually relevant, on the basis of simple functional inter-relationships, and thus represents a virtual sensor technology.

The characteristics of modern technical systems result in a variety of challenges for these state estimators.

**Real-Time Capability** For many applications requiring interventions to control and regulate the system, it is essential to deliver the needed system state estimates in what amounts to real time. Particularly for highly dynamic processes, this is a true challenge that requires the combined use of dimensionally restricted system models and correspondingly powerful (i.e., fast) hardware. Here, one must also ensure that the sensors being used are up to the dynamic challenges of the process in question.

**Robustness** Because state estimators often deliver the basis information for associated system control algorithms, a certain level of robustness must be guaranteed in the face of changes in system specifications. This applies both to short-term variations in the parameter values of particular components due to changing ambient conditions in the process environment and also to permanent variations in parameter values due to aging processes. Beyond this, most system parameters are initially specified with only limited exactness by the equipment suppliers. Nonetheless, one is interested in having the most exact information possible about the true dynamics of the system. One also wants to guarantee the robustness of the state estimations in the face of disturbances arising from outside the system—which are often only partially known. The reliability of the sensor technology is another important consideration; occasional faulty measurements or even the complete loss of a particular sensor signal cannot lead to a collapse of the overall state estimation procedure. Here, under some circumstances, redundancy concepts must be incorporated to rule out such scenarios. In this context, the problem of non-synchronized sensor technology must also be managed in an appropriate fashion.

## 2 Challenges for Mathematics

In many technical, medical, and biological processes, mathematical state estimation is an important tool for determining process states that are hidden or not directly measureable, based on the synergetic combination of information from a system simulation and real measurements of various system quantities. When preparing state estimators, the following challenges present themselves:

**System Model** On the basis of a suitably defined system state that includes all information needed for the further dynamic development of the system, one uses the existing technical and/or biological understanding of the system, along with the relevant, available process data, to prepare a model that accurately predicts the future development of the system state. In many cases, this modeling leads to a state dynamic in the form of an ordinary differential equation system or a differential algebraic system. When one uses a purely knowledge-driven modeling approach, the result is a so-called white box model. As the proportion of data used in the modeling approach increases, the model is then referred to as a gray box and, ultimately, a black box model. The model of the state dynamic is supplemented by equations that permit calculation of the system quantities that are actually to be monitored on the basis of the system state. The relationships to the measured system quantities must also be captured appropriately. In particular, when designing a state estimator, the information content of the possible variants can be appraised and compared on the basis of the measurements. This supports the selection of the best possible measurement configuration.

Depending on the characteristics of the underlying application, one must ensure that the complexity of the model being developed is compatible with the time available for executing the state estimation. Highly dynamic applications, for example, demand state estimation in close to real time. If the dimension of the resulting model is too large, model reduction techniques can be used to generate an error-controlled approximation by means of a smaller system model. A variety of model reduction methods is available, depending on the type of state-space model being used.

Any special requirements for preparing the model and the associated state estimator result primarily from significant nonlinearities in the dynamic behavior of the underlying system. When dealing with large, networked systems, one must also decide whether to use a centralized or localized design for the state estimator(s).

**Uncertainties** The appropriate treatment of uncertainties during the modeling process is a key concern. These can be uncertainties in one's understanding of the physical or biological relationships that dominate the process or system. Or they can be uncertainties about parameters used in the model, which may be known with only limited accuracy and are often subject to short-term fluctuations caused by the process environment. Aging processes, such as wear and corrosion, also cause parameter values to drift over longer time periods. If important parameters are not known at all, a suitable parameter estimation process can be incorporated into an extended state estimation problem.

Along with the uncertainties in parameters, the unavoidable errors associated with measurements also play an important role and must be treated appropriately. The technology of the sensors being used and the associated signal processing chain offer clues about how to model the system. Analyzing a sufficient number of measurements is of central importance for establishing appropriate distribution functions. The characteristics of the individual sensors, as well as knowledge about the time-points of the measurements and the relationships between these time-points for the different sensors, are both of central importance for designing the state estimator. Uncertainties in these quantities must also be suitably accounted for in the model.

Furthermore, there are almost always external effects or phenomena impacting the system that cannot be explicitly accounted for in the model due to a lack of detailed information. Here, rough disturbance models are the best one can do to treat these impacts. Depending on whether deterministic or stochastic phenomena predominate in the model, the state estimation problem tends to also be viewed in either a deterministic or stochastic light.

**Performance Criteria** The appropriate specification of the performance criteria depends on the desired characteristics of the state estimator being designed. Here, of course, the expected estimation errors play a central role, and special emphases result from the specifically chosen error norms and signal classes, for which the appropriate optimization is carried out. In many cases, the estimation errors are not weighted uniformly, since it proves advantageous to weight by specifying time horizons.

On the basis of the prepared model and the selected performance criterion, the weighting matrices for the combination of model simulation and measurement value can be defined explicitly in advance by solving the appropriate Riccati equations. This applies especially in the context of linear, time-invariant system models. Important, well-known variants here are first the Kalman filter and then the $H_\infty$ filter. For nonlinear systems, linearization around certain operating points makes it possible to apply the linear concepts to a certain degree within the framework of the extended Kalman filter. Here, however, no optimality characteristics can be shown and the derived confidence intervals are not valid. Although, in the general nonlinear case, an optimal state estimator can indeed be specified in theory, direct calculation—as in the case of the Kalman filter—is not possible. One must therefore rely on approximations. The particle filter accomplishes one such approximate calculation with the help of a sequential Monte Carlo approach. In principle, this amounts to simulating in parallel numerous possible system trajectories across an appropriate proposal distribution and weighting them by the measurement values with an importance sampling approach. An additional resampling step prevents degeneration of this method over time. The filter distribution is then given at every point in time as an empirical distribution (weighted samples), which is used to calculate approximation values for further quantities of interest, such as averages, variances, or confidence intervals. In contrast to the extended Kalman filter, this approach ensures mathematical convergence. On the other hand, the efficiency of the implemented algorithm depends strongly on the selection of suitable proposal distributions. Finding them is a problem that must be solved specifically for the application in question when preparing the model.

## 3 Previous Studies on This Subject

For many years, the System Analysis, Prognosis, and Control Department has been working in various application contexts with the subject of state estimations. Here, in many cases, existing methods have been adapted to the specific applications. Extensions and brand-new solutions for special problems have also been developed, however.

**Robust Observers for Elastomechanical Systems** The Department's first studies of state estimation came in connection with a project to develop an observer for turbo generator sets in power plants. These turbo generator sets consist of a long shaft on which the generator and, in general, several turbines are mounted. They are vulnerable to torsional vibrations, which can be induced by disturbances in the electrical grid. These vibrations can reach considerable amplitudes due to weak system damping, and the resulting material fatigue can substantially decrease the turbo generator set's life expectancy. Thus, the need arose for a monitoring system to permanently estimate and track the system's expected remaining operating life. Because the structures surrounding the turbo generator set limit access to the actual shaft to only a few places, a state estimator was developed on the basis of torque measurements at a single shaft position. Here, the starting point for describing the system dynamic is a high-dimensional, second order, linear state-space model, where the states describe the torsion angle of the shaft sections relative to the zero position. Whereas the resulting matrices for the moments of inertia and stiffness are quite well known from the finite element model, the damping matrix is generally subject to large uncertainties. At best, one has merely rough estimates for the modal damping. These boundary conditions meant that the project focused on adapting known approaches from the field of robust state estimation, with regard, in particular, to the high dimensionality. Especially for weakly damped systems, however, the required solution of certain Riccati equations is very poorly conditioned and presents problems for traditional solution methods. On the other hand, explicit formulas for approximation solutions can be specified for special system representations resulting from modal state-space transformations. Error estimates for the approximation quality can be derived from familiar matrix inequalities.

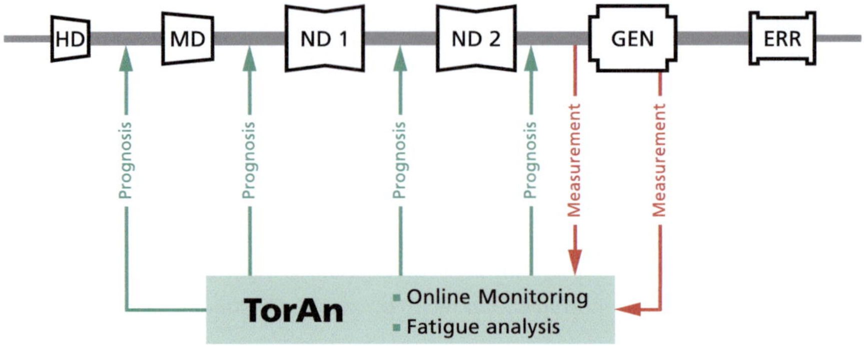


**Fig. 1** Schematic drawing of an observer for power plant turbo generator sets with high-pressure turbine (HD), intermediate-pressure turbine (MD), low-pressure turbine (ND), Generator (GEN), and excitation machine (ERR)

To develop the state estimator for turbo generator shaft lines, the Department collaborated for many years on the industrial side with Siemens AG, in Mülheim, and service providers such as E.ON Anlagenservice. On the scientific side, we should mention our cooperation with the Electrical Drives and Mechatronic Chair, headed by Professor Dr. Stefan Kulig at the University of Dortmund—a cooperation that is still active today. Together with staff from Professor Kulig's department, we developed the torque monitoring and analysis system TorAn (see Fig. 1) introduced in Sect. 6. On the basis of the developed state estimators, this system delivers on-line predictions of the torsional vibration behavior of turbo generator sets at critical shaft components and determines the resulting material fatigue in the case of disturbances [5, 6]. To measure the torques needed to determine the correction term, a contact-free magnetostrictive sensor was further developed on behalf of the ITWM. In the course of further developments, TorAn was supplemented with the monitoring systems TorFat and TorStor. Unlike TorAn, however, these systems do not generate prognoses of the torsional vibrations by means of a state estimator. The focus of TorFat was to develop methods for rapid detection of highly critical torsional vibrations, such as sub-synchronous resonances. TorStor was designed to record torques within experiments and determine as precisely as possible such relevant system quantities as damping parameters or the resonant frequencies of the shaft line. To accomplish this, a filter had to be developed to allow for compensation of the periodic disturbances—the so-called run-out—resulting from the magnetostrictive measurement principle used by the contact-free torque sensor. The phenomenon of run-out and the possibilities for using a state estimator to filter these disturbances are described in Sect. 6. The torsional recording and analysis system developed at the ITWM have been adopted by power plants and large-scale industrial installations and are now in service around the world.

**Controller Design for Active Vibration Damping of Elastomechanical Systems** Many model-based control approaches work on the principle of state feedback; that is, the estimation of the system state is part of the control algorithm. Thus, a close relationship exists between state estimation and controller design. In this regard, the expertise in state estimation of elastomechanical systems described in the previous section was further pursued in projects involving active vibration damping. Only an optimal interplay between system structure and system control can produce the best-possible damping of oscillation behavior in relation to vibrations or noise reverberation, for example. Here too, for model-based controller design, one starts with the second order differential equation systems for describing the system dynamics of elastomechanical systems. Based on either an explicit estimate of the system states—as, for example, in the context of model predictive control—or an implicit state estimate—as, for example, in the context of optimal $H_2$ controlling or robust $H_\infty$ controlling—the regulating variables for the actuators are defined in relation to the selected performance goals. Here, the model parameters are adjusted to "reality" by means of the state estimations. The estimated states then form the starting point for calculating the control input needed to achieve the desired performance behavior. Thus, in comparison with such classical control approaches as PID, model-based control also permits adjustment of performance quantities that are not directly measurable, but must be optimized nonetheless.

On behalf of Volkswagen AG, a MATLAB Toolbox was developed for automated controller design of active vibration damping in the drive train of a motor car, taking due consideration of nonlinear actuator behavior [39]. In other projects, we investigated the use of new "smart actuators" with nonlinear behaviors, such as hysteresis and saturation, and developed controller concepts for compensating these effects. Active noise reduction in vehicle interiors by means of smart actuators was one of our studies in this area [7].

**Particle Filters** The problem of state estimation in nonlinear system models having non-Gaussian disturbance processes can be solved approximately with the help of sequential Monte Carlo methods. Worth mentioning in particular is the particle filter algorithm, which works on a set of weighted samples (particles). Parameter estimation problems can also be addressed by including parameters in the state set or by means of additional Markov Chain Monte Carlo (MCMC) approaches.

Within the Department, the methodology of particle filters was initially investigated and adapted for state estimation on the basis of hysteresis-prone, nonlinear component models from the automobile industry. Subsequently, these techniques were then used primarily in the context of state and parameter estimation in biological systems. New methodological developments took place in connection with the explicit treatment of uncertainties in the measurement time-points in the particle filter approach. It was also shown that a model predictive controller (MPC) can be realized by suitably coupling two particle filters.

In this field of endeavor, the Department has worked together for years with the System Biology Department of Professor Mats Jirstrand, from the Fraunhofer Chalmers Center in Gothenburg. Here, the particular focus of activity was the development of a Mathematica-based system biology toolbox.

## 4 Modeling Principles

In this section, we will present filtering theory from the standpoint of a very general stochastic approach. Stochastic state-space models form the basis for these reflections. They separate the modeling of non-observable, internal system states from those system quantities observed by means of measurements. Both are modeled using coupled stochastic processes. Here, the stochastics is used to model disturbances and uncertainties, as well as intrinsic system variability. The state process models the dynamics of the system, while the measuring process describes the measurement/observation procedure. The underlying spaces for state and measuring processes are kept very general: they are not restricted to discrete or real spaces, and mixtures are also possible. Nor are there restrictions on the time-points at which measurements may take place: multiple measurements with different sampling times can be modeled, along with uncertainties in measurement time-points. The advantage of this approach is that it makes possible a very flexible mathematical modeling; models do not have to be restricted to a particular model class in advance. The disadvantage is that drawing inferences about internal states on the basis of observed measurements on real systems becomes very difficult and can only be made in this generality using Monte Carlo approaches.

### 4.1 Inference in Complex Systems

Real systems always exhibit a certain variability. Whereas, in technical systems, one uses design and control mechanisms to try and keep this variance small—or at least under control—in biological systems, this is not feasible in most cases. Living cells, for example—even those of the same type and age—differ greatly from one another in their characteristics: they are different in size and shape, or they are at different developmental stages. When using such microorganisms in bioreactors to produce pharmaceutical ingredients, this variability is also ultimately transmitted to the technical systems involved. But even in non-biological technical systems, (mostly undesired) variabilities can also arise, through aging and wear, for example, or through faulty system behavior.

The appropriate mathematical tool for dealing with these uncertainties is probability theory. While it often suffices in simple (mostly technical) systems to calculate using averages, and one can therefore limit oneself to deterministic calculations (i.e., the solution of differential equations), in many naturally-arising complex systems, this is not justified. In these cases, therefore, we choose a stochastic approach right from the start.

We obtain qualitative information about technical or natural systems by observing them; we obtain quantitative information by making measurements. The measurement process itself should always be viewed independently from the actual system (see Fig. 2). The actual state of the system at any point in time is not apparent to us. In this sense, the process that describes the state of the system is hidden from our view. The measurements serve to nonetheless make indirect quantitative statements about the current state of the

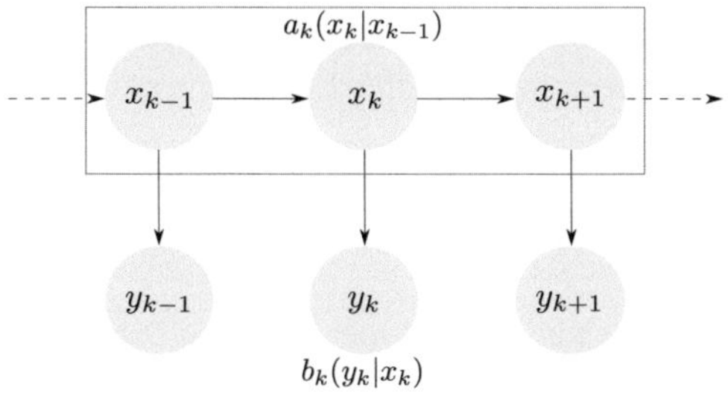


**Fig. 2** A state-space model in discrete time. Here, $x_k$ is the state vector at time $k$, $y_k$ is the measurement at time $k$, dependent only on state $x_k$ at the same time. The stochastic dependencies are given here by conditioned densities: $a_k(x_k \mid x_{k-1})$ describes the dependency of the current state $x_k$ exclusively on the basis of the previous state $x_{k-1}$ (Markov property of the state process) and $b_k(y_k \mid x_k)$, the dependency of the measurement $y_k$ exclusively from the current state $x_k$. Only the measurements are observable; the states $x_k$ themselves are hidden (non-observable)

system we are studying. Both processes—the hidden state process and also the observation process—must be viewed as having uncertainties: the state process, due to internal variabilities in the system or unobservable and/or unmodeled external disturbances; the measurement process, due to measurement errors or inaccuracies.

As a result of these multiple stochastic dependencies, the measurement results collected over time for a dynamic, changing system exhibit complicated correlations among one another, so that simple statistical evaluations of the measurements are not adequate.

The key that allows us to draw any inferences at all from noisy measurements about the internal system states lies in the stochastic dependencies of both processes. These dependencies affect, first, the time dependencies of the system states among each other and, second, the stochastic dependencies of the measurement process on the system process. Here, as well, probability theory offers a self-contained and extremely efficient tool that allows us to both model such systems and also draw inferences in a rigorous and unambiguous manner.

The Bayesian approach, in particular, which allows each quantity to be equipped with distributions, delivers here via Bayes's law a universal tool with which any inference problem subject to uncertainty can be at least theoretically solved in a transparent and simple fashion. This last trait—the simplicity of the theoretical solution—does not necessarily transfer to practical calculations. Here, one finds analytical and, thus, easily calculable solutions in only very few instances. In the case of state filtering, there are exactly two: systems with a finite number of discrete states and linear systems with Gaussian disturbances. For the latter, the solution is given by the familiar Kalman filter.

In all other cases, the calculation proves difficult. Only two developments in the second half of the 20th century—powerful computers and Monte Carlo methods—finally made it possible to execute these calculations for complex cases as well. This advance has not nearly run its course. Many algorithms, particularly in the area of state filtering or parameter estimation in dynamic systems, are new. The particle filter for state estimation in nonlinear systems, for example, is not yet 20 years old, and a promising method for joint Bayes estimation of states and parameters—supported by convergence proofs—seems only to have recently been found in the form of the new PMCMC approach [16].

### 4.2 *A Posteriori* Path and Filter Distributions

The question arises as to what the existing measurement data allows one to claim, in the best case, about a system's internal states. In dynamic systems that are mathematically represented by means of state-space models, the trajectories (paths) of the internal (current) system states play an important role. Because we are considering stochastic systems, the paths are not defined deterministically, but are subject to random distributions. These random distributions are initially specified by the system model and define the possible temporal developments of the system. One sometimes speaks of the prior or *a priori* distributions of the system trajectories (paths), since these are the probability distributions that are valid *before* the measuring process begins. In systems with a large proportion of stochastic disturbances, the system's range of possibilities is typically very broad, that is, the prior probability distribution of the path is very wide. It is the task of the filter to modify the system's prior probabilities with the help of the likewise randomly disturbed measurement data, so that system paths that do not fit the data become less probable and system paths that explain the measurements satisfactorily become more probable. These probability distributions, which describe the system states and/or trajectories *after* measurement data collection, are then referred to as posterior or *a posteriori* distributions. These are conditional probabilities that are dependent on the measurements. The selection of both the prior distributions for the system's state trajectories and the distributions for the measurements as functions of the state trajectories belongs to the model design process. Once these distributions have been defined, the posterior distribution of the states is—from a probability theory standpoint—the best possible information that one can obtain about the system's development on the basis of the measurement data. The mathematical result delivering the posterior distribution is Bayes's Theorem.

Thus, we are actually interested in the posterior distributions of the state trajectories. Although these path distributions are often very difficult to treat, under certain circumstances, it is not even necessary to consider complete paths. This is so when the current system states at each point in time already contain all the information about the future development of the system. In this instance, it is no longer necessary to consider past states (or entire past paths), since this would deliver no additional information. One then describes the system as having the Markov property. In systems having the Markov property, it therefore suffices to consider the current distributions of the states over time, instead of the path distributions. If one now calculates the corresponding posterior distributions of the current system states at a given time $t$, taking only into consideration those measurements made before or at time $t$, then one obtains exactly the filter distributions. The filter distribution at each time $t$ is therefore the posterior distribution of the states at time $t$, given the measurements up to time $t$. It turns out that the filter distributions for consecutive measurement time-points can be calculated recursively. At this level of generality, the state filtering can then be performed with sequential Monte Carlo methods, the most important of which is the particle filter, a combination of importance sampling and re-sampling over time. Here, theoretical convergence results are available. The filter distribution serves as the foundation for further important applications, such as parameter estimation and control.

## 4.3 Parameter Estimation with the Maximum Likelihood Approach

Parameter estimation in stochastic state-space systems is an extremely difficult problem. In cases where the system dynamics can simply be modeled using ordinary differential equations—that is, without stochastic noise in the states and/or correlated noise in the measurements—the problem is often considered as a deterministic optimization problem on the basis of a Maximum Likelihood (ML) approach. An overview of these approaches having a focus on biological applications can be found in [38] and [22]; see [17] also, where other aspects are considered, such as identifiability. A generalization of the ML approach via introduction of more flexible cost functions is offered by the Prediction Error Estimation methods [20]. In contrast, if one takes as a basis a model that assumes additional stochastic disturbances in the state dynamics, then one arrives at an optimization problem with constraints, where these constraints are given by stochastic differential equations (SDEs). In this case, the internal system states can no longer be directly observed or calculated and must therefore be estimated, along with the parameters, on the basis of existing measurement data. Toward this end, the method used for parameter estimation must be supplemented by the appropriate state filter methods. An overview of ML estimation for this case is found in [40]. If the underlying SDEs are linear, then the Kalman filter delivers an exact solution of the filter distribution. If the SDEs are nonlinear, then one typically relies on linearized versions of the Kalman filter, such as the Extended Kalman Filter (EKF) or the Unscented Kalman Filter (UKF), in order to obtain approximations of average values and co-variances of the filter distribution over time. All these approximations based on the Kalman filter have a crucial disadvantage, however: they approximate the filter distribution over time, in the best case, only with a Gaussian normal distribution. Therefore, they cannot properly approximate multi-modal distributions (that is, those with multiple local maxima in the probability density) or skewed distributions. Better approximations are given by simulation-based methods (sequential Monte Carlo, SMC), to which the particle filter also belongs. Good convergence results have been achieved here [24]. However, these algorithms still exhibit significant problems when applied to simultaneous estimation of dynamic states and fixed parameters ([15, 31, 45]; see [16] also).

## 4.4 Parameter Estimation with the Bayesian Approach

The Bayesian context differs from the "classical" ML approach in that a prior probability distribution is assigned to the parameter vector. The parameters are thus treated as random variables, just like the state and measurement variables. The prior distribution reflects knowledge about the parameters before considering the measurements. The estimation problem thus consists of determining the posterior distribution, that is, the probability distribution that describes knowledge about the parameters after incorporation of the measurement results (observations). At least theoretically, this can be calculated with the aid of Bayes's Theorem, if the prior distribution and the observations are given. For non-trivial problems, however, this requires calculating high-dimensional integrals, for which there are no analytical solutions. In practice, then, calculation presents great difficulties. Simulation-based methods once again offer a remedy—in this case, Markov Chain Monte Carlo (MCMC) methods. They represent a generally applicable tool for approximating posterior distributions. Here as well, however, problems arise with the joint estimation of dynamic states and fixed parameters. For example, the design of good distribution proposals for standard MCMC methods, such as the Metropolis-Hastings sampler, is practically impossible. Therefore, these methods cannot be used profitably for estimations in stochastic state-space models.

It would therefore be desirable to find an approach that combines the dynamic SMC method with the static MCMC method—with SMC as a suitable tool for estimating the states, and MCMC as a suitable tool for estimating the posterior distribution of the parameters. One would then have at one's disposal a general tool for estimating parameters in stochastic state-space models. For a long time, this combination approach remained unattainable, since calculating the acceptance probabilities of the MCMC methods presupposes knowledge of the density function of the particle distributions. Andrieu et al. [16] were the first to succeed, when they used an auxiliary variable approach (extension of the state-space by the ancestral path distributions) to show that knowledge of the approximate data likelihoods alone suffices; the particle filter delivers this knowledge for free, so to speak. Their promising approach, known as Particle Markov Chain Monte Carlo (PMCMC), is generally applicable and is backed up by good convergence results.

An alternative to PMCMC remains an established approach in which the fixed states are provided with an artificial dynamic by allowing the parameters to change their values slightly over time in a stochastic way. Thus, in the Bayesian context, states and parameters are placed on the same level conceptually: parameters can be added to the system states (augmented states) and estimated jointly via filter methods. In order for this approach to work well, however, it is important for the variances of the artificial parameter dynamics to be well chosen, a task that is quite difficult in many instances.

## 4.5 Nonlinear Mixed Effects Models

Estimating in Nonlinear Mixed Effects Models (NLME) requires estimating both global and individual parameters. With classical Maximum Likelihood estimations, there is a large conceptual difference between these two types of parameters. Whereas the individual parameters are conceived as random variables that are appropriately outfitted with probability distributions, the global parameters remain pure constants whose "true" values are simply unknown. If the equations underlying the model are nonlinear, this leads to likelihood functions that can no longer be directly evaluated. In this case, one must work with approximations. The tool NONMEM [18] has become established for certain application areas in cases where the state dynamics are modeled using deterministic ODE. In [41], in contrast, for NLME models based on stochastic differential equations (SDE), an estimation algorithm is proposed that relies on the Extended Kalman Filter (EKF) for filtering SDE. This approach was added to NONMEM [46]. In [19], a comparison was performed between ODE-based and SDE-based methods for parameter estimation in NLME models. The result here was that the estimations of the variabilities between the individual parameters generally assume smaller values for the SDE model. Donnet and Samson [27] propose combining a stochastic version of the Expectation Maximization Algorithm (for estimating global parameters) with MCMC methods (for estimating the states and the individual parameters). However, because MCMC methods exhibit problems regarding the use of joint state and parameter estimation (as mentioned above), the MCMC approach was replaced in [28] by the more suitable PMCMC method of Andrieu et al. [16].

In contrast to the Maximum Likelihood approach, in the Bayesian context, the global parameters are also supplied with (prior) probabilities, and the conceptual differences between global and individual parameters do not exist. The Mixed Effects model can be understood here simply as a hierarchical stochastic model with independent and dependent parameters [14, 43, 44]. Simulation-based (Monte Carlo) methods can be easily adapted to this case. However, the above-mentioned difficulties and requirements for the SMC and MCMC methods (or combinations of the two) become even more pronounced due to the correspondingly larger number of states and parameters in NLME models, since the number of states and individual parameters must be multiplied by the number of individual parameters.

## 4.6 The State-Space Model

We now want to provide the mathematical foundations that give a concrete form to our descriptions of the previous sections, and do so in a very generalized context. We consider the state-space models in continuous time, that is, we take as given that the state process, in particular, is a continuous-time Hidden Markov process with corresponding continuous-time transition kernels.

### 4.6.1 The State Process

Let $(\Omega, \mathscr{A}, \mathsf{P})$ be a probability space, and for each $t \in [t_0, \infty)$ with $t_0 \in \mathbf{R}$, let $(\mathscr{X}_t, \mathscr{B}_{\mathscr{X}_t})$ be an arbitrary measurable space. Furthermore, for each $t \in [t_0, \infty)$, let $X_t : \Omega \to \mathscr{X}_t$ be a $\mathscr{A} - \mathscr{B}_{\mathscr{X}_t}$ measurable random variable, such that $X_{[t_0,\infty)} := (X_t)_{t\in[t_0,\infty)}$ is a continuous-time Markov process with general state-space

$$\mathscr{X}_{[t_0,\infty)} := \prod_{t_0 \le s} \mathscr{X}_s.$$

For each $t \in [t_0, \infty)$, $\mathscr{L}_{X_t}$ denotes the pushforward measure of $\mathsf{P}$ under $X_t$, that is, $\mathscr{L}_{X_t}(B) := \mathsf{P}(X_t^{-1}(B))$ for all $B \in \mathscr{B}_{\mathscr{X}_t}$. Moreover, $\mathscr{L}_{X_{[t_0,\infty)}}$ denotes the pushforward measure of $\mathsf{P}$ under $X_{[t_0,\infty)} := (X_s)_{s\in[t_0,\infty)}$ (with the corresponding product algebra). Analogously,

$$\mathscr{X}_{[t_0,t]} := \prod_{t_0 \le s \le t} \mathscr{X}_s \quad \text{for each } t \ge t_0$$

denotes the state-space restricted to the interval $[t_0, t]$, and $\mathscr{L}_{X_{[t_0,t]}}$ denotes the corresponding pushforward measure. For each $s$ and $t$, with $t > s \ge t_0$, let $K_{s,t}(x_s, \mathrm{d}x_t)$ be the Markov kernel of the process $X_{[t_0,\infty)}$ from time $s$ to time $t$.

An important special case for $X_{[t_0,\infty)}$ is given by a multi-dimensional Itô process on $\mathscr{X}_t = \mathbf{R}^n$ (equipped with the corresponding Borel $\sigma$-algebra), defined by a stochastic differential equation (SDE)

$$\mathrm{d}X_t = a(X_t, t)\mathrm{d}t + B(X_t, t)\mathrm{d}\mathscr{W}_t,$$

with drift $a(x, t)$, diffusion matrix $B(x, t)$, multi-dimensional standard Wiener process $\mathscr{W}_t$, and initial value given by the random variable $X_{t_0}$. In this case, it is possible to sample directly (at least approximately) from the kernels $K_{s,t}$, when a suitable discretization method is applied, for instance, the Euler–Maruyama method.

### 4.6.2 Observations/Measurements

Let the process $X_{[t_0,\infty)}$ be observed via $M$ random variables $Y_{1:M}$ with values in the measurable spaces $(\mathscr{Y}_j, \mathscr{B}_{\mathscr{Y}_j})$. Each single observation (measurement) $Y_j$ depends on the state variable $X_{t_j}$ at some time $t_j$ and on the observation time (measurement time) $t_j$ itself. We assume that, given the observation time $t_j$ and the state $X_{t_j} = x_{t_j}$, the variable $Y_j$ is independent of all other variables, and that the conditional probability measure can be expressed via some conditional probability density $g_j(y_j \mid x_{t_j}, t_j)$ with respect to a given reference measure $\mu_{\mathscr{Y}_j}$ on $(\mathscr{Y}_j, \mathscr{B}_{\mathscr{Y}_j})$. We place no further conditions on $g$, such as linear dependence on the states, a normal distribution, or the like.

### 4.6.3 Observation Times/Measurement Times

The observation times (measurement times) $t_j$ for $j = 1, \dots, M$ are typically assumed to be deterministically given and known. At the ITWM, a variant of the particle filter was developed that is able to directly account for uncertainties in the measurement times themselves [1, 8]. Here, it is assumed that the observation times $t_j$ are realizations of the random variables $T_j$. These variables thus model the uncertainty about the exact measurement times.

We will consider initially the standard case, which presupposes that all $t_j$ are deterministically given and known. Formally, this corresponds to the case in which all $t_j$ are random, but observed, so that all other emerging probabilities can be seen as conditionally depending on them. Therefore, we will always express this dependence on the time-points in our notation $g_j(y_j \mid x_{t_j}, t_j)$ for the observation density. For simplicity's sake, we also presuppose that the observation times $t_{1:M}$ are strictly arranged in ascending order, so that $t_0 < t_1 < \dots < t_M$.

The standard particle filter is usually formulated for discrete-time Markov processes $X_{t_{0:M}} := (X_{t_j})_{j \in \{0,\dots,M\}}$ with general state-space, so that the state variables are only defined for the initial time $t_0$ and those time-points $t_1, \dots, t_M$ for which measurements exist. This case is included as a special case in a more generalized framework, however, in which the state variable $X_t$ is defined for all times $t \geq t_0$ (one only has to pick out the states at the discretely given measurement times and ignore the others).

### 4.6.4 Full Model and Filter Model

The full model is given by the joint density of the variables $X_{t_{0:M}}$ and $Y_{1:M}$ (conditioned on the observation times $T_{1:M} = t_{1:M}$) with respect to the product measure $\mathscr{L}_{X_{t_{0:M}}} \prod_{j=1}^{M} \mu_{\mathscr{Y}_j}$:

$$f^{X_{t_{0:M}}, Y_{1:M} | T_{1:M}}(x_{t_{0:M}}, y_{1:M} \mid t_{1:M}) := \prod_{j=1}^{M} g_j(y_j \mid x_{t_j}, t_j). \tag{1}$$

The filter distribution at time $t_k$, in contrast, is based on a reduced model.

This filter model is given by the joint distribution density of the variables $X_{t_{0:k}}$ and $Y_{1:k}$ (given that $T_{1:M} = t_{1:M}$) with respect to the product measure $\mathscr{L}_{X_{t_{0:k}}} \prod_{j=1}^{k} \mu_{\mathscr{Y}_j}$:

$$f^{X_{t_{0:k}},Y_{1:k}|T_{1:M}}(x_{t_{0:k}}, y_{1:k} \mid t_{1:M}) := \prod_{j=1}^{k} g_j(y_j \mid x_{t_j}, t_j). \tag{2}$$

This probability density is based on the state sequence $X_{t_{0:k}}$. In contrast, we can concentrate on the single state $X_{t_k}$ by considering the joint density of the variables $X_{t_k}$ and $Y_{1:k}$ (given that $T_{1:M} = t_{1:M}$) with respect to $\mathscr{L}_{X_{t_k}} \prod_{j=1}^{k} \mu_{\mathscr{Y}_j}$. This density can be calculated by marginalization as follows:

$$\begin{aligned} &f^{X_{t_k},Y_{1:k}|T_{1:M}}(x_{t_k}, y_{1:k} \mid t_{1:M}) \\ &:= \int_{\{\tilde{x}_{t_{0:k}} \in \mathscr{X}_{t_{0:k}} \,:\, \tilde{x}_{t_k} = x_{t_k}\}} f^{X_{t_{0:k}},Y_{1:k}|T_{1:M}}(\tilde{x}_{t_{0:k}}, y_{1:k} \mid t_{1:M}) \mathrm{d}\mathscr{L}_{X_{t_{0:k}}}(\tilde{x}_{t_{0:k}}). \end{aligned} \tag{3}$$

The filter density at time $t_k$ with respect to $\mathscr{L}_{X_{t_k}}$ can be calculated by means of Bayes's Theorem:

$$f^{X_{t_k}|Y_{1:k},T_{1:M}}(x_{t_k} \mid y_{1:k}, t_{1:M}) := \frac{f^{X_{t_k},Y_{1:k}|T_{1:M}}(x_{t_k}, y_{1:k} \mid t_{1:M})}{f^{Y_{1:k}|T_{1:M}}(y_{1:k} \mid t_{1:M})} \tag{4}$$

with

$$f^{Y_{1:k}|T_{1:M}}(y_{1:k} \mid t_{1:M}) := \int_{\mathscr{X}_{t_{0:k}}} f^{X_{t_{0:k}},Y_{1:k}|T_{1:M}}(x_{t_{0:k}}, y_{1:k} \mid t_{1:M}) \mathrm{d}\mathscr{L}_{X_{t_{0:k}}}(x_{t_{0:k}}). \tag{5}$$

For general (nonlinear) models, the practical calculation of the filter density is extremely difficult. Nonetheless, a Monte Carlo approximation can be calculated with the help of the particle filter. This is based on the crucial fact that filter densities $f^{X_{t_k}|Y_{1:k},T_{1:M}}$, in contrast to the density of the full model, can be calculated recursively over time. This takes place in two steps. First, we consider the filter distribution at time $t_{k-1}$ given by the probabilities

$$\begin{aligned} &\mathsf{P}(X_{t_{k-1}} \in B \mid Y_{1:k-1} = y_{1:k-1}, T_{1:M} = t_{1:M}) \\ &= \int_B f^{X_{t_{k-1}}|Y_{1:k-1},T_{1:M}}(x_{t_{k-1}} \mid y_{1:k-1}, t_{1:M}) \mathrm{d}\mathscr{L}_{X_{t_{k-1}}}(x_{t_{k-1}}) \end{aligned} \tag{6}$$

for each set $B \in \mathscr{B}_{\mathscr{X}_{t_{k-1}}}$. We initially obtain the prediction distribution, that is, the distribution of $X_{t_k}$ given by the data $Y_{1:k-1}$ and $T_{1:M}$ by using the Markov kernel $K_{t_{k-1},t_k}$:

$$\begin{aligned} &\mathsf{P}(X_{t_k} \in B \mid Y_{1:k-1} = y_{1:k-1}, T_{1:M} = t_{1:M}) \\ &= \int_B \int_{\mathscr{X}_{t_{k-1}}} f^{X_{t_{k-1}}|Y_{1:k-1},T_{1:M}}(x_{t_{k-1}} \mid y_{1:k-1}, t_{1:M}) \mathrm{d}\mathscr{L}_{X_{t_{k-1}}}(x_{t_{k-1}}) K_{t_{k-1},t_k}(x_{t_{k-1}}, \mathrm{d}x_{t_k}) \end{aligned} \tag{7}$$

for each set $B \in \mathscr{B}_{\mathscr{X}_{t_k}}$.

In the second step, we then use Bayes's Theorem to obtain the filter distribution at time $t_k$:

$$\begin{aligned}
&\mathsf{P}(X_{t_k} \in B \mid Y_{1:k} = y_{1:k}, T_{1:M} = t_{1:M}) \\
&= \int_B \frac{g_k(y_k \mid x_{t_k}, t_k)}{f^{Y_k|Y_{1:k-1},T_{1:M}}(y_k \mid y_{1:k-1}, t_{1:M})} \\
&\quad \times \int_{\mathscr{X}_{t_{k-1}}} f^{X_{t_{k-1}}|Y_{1:k-1},T_{1:M}}(x_{t_{k-1}} \mid y_{1:k-1}, t_{1:M}) \mathrm{d}\mathscr{L}_{X_{t_{k-1}}}(x_{t_{k-1}}) \\
&\quad \times K_{t_{k-1},t_k}(x_{t_{k-1}}, \mathrm{d}x_{t_k})
\end{aligned} \tag{8}$$

for each set $B \in \mathscr{B}_{\mathscr{X}_{t_k}}$, with the normalizing constant

$$\begin{aligned}
&f^{Y_k|Y_{1:k-1},T_{1:M}}(y_k \mid y_{1:k-1}, t_{1:M}) \\
&:= \int_{\mathscr{X}_{t_k}} g_k(y_k \mid x_{t_k}, t_k) \\
&\quad \times \int_{\mathscr{X}_{t_{k-1}}} f^{X_{t_{k-1}}|Y_{1:k-1},T_{1:M}}(x_{t_{k-1}} \mid y_{1:k-1}, t_{1:M}) \mathrm{d}\mathscr{L}_{X_{t_{k-1}}}(x_{t_{k-1}}) \\
&\quad \times K_{t_{k-1},t_k}(x_{t_{k-1}}, \mathrm{d}x_{t_k}).
\end{aligned} \tag{9}$$

## 4.7 Particle Filter Algorithms for State Estimation

Particle filters [21, 30, 32] belong to the class of SMC methods used for state filtering in state-space models. Thus, with the appropriate adaptations and/or extensions, they also form the basis for parameter estimations. The standard particle filter works on discrete-time, nonlinear and non-Gaussian models and can be easily adapted for use on continuous-time systems with discrete-time measurements. The idea of the particle filter is to store a representation of the current filter distribution at each time-point by means of a set of weighted realizations (weighted samples or particles). This particle set is propagated through time in a suitable manner by adapting the realizations and the particle weights via the system dynamics and/or the measurements available at each time-point.

### 4.7.1 Importance Sampling

A key element of the particle filter is sequential importance sampling. We assume that a second Markov chain $\tilde{X}_{t_{0:M}}$ is given for the same state-space with pushforward measure $\mathscr{L}_{\tilde{X}_{t_j}}$ and Markov kernels $\tilde{K}_{t_{j-1},t_j}(x_{t_{j-1}}, \mathrm{d}x_{t_j})$ for $j = 1, \dots, M$. We also assume that for each $x_{t_{j-1}} \in \mathscr{X}_{t_{j-1}}$, the measure $K_{t_{j-1},t_j}(x_{t_{j-1}}, \cdot)$ is absolutely continuous with respect to the measure $\tilde{K}_{t_{j-1},t_j}(x_{t_{j-1}}, \cdot)$.

It follows that the Radon–Nikodym derivative (written as a conditional probability density)

$$\varrho_{t_j|t_{j-1}}(x_{t_j} \mid x_{t_{j-1}}) := \frac{K_{t_{j-1},t_j}(x_{t_{j-1}}, \mathrm{d}x_{t_j})}{\tilde{K}_{t_{j-1},t_j}(x_{t_{j-1}}, \mathrm{d}x_{t_j})}$$

exists. We also require that the pushforward measure $\mathscr{L}_{X_{t_0}}$ under $X_{t_0}$ is absolutely continuous with respect to the corresponding pushforward measure $\mathscr{L}_{\tilde{X}_{t_0}}$ under $\tilde{X}_{t_0}$ with the Radon–Nikodym derivative

$$\varrho_{t_0}(x_{t_0}) := \frac{\mathrm{d}\mathscr{L}_{X_{t_0}}(x_{t_0})}{\mathrm{d}\mathscr{L}_{\tilde{X}_{t_0}}(x_{t_0})}.$$

Sequential importance sampling can be performed under those circumstances in which we are able to draw random realizations from both the initial distribution $\mathscr{L}_{\tilde{X}_{t_0}}$ and the kernels

$$\tilde{K}_{t_{j-1},t_j}(x_{t_{j-1}}, \cdot)$$

for each $x_{t_{j-1}} \in \mathscr{X}_{t_{j-1}}$, and when we can calculate $\varrho_{t_0}(x_{t_0})$ and $\varrho_{t_j|t_{j-1}}(x_{t_j} \mid x_{t_{j-1}})$ pointwise.

Using

$$K_{t_{k-1},t_k}(x_{t_{k-1}}, \mathrm{d}x_{t_k}) = \varrho_{t_k|t_{k-1}}(x_{t_k} \mid x_{t_{k-1}})\tilde{K}_{t_{k-1},t_k}(x_{t_{k-1}}, \mathrm{d}x_{t_k}),$$

we can then rewrite the recursive formula (8) for the filter distribution at time $t_k$ as follows:

$$\begin{aligned}
&\mathsf{P}(X_{t_k} \in B \mid Y_{1:k} = y_{1:k}, T_{1:M} = t_{1:M}) \\
&= \int_B \frac{g_k(y_k \mid x_{t_k}, t_k)}{f^{Y_k|Y_{1:k-1},T_{1:M}}(y_k \mid y_{1:k-1}, t_{1:M})} \\
&\quad \times \int_{\mathscr{X}_{t_{k-1}}} f^{X_{t_{k-1}}|Y_{1:k-1},T_{1:M}}(x_{t_{k-1}} \mid y_{1:k-1}, t_{1:M}) \\
&\quad \times \varrho_{t_k|t_{k-1}}(x_{t_k} \mid x_{t_{k-1}})\mathrm{d}\mathscr{L}_{X_{t_{k-1}}}(x_{t_{k-1}}) \\
&\quad \times \tilde{K}_{t_{k-1},t_k}(x_{t_{k-1}}, \mathrm{d}x_{t_k})
\end{aligned} \tag{10}$$

for each $B \in \mathscr{B}_{\mathscr{X}_{t_k}}$.

The direct calculation of the normalizing constants

$$f^{Y_k|Y_{1:k-1},T_{1:M}}(y_k \mid y_{1:k-1}, t_{1:M})$$

(while the values $y_{1:M}$ are considered to be fixed) is unnecessary.

Sequential importance sampling is then performed as follows: we randomly draw a number $N$ of realizations $x^i_{t_0}$ from $\mathscr{L}_{\tilde{X}_{t_0}}$ and calculate the corresponding unnormalized weights

$$w^i_{t_0} := \varrho_{t_0}(x^i_{t_0}) \quad \text{for all } i = 1, \dots, N.$$

We then randomly draw for all $k = 1, \dots, M$ realizations $x^i_{t_k}$ from the kernel

$$\tilde{K}_{t_{k-1},t_k}(x^i_{t_{k-1}}, \mathrm{d}x_{t_k})$$

for each $i = 1, \dots, N$ and calculate the unnormalized weights

$$w^i_{t_k} := \varrho_{t_k|t_{k-1}}(x^i_{t_k} \mid x^i_{t_{k-1}}) g_k(y_k \mid x^i_{t_k}, t_k) w^i_{t_{k-1}} \quad \text{for all } i = 1, \dots, N.$$

For suitable integrable functions $h$ (for example, when certain restrictions are fulfilled on the rate with which $h$ may increase relative to $x$; see [34] for details), one can approximate the expected value of $h$ with respect to the filter density conditioned on the observations $Y_{1:k} = y_{1:k}$, given by

$$\begin{aligned} &\mathsf{E}\big[h(X_{t_k}) \bigm| Y_{1:k} = y_{1:k}, T_{1:M} = t_{1:M}\big] \\ &:= \mathsf{E}_{f^{X_{t_k}|Y_{1:k}=y_{1:k},T_{1:M}=t_{1:M}}(\cdot|y_{1:k},t_{1:M})}\big[h(X_{t_k})\big] \\ &= \int f^{X_{t_k}|Y_{1:k},T_{1:M}}(x_{t_k} \mid y_{1:k}, t_{1:M}) h(x_{t_k}) \mathrm{d}\mathscr{L}_{X_{t_k}}(x_{t_k}), \end{aligned} \tag{11}$$

by means of

$$\mathsf{E}_{t_k,N}\big[h(X_{t_k}) \bigm| Y_{1:k} = y_{1:k}, T_{1:M} = t_{1:M}\big] := \frac{\sum_{i=1}^N w^i_{t_k} h(x^i_{t_k})}{\sum_{i=1}^N w^i_{t_k}} \tag{12}$$

where $N$ is the number of particles. It can be shown that as $N$ approaches infinity, these empirical expected values converge to the expected values from the filter distribution:

$$\lim_{N\to\infty} \mathsf{E}_{t_k,N}\big[h(X_{t_k}) \bigm| Y_{1:k} = y_{1:k}, T_{1:M} = t_{1:M}\big] = \mathsf{E}\big[h(X_{t_k}) \bigm| Y_{1:k} = y_{1:k}, T_{1:M} = t_{1:M}\big]. \tag{13}$$

If we are in a position to sample from the Markov kernels $X_{t_j}$ of the states themselves, then we can select $\tilde{X}_{t_j} = X_{t_j}$ (at least in distribution), from which $\varrho_{t_0}(x_{t_0}) \equiv 1$ and $\varrho_{t_j|t_{j-1}}(x_{t_j} \mid x_{t_{j-1}}) \equiv 1$ follow. This selection is indeed standard, but it is not always the best choice with regard to the effectiveness of the particle filter algorithm. Finding a Markov chain $\tilde{X}_{t_{0:M}}$ different than $X_{t_{0:M}}$ that can improve the effectiveness of the algorithm, however, is an application-specific, and not always simple, task.

### 4.7.2 Resampling

Sequential importance sampling converges when the number of samples (particles) increases exponentially over time. This is not practicable; typically, $N$ is even held constant over time. However, when the number $N$ of particles remains constant over time, the particles propagated by sequential importance sampling quickly degenerate, since most of the normalized weights converge rapidly toward 0.

The degree of degeneracy of the particle set is often measured by an estimator for the so-called Effective Sample Size (ESS). This estimator at time $t$ is given by

$$n_{\text{ESS}} := \frac{1}{\sum_{i=1}^{N}(\tilde{w}_t^i)^2}, \tag{14}$$

where

$$\tilde{w}_t^i := \frac{w_t^i}{\sum_{i=1}^{N} w_t^i} \tag{15}$$

refer to the normalized weights.

The ESS estimator assumes its maximum value $N$ (number of particles), when all weights are equal, and it approaches 1 when the variance of the weights, and thus the degree of degeneracy, becomes large. To avoid this degeneration, one must insert a resampling step in the algorithm, to be performed when the ESS drops below a certain threshold $N_{\text{Threshold}}$ (usually selected to be $N/2$).

Resampling at time-point $s_\ell$ is based on given, non-negative (unnormalized) selection weights $v_{s_\ell}^i$ for each particle index $i$. One repeats random selections (with replacement) of particles having probabilities $p_\ell^i$ given by the normalized selection weights

$$p_\ell^i := \frac{v_{s_\ell}^i}{\sum_{\nu=1}^{N} v_{s_\ell}^\nu}. \tag{16}$$

This is referred to as multinomial resampling. There are also procedures in which each individual particle continues to be selected with probability $p_\ell^i$, but which exhibit a reduced overall variance, such as Stratified Resampling or Systematic Resampling, and these should be chosen in preference to multinomial resampling (see [29, 33]). In any case, resampling defines a (random) selection function $\iota_\ell : I \to I$ on the index set $I := \{1, \dots, N\}$.

The resampling step is then performed in two phases:

- Replacement of the state samples $(x^i_{s_\ell})_{i=1,\dots,N}$ by the selected state samples $(x^{\iota_\ell(i)}_{s_\ell})_{i=1,\dots,N}$.
- Replacement of the unnormalized weights $(w^i_{s_\ell})_{i=1,\dots,N}$ by the corrected unnormalized weights $(w^{\iota_\ell(i)}_{s_\ell}/v^{\iota_\ell(i)}_{s_\ell})_{i=1,\dots,N}$.

It is necessary to correct the weights in the final step in order to compensate for the bias introduced in the particle distribution by the selection process. This bias results from the following consideration: Before sampling, the selection probability for particle $i$ (at each draw) is given by $p^i_\ell$. The expected value for the number of times that particle $i$ will actually be drawn after $N$ samplings is therefore $Np^{\iota_\ell(i)}_\ell$.

As a result, each normalized weight $\tilde{w}^i_{s_\ell}$, for each selected particle $i$, must be corrected by replacing it with the weight

$$\frac{\tilde{w}^{\iota_\ell(i)}_{s_\ell}}{Np^{\iota_\ell(i)}_\ell} \Big/ \sum_{\nu=1}^{N} \frac{\tilde{w}^{\iota_\ell(\nu)}_{s_\ell}}{Np^{\iota_\ell(\nu)}_\ell} = \frac{w^{\iota_\ell(i)}_{s_\ell}}{v^{\iota_\ell(i)}_{s_\ell}} \Big/ \sum_{\nu=1}^{N} \frac{w^{\iota_\ell(\nu)}_{s_\ell}}{v^{\iota_\ell(\nu)}_{s_\ell}} \tag{17}$$

(using (16)).

Note that in the original particle filter, the selection weights $v^i_{s_\ell}$ at time $s_\ell$ are chosen so that they are given by the particle weights (before the replacement), that is,

$$v^i_{s_\ell} = w^i_{s_\ell} \quad \text{for } i = 1, \dots, N,$$

so that after the resampling step, the unnormalized weights are all equal to 1. Nonetheless, in general, their choice is free and may be influenced by the system observations (measurements), for example (as used in the so-called Auxiliary Particle Filter [42]).

### 4.7.3 Particle Filter Algorithm

The particle filter calculates the state realizations and weights recursively through time. In its standard form, the particle filter can be specified as pseudo code, as in Algorithm 1.

**Algorithm 1** Standard particle filter

{*At time* $t_0$*:*}

Randomly sample $N$ state realizations $(x^i_{t_0})_{i=1,\ldots,N}$ of $\tilde{X}_{t_0}$ with large $N$.

**for all** $i = 1, \ldots, N$ **do**

Set the weight $w^i_{t_0} = \varrho_{t_0}(x^i_{t_0})$.

**end for**

**for all** times $t_k$, $k = 1, \ldots, M$ **do**

{*Resample the particles* $(x^i_{t_{k-1}})_{i=1,\ldots,N}$, *if necessary (e.g., if the ESS drops below a threshold):*}

Randomly generate a selection function $\iota$ according to certain selection weights $(v^i_{t_{k-1}})_{i=1,\ldots,N}$.

**for** $i = 1, \ldots, N$ **do**

Replace the state realization $x^i_{t_{k-1}}$ by the selection $x^{\iota(i)}_{t_{k-1}}$.

Replace the unnormalized weight $w^i_{t_{k-1}}$ by the corrected weight $w^{\iota(i)}_{t_{k-1}} / v^{\iota(i)}_{t_{k-1}}$.

**end for**

**for** $i = 1, \ldots, N$ **do**

Randomly sample a realization $x^i_{t_k}$ from the Markov kernel

$$\tilde{K}_{t_{k-1},t_k}(x^i_{t_{k-1}}, \cdot).$$

Update the weight by:

$$w^i_{t_k} = \varrho_{t_k|t_{k-1}}\left(x^i_{t_k} \mid x^i_{t_{k-1}}\right) g_k\left(y_k \mid x^i_{t_k}, t_k\right) w^i_{t_{k-1}}.$$

**end for**

For given suitable integrable functions $h$, calculate the estimates

$$\mathsf{E}_{t_k,N}\left[h(X_{t_k}) \mid Y_{1:k} = y_{1:k}, T_{1:M} = t_{1:M}\right] := \frac{\sum_{i=1}^N w^i_{t_k} h(x^i_{t_k})}{\sum_{i=1}^N w^i_{t_k}}.$$

**end for**

Note that, in choosing $\tilde{X}_{[t_0,\infty)} = X_{[t_0,\infty)}$ (in distribution), the identity

$$\varrho_{t_k|t_{k-1}}\left(x^i_{t_k} \mid x^i_{t_{k-1}}\right) \equiv 1$$

holds, and the updating of the weights simplifies to

$$w^i_{t_k} = g_k\left(y_k \mid x^i_{t_k}, t_k\right) w^i_{t_{k-1}}.$$

### 4.7.4 Data Likelihood

The model validation and discrimination are generally based on the data likelihood

$$\begin{aligned} Z_{t_k}(t_{1:M}) := f^{Y_{1:k}|T_{1:M}}(y_{1:k} \mid t_{1:M}) &= \int_{\mathscr{X}_{t_{0:k}}} f^{X_{t_{0:k}},Y_{1:k}|T_{1:M}}(x_{t_{0:k}}, y_{1:k} \mid t_{1:M}) \mathrm{d}\mathscr{L}_{X_{t_{0:k}}}(x_{t_{0:k}}) \\ &= \mathsf{E}\left[f^{X_{t_{0:k}},Y_{1:k}|T_{1:M}}(\cdot, y_{1:k} \mid t_{1:M})\right] \end{aligned} \tag{18}$$

for given observations $y_{1:k}$. Without resampling, the data likelihood could be approximated by the empirical average of the unnormalized weights, that is, by

$$\hat{Z}_{t_k}(t_{1:M}) := \frac{1}{N} \sum_{i=1}^{N} w_{t_k}^i, \tag{19}$$

since this is the empirical estimate of the above expected value. After a resampling step, this no longer holds true.

In any case (with or without resampling), the ratio estimator

$$Z_{t_k}(t_{1:M}) / Z_{t_{k-1}}(t_{1:M})$$

can be used to recursively approximate the data likelihood:

$$\frac{\widehat{Z_{t_k}(t_{1:M})}}{Z_{t_{k-1}}(t_{1:M})} := \frac{\sum_{i=1}^{N} \varrho_{t_k|t_{k-1}}(x_{t_k}^i \mid x_{t_{k-1}}^i) g_k(y_k \mid x_{t_k}^i, t_k) w_{t_{k-1}}^i}{\sum_{i=1}^{N} w_{t_{k-1}}^i}, \tag{20}$$

with the initial estimator $\hat{Z}_{t_0}(t_{1:M}) = 1$ (see [25], for example).

## 4.8 Kalman Filter

As mentioned, the explicit formulaic calculation of the filter distributions is only possible in a very few instances. This is due to the recursively nested, high-dimensional integrals, which, in general, cannot be analytically solved. In two cases, however, this is still possible and practicable. In the first case, one assumes that the state-space can only jump between a finite number of discrete states (here, one is also usually working with a discrete-time model). The measurement disturbances have a normal distribution. One speaks here of a hidden Markov model in the narrower sense. One is dealing with a finite number of transitional probabilities, which can be directly calculated, and the filter distributions can be determined accordingly via a recursive procedure by means of direct computation. The second case, which is far more important for modeling, is given by linear systems with exclusively Gaussian disturbances. Gauss distributions are characterized solely by the first two moments (average and variance). Moreover, in linear systems, the Gaussian form is retained for all other relevant distributions. The appropriate state filter is the Kalman filter: The average and variance can be recursively calculated, simply and directly, using matrix operations.

Based on the terminology developed in the previous section, the Kalman filter can be derived as a special case without much effort. The Kalman filter is only correct as a state filter when we subject our system to certain restrictions: linearity and Gaussian normality in the state and measurement processes. The assumption of linearity in the entire system means that all system disturbances remain normally distributed, regardless of whether one propagates them forward or backward through the system. This is evident by the corresponding property of the multivariate Gauss distribution.

Here, we consider only the discrete-time case.

### 4.8.1 Multivariate Normal Distribution and Linearity

A random variable $X$ has a multivariate normal distribution (multivariate Gauss distribution), denoted $X \sim \mathcal{N}_d(\mu, \Sigma)$, with average $\mu \in \mathbf{R}^d$ and positive-semi-definite covariance matrix $\Sigma \in \mathbf{R}^{d\times d}$, when there is a random $\ell$-vector $Z$ with standard, normally-distributed coefficients and a matrix $A \in \mathbf{R}^{k\times\ell}$ with $AA^\top = \Sigma$, such that

$$X = AZ + \mu.$$

If the covariance matrix $\Sigma$ is positive-definite rather than positive-semi-definite, then there exists a corresponding probability density, and it is given by

$$\frac{1}{(2\pi)^{d/2}|\Sigma|^{1/2}} \exp\left(-\frac{1}{2}(x-\mu)^\top \Sigma^{-1}(x-\mu)\right).$$

An affine-linear transformation $Y = BX + c$, with $B \in \mathbf{R}^{m\times d}$ and $c \in \mathbf{R}^m$, leads to a variable $Y$, which also has a multivariate normal distribution:

$$Y \sim \mathcal{N}_m\big(B\mu + c, B\Sigma B^\top\big).$$

Special cases are given by:

- Marginalization: Let $X = (X_1, X_2)^\top$. Then $X_1$ (and $X_2$) are normally distributed, since

$$X_1 = BX \quad \text{with } B = \begin{pmatrix} I & 0 \\ 0 & 0 \end{pmatrix}.$$

- Let $Y = BX + V$, with $V \sim \mathcal{N}_m(0, R)$. It then follows that $Y \sim \mathcal{N}(BX, B\Sigma B^\top + R)$, since

$$Y = \begin{pmatrix} B & I \end{pmatrix} \begin{pmatrix} X \\ V \end{pmatrix}.$$

Note that, when

- $p(x)$ is normally distributed,

$$X \sim \mathcal{N}_d(\mu, \Sigma),$$

- $p(y \mid x)$ is conditionally normally distributed to a given $x$,

$$Y \mid (X = x) \sim \mathcal{N}_m\big(\hat{y}(x), R\big),$$

*and*

- the average $\hat{y}$ is affine-linearly dependent on $x$,

$$\hat{y}(x) = Bx + c,$$

then

- the joint distribution density $p(x, y) = p(y \mid x)p(x)$ is normal and
- the marginal distribution density $p(y) = \int p(x, y)dx$ is normal.

Here, the joint distribution density $p(x, y)$ is given by

$$\begin{pmatrix} X \\ Y \end{pmatrix} \sim \mathcal{N}_{d+m}\left(\begin{pmatrix} 1 \\ B \end{pmatrix}\mu + \begin{pmatrix} 0 \\ c \end{pmatrix}, \begin{pmatrix} 1 \\ B \end{pmatrix}\Sigma\begin{pmatrix} 1 \\ B \end{pmatrix}^{\top} + \begin{pmatrix} 0 & 0 \\ 0 & R \end{pmatrix}\right)$$

$$= \mathcal{N}_{d+m}\left(\begin{pmatrix} \mu \\ B\mu + c \end{pmatrix}, \begin{pmatrix} \Sigma & \Sigma B^{\top} \\ B\Sigma & B\Sigma B^{\top} + R \end{pmatrix}\right),$$

and the marginal distribution density $p(y)$ is given by

$$Y \sim \mathcal{N}_m\big(B\mu + c, B\Sigma B^{\top} + R\big).$$

In the following treatment, we denote the average and covariance matrix of $Y$ as

$$\mu_y = B\mu + c, \quad \text{and} \quad \Sigma_y = B\Sigma B^{\top} + R.$$

### 4.8.2 Bayes's Theorem for Normal Distributions: Kalman Gain

Let us consider Bayes's Theorem

$$p(y \mid x)p(x) = p(x, y) = p(x \mid y)p(y)$$

under the prerequisites listed above. All distribution densities that occur are thus normal. It remains to be shown that this is also correct for the posterior distribution $p(x \mid y)$. Let us take the approach

$$X \mid (Y = y) \sim \mathcal{N}_d(Ky + e, G),$$

and consider both the left and right sides of the previous equation. In so doing, we obtain the equation

$$\mathcal{N}_{d+m}\left(\begin{pmatrix} \mu \\ B\mu + c \end{pmatrix}, \begin{pmatrix} \Sigma & \Sigma B^{\top} \\ B\Sigma & B\Sigma B^{\top} + R \end{pmatrix}\right)$$

$$= \mathcal{N}_{d+m}\left(\begin{pmatrix} K\mu_y + e \\ \mu_y \end{pmatrix}, \begin{pmatrix} K\Sigma_y K^{\top} + G & K\Sigma_y \\ \Sigma_y K^{\top} & \Sigma_y \end{pmatrix}\right).$$

Solving for $K$, $G$ and $e$ leads to:

$$K = \Sigma B^\top \Sigma_y^{-1} \quad \text{(Kalman gain)},$$
$$G = \Sigma - K\Sigma_y K^\top = \Sigma - K\Sigma_y \Sigma_y^{-1} B\Sigma = (I - KB)\Sigma,$$
$$e = \mu - K\mu_y = \mu - K(B\mu + c).$$

The last equation yields:

$$Ky + e = \mu + K\big(y - (B\mu + c)\big).$$

### 4.8.3 Application to Recursive State Filtering: the Kalman Filter

As just shown, the posterior distribution $p(x \mid y)$ is given by

$$\mathcal{N}_d\big(\mu + K\big(y - (B\mu + c)\big), (I - KB)\Sigma\big) \quad \text{with } K = \Sigma B^\top \Sigma_y^{-1}.$$

We now consider the linear, normal, dynamic model:

$$x_0 \sim \mathcal{N}_d(\hat{x}_0, Q_0), \qquad x_t \sim \mathcal{N}_d\big(A_t x_{t-1} + b(u_{t-1}), Q_t\big), \qquad y_t \sim \mathcal{N}_m(C_t x_t, R_t).$$

We further assume that $p(x_{t-1} \mid y_{1:t-1})$ is recursively given by

$$\mathcal{N}_d(\hat{x}_{t-1|t-1}, P_{t-1|t-1}),$$

starting with $p(x_0)$, that is, $\hat{x}_{0|0} = \hat{x}_0$ and $P_{0|0} = Q_0$. We must now show that $p(x_t \mid y_{1:t})$ is also normally distributed, that is, it is given by

$$\mathcal{N}_d(\hat{x}_{t|t}, P_{t|t}).$$

The Kalman filter is calculated in two steps:

- Prediction:

$$p(x_t \mid y_{1:t-1}) = \int p(x_t \mid x_{t-1}) p(x_{t-1} \mid y_{1:t-1}) dx_{t-1}$$

  is given by $\mathcal{N}_d(\hat{x}_{t|t-1}, P_{t|t-1})$, with

$$\hat{x}_{t|t-1} = A_t \hat{x}_{t-1|t-1} + b(u_{t-1}), \quad \text{and} \quad P_{t|t-1} = A_t P_{t-1|t-1} A_t^\top + Q_t.$$

- Update:

$$p(x_t \mid y_{1:t}) = \frac{p(y_t \mid x_t)p(x_t \mid y_{1:t-1})}{p(y_t \mid y_{1:t})}$$

is given by $\mathcal{N}_d(\hat{x}_{t|t}, P_{t|t})$, with

$$S_t = C_t P_{t|t-1} C_t^\top + R_t, \qquad K_t = P_{t|t-1} C_t^\top S_t^{-1},$$

$$\hat{x}_{t|t} = \hat{x}_{t|t-1} + K_t\big(y_t - (C_t \hat{x}_{t|t-1})\big), \qquad P_{t|t} = (I - K_t C_t) P_{t|t-1}.$$

In the continuous-time case, we have a similar situation for the solution of the corresponding differential equations.

## 4.9 Extended Kalman Filter

As mentioned above, in linear systems with normally distributed disturbances, the normal distribution is also transferred to all other relevant distributions, including the filter distribution. This is not so for nonlinear systems, even when all disturbances are assumed to be normally distributed. The filter distribution, in particular, can be arbitrarily complex in these cases. Except for the simple cases, in which merely an additional asymmetry appears in the distribution, there can also be filter distributions with multiple local maxima (modes). Although such distributions can only be very poorly approximated by the single-mode Gauss distribution, this type of approximation is standard and is applied in the vast majority of cases. One does so by linearizing the nonlinear system for the current state values and then applying the Kalman filter to this linearized system. The resulting filter is then called an Extended Kalman Filter (EKF). However, due to the aforementioned poor approximation of Gauss distributions for multi-mode filter distributions, general convergence results for this filter are not to be expected. One often tries to improve at least the covariance values of the EKF estimator by increasing computational efforts, which occurs in the case of the Unscented Kalman Filter, for example. However, the fundamental problem of approximating complex filter distributions using unimodal Gauss distributions still remains.

## 4.10 MTU-PF: Accounting for Uncertainties in the Measurement Time-Points when State Filtering with the Particle Filter

We now return to the general case of the stochastic state-space and want to dispense with the assumption that the observation time-points (measurement time-points) $t_j$ for $j = 1, \dots, M$ are given and known deterministically. Instead, we want to assume that the observation times $t_j$ are realizations of random variables $T_j$. These variables thus model the uncertainty about the exact measurement time-points. In contrast to the observation variables $Y_j$ themselves, the observation times $T_j$ are never directly observed (measured). Instead, we assume that the only information we have at our disposal is their probability distribution on the half-axis $[t_0, \infty)$ itself, whereas, for the observations $Y_j$, we know both the densities $g_j(y_j \mid x_{t_j}, t_j)$ *and* the observed values $y_j$ themselves. Consequently, we have here a significant conceptual difference.

We consider here only the simplest case, in which each time variable $T_j$ is independent from each other time variable. Indeed, this contradicts the fact that measurement values typically follow a prescribed chronology, such as $T_1 < T_2 < T_3 < \cdots$, which would imply a stochastic dependency between the variables $T_j$. However, this would lead to substantially more complicated algorithms. Moreover, dependencies in the chronology can also be simply introduced via appropriate restrictions to the support $\operatorname{supp} T_j$ of the random variables $T_j$, for example, by requiring that all elements of $\operatorname{supp} T_j$ are smaller than all elements of $\operatorname{supp} T_{j+1}$. In this way, the independence of the variables from one another is preserved. In general, the probability distribution of each individual variable $T_j$ should be given by a density $\gamma_j(t_j)$ relative to the Lebesgue measure $\lambda_{[t_0,\infty)}$ on the interval $[t_0, \infty)$.

Normally, the uncertainty in the measurement time-points is incorporated into the uncertainty of the measurement value by increasing the latter's variance (lumped measurement disturbances; see Fig. 3). This generally leads to parameters that can only be very conservatively estimated (with large uncertainties); for rapidly changing states, however, this procedure can also lead to really erroneous estimates (see Fig. 4(b)–(d)).

The standard particle filter can be extended appropriately (see [1, 8]). If there really are uncertainties in the measurement time-points, the resulting Measurement Time Uncertainty-Particle Filter (MTU-PF) delivers substantially better estimates than the standard particle filter (see Fig. 4 (a)).

The main difference between the MTU and the standard particle filter is that the weights are not just updated at discrete time-points (in the standard filter, at the exact measurement time-points); in principle, they are updated continuously, over all time-points. This results from the fact that the measurement time-points themselves are "smeared" across the time axis due to the densities $\gamma_j(t_j)$. This opens up a much broader range of possibilities for stabilizing the algorithm—for example, by choosing an adaptive increment control based on the development of the ESS estimator. With strongly decreasing ESS (i.e., high risk of algorithm degeneration), a smaller time increment (step size) can be selected so that early, repeated resampling can keep the particle set in good condition (at least from the standpoint of a high ESS value) (see Fig. 5). This is not possible in the standard case, since here, the algorithm's step size is fixed by the measurement intervals. More details can be found in [1] and [8]. In Sect. 7, we will introduce a biomedical application of this MTU particle filter and compare it to the standard particle filter.

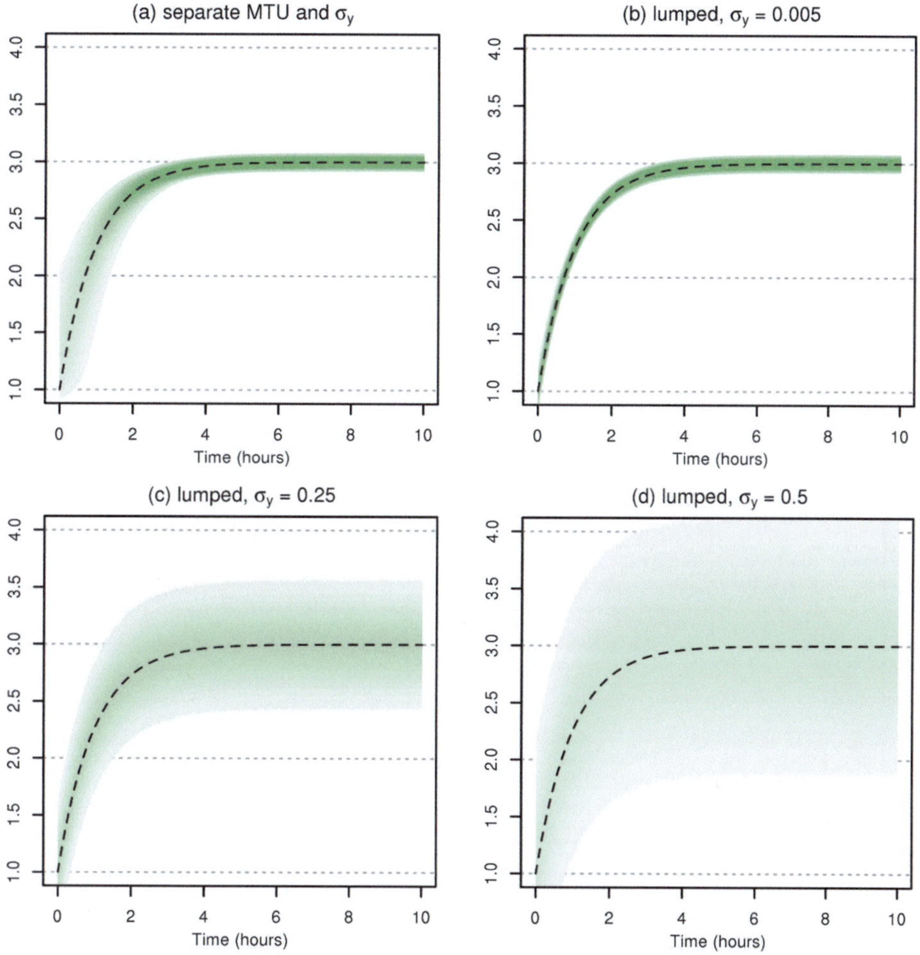


**Fig. 3** Monte Carlo simulation of measurement values for a simple model consisting of one state with exponential growth and normally-distributed disturbances, as well as normally-distributed disturbances in the measurement values. The *dashed green lines* represent the nominal development (average) of the state over time. The *green shaded areas* show the distribution of the measurement times and values. (**a**) Separate modeling of uncertainty in measurement time-points (*horizontal*) and values (*vertical*); variance of the measurement values $\sigma_y = 0.005$. (**b**)–(**d**) Without special modeling of the uncertainty in the measurement time-points; scattering only in the measurement values (*vertical*). Here, to compensate, the variance $\sigma_y$ of the measurement values is gradually increased. It shows that this compensation in (**b**)–(**d**) cannot adequately reproduce the distribution of the measurement values in case (**a**). This is especially visible for areas with steep increases (early time-points). Whereas in (**a**), the distributions in the $y$-direction scatter more here than where the state is constant (later time-points), in (**b**)–(**d**), the scattering in the $y$-direction is equally pronounced everywhere, as dictated by the model

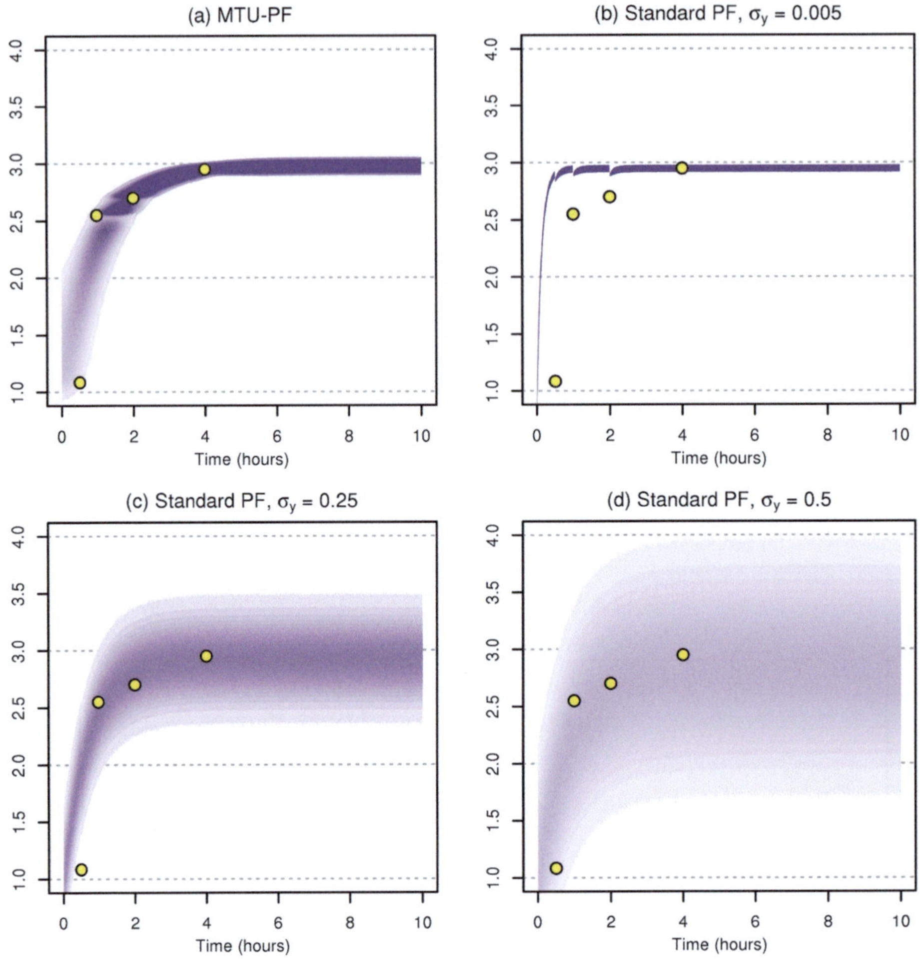


**Fig. 4** Simulated distributions of measurement values after estimating parameters and states in the sample model with various filters. The *yellow points* correspond to the measurements used for the estimates, which were generated as random realizations of the distribution shown in Fig. 3(a). (**a**) MTU particle filter; (**b**)–(**d**) standard particle filter with different measurement value variances $\sigma_y$. The *purple shaded areas* show the distributions of the measurements that would result from the various estimates of the particle filters. The actual distribution of the measurement values can be seen in Fig. 3(a). The estimates with the MTU particle filter deliver clearly better matching distributions than the standard particle filter

## 4.11 PF-MPC as a Particle Filter-Based MPC Approach

At the ITWM, we developed a generalized, stochastic, nonlinear Model Predictive Control (MPC) approach based on a double application of the particle filter [3]. Model Predictive Control refers to a class of model-based controllers whose development began in the

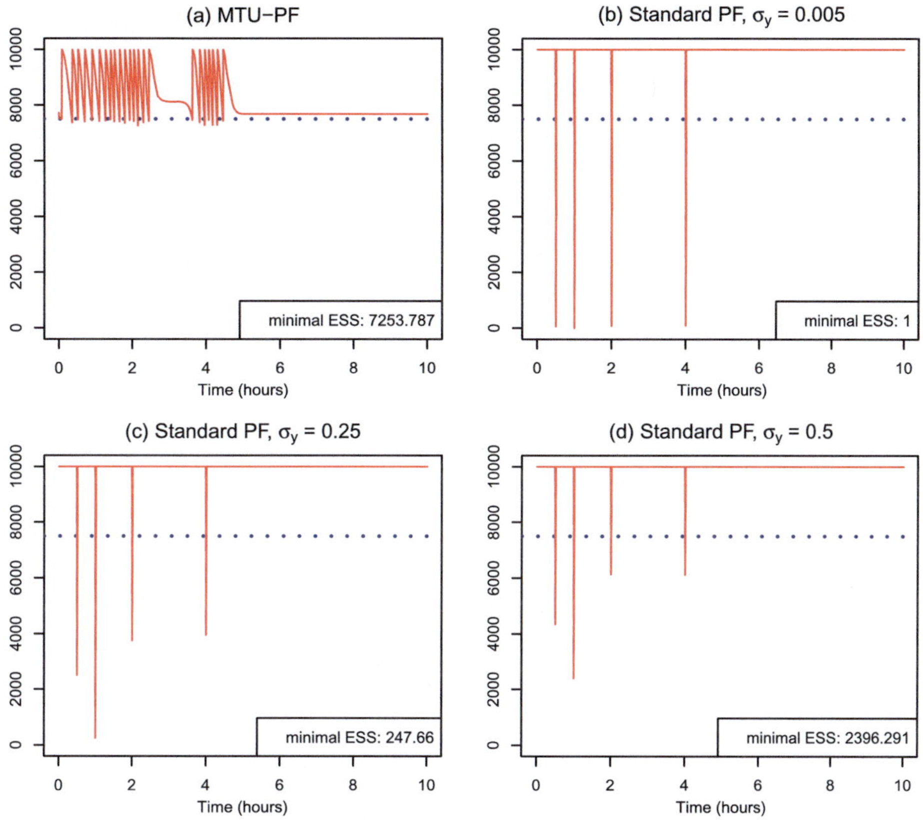


**Fig. 5** Comparison of ESS estimates during filtering. (**a**) MTU particle filter; (**b**)–(**d**) standard particle filter with different measurement value variances $\sigma_y$. Clearly, the ESS in the standard filters (**b**)–(**d**) falls off significantly at the discrete time-points for which a measurement is available. In (**a**), this is prevented by early resampling

1970s. Unlike traditional control approaches, such as PID controllers, the control signal is not determined solely by the current state (or an estimate thereof). Instead, the MPC controller makes use of a system model to enable predictive calculation of the system's development under the influence of the control signal $u_j$. Based on these predictions, the control signal is defined over a particular time period (horizon) $T_p$, so as to minimize a given target function $J$. Then, the first value of this calculated control signal is delivered to the system as a control input. This procedure is continually repeated over time.

In the past, the particle filter was often used for state estimation in the context of an MPC approach. Our approach is new in the sense that our controller not only uses the particle filter for state estimation, but also for solving the optimization problem. This is accomplished by considering the control targets as virtual measurements. After adding the control variable to the state variables, the optimization problem reduces to a filter problem: the conditioned distributions of the control signal under given targets can thus

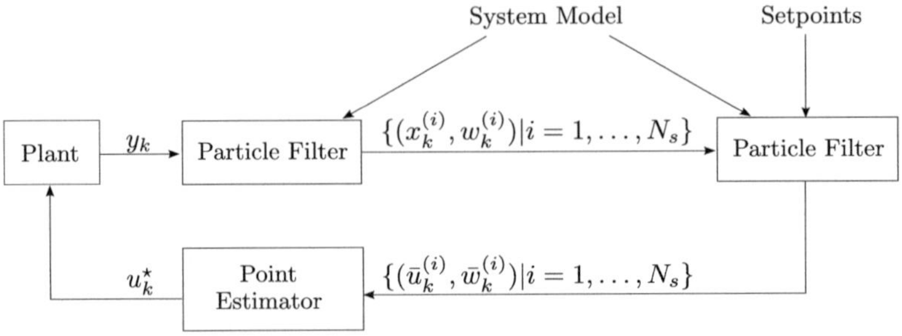


**Fig. 6** Schematic diagram of the PF-MPC controller. Here, $x_k^{(i)}$ is the state vector of the $i$-th particle with weight $w_k^{(i)}$ at time $k$ in the first particle filter for state estimation; $y_k$ is the measurement at time $k$; $\bar{u}_k^{(i)}$ is the value of the control variable of the $i$-th particle with weight $\bar{w}_k^{(i)}$, as output of the second particle filter. This particle set serves to calculate the optimal control value $u_k^*$; $N_s$ is the number of particles in each filter

be understood as a filter distribution and can therefore be calculated with a second particle filter. On the basis of these filter distributions for the control signal, the filtering trajectory most likely to lead to good system behavior can then be selected (see Fig. 6). The first value of this trajectory then serves as the next control input.

In the standard MPC approach, the target function $J = J(x_k, \bar{u}_{k:(k+T_p)}, T_p)$ is usually of the form

$$J = \sum_{j=k}^{k+T_p} \|\bar{u}_j - \bar{u}_{j-1}\|_Q^2 + \sum_{j=k+1}^{k+T_p} \|s_j - x_j\|_R^2.$$

Here, the norms refer to weighted Euclidian norms with the weighting matrices $Q$ and $R$. The first term ensures that the difference between consecutive control values $\bar{u}_{j-1}$ and $\bar{u}_j$ remains small. The second term penalizes deviations of the system states $x_j$ from the target states $s_j$; these target states describe those state trajectories that the system is to preferentially follow. The trick is that a minimization of $J$ corresponds to a maximization of $\exp^{-1/2J}$, that is—except for a normalizing constant—to a multivariate Gauss distribution density. For given states $x_j$, this can be viewed as a joint distribution of the control signal transition probabilities and the observation probabilities of $s_j$. But the variant of the particle filter introduced above realizes just that. The treatment of the target function as, in its essence, a distribution density immediately opens the possibility of lifting the restriction to Gauss distributions and permitting general, complex probability densities. In this way, one can incorporate very complex strategies in the control system, along with, in a very free manner, restrictions and constraints. These only have to be appropriately reproduced in probability distributions, which can then be treated, so to speak, as the preferred distributions of the system states. Here, however, problems can arise from the possible degeneration of the particle filter algorithm; the control strategy must be well designed in order to keep the degeneration as small as possible. Real-time controlling on the basis of nonlinear models is then quite feasible. As an example, Fig. 7 shows the control system

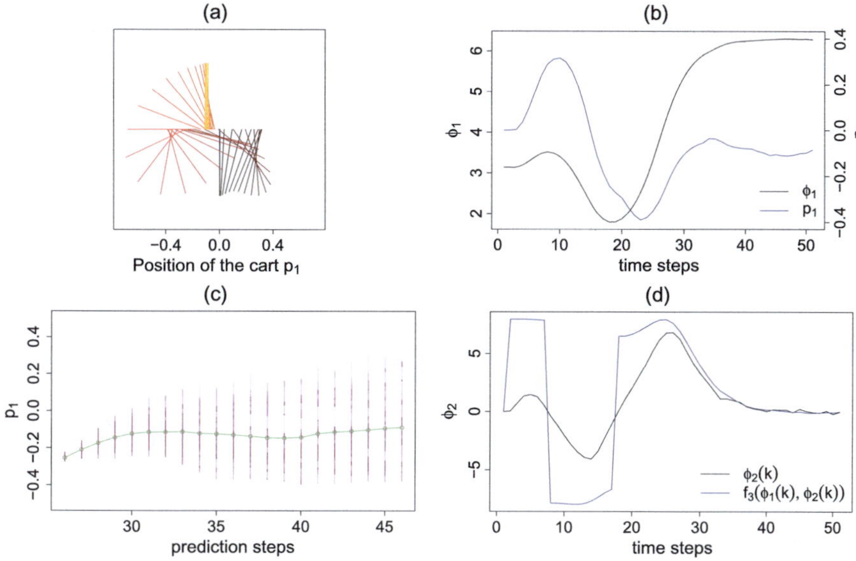


**Fig. 7** Controlling a simulated inverted pendulum. (**a**) Movement of the pendulum over time, as indicated by color transition from black to red to yellow. *Black*: Start, with pendulum hanging below. *Red*: Swinging the pendulum into inverted position. *Yellow*: Inverted, balanced pendulum. The movement of the pendulum in the $x$-direction (cart movement) remains within narrow bounds. (**b**) Progression of the states $p_1$ (position, *blue*) and $\phi_1$ (deflection angle, *black*) of the pendulum over time. (**c**) Prediction of the marginal distribution of the position $p_1$ (*magenta*) across the time horizon ($j = k, \dots, k + T_p$ with $T_p = 21$) in the second particle filter at time-point $k = 25$. The *green line* describes the estimated average. (**d**) Progression of the state $\phi_2$ (velocity of the deflection angle, *black*) in comparison with the specified progression of the corresponding set-point $f_3$ (*blue*), as a function of the deflection angle $\phi_1$ and its velocity $\phi_2$

for a nonlinear, inverted pendulum mounted on a cart that is sitting on a track. By moving the cart along the track, the pendulum is to be first swung into a vertical position and then held in this balanced state. As a supplementary constraint, the cart is not to drive across specified boundaries as it is moved along the track. It was possible to design the controls so simply that the computer-simulated system could be controlled in real-time (on a normal PC). Because the original, non-linearized model is used, both control tasks—swinging the pendulum into the vertical position and keeping it balanced there—can be accomplished with a single controller. More details can be found in [3].

## 5 Relationship to Simulation

As shown in the previous sections, one can estimate the hidden states of a process at prescribed measurement time-points by combining just a small amount of measurement data and a suitable mathematical system model. To do so, one performs a single system simulation step and then appropriately adapts the resulting calculated system state on the basis of the new measurement information that has arrived during this time step. In addition to the actual process dynamic, another significant factor in choosing and implementing a state estimator is the computation time available for the algorithm and/or the level of real-time capability required for the application.

## 5.1 Requirements Relating to the Application

The amount of computation time available depends primarily on the application context of the state estimation. One must first decide whether it is to be performed online or offline, that is, in real-time or not.

For offline state estimations, the amount of computation time required is usually not a critical factor; the differential equation systems to be solved in the course of the simulation "simply" need to be solvable. The required computation time plays a subordinate role, since data acquisition and state estimation are not temporally linked to one another. This approach is frequently used when performing state estimations in connection with the identification of process parameters.

For online applications, however, such as safety systems, process monitoring and/or diagnosis, and process control, real-time capability is required for state estimating. Here, the real-time capability is assessed in relation to the updating time required for the state estimation. Depending on the process dynamic and the application, this can range from milliseconds to minutes. The requirements resulting from the three above-mentioned applications are discussed in the following sections:

- For critical situations in safety systems, the state estimation, the simultaneous process analysis for risk assessment, and the protective response trigger must be accomplished on the order of milliseconds. Even with very fast hardware systems, this speed is frequently impossible. Therefore, when developing such a system, the algorithms for risk analysis and protective response should be implemented on their own very high-speed hardware platform. For storage and diagnosis of events categorized as relevant by the safety system, one can then use a downstream monitoring system based on a state estimator on independent hardware. One can justify this separation, since, for system diagnosis, one is generally interested only in conspicuous and/or critical events and the upstream safety mechanism functions as a corresponding event detector.
- In addition to their downstream use in safety systems, independent monitoring systems are also frequently used for process behavior analysis or system diagnosis. Here as well, slight time delays in the analysis of the results are often permissible. In this case, one performs a delayed execution of the state estimation for measurements collected within a specified time window, while, in parallel, data for the next time window is being collected and stored. To prevent data loss, however, execution of both the state estimation and the diagnosis or evaluation algorithms must

be completed within the time required to store the measurement data. The ITWM's torque monitoring systems introduced in Sect. 6 also work according to this principle.

- In the course of process management and control, a controller must optimally adjust the performance of the process and, in particular, maintain process stability. In cases where there are no directly measureable performance variables, the optimal control inputs are calculated on the basis of the system state determined by a model-based state estimator. The execution of the state estimation must take into account the updating rate of the controller. Here, one must allow for the fact that additional computing capacity is needed for the controller to calculate the control inputs for the system. State estimation and calculation of the control signals must both take place within the available updating time. In connection with the BMWI project "Development of an energy-efficient furnace concept for the heat treatment of glass," this concept was used at the Fraunhofer ITWM on behalf of Schott AG to design a controller for the energy-efficient management of the glass cooling process. Here, in each time step, measurement information from a few air temperature sensors was used to estimate the temperature distribution in the entire furnace with a Kalman filter.

The computing time required for the state estimation depends on the time needed to perform the system simulation step and on the subsequent adaptation of the state. In many control engineering applications, system behavior is modeled by means of finite element approaches. Even at moderate resolution, these lead to high-dimensional models and, thus, to correspondingly long simulation times. Therefore, in many applications, regardless of the state estimator being used, one must initially reduce model complexity with mathematical model order reduction methods, so that the model-based state estimation is possible within the available computing time. The challenge with model order reduction is generating a less complex model that still approximates the dynamic of the real process in the relevant working areas as well as possible. The errors resulting from the model order reduction must then be accounted for in the state estimation in the form of process uncertainties. In this field, the Fraunhofer ITWM develops symbolic and numerical model reduction algorithms for parametric nonlinear systems.

Another component with a comparable impact on simulation time is the computational and storage capacity of the hardware platform being used. Along with the available working memory, the processing power must also be taken into consideration when implementing the selected filter. While today's typical PC processors or modern embedded systems exhibit no computation time problems for many applications when suitably reduced models are used, low cost processors with very low computing power and storage capacity are still often utilized for industrial mass-produced goods to save money.

## 5.2 Implementations at the ITWM

### 5.2.1 Linear State Estimators

Various linear state estimators have already been put to use at the Fraunhofer ITWM for diverse industrial projects and products. The starting point for the following treatment is a linear, time-variant state-space model of the form

$$\begin{aligned}\dot{x}(t) &= A(t)x(t) + B(t)u(t) + Q(t)w(t)\\ z(t) &= C_1(t)x(t)\\ y(t) &= C_2(t)x(t) + M(t)v(t)\end{aligned} \tag{21}$$

for the technical or biological process under consideration. Here, $A(t) \in \mathbb{R}^{n\times n}$ is the state matrix and $B(t) \in \mathbb{R}^{n\times q}$ is the input matrix, with which the measured system inputs $u(t)$ are assigned to the states and, where necessary, also converted into the needed physical quantities. The outputs of actual interest $z(t)$ are calculated from the states with the matrix $C_1(t) \in \mathbb{R}^{k\times n}$. These outputs can be any of the states themselves, that is, $k = n$ and $C_1 = I_{n\times n}$, or physical quantities converted from one or more states, such as torque calculated from the twisting angles (states $x(t)$) for a shaft. The outputs $z(t)$ thus calculated are frequently used as virtual sensors for process analysis or control. State estimation also requires a comparison of the simulation result with the real measurement information in order to calculate the correction terms. Therefore, one must determine from the model the physical quantities corresponding to the sensor measurement values. This is done by transforming the states $x(t)$ with the matrix $C_2(t) \in \mathbb{R}^{p\times n}$. Moreover, in (21), $w(t)$ and $v(t)$ are stochastic disturbances that are modeled by the matrices $Q(t)$ and $M(t)$ and which impact the system and/or the measurements. Selection of the appropriate state estimator now depends on the assumptions made and/or on the process characteristics.

**Discrete Kalman Filter** Due to its simple iterative calculation scheme, the linear discrete Kalman filter, along with its diverse variations, is the most widely used algorithm for state estimations of linear systems (see Sect. 4.8). In particular, it can also be used in cases of non-steady-state noise or with time-variant systems. In order to apply it to the continuous model being treated here (21), one must first discretize it, reduce it to an appropriate dimension, and implement the iterative, discrete Kalman filter specified in Sect. 4.8. One such application at the ITWM involved the aforementioned BMWI project "Development of an energy-efficient furnace concept for the heat treatment of glass," where we estimated the temperature distribution in a passive glass annealing furnace on the basis of the time variance of the underlying linear model.

**Continuous-Time Filter** The Kalman-Bucy filter is the continuous form of the Kalman filter. Analogously to the discrete Kalman filter, the disturbances are handled purely by means of the expectation value and covariances, so that one can assume $M(t) = I$ for the continuous state-space system (21) without any restrictions. Here, we let $w(t)$ represent normally-distributed process noise with expectation value 0 and covariance matrix $R_w(t)$, and we let $v(t)$ represent normally-distributed measurement noise with expectation value 0 and covariance matrix $R_v(t)$.

Using the previous assumptions, the continuous Kalman filter is given by

$$\dot{\hat{x}}(t) = A(t)\hat{x}(t) + B(t)u(t) + K(t)\big(y(t) - C_2\hat{x(t)}\big) \tag{22}$$

where $K(t) = P(t)C_2(t)^T R_v^{-1}(t)$ and $P(t)$ is the solution of the differential equation

$$\dot{P} = A(t)P + PA(t)^T - PC_2^T R_v^{-1}(t)C_2(t)P + QR_w(t)Q^T \tag{23}$$

with $P(0) = E\{x(0)x^T(0)\}$.

Due to the time-dependency of the noise and the time-variance of the model, calculating the solution of the differential equation (23) for each time-step is very computationally intensive and, in many cases, cannot be done online. This is not so for a time-invariant system, that is, one in which the state matrices $A$, $B$ and $C_2$ in (21) are constant. For the underlying process, one frequently assumes steady-state measurement and process noise, along with time-invariance. In this case, the Kalman gain $K$ converges to a constant matrix and is given by

$$K = PC_2^T R_v^{-1},$$

where $P$ is the stabilizing solution of the Riccati equation

$$AP + PA^T - PC_2^T R_v^{-1}C_2P + QR_wQ^T = 0. \tag{24}$$

The resulting filter can then be represented as a linear state-space system, with $(A_{KF}, B_{KF}, C_{KF})$ given by

$$\begin{aligned} A_{KF} &= A - PC_2^T R_v^{-1}C_2 \\ B_{KF} &= \big[PC_2^T R_v^{-1}\ B\big] \\ C_{KF} &= C_1. \end{aligned} \tag{25}$$

To implement this filter, one can now perform a single *a priori* offline calculation of the Kalman gain and the error covariance estimation before the actual state estimation. Then, one determines the desired states online for each time-step by solving the differential equation system. Here, one can either discretize the system *a priori* or calculate the solution of the differential equation system stepwise using a suitable algorithm. This separation into offline and online calculation steps makes implementation possible even for short updating times.

As described in Sect. 4, the Kalman filter offers, in the form of the error covariance matrix, a confidence measure for the estimated states under the assumed stochastic influences. From a systems theory perspective, the Kalman filter is the state estimator that delivers optimal estimates, in the sense that it averages across all frequencies. An estimator that focuses on critical frequencies is the $H_\infty$-filter [48], which is based on the $H_\infty$-norm. For a well-defined, stable, time-invariant system $G$, this is defined by

$$\left\| G(s) \right\|_\infty := ess \sup_\omega \bar{\sigma}\left(G(j\omega)\right)$$

with the maximum singular value

$$\bar{\sigma}\left(G(j\omega)\right) := \max_{u(\omega)\neq 0} \frac{\|z(\omega)\|_2}{\|u(\omega)\|_2}$$

and $z(\omega) = G(j\omega)u(\omega)$. For linear systems with one input and one output, the norm describes the maximum gain factor across all frequencies. Choosing the $\infty$-norm shifts the focus from the simultaneous minimization of the energy of the transfer functions for all frequencies to the most critical frequency of the system. In other words, it deals with a worst-case scenario.

As the starting point for calculating the filter, we take a time-invariant linear state-space system; that is, the matrices $A$, $B$, $C_1$, and $C_2$ in (21) are constant. With $H_\infty$-filter problems, one assumes that the disturbances have limited energy. The actual information about the intensity of the disturbances is captured by the time-invariant matrices in (21), $Q$ and $M$ [2]. On the basis of the $H_\infty$-norm, the $H_\infty$-filter problem can be formulated as follows [48]:

**$H_\infty$-Filter Problem** For a given $\gamma > 0$, find a causal filter $F(s) \in \Re H_\infty$, where $\Re H_\infty$ is the set of all well-defined and real-rational, stable transfer functions, so that

$$\sup_{w\in L^2[0,\infty)} \frac{\|\tilde{z} - \hat{z}\|_2^2}{\|w\|_2^2} < \gamma^2.$$

Here $\tilde{z}$ denotes the real system output and $\hat{z}$, the estimated filter output.

One then obtains the desired gain-matrix for a linear, robust $H_\infty$-filter by solving the following algebraic Riccati equation:

$$PA^T + AP + P\left(\gamma^{-2}C_1^T C_1 - C_2^T M M^T C_2\right)P + QQ^T = 0. \qquad (26)$$

If the positive, semi-definite stabilizing solution $P$ exists, then the desired filter $F(s) \in \Re H_\infty$ can be represented in the state-space representation and the system matrices $(A_{HF}, B_{HF}, C_{HF})$ are given by

$$\begin{aligned} A_{HF} &= A - PC_2^T (MM^T)^{-1} C_2 \\ B_{HF} &= \left[ PC_2^T (MM^T)^{-1} \; B \right] \\ C_{HF} &= C_1. \end{aligned} \tag{27}$$

The difference between the Riccati equation (26) and the Riccati equation (24) for calculating the Kalman gain is essentially the additional term $\gamma^{-2} C_1^T C_1$ resulting from the robustness requirement. Here, the existence of a solution to the Riccati equation (26) is not guaranteed for each $\gamma$. To obtain the most robust estimator possible, $\gamma$ is iteratively reduced until the solution of the algebraic Riccati equation exists. Therefore, as with the Kalman filter (25), implementation of the robust linear $H_\infty$-filter (27) also involves first solving an algebraic Riccati equation offline for the underlying linear, time-invariant state-space model. The resulting $H_\infty$-filter is also given in the form of a dynamic continuous-time state-space system (27). However, minimizing the $H_\infty$-norm results in more robustness relative to unstructured disturbances and/or model uncertainties than exists for the Kalman filter. The $\mu$-synthesis, also used at the ITWM, delivers extensions relating to structured uncertainties. The decision whether to use the $H_\infty$-filter or the Kalman filter depends on the model uncertainties and the resulting robustness requirements.

The torque detection and analysis system TorAn described in Sect. 6 is an ITWM monitoring system based on an online-capable, robust $H_\infty$-filter or the continuous Kalman filter. On the basis of the measured mechanical torque signals of the energizing drive train components, such as the motor or generator torque, and a direct torsion measurement with a torque sensor, TorAn uses the selected filter to estimate the states given in the form of the twisting angle of the rotating shaft. TorAn also uses the estimated states to predict and analyze the torque characteristics for other critical, inaccessible shaft components. The algorithms for data acquisition and state estimation and the criteria monitoring and fatigue analysis were implemented as a real-time-capable C-Library with a link to an analog-digital converter.

### 5.2.2 Nonlinear State Estimator

**Constrained Extended Kalman Filter** The extended Kalman filter is the extension of the discrete Kalman filter to nonlinear systems. As with the previously described linear state estimator, these filters do not initially allow for consideration of physical constraints. However, there are some approaches that do make this possible, such as the Moving Horizon Estimation or the Constrained Extended Kalman Filter (CEKF) proposed in [47]. The basic idea of the CEKF is to initially perform a general state estimation with a first extended Kalman filter. Then, in a second extended Kalman filter, a correction to the first estimation is undertaken so that the states lie within the permissible value range. Particularly for nonlinear systems, one can frequently limit the states' solution space by means of physically motivated restrictions and, thus, prevent a possible divergence in the state estimation. The ITWM uses this filter to compensate rpm-dependent, periodic disturbances in connection with torque measurements using an inductive torque sensor based on the magnetostrictive effect (see Sect. 6.2).

**Particle Filter Algorithm** As described in detail in Sect. 4, the standard particle filter works on discrete-time, nonlinear, non-Gaussian models and can be easily adapted for use with continuous-time systems with discrete-time measurements. The particle filter algorithm has already been used in diverse application areas in the form of prototype implementations. Among the applications are the measurement-based model identification of a shock absorber with hysteresis effects, the estimation of gene copies (copy number) in tumor DNA, and the estimation of parameters in the biomedical field (see Sect. 7). There are implementations in Java and in the statistical programming language R, which is often used in biomedicine. In addition, the particle filter algorithm has been implemented as an element of a system-biology toolbox being developed at the Fraunhofer Chalmers Centre (FCC) in Gothenburg on the basis of the symbolic programming environment Mathematica. Extensions to the particle filter algorithm have also been developed at the ITWM, in particular, the adaptation of the algorithm to uncertainties in measurement time-points and its double-use in a new nonlinear Model Predictive Control (MPC) approach.

## 6 Online Monitoring of Torsional Vibrations in Power Plant Turbine Generator Shaft Lines

### 6.1 Problem Description

The first of what have become many endeavors involving state estimation in the System Analysis, Prognosis, and Control Department began with a project to develop a state observer for power plant turbine sets. These turbine sets consist of a long shaft on which are mounted a generator for electricity generation and one or more turbines to drive the shaft (see Fig. 8). Grid malfunctions or operational errors can trigger torsional vibrations in a turbine set, which lead to fatigue in the shaft components and may even result in serious mechanical damage. In some cases, the latter can induce additional, permanent torsional vibrations at the rotational frequency and its harmonics. It is therefore necessary to ensure continuous monitoring of turbine generator shaft lines for torsional vibrations. For many years, the Fraunhofer ITWM has worked to develop methods for online monitoring of torsional vibrations and has developed procedures for model-based prognosis of torsional vibrations and run-out compensation of inductive magnetostrictive sensors. These results

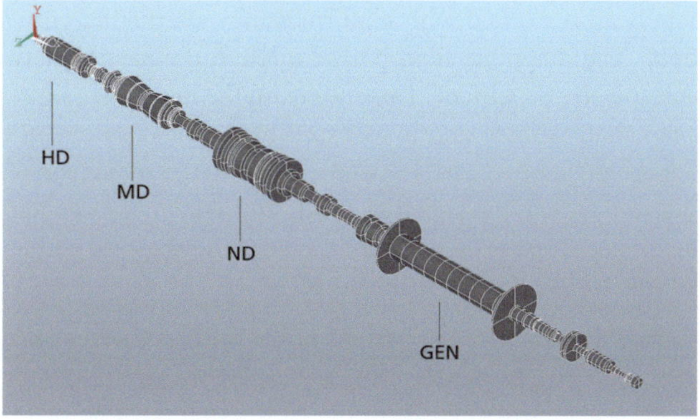


**Fig. 8** Schematic of a power plant turbine set with generator (GEN), low-pressure turbine (ND), intermediate-pressure turbine (MD), and high-pressure turbine (HD)

have been put into service around the world by such customers as E.ON Anlagenservice, Siemens Energy, and ABB Utilities.

The challenges associated with the torsion monitoring of power plant turbine generator shaft lines are diverse. The methods and products developed at the Fraunhofer ITWM to meet these challenges include systems for the following:

- Experimental torsional analysis with which, for example, torsional natural frequencies from turbine generator shaft lines can be determined (TorStor),
- Measurement-data-based detection and assessment of critical torsional vibrations, such as sub-synchronous oscillations (TorFat),
- Targeted monitoring of at-risk locations in a turbine set—the shaft couplings, for example—using model-based state estimators (TorAn), and
- Detection and classification of shaft damage.

All systems require torque measurements from at least one position on the drive train. Here, one can make use of the magnetostrictive effect, which describes the relationship between the magnetic permeability change and a change in strain when external loads are applied. The measurement systems function with no shaft contact and can be positioned flexibly. On the basis of the inverse magnetostrictive effect of ferromagnetic shafts, they detect changes in the torsional stress on the shaft surface by means of induction and magnetic field measurements. Thus, they represent a useful alternative to traditional torque sensing technology. The great advantage of these sensors is that their use requires no structural modifications to the shaft itself. They can therefore be flexibly implemented without influencing the dynamics of the shaft. The magnetostrictive sensor developed on behalf of the Fraunhofer ITWM for torsion monitoring in power plants inductively measures the difference in magnetic permeability, which is proportional to the torsional stress on the shaft surface across a large measurement range. Here, a primary coil in the center of the measurement head is excited with high frequency alternating current, thus producing a magnetic field. The magnetic field passes through the air gap between sensor and shaft and penetrates the shaft's surface. Depending on the magnetic permeability, the field spreads out over the shaft surface and is assessed by four measurement coils within the sensor head, which are positioned at a 45-degree angle to the main axis of the shaft. With this measurement arrangement, the signal resulting from the measurement coil circuitry is proportional to the torsional stress on the shaft surface. The sensor's output voltage/current $S_V(t)$ is converted into torque $S_M(t)$ by means of

$$S_M(t) = gain * \left(S_V(t) - offset\right).$$

The quantities *offset* and *gain* needed for the conversion must be determined in a calibration step using measurements from two known load points. After calibration, the sensor can then be used for torque measurements.

## 6.2 Run-out Compensation Using the Constrained Extended Kalman Filter

In addition to the torsional stresses resulting from external loads, there are always permanent, frozen stresses on the shaft surface that arise during the manufacturing process. These so-called inhomogeneities vary locally, and it is impossible to make any general *a priori* statement about their shape and size. Therefore, when performing torsional measurements on a rotating shaft with an inductive magnetostrictive sensor, one must take into consideration that these inhomogeneities along the measurement track lead to varying magnetic flows around the entire circumference. However, the rotation of the shaft causes the signals resulting from the inhomogeneities along the measurement track around the shaft circumference to always recur in the same sequence. Thus, one obtains a deterministic, periodic disturbance signal $y_{Inhom}(t)$, referred to as the run-out signal. Because one is dealing with localized characteristics distributed around the circumference, the frequencies of the various elements of the disturbance signal correspond to whole-numbered multiples of the shaft's rotational frequency $f(t)$. However, for state evaluations of industrial turbines, for example, it is often exactly these frequencies that are of interest. Thus, the run-out masks the significant system information, and determinations of the torsional load made without signal correction always contain errors. Due to their characteristics, run-out signals can be modeled with the time-dependent rotational frequency $f(t)$ as the basic frequency of a Fourier sum at time $t_k$ as follows:

$$y_{Inhom}(t_k) = \sum_{l=1}^{n} a_l(t_k) \sin\left(l 2\pi f(t_k) t_k\right) + b_l(t_k) \cos\left(l 2\pi f(t_k) t_k\right). \tag{28}$$

The amplitudes $a_l(t_k)$ and $b_l(t_k)$ do not change relative to a fixed reference point on the shaft's circumference, even when the rotational speed varies. For a non-changing measurement track, they can be assumed to be constant. When the shaft is loaded with an external torque $y_T(t_k)$, the disturbance signal is then superimposed on the torque signal. The measurement noise of the measurement system $v(t)$ must also be considered, so that the measurement signal $y_{Mess}(t_k)$ of the torque sensor at time $t_k$ thus becomes

$$y_{Mess}(t_k) = y_T(t_k) + y_{Inhom}(t_k) + v(t_k). \tag{29}$$

In some applications, one can resort to classical signal filters, such as high-pass, low-pass, or notch filters, to compensate the run-out signal. In the context of detecting critical torsional frequencies for turbine sets, for example, one encounters mostly an excitation of the critical excitable torsional natural frequencies, which do not correspond to the shaft rpm and its harmonics for standard grid operation. By means of frequency-selective analysis or band-pass filtering of the relevant frequency zones, one can therefore isolate the torsional vibrations of interest $y_T(t_k)$ from the measurement signal $y_{Mess}(t_k)$ for subsequent analysis. If a disturbance frequency should correspond to one of the critical torsional frequencies of interest, however, these torsional vibration signals would either be completely eliminated with the aforementioned filter, or be at least strongly distorted. Therefore, one should use a filter that eliminates the run-out signal, but leaves the actual relevant vibration information unchanged also for the shaft rpm and its whole-numbered multiples.

In a compensation method developed at the ITWM, the run-out signal $y_{Inhom}(t_k)$ is estimated online relative to the circumference for each time-step $t_k$ and subtracted from the measurement value $y_{Mess}(t_k)$. Here, however, the influences of all transfer functions, such as phase shifts and amplitude attenuations from signal filters, must be taken into consideration within the measurement chain.

The parameters to be determined in the run-out function (28) are described by the dynamic discrete state-space model

$$x(t_{k+1}) = Ax(t_k) + w(t_k) \tag{30}$$

$$y(t_k) = h\big(x(t_k)\big) + v(t_k) \tag{31}$$

with $x(t_k) = [a_0(t_k), \dots, a_n(t_k), b_1(t_k), \dots, b_n(t_k), f(t_k), \Theta(t_k)]^T \in \mathbb{R}^{2n+3}$ and

$$A = \begin{bmatrix} I_1 & 0 & 0 & 0 \\ 0 & I_2 & 0 & 0 \\ 0 & 0 & 1 & 0 \\ 0 & 0 & 2\pi\Delta t & 1 \end{bmatrix} \in \mathbb{R}^{(2n+3)\times(2n+3)}$$

where $I_1 \in \mathbb{R}^{(n+1)\times(n+1)}$, $I_2 \in \mathbb{R}^{n\times n}$ are unit matrices. Moreover, $w(t_k)$ and $v(t_k)$ in Eqs. (30) and (31) are normally-distributed, white process and measurement noise, respectively, while $h(x(t_k))$ represents the use of the states $x(t_k)$ in Eq. (28) and the subsequent signal filtering. The states $x(t)$ are estimated by a mathematical state observer online from the measurements by comparing $y_{Mess}(t_k)$ and $y(t_k)$. Here, in particular, constraints resulting from the technical and physical boundary conditions of the measurements are taken into account in the form of upper and lower bounds for each physical quantity. As

**Fig. 9** Run-out compensation result with Constrained Extended Kalman Filter (CEKF)

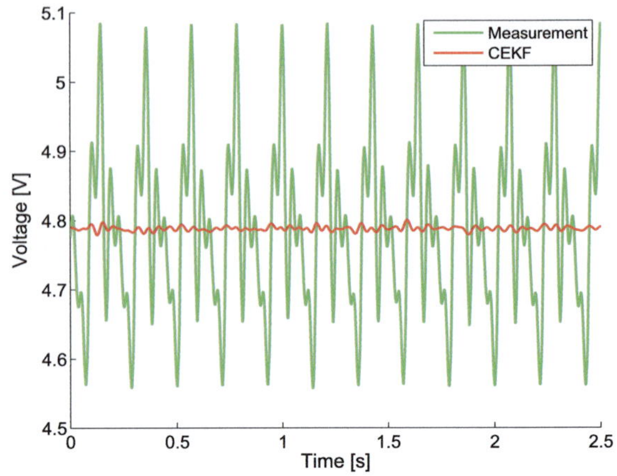


the state estimator, the Constrained Extended Kalman Filter (CEKF) proposed in [47] was adapted to the given run-out compensation problem. The CEKF allows one to distinguish between the run-out signal and torsional vibrations, even when the torsional natural frequency corresponds to the rotational frequency. Figure 9 shows the results of a CEKF run-out compensation for a measurement with constant load. After the compensation, one obtains the load value with sensor noise, which in this case is less than 0.5 %. Along with the run-out compensation, the method also estimates the rotational frequency $f(t_k)$ and the current position $\Theta(t_k)$ of the shaft in relation to the sensor. Both this and other methods have been used successfully for run-out filtering on magnetostrictive torque measurements in industrial installations.

## 6.3 Prognosis of Torsional Vibrations for Inaccessible Components on a Turbine Generator Shaft Line

During state monitoring of torsional vibrations on a power plant turbine generator shaft line, one must be able to guarantee uninterrupted surveillance of the drive train's critical components. However, technical restrictions and cost concerns sometimes prevent placement of torque sensors on all the critical shaft components one would like to observe. In order to nonetheless be able to make a statement about the torsional oscillations and their influence on any given shaft component, one must use a suitable, model-based prognosis system. Older torsional oscillation monitoring systems are based on a pure system simulation, in which modeling errors, estimated initial conditions, and model uncertainties during the simulation can lead unavoidably to deviations from true system behavior. An overview of the existing systems can be found in [35]. Use of a mathematically robust, online-capable state estimator represents an extension of the pure simulation approach. This approach was first introduced by the Fraunhofer ITWM in collaboration with the Electrical Drives and Mechatronics Chair of the Technical University of Dortmund, under the supervision of Professor Stephan Kulig, for the torsion monitoring of power plant turbine generator shaft lines. Along with the measurement data needed for the simulation, the state estimator receives a torque measurement for one component as an additional input quantity. The mathematical state estimator then implicitly compares the real, measured data with the time-series predicted by the simulation for the measurement site. Information obtained on the difference between the measured torque signal and the system simulation is integrated—as described in Sect. 5—into the prediction of the torsional vibration behavior of the remaining components in the form of a correction term for error compensation. Depending on the quality of the available physical information, either a Kalman filter or a robust $H_\infty$-filter (see Sect. 5.2.1) is used for the torsion monitoring. The filter design is accomplished at the ITWM by following the steps outlined below.

On the basis of the geometric and physical information available for the given drive train, the finite element method is used to generate the Newtonian equations of motion—a system of 2nd order ordinary differential equations—for the torsional behavior of the drive train:

$$J\ddot{\varphi}(t) + K\varphi(t) = \bar{B}u(t); \qquad y(t) = \bar{C}\varphi(t). \tag{32}$$

Here, $0 < J^T = J \in \mathbb{R}^{n\times n}$ is the matrix of the mass moment of inertia and $0 \preceq K^T = K \in \mathbb{R}^{n\times n}$, the torsional stiffness matrix of the system. Moreover, $\bar{B} \in \mathbb{R}^{n\times q}$ is the input matrix, which contains information about the position and conversion factors for the externally applied torques of the generator and the turbines $u(t)$. The matrix $\bar{C} \in \mathbb{R}^{p\times n}$ transforms the angular displacements $\varphi(t)$ into torques $y(t)$ for the targeted system components, for example, the coupling between the turbine trains. The torque at the sensor's measurement site is another output, which is then used in the state estimators to compare simulation and measurement. One knows that the matrices $J$ and $K$ can be diagonalized by an equivalence transformation with the modal matrix $V = [v_1 \cdots v_n]$ and a suitable norming of the modes $v_i \in \mathbb{R}^n$, $i = 1, \ldots, n$. This then yields:

$$V^T J V = I, \qquad V^T K V = \Lambda, \tag{33}$$

where $I \in \mathbb{R}^{n\times n}$ is the unit matrix. The diagonal matrix $\Lambda \in \mathbb{R}^{n\times n}$ contains the generalized eigenvalues of the undamped system, that is, $KV = JV\Lambda$.

To account for the damping neglected up to this point in Eq. (32), one usually has only rough approximations regarding the modal damping. Therefore, Eq. (32), as will be now demonstrated, is first modally transformed and then augmented by a modal damping term $D\dot{x}(t)$ with the modal damping matrix $D \in \mathbb{R}^{n\times n}$. Because of this simplification, and also because the modal damping coefficients are usually not known exactly for turbine generator shaft lines, one must treat these as model uncertainties during the subsequent filter design. With these damping assumptions, and with the substitution of $\varphi(t) = Vx(t)$, one obtains, taking into account the specified orthogonality conditions, the following system of decoupled 2nd order differential equations:

$$\ddot{x}(t) + D\dot{x}(t) + \Lambda x(t) = V^T \bar{B}u(t); \qquad y(t) = \bar{C}Vx(t) \tag{34}$$

If one now also substitutes $z(t) = \left[\begin{smallmatrix} x(t) \\ \dot{x}(t) \end{smallmatrix}\right]$ in (34), one obtains the following state-space model:

$$\dot{z}(t) = Az(t) + Bu(t);$$
$$y(t) = Cz(t); \tag{35}$$

with

$$A = \begin{bmatrix} 0 & I \\ -\Lambda & -D \end{bmatrix} \in \mathbb{R}^{2n\times 2n}, \qquad B = \begin{bmatrix} 0 \\ V^T\bar{B} \end{bmatrix} \in \mathbb{R}^{2n\times q} \quad \text{and}$$
$$C = \begin{bmatrix} \bar{C}V & 0 \end{bmatrix} \in \mathbb{R}^{p\times 2n}.$$

In order for the state estimator to achieve real-time capability, that is, in order to be able to calculate one time-step of the state estimation within the real time sampling interval, the dimension of the state-space model (35) is reduced using a model order reduction method. Here, the system model is transformed using the appropriate projection matrices $T_R, T_L \in \mathbb{R}^{n\times s}$ with $s \ll n$. This yields

$$\dot{z}_r(t) = A_r z(t) + B_r u(t)$$
$$y(t) = C_r z_r(t); \tag{36}$$

with

$$z(t) = T_R z_r(t), \qquad A_r = T_L^T A T_R, \qquad B_r = T_L^T B, \qquad C_r = C T_R.$$

Especially for the reduction of weakly damped, 2nd order systems—as we have with the torsion model for power plant turbine shaft lines —the Fraunhofer ITWM has developed efficient methods for an approximated, frequency-weighted, balanced reduction [9]. With regard to the quality of the approximation, special emphasis is placed here on specifying a frequency range in advance. For the torsion monitoring of turbine generator shaft lines, this is the low-frequency range from 0 to 200 Hz, since most malfunctions in the electrical grid result primarily in an excitation of the torsional natural frequency of the power plant turbine shaft line below the grid frequency.

When designing the state estimator, one must then consider both the above-mentioned modeling assumptions and also the uncertainties resulting from the model reduction. Therefore, one adapts the model using suitable methods based on measurements, such as the modal data of the real system. Here, one must consider that the torsional natural frequencies cannot be excited experimentally at will for power plant turbine generator shaft lines. Thus, the modal data must often be determined from actual grid disturbances by means of permanent monitoring and analysis. Moreover, one only has torsion measurements for a few shaft components. All told, one has usually measured only a few natural frequencies from the low-frequency range, and measurement values are available

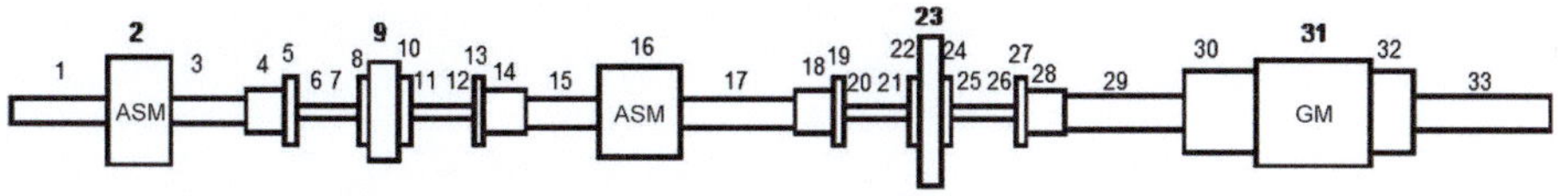


**Fig. 10** Schematic of the test rig showing the division into finite elements

**Fig. 11** Torque test rig (©Chair Electrical Drives and Mechatronics, TU Dortmund)

for few of the nodes for the corresponding modes. In contrast, the analytical model contains significantly more natural frequencies, depending on the model dimension. However, many of these "extra" natural frequencies are in the high-frequency range. To the extent that one has modal information, that is, measurements of natural frequencies and modes, one then uses model updating methods to improve model quality. Using the iterative model updating methods [10] developed for this problem at the Fraunhofer ITWM, the model parameters are adapted so as to achieve the desired correspondence between the model's modal data and the measured data. The reduced and adapted model then serves as the basis for the filter design (e.g., Kalman filter or $H_\infty$-filter; see Sect. 5.2.1), which is then used in the monitoring system TorAn for the state estimation and torsional vibration prognosis. Extensive descriptions of the design steps outlined here can be found in [4–6].

The functionality of the prediction of the mathematical, robust state observer for torsional vibration prognosis is demonstrated using the example of the test stand from the Electrical Drives and Mechatronics Chair of the TU Dortmund. In contrast to a real power plant turbine generator shaft line, it was a simple procedure to install extra sensors here to assess the quality of the state estimator. Figure 10 shows a schematic diagram of the test rig; Fig. 11, a photo; and Fig. 12, the torque sensor positioned at node 7. The test rig was designed to exhibit the typical natural frequency and mode data of a turbine generator

**Fig. 12** Contact-free torque sensor on element 7 (©Chair Electrical Drives and Mechatronics, TU Dortmund)

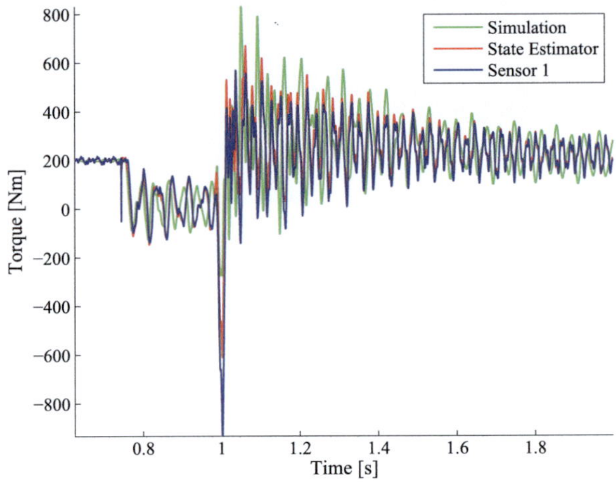


**Fig. 13** Comparison of measurement (*blue*), state estimation (*red*), and pure simulation (*green*) at node 7

shaft line, and thus a similar dynamic. According to the previously described modeling steps, an online-capable, robust $H_\infty$-filter was designed for the test rig. As measurement quantities, that is, as input for the filter, we used the mechanical moment of the asynchronous drive machine (obtained via power and rpm measurements)—node 2—and the DC machine—node 31—along with a measurement from the contact-free torque sensor—node 7. As explained earlier, the state observer, unlike a pure simulation, uses the comparison of measurement signal and simulation as a central quantity for determining the torque estimates.

For the disturbance scenario referred to as a "short interruption," the torsional vibrations for nodes 7 (Fig. 13) and 21 (Fig. 14) were predicted on the basis of the pure finite element model and also using the filter generated from this model. In order to be able to assess the results of the pure system simulation and the state estimation, the torsional vibrations on node 21 were measured with another sensor. Unlike the measurement from node 7,

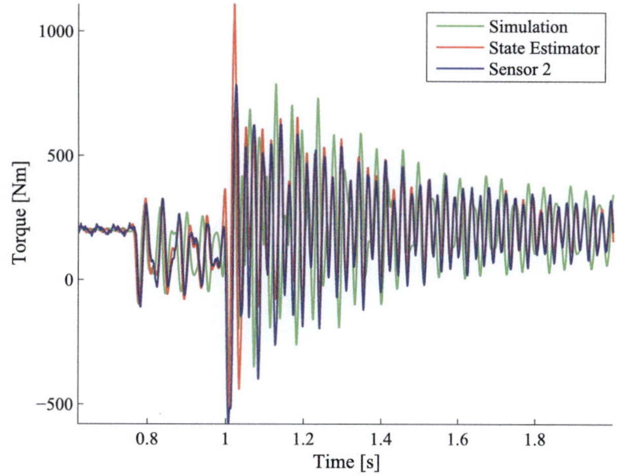


**Fig. 14** Comparison of measurement (*blue*), state estimation (*red*), and pure simulation (*green*) at node 21

the measurement from node 21 is not used as an input quantity for the state estimator. The correspondence between measurement and state estimation is, as expected, clearly better than that between system simulation and measurement. The cause of the overall poor correspondence for the system simulation is the erroneous modeling of the drive train damping. With the state estimation, this leads to short-lived errors at the beginning of the short interruption. However, this prediction error is corrected by the filter within one to two cycles, which highlights the capability of a robust mathematical state estimator to compensate for model uncertainties.

## 7 Application of the MTU Particle Filter to a Plasma-Leucine Model with Population Data

In this section, we describe an application of the MTU-PF algorithm, a version of the particle filter developed at the ITWM that allows for inclusion of uncertainties in the measurement time-points (MTU stands for Measurement Time Uncertainties; see Sect. 4.11). The results are based on a collaboration between the ITWM and Mats Jirstrand, from the Fraunhofer Chalmers Center (FCC) in Gothenburg, Sweden, Martin Adiels, from the Sahlgrenska Center for Cardiovascular Research in Gothenburg, and Marja-Riitta Taskinen, from the medical faculty of the University of Helsinki, Finland [1, 8]. In this project, we apply the MTU-PF to a study that analyzes the kinetics of the amino acid leucine in blood plasma by means of so-called tracer/tracee experiments. This plasma-leucine is a component of certain lipoproteins, which serve as fat transporters in blood and play an important role in cardiovascular disorders. Specifically, in the course of our project, we were asked to confirm a hypothesis about the deviation of a rate parameter for diabetes patients, in comparison with test subjects from a control group. The difficulties in performing the corresponding estimates result, on the one hand, from the assumption of hierarchically arranged parameters (global, group-specific, individual parameters), which lead to so-called mixed effects in the models, and, on the other hand, from uncertainties in the measurement values (blood samples), missing measurements, and, in particular, from uncertainties in the measurement time-points.

Tracer/tracee experiments were carried out in the study to analyze the plasma-leucine kinetics. The kinetics of the actual substance of interest—plasma-leucine—in this case referred to as the tracee, are determined by observing a labeled leucine added in the experiment, referred to as the tracer. The underlying model for the plasma-leucine kinetics comes from Demant et al. [26] and is based, in turn, on Cobelli et al. [23]. The data is taken from a clinical study on diabetes patients [12, 13]. In [1] and [8], both model and data are used to perform a Bayesian population-based parameter estimation, and were already used earlier for a maximum likelihood estimation [19]. In this case, the original model, based on ordinary differential equations (ODE), had to be supplemented with a stochastic element, with the result that the kinetics are now modeled using stochastic differential equations (SDE). The approach in [19] differs from ours also in that it assumes the stochastic fluctuations to be in the plasma-leucine (tracee), whereas we place the variability in the labeled leucine (tracer). In point of fact, stochastic variability should be assumed for both concentrations; for simplicity's sake, however, we limit ourselves to just one.

The negative effects on the estimates of uncertainties or inaccuracies in determining the measurement time-points are to be expected primarily at the beginning of the measurement series, since it is here, directly after addition of the tracer, that the concentrations change most abruptly. Our algorithm has the ability to counteract this problem.

## 7.1 The Leucine Model

In [13] (see [11] also), a new, combined multi-compartmental model for apolipoprotein B-100 (apoB) and triglyceride metabolism in very low density lipoprotein (VLDL) subfractions was developed (see Fig. 15). VLDL serve as transporters of triglycerides and cholesterol from the liver to the periphery. Elevated values are associated with an increased risk of cardiovascular disorders. Each VLDL particle contains exactly one apoB molecule, which makes apoB a suitable marker for triglyceride transport. The secreted particles become denser and denser as more triglycerides are delivered to target sites, such as muscles and adipose tissue, so that the relative protein content increases. As the density increases, the VLDL becomes an intermediate density lipoprotein (IDL) and, finally, a low density lipoprotein (LDL).

For our purposes, we use only the portion of the model that concerns the leucine pool, that is, compartments 1–4 (see Fig. 16). The fluxes exiting the subsystem are located in compartments 1 and 2. The flux entering compartment 1 is designated $U_1$.

The data is obtained from tracer/tracee experiments. Here, the tracee (i.e., the concentration we are actually interested in) consists of the leucine amino acids as components of the apoB molecule. Additional, labeled leucine (the tracer) is injected as a bolus infusion. Knowledge about the kinetics (fluxes between the compartments) of the tracee can be gained by studying the kinetics of the tracer.

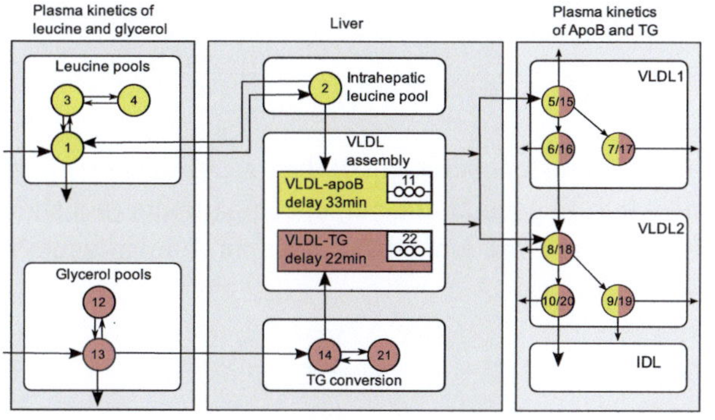


**Fig. 15** Multi-compartmental model for the metabolism of apolipoprotein B-100 (apoB) and triglycerides (TG) in very low density lipoprotein (VLDL) subfractions. This multi-compartment model was developed in [13]

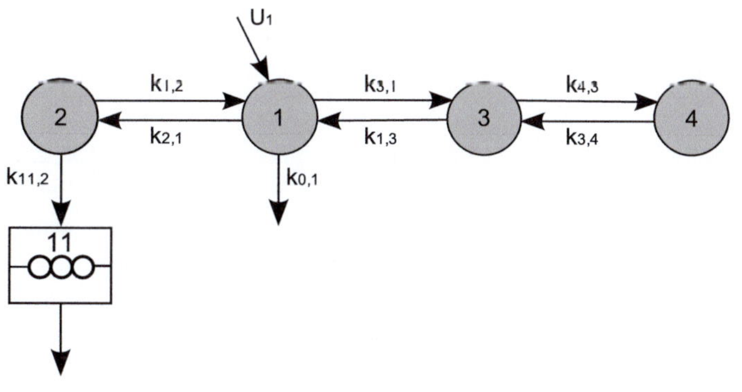

**Fig. 16** Schematic depiction of the restricted model (leucine pool) [11]. This scheme is a sub-scheme of Fig. 15. *Circles* depict compartments. *Arrows* depict fluxes between compartments and are labeled with the corresponding fractional transfer coefficients. Compartment 1 is the plasma-leucine compartment, into which the leucine is injected. Compartment 2 is an intrahepatic compartment and source of the apoB synthesis. Compartments 3 and 4 are body protein pools. The output is from compartment 1. Compartment 11 is a delay compartment, used here only as an output from compartment 2

For each compartment $i$, with $i = 1, \ldots, 4$, $Q_i$ and $q_i$ now refer to the concentrations of tracee and tracer, respectively. Similarly, $U_i$ and $u_i$ refer to the input of tracee and tracer, respectively. For tracer/tracee experiments, a steady-state is generally assumed for the concentration $Q_i$ of the tracee. If the concentration of the labeled injection is small compared with the overall concentration levels, and if the model is linear, than the following holds approximately:

$$\frac{\mathrm{d}q(t)}{\mathrm{d}t} = K(t)q(t) + u(t)$$

with $q(t) = (q_i(t))^{\mathrm{T}}_{i=1,2,3,4}$, $u(t) = (u_1(t), 0, 0, 0)^{\mathrm{T}}$, and

$$K(t) = (k_{j,i})_{j,i=1,2,3,4}.$$

Here, $k_{j,i}$ for $i \neq j$ is the transfer coefficient of the tracer from compartment $i$ to compartment $j$. Compartment 0 is, in general, the output compartment (not shown in the figures). Here, compartment 11 is also considered to be an output compartment. Moreover, for each $i = 1, \ldots, 4$,

$$k_{i,i} := - \sum_{\substack{j=0,1,2,3,4,11 \\ j \neq i}} k_{j,i}.$$

As time unit, we always assume 1 hour (h); all transfer coefficients are given in the unit $\mathrm{h}^{-1}$, and the amount of material in the compartments, in mg. In our model, only $k_{0,1}$, $k_{1,2}$, $k_{1,3}$, $k_{2,1}$, $k_{3,1}$, $k_{3,4}$, $k_{4,3}$, and $k_{11,2}$ are assumed to be non-zero, while the following dependencies between the transfer coefficients are also assumed to be valid:

$$k_{1,2} = k_{2,1},$$
$$k_{3,4} = 0.1 \cdot k_{4,3}.$$

The transfer coefficient $k_{11,2}$ must be fixed and specified in order for the system to be identifiable. We set $k_{11,2} = 0.01\ \mathrm{h}^{-1}$ as an estimated average from earlier measurements. We generate stochastic differential equations (SDE) based on the resulting ordinary differential equations by adding stochastic noise terms. These are given by standard Wiener processes $\mathscr{W}_{1,t}, \ldots, \mathscr{W}_{4,t}$, multiplied by the corresponding diffusion parameters $\sigma_1, \ldots, \sigma_4$. All fluxes that occur within the subsystem should follow the principle of mass conservation. We therefore depart from the usual procedure and add the stochastic terms in the following manner:

$$\mathrm{d}q(t) = K(t)q(t)(\mathrm{d}t + \Sigma \mathrm{d}\mathscr{W}_t) + u(t)\mathrm{d}t$$

with $\Sigma = \mathrm{diag}(\sigma_1, \sigma_2, \sigma_3, \sigma_4)$ and $\mathscr{W}_t = (\mathscr{W}_{1,t}, \ldots, \mathscr{W}_{4,t})^{\mathrm{T}}$. We fix the diffusion parameters thus: $\sigma_1 = \sigma_2 = \sigma_3 = \sigma_4 = 3$. The initial conditions are given by

$$q_2(0) = q_3(0) = q_4(0) = 0.$$

The test subjects receive a bolus injection with labeled leucine, so that we can fix the initial condition

$$q_1(0) = u_{1,0}$$

and simultaneously assume $u_1(t) = 0$ in the differential equation.

We assume identical differential equations, without the stochastic noise terms, however, for the states $Q_i$ and the input $U_1$ of the tracee:

$$\frac{\mathrm{d}Q(t)}{\mathrm{d}t} = K(t)Q(t) + U(t)$$

with $Q(t) = (Q_i(t))_{i=1,2,3,4}^{\mathrm{T}}$, $U(t) = (U_1(t), 0, 0, 0)^{\mathrm{T}}$. Here, the tracee input $U_1(t) = U_1$ is presumed to be constant, but unknown. We therefore want to estimate this value along with the transfer parameters. Because a steady-state is presumed for the tracee (i.e., $\mathrm{d}Q_i(t)/\mathrm{d}t = 0$), we can solve the equations for $Q_1(t)$ and thus obtain:

$$Q_1(t) = \frac{(k_{11,2} + k_{1,2})U_1}{k_{0,1}(k_{11,2} + k_{1,2}) + k_{11,2}k_{1,2}}.$$

Each measurement is given by a value that is proportional to the ratio of tracer and tracee, with additional log-normal disturbances:

$$y_1(t) = p_1 \frac{q_1(t)}{Q_1(t)} \xi_t, \quad \xi_t \sim \text{Log-}\mathscr{N}\left(0, \sigma_{y_1}^2\right) \text{ independently for each } t,$$

where we assume the value of the variance parameter (this denotes the variance of $\log \xi_t$) to be $\sigma_{y_1}^2 = 0.5^2$. The parameter $p_1$ denotes the unknown proportion of plasma-leucine that is actually in the plasma. Since the parameters $p_1$ and $U_1$ are not jointly identifiable, we specify $p_1 = 0.65$ (on the basis of previous knowledge). More details concerning the deterministic model (without stochastic disturbances) can be found in [13] and [11]. Note that the stochastic disturbances are not part of the original model, but are our subsequent enhancements.

## 7.2 The Mixed-Effects Model

The model, as it was presented in the previous paragraphs, contains only flux parameters $k_{j,i}$ that are the same for each individual. In this form, the model does not account for individual differences between the various persons, nor does it consider group-specific differences between the patients and the control group. In the latter case, differences in flux parameters may arise when the persons examined belong in part to a group whose members are affected by a disease or have received a special treatment, while other persons belong to a control group. In order to account for these differences, we now introduce group-specific and patient-specific parameters into the model. Specifically, we split the transfer coefficients $k_{0,1}$ into a group-dependent and a patient-dependent part. In this fashion, we introduce so-called mixed effects into the model. Mixed effects generally make it more difficult to perform the estimates, since they not only increase the number of parameters to be estimated, but also result in a hierarchical ranking among the parameters. For the following estimates, we use measurement data collected in a study involving 34 persons—data that was already used in another context [12, 13]. From these 34 persons, 15 belong to the group of diabetes patients, while the other 19 belong to the control group. From earlier experiments, one sees that the degradation rate $k_{0,1}$ of the plasma-leucine differs significantly for persons with and without diabetes. We therefore assume that the expected value of $k_{0,1}$ is different in each group, that is, has either a value $k_{0,1}^d$ or a value $k_{0,1}^c$, depending on whether the person belongs to the diabetes or the control group. Moreover, we also assume patient-dependent random factors $\zeta_p$ that reflect the parameter uncertainties among the individuals. In the end, we obtain

$$k_{0,1}^{(p)} = \begin{cases} \zeta_p k_{0,1}^{\mathrm{d}} & \text{if patient } p \text{ belongs to the diabetes group,} \\ \zeta_p k_{0,1}^{\mathrm{c}} & \text{if patient } p \text{ belongs to the control group,} \end{cases}$$

where all $\zeta_p$ are assumed to be static and independently log-normally distributed:

$$\zeta_p = \exp(\eta_p) \quad \text{with } \eta_p \sim \mathcal{N}\left(0, \sigma_{\eta_p}^2\right) \text{ independently for all } p$$

for $p = 1, \ldots, 34$. Consequently, each state $q_1, \ldots, q_4$ must be considered separately for each patient $p$. We indicate this in the notation by means of indices $q_1^{(p)}, \ldots, q_4^{(p)}$, $p = 1, \ldots, 34$.

The goal of our investigations, apart from estimating the remaining parameters, is thus to show that the group-dependent parameters $k_{0,1}^{\mathrm{d}}$, $k_{0,1}^{\mathrm{c}}$ are indeed different. To do so, we use the Bayesian approach for parameter estimation. For this reason, we treat the parameters essentially like state variables in the particle filter. Because the estimation of constant parameters is problematic with particle filter methods, it is standard to introduce a small, artificial stochastic dynamic so that the parameters can change slightly over time. This is done by allowing normally or log-normally distributed increments with decaying variances for the parameters in each time-step [36]. We also introduce corresponding dynamics for the static individual parameters $\eta_p$, which must also be estimated. Our process $X_t$ is therefore given as an augmented state vector

$$X_t = \left(q_{1:4}^{(1:34)}(t), k_{0,1}^{\mathrm{c}}(t), k_{0,1}^{\mathrm{d}}(t), k_{1,2}(t), k_{1,3}(t), k_{3,1}(t), k_{4,3}(t), U_1(t), \eta_{1:34}(t)\right)^{\mathrm{T}}.$$

The complete model is thus a nonlinear mixed-effects model with three levels of effects (parameters), namely, global parameters, group-dependent parameters $k_{0,1}^{\mathrm{d}}$, $k_{0,1}^{\mathrm{c}}$, and individual parameters $\zeta_p$.

## 7.3 Estimation Results

In this section, we compare the results of parameter estimations with the MTU particle filter and the standard particle filter. We performed estimations and subsequent test runs with the estimated parameters, using the data from all 34 patients (19 in the control group and 15 in the diabetes group). Specifically, we carried out the following computer experiments: In the first phase, we estimated the parameters with the MTU particle filter; for comparison purposes, we also separately estimated the parameters with the standard particle filter—under the same conditions and with the same seed value for the random number generator. The initial distribution of the particles was thus the same in both cases. For each run, we also calculated estimators for the effective sample size (ESS) and the data likelihood over time. These estimators allow a performance comparison for the MTU and the standard particle filters. In a second phase, we then used the empirical medians of the final parameter

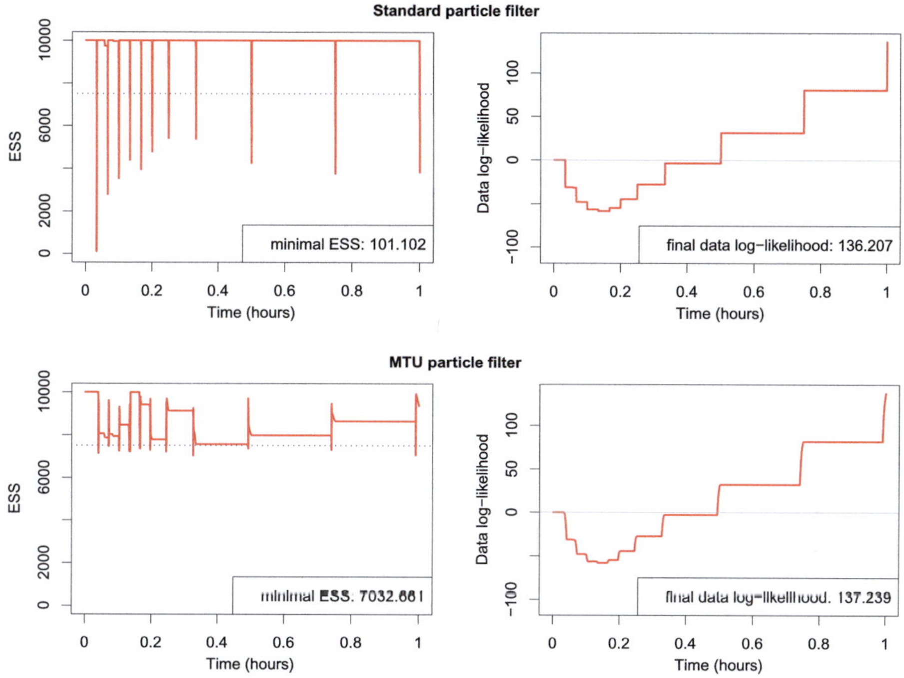


**Fig. 17** Development of the Effective Sample Size (ESS) and the data likelihood over time during the parameter estimation. Standard particle filter (*top*) and MTU particle filter (*bottom*)

distributions in test runs, separately for each case. Both versions of the particle filter were used in these runs for state filtering and calculating the data likelihood—this time with fixed parameters given by the estimated values. In this manner, the resulting simulated distributions of the measurement values can be compared with the actual measurement values, both visually and quantitatively, by examining the data likelihood.

We performed our calculations with 10 000 particles and a resampling threshold of 7500. The step-size in the MTU filter was between $10^{-7}$ h and $10^{-3}$ h, adaptively calculated on the basis of the ESS estimate. In the standard filter, we used a fixed step-size of $10^{-3}$ h. Although the data contains measurements up to time $t = 8$ h, we only used values up to time $t = 1$ h for our estimates and test runs, mainly to save computing time. In any event, after time $t = 1$ h, the tracer concentrations are quite small and almost static, so it is not to be expected that including the later data would alter the estimates significantly. In our implementation of the particle filters, we sample directly from the (augmented) states $X_t$ (i.e., $\tilde{X}_{[t_0,\infty)} = X_{[t_0,\infty)}$ in distribution), with the aid of the Euler–Maruyama method for discretizing the SDE over time [37].

Figure 17 shows the development over time $t$ of the estimated value of the effective sample size ESS and the estimated data likelihood. Figures 18 and 19 are box plots showing the posterior distributions of the global/group parameters and the individual parameters, respectively, in the final time-step of the estimation.

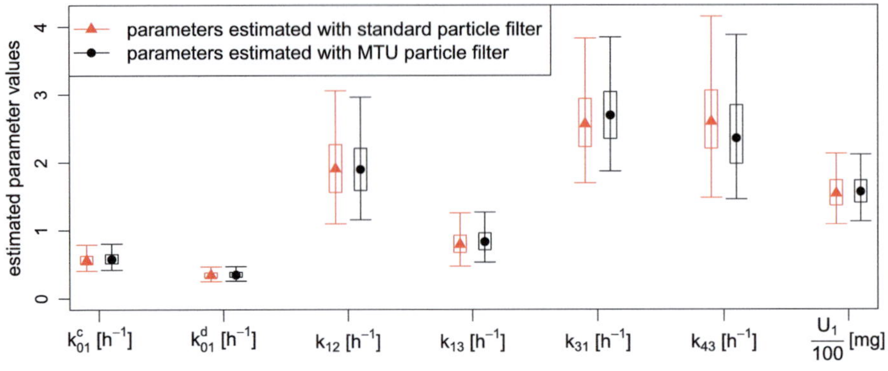


**Fig. 18** Estimated global and group parameters. Box plots of the estimated posterior distributions for the global parameters and the group-dependent parameters. The medians are depicted with *triangles* (standard particle filter) or *circles* (MTU particle filter). The bottom and top of the box are the 0.25-quantile and the 0.75-quantile; i.e., 50 % of the values lie within the box. The whiskers mark the 0.025-quantile and the 0.975-quantile; i.e., 95 % of the values lie between the whiskers. The values for $U_1$ have been scaled by a factor of 0.01

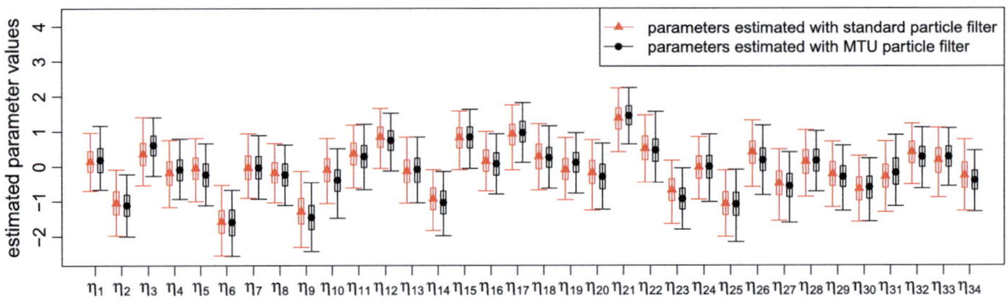


**Fig. 19** Estimated individual parameters. Box plots of the estimated posterior distributions for the individual parameters. The medians are depicted with *triangles* (standard particle filter) or *circles* (MTU particle filter). The *bottom* and *top* of the box are the 0.25-quantile and the 0.75-quantile; i.e., 50 % of the values lie within the box. The whiskers mark the 0.025-quantile and the 0.975-quantile; i.e., 95 % of the values lie between the whiskers

A comparison of the results of the MTU-PF and the standard particle filter shows that both algorithms deliver very similar performance with regard to the quality of the estimated parameters; in each case, the development of the data likelihood is very similar, both during the estimation and the test run. The estimated log-likelihood of the data in the final estimation step is 137.239 for the MTU particle filter and 136.207 for the standard case; in other words, for all practical purposes, they are equal. The computation time for the MTU-PF is only slightly longer than that of the standard filter. A visual inspection of the test runs shows that the predicted distribution of the measurement values based on parameters estimated by both filters fits the data equally well. This impression is reinforced by the values of the estimated data likelihood. In the final step, the MTU particle filter delivers a log-likelihood value of 157.622, which is very similar to the 155.952 value delivered by the standard filter. The difference is insignificant; the uncertainty in the measurement time-points thus appears not to lead to differences in the actual estimation results, at least in this case.

In contrast to the insignificant differences in the likelihoods between the MTU-PF and the standard PF, the development of the ESS estimate in the estimation runs differs remarkably. The runs with the MTU particle filter deliver an ESS estimate with very high values throughout the estimation, and a minimum of 7032.661. This value lies just slightly under the resampling bound of 7500 (see Fig. 17, top). In contrast, the standard particle filter shows a substantially worse performance. One can see from the bottom of Fig. 17 that the ESS drops repeatedly to very low values, with a minimum of 101.102. Here, the MTU particle filter avoids degeneration of the particle cloud by holding the ESS at a high value at all time-points. This indicates that these results have been obtained on a sound basis and may be considered more reliable than those delivered by the standard algorithm.

A glance at the estimated values of the group parameters $k_{0,1}^c$ and $k_{0,1}^d$ (see Fig. 18) shows that, in both estimation cases (standard PF and MTU-PF), the rate $k_{0,1}^d$ for diabetes patients is only about 60 % of the rate $k_{0,1}^c$ for the control group (standard PF: 0.337 $h^{-1}$ vs. 0.557 $h^{-1}$; MTU-PF: 0.346 $h^{-1}$ vs. 0.577 $h^{-1}$). The good performance of the MTU particle filter, in particular, strengthens one's confidence in the results obtained and leads to the conclusion that the secretion rate $k_{0,1}$ is, in fact, lower for the group of diabetes patients than for the control group.

## References

### Publications of the Authors


1. Krengel, A., Hauth, J., Taskinen, M.R., Adiels, M., Jirstrand, M.: A continuous-time adaptive particle filter for estimations under measurement time uncertainties with an application to a plasma-leucine mixed effects model. BMC Syst. Biol. **7**, 8 (2013). doi:10.1186/1752-0509-7-8
2. Lang, P., Prätzel-Wolters, D., Kulig, S.: Modellreduktion und dynamische Beobachter für Torsionsschwingungen in Turbosätzen. In: Hoffmann, K.H., Jäger, W., Lohmann, T., Schunk, H. (eds.) Mathematik-Schlüsseltechnologie für die Zukunft, pp. 491–501. Springer, Berlin (1997)
3. Stahl, D., Hauth, J.: PF-MPC: Particle Filter-Model Predictive Control. Syst. Control Lett. **60**(8), 632–643 (2011)
4. Wirsen, A.: Monitoring von Torsionsschwingungen in Kraftwerksturbosätzen. In: Turbogeneratoren in Kraftwerken, Technik–Instandhaltung–Schäden. Haus der Technik, Essen (2011)
5. Wirsen, A., Humer, M.: Online Monitoring von Torsionsschwingungen in Wellensträngen von Kraftwerksturbosätzen. In: Symposium Schwingungsdiagnose–Schwingungsdiagnostische Überwachung von Kraftwerksturbosätzen–Methoden, Nutzen, Erfahrung. Potsdam Sanscoussi, Germany (2006)
6. Wirsen, A., Lang, P., Humer, M.: Systems for monitoring and analysing torsional vibrations in turbine generator shaft lines. In: Conference Proceedings of the 16th International Conference on Electrical Machines, Krakau, Polen (2004)

7. Wirsen, A., Mohring, J.: Methods for $H_2$ optimal actuator placement and controller design based on high dimensional parametric models of mechanical structures. In: Conference Proceedings IV European Conference on Computational Mechanics (ECCM) (2010)

## Dissertations on the Topic at the Fraunhofer ITWM

8. Krengel, A.: A Modified Particle Filter with Adaptive Stepsize for Continuous-Time Models with Measurement Time Uncertainties (2013). Verlag Dr. Hut
9. Lang, P.: Model Reduction, Sensor Placement and Robust $H_\infty$-Filter Design for Elastomechanical Systems (1998). Shaker Verlag
10. Wirsen, A.: Sensitivitätsanalyse und modaldatenbasierte Modelladaption bei elastomechanischen Systemen. dissertation.de (2002)

## Further Literature

11. Adiels, M.: A compartmental model for kinetics of apolipoprotein B-100 and triglycerides in $VLDL_1$ and $VLDL_2$ in normolipidemic subjects. Licentiate thesis, Chalmers University of Technology, Göteborg (2002)
12. Adiels, M., Borén, J., Caslake, M.J.J., Stewart, P., Soro, A., Westerbacka, J., Wennberg, B., Olofsson, S.O.O., Packard, C., Taskinen, M.R.R.: Overproduction of VLDL1 driven by hyperglycemia is a dominant feature of diabetic dyslipidemia. Arterioscler. Thromb. Vasc. Biol. **25**(8), 1697–1703 (2005). doi:10.1161/01.ATV.0000172689.53992.25
13. Adiels, M., Packard, C., Caslake, M.J., Stewart, P., Soro, A., Westerbacka, J., Wennberg, B., Olofsson, S.O., Taskinen, M.R., Borén, J.: A new combined multicompartmental model for apolipoprotein B-100 and triglyceride metabolism in VLDL subfractions. J. Lipid Res. **46**, 58–67 (2005)
14. Andersen, K.E., Hojbjerre, M.: A population-based Bayesian approach to the minimal model of glucose and insulin homeostasis. Stat. Med. **24**(15), 2381–2400 (2005)
15. Andrieu, C., De Freitas, N., Doucet, A.: Sequential MCMC for Bayesian model selection. In: Proceedings of the IEEE Signal Processing Workshop on Higher-Order Statistics, pp. 130–134. IEEE, Caesarea (1999). doi:10.1109/HOST.1999.778709
16. Andrieu, C., Doucet, A., Holenstein, R.: Particle Markov chain Monte Carlo methods. J. R. Stat. Soc., Ser. B, Stat. Methodol. **72**(3), 269–342 (2010)
17. Ashyraliyev, M., Fomekong-Nanfack, Y., Kaandorp, J.A., Blom, J.G.: Systems biology: parameter estimation for biochemical models. FEBS J. **276**(4), 886–902 (2009). doi:10.1111/j.1742-4658.2008.06844.x
18. Beal, S., Sheiner, L.: NONMEM User's Guides. NONMEM Project Group. University of California, San Francisco (1994)
19. Berglund, M., Sunnåker, M., Adiels, M., Jirstrand, M., Wennberg, B.: Investigations of a compartmental model for leucine kinetics using non-linear mixed effects models with ordinary and stochastic differential equations. Math. Med. Biol. **29**(4), 361–384 (2011)
20. Bohlin, T.: Practical Grey-Box Process Identification: Theory and Applications. Advances in Industrial Control. Springer, London (2010)
21. Cappé, O., Godsill, S., Moulines, E.: An overview of existing methods and recent advances in sequential Monte Carlo. Proc. IEEE **95**(5), 899–924 (2007)
22. Chou, I.C.C., Voit, E.O.: Recent developments in parameter estimation and structure identification of biochemical and genomic systems. Math. Biosci. **219**(2), 57–83 (2009). doi:10.1016/j.mbs.2009.03.002

23. Cobelli, C., Saccomani, M.P., Tessari, P., Biolo, G., Luzi, L., Matthews, D.E.: Compartmental model of leucine kinetics in humans. Am. J. Physiol. **261**(4 Pt 1), E539–50 (1991)
24. Crisan, D., Doucet, A.: A survey of convergence results on particle filtering methods for practitioners. IEEE Trans. Signal Process. **50**(3), 736–746 (2002)
25. Del Moral, P., Doucet, A., Jasra, A.: Sequential Monte Carlo samplers. J. R. Stat. Soc., Ser. B, Stat. Methodol. **68**(3), 411–436 (2006)
26. Demant, T., Packard, C.J., Demmelmair, H., Stewart, P., Bedynek, A., Bedford, D., Seidel, D., Shepherd, J.: Sensitive methods to study human apolipoprotein B metabolism using stable isotope-labeled amino acids. Am. J. Physiol. **270**(6 Pt 1), E1022–36 (1996)
27. Donnet, S., Samson, A.: EM algorithm coupled with particle filter for maximum likelihood parameter estimation of stochastic differential mixed-effects models. http://hal.archives-ouvertes.fr/hal-00519576. Preprint version 2–21 Jul. 2011
28. Donnet, S., Samson, A.: Parametric inference for mixed models defined by stochastic differential equations. ESAIM Probab. Stat. **12**, 196–218 (2008)
29. Douc, R., Cappé, O., Moulines, E.: Comparison of resampling schemes for particle filtering. In: Proceedings of the 4th International Symposium on Image and Signal Processing and Analysis, 2005 ISPA, pp. 64–69. IEEE, Zagreb (2005)
30. Doucet, A., de Freitas, N., Gordon, N. (eds.): Sequential Monte Carlo Methods in Practice. Statistics for Engineering and Information Science. Springer, New York (2001)
31. Fearnhead, P.: MCMC, sufficient statistics and particle filters. J. Comput. Graph. Stat. **11**(4), 848–862 (2002)
32. Gordon, N., Salmond, D., Smith, A.: Novel approach to nonlinear/non-Gaussian Bayesian state estimation. IEE-Proc.-F **140**(2), 107–113 (1993)
33. Hol, J.D., Schön, T.B., Gustafsson, F.: On resampling algorithms for particle filters. In: Proceedings of Nonlinear Statistical Signal Processing Workshop (NSSPW), pp. 79–82. IEEE, Cambridge (2006)
34. Hu, X.L., Schön, T., Ljung, L.: A basic convergence result for particle filtering. IEEE Trans. Signal Process. **56**(4), 1337–1348 (2008)
35. Humer, M.: Erfassung und Bewertung von Torsionsschwingungen in Wellensträngen von Kraftwerksturbosätzen. Ph.D. thesis, TU Dortumund (2004)
36. Hürzeler, M., Künsch, H.R.: Approximating and maximizing the likelihood for a general state-space model. In: Sequential Monte Carlo Methods in Practice. Springer, New York (2001)
37. Kloeden, P.E., Platen, E.: Numerical Solution of Stochastic Differential Equations. Springer, Berlin (1999)
38. Moles, C.G., Mendes, P., Banga, J.R.: Parameter estimation in biochemical pathways: a comparison of global optimization methods. Genome Res. **13**(11), 2467–2474 (2003). doi:10.1101/gr.1262503
39. Neubauer, M.: Aktive Bedämpfung von Drehschwingungen im Fahrzeugantriebstrang. Ph.D. thesis, TU, Braunschweig (2011)
40. Nielsen, J., Madsen, H., Young, P.: Parameter estimation in stochastic differential equations: an overview. Annu. Rev. Control **24**, 83–94 (2000)
41. Overgaard, R.V., Jonsson, N., Tornøe, C.W., Madsen, H.: Non-linear mixed-effects models with stochastic differential equations: implementation of an estimation algorithm. J. Pharmacokinet. Pharmacodyn. **32**(1), 85–107 (2005)
42. Pitt, M.K., Shephard, N.: Filtering via simulation: auxiliary particle filter. J. Am. Stat. Assoc. **94**, 590–599 (1999)
43. Racine-Poon, A., Wakefield, J.: Statistical methods for population pharmacokinetic modelling. Stat. Methods Med. Res. **7**(1), 63–84 (1998)
44. Sheiner, L., Wakefield, J.: Population modelling in drug development. Stat. Methods Med. Res. **8**(3), 183 (1999)

45. Storvik, G.: Particle filters for state-space models with the presence of unknown static parameters. IEEE Trans. Signal Process. **50**(2), 281–289 (2002)
46. Tornøe, C.W., Overgaard, R.V., Agersø, H., Nielsen, H.A., Madsen, H., Jonsson, E.N.: Stochastic differential equations in nonmem: implementation, application, and comparison with ordinary differential equations. Pharm. Res. **22**(8), 1247–1258 (2005)
47. Ungarala, S., Dolence, E., Li, K.: Constrained extended Kalman filter for nonlinear state estimation. In: 8th International IFAC Symposium on Dynamics and Control of Process Systems, vol. 2. Cancun, Mexico (2007)
48. Zhou, K., Doyle, J.C., Glover, K.: Robust and Optimal Control. Prentice Hall, New York (1996)

# Option Pricing in Practice—Heston's Stochastic Volatility Model

Sascha Desmettre, Ralf Korn, and Tilman Sayer

## 1 The Finance Industry as Employer, Supplier of Mathematical Challenges, and Risk Factor

Over the past 50 years, the growth in complexity of the products offered in the financial markets has opened up a completely new sub-domain in mathematics—(modern) financial mathematics—which, in terms of its mathematical sophistication goes far beyond what used to be understood as financial mathematics, which would be better referred to as business accounting. The significant role of mathematics in the financial world can be demonstrated emphatically in a number of ways. The finance and insurance industries are today, more than ever, the most important employers of graduates from mathematics curricula, although thorough training in applied mathematics (especially stochastics, statistics, optimization, and numerics) is an advantage. In the financial field, mathematicians are sometimes employed as traders, as so-called *quants* (mathematics experts who implement quantitative methods and models in the investment banking sector), or as external staff for large consulting firms, who offer their competences to banks and insurance companies. The insurance branch even has its own professional designation—*actuary*—awarded by the Deutsche Aktuarvereinigung DAV after completion of an intensive training and examination.

The new mathematical challenges can be illustrated by means of four significant domains in financial mathematics:

S. Desmettre · R. Korn (✉) · T. Sayer
Abteilung Finanzmathematik, Fraunhofer ITWM, Fraunhofer Platz 1, 67663 Kaiserslautern, Germany
e-mail: ralf.korn@itwm.fraunhofer.de



H. Neunzert, D. Prätzel-Wolters (eds.), *Currents in Industrial Mathematics*,
DOI 10.1007/978-3-662-48258-2_10

- *Modeling*—This is the simulation of price movements of all kinds (stock prices, interest rates, currency exchange rates, etc.). Here, stochastic processes are applied, based essentially on so-called *Itô processes*, developed in the middle of the last century. As functions of time, they are non-differentiable and require their own calculus, the *Itô calculus* (see [33] or [4] for an introduction).
- *Portfolio optimization*—The goal here is to determine an optimal trading strategy for an investor. Today, this represents one of the main fields of dynamic optimization and stochastic control (see [1] or [3]).
- *Risk management*—This deals with measuring and managing the risks associated with unforeseeable developments in investments. In the past years, the theory of risk measures has opened up a new field of theoretical research (see [15], the standard reference).
- *Option pricing*—This is the showpiece of financial mathematics. It deals with determining the prices of option contracts and has led to numerous new theoretical problems and results in such fields as martingale theory and the numerics of stochastic processes (see [4, 34], and [5]). In the following discussion, we will consider this field in greater depth.

Finally, due to the financial industry's intrinsic uncertainties, it represents one of the largest risk factors in modern economies, possibly *the* largest. The role of mathematics in this regard can indeed be viewed with some ambivalence. On the one hand, the methods of mathematical modeling and financial mathematics provide significant support in recognizing, understanding, and managing risks. On the other hand, the technology of mathematics can tempt its users into creating new, ever more complicated products, in the belief that mathematical methods can make their characteristics and risks understandable and controllable. Here, it is essential for mathematicians to emphasize the limits of modeling and, in particular, to point out the enormous dependency of the models on their input parameters—a risk that is often largely ignored.

Moreover, it is important to make clear the difference between prediction and simulation. Whereas stock price simulations can indeed be used to calculate risk measures or options prices by means of Monte Carlo methods, these simulations often have about as much utility in predicting actual stock prices as the choice of a number $2, \ldots, 12$ has in predicting the toss of a pair of dice. The financial crisis of 2007–2012 should be taken as a strong warning to sharpen the awareness of the limits of mathematical modeling and simulation. Conversely, however, one should also be aware that mathematical modeling was not the cause of the financial crisis. This fig leaf—so happily put to use by investment bankers—is not large enough to cover up the actual causes, such as egregious misjudgments of creditworthiness, the taking of tremendously risky market positions, and, in many cases, plain ignorance in acquiring products that were not understood.

In the course of this chapter, we will concentrate on the field of option pricing, which is a central concern of both modern investment banking and financial mathematics theory, but which is treated quite differently in these two fields. We want to demonstrate that the popular Black–Scholes model is no longer adequate for many practical purposes and introduce instead the Heston model (see [29]), an alternative often used in practice that represents a compromise between practicability and theoretical generality. In so doing, we will encounter mathematical challenges in the areas of modeling, theoretical stochastics, and the numerical computation of options prices.

Along the way, we will first introduce the reader to the terminology of the world of options, describe project collaborations between the financial industry and the Financial Mathematics Department of the ITWM, and then delve into the theoretical foundations of stock price modeling and options pricing.

## 2 Options as Modern Ingredients of the Financial Markets

The terms *option* and *derivative* have developed negative connotations in the aftermath of the great financial crisis of 2007–2012. This is not entirely unjustified, given that many credit derivatives, in both their form and ultimate impact, were so complicated and opaque—not only for lay persons—that their trading contributed significantly to the outbreak of the crisis. Ironically, the designation *exotic option*, applied to many of these securities, seems very aptly chosen, given their mysterious and arcane nature.

In the following section, we first present the basic features of options and option trading by using the simplest options as examples. We then take a look at more complicated types, in order to illustrate the necessity of mathematical modeling.

### 2.1 Simple Options—Call and Put

The word *option*, in its colloquial sense, stands for an opportunity that one is not compelled to take, but which one may indeed take, if so desired. Having such an opportunity is always a good thing. Options contracts on the financial market, which represent a similar opportunity, are therefore to be had only at a price.

*Option contracts* are securities derived from underlying assets, which explains why they are also known as *derivatives*. These securities have been traded for centuries in one form or another, but they only achieved great economic significance at the beginning of the 1970s. They are used to secure market positions and design special payoff profiles, as well as for purposes of pure speculation. As the term itself implies, possession of an option includes a right of choice that the owner can, but does not have to, exercise.

With the simplest option, the *European call* on a stock, the buyer has the right (but not the obligation!) to acquire from the seller a share of a given company at time $T$ (the *maturity*) at a fixed price $K$ (the *exercise price* or *strike*). He will only do so if it is to his advantage, that is, if the stock price $S(T)$ lies above the strike $K$, since he could otherwise obtain the same share more cheaply by acquiring it directly in the market. Consequently, possession of a European call is equivalent to the payment

$$Y_{\text{call}} = \big(S(T) - K\big)^+ = \max\big(S(T) - K, 0\big)$$

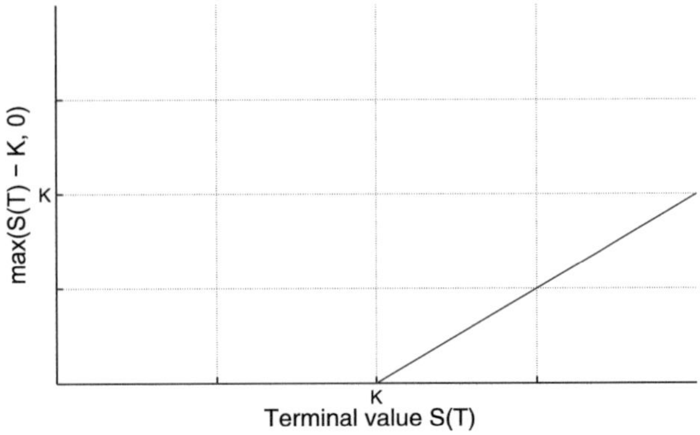


**Fig. 1** Final payoff of a European call with strike price $K$

at time $T$ and is often depicted graphically by means of its payment profile, as in Fig. 1. Because the amount of the final payment is non-negative and may be positive, the possession of the option today must have a positive value. Determining this value is the object of option pricing.

The direct counterpart to the European call is the *European put*, which is given by the final payment

$$Y_{\text{put}} = \left(K - S(T)\right)^{+}$$

at time $T$, and which gives the owner the right to sell a share at price $K$ at time $T$ to the seller of the European put. Here, the owner will only exercise his right if the current market price of the stock is lower than $K$. Such simple options as the European call and put are often referred to in the market as *vanilla options*.

L.F. Bachelier's dissertation *Théorie de la Spéculation* (see [16]), from 1900, may be regarded as the long-forgotten starting point for option pricing. Bachelier's idea was to use asset price modeling to derive theoretical values for different types of options on particular assets and to compare these values with the actual market prices. As the option price, he proposed using the expected value of the future payment arising from the option. In so doing, he used implicitly, for the first time, the so-called Brownian motion (with drift) as an asset price model (see Sect. 4.1)—albeit, without designating it as such.

One consequence of this modeling, however, was that the prices in the model could fall below zero. His ideas were taken up again in the 1960s with the introduction of geometric Brownian motion (see again Sect. 4.1) as a price model. In 1973, Fischer Black and Myron Scholes achieved the first crucial breakthrough in the field of option pricing with the derivation of explicit price formulas for European calls and puts (see [20] and Sect. 4.3).

## 2.2 Exotic Options—The Next Stage

For the simple options described above, the payment resulting from ownership depends only on the stock price $S(T)$ at the maturity of the option. This need not be the case, however. The general form of European options is given by a final payment of the form

$$Y = f\big(S(t), t \in [0, T]\big),$$

where $f$ (and thus also $Y$) is a function describing the entire price development of the stock over the time interval $[0, T]$. Here, we identify time $t = 0$ as the present moment and $S_0 > 0$ as the initial stock price. One then speaks of a *path-dependent option*. For the European call and put, this dependency was given simply by the stock price at $t = T$. Examples of European options with genuine path dependency are:

- the *lookback* or *maximum option*, for which the maximum of the stock price appears in the final payment as follows:

$$Y_{\text{lb}} = \Big(\max_{t\in[0,T]} S(t) - K\Big)^{+}.$$

- the *Asian option*, which is given by a final payment in the form

$$Y_{\text{ao}} = \left(\frac{1}{T}\int_0^T S(t)\text{d}t - K\right)^{+}$$

  for example, and similar variants (e.g., as discrete mean).
- or the *barrier option* class, for which the stock price in the interval $[0, T]$ may not exceed or must exceed (depending on the variant) one or more specified barriers $H_i$, so that at $T$, a positive payment flows to the option's owner. One example is the *double barrier knock out call*, with strike $K$, barriers $0 \leq H_1 < H_2$, and final payment

$$Y_{\text{dbkoc}} = 1_{\{H_1 < S(t) < H_2 \forall t\in[0,T]\}}\big(S(T) - K\big)^{+}.$$

  Here, as seen in Fig. 2, there may indeed be price paths that end above the strike $K$, but violate a barrier condition beforehand and thus do not lead to a final payoff.

The pricing of such options that are strongly dependent on the stock price path—options that are often grouped together under the title of exotic options—requires a specialized and highly efficient numerical method and also demonstrates empirically the need for more sophisticated stock price models than the geometric Brownian motion (cf. here also Sect. 4.6).

*Remark 1* It is certainly justified to ask why such complicated options are traded at all. The answer is multi-faceted.

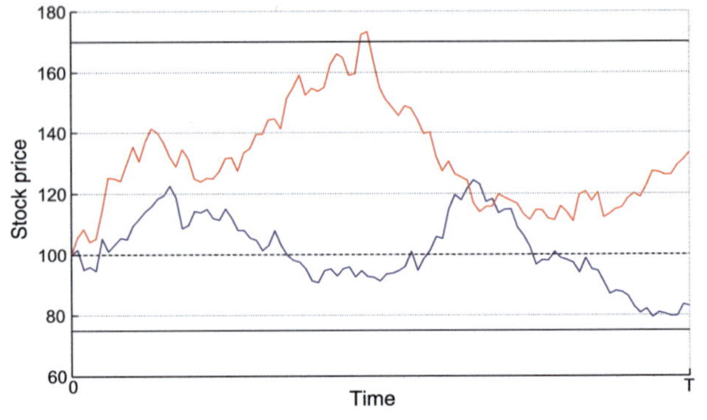


**Fig. 2** Two possible stock price paths and barriers $H_1 = 75$ and $H_2 = 170$, as well as strike $K = 100$ for a double barrier knock out call

A barrier option must obviously be cheaper than its counterpart without barriers since, in order to receive the same payment at the maturity of the option, the barrier conditions must be fulfilled, which is not always the case. If they are not fulfilled, the buyer of the barrier option receives nothing, whereas the buyer of the common European variants receives the full payment. Despite this risk, barrier options are happily bought in preference to the common variants, due to their lower price.

In the case of the maximum option, the option's owner wants to secure the largest possible difference between the option price and the execution price—an advantage for which he must then also be ready to pay a higher price than for the simple European call.

Finally, the Asian option represents a kind of insurance against short-term market manipulation. It is quite conceivable that large market players might use their trading power at the option's maturity to drive the price of the option's associated stock in a direction that works to their own advantage, but this tactic is not possible over a long time period. Therefore, Asian options typically use an average of stock prices over specified time points to determine the size of the option payment.

## 2.3 American Options and More—Free Choice

Another natural variant is the so-called *American option*, in which the option's holder can decide at what time-point $t \in [0, T]$ he exercises his option right. Here, the option can be identified by a whole family $Y(t)$, $t \in [0, T]$ of possible final payments, from which the option's owner can choose one. For each of the above-mentioned European option types, there is also a corresponding American variant, which, for a call, for example, is given by the following family of possible final payments:

$$Y_{\text{call}}(t) = \big(S(t) - K\big)^+, \quad t \in [0, T].$$

Generally speaking, there is a tremendous variety of options available in the financial marketplace. Along with options on stocks, one finds options on bonds, commodities, loans, options, contracts, foreign currencies, electricity, and virtually any asset that is traded. Each of these option classes often generates its own mathematical problems for price calculation, not least because the underlying assets exhibit completely different characteristics. For example, electricity is, in general, non-storable, whereas commodities generate inventory costs but some may offer compensating strategic advantages, etc.

For those interested in more background on options trading, including historical background, we recommend [4] or [2].

### 2.4 How Much Do Options Cost?

This fundamental question regarding option pricing cannot be answered without two significant ingredients:

- a mathematical model of the price development of the asset underlying each option and
- the insight that, since an option is a derivative, its price must always depend on the current market price of the underlying asset.

Both ingredients are considered in detail in Sect. 4 and lead to a surprising result—one that was honored with the Nobel Prize for Economics.

## 3 Options at the ITWM

Because option pricing is one of the central domains of financial mathematics and so important for trading in modern financial markets, it also plays a central role in the work of the Financial Mathematics Department at the Fraunhofer ITWM. Our non-disclosure agreements with customers in the finance and insurance industries require us to remain somewhat vague in our descriptions of the associated projects, but we do want to give our readers an impression of what sort of work a mathematics research institute can undertake in this field.

The essential components underlying all the projects in the field of option pricing are:

- the development of new, and appropriate modification of existing, dynamic stock price models, toward the end of achieving realistic modeling (do the price movements in the model have the same characteristics as the empirically observed ones?) and numerical tractability (can parameters be stably and efficiently calculated?),
- the derivation of explicit analytical pricing formulas for special, exotic options,
- the development of numerical algorithms for pricing exotic options without explicit price formulas, for which new Monte Carlo methods, tree methods, and solution methods for differential equations are put to use,
- and the implementation of the developed algorithms in modern software for direct application by the trader or for risk management.

The years 2000 to 2013 witnessed the successful completion of many industrial projects. Among our clients were banks, such as the Hypovereinsbank and the Landesbank Baden-Württemberg; insurers, such as the R+V Versicherung; and financial services providers, such as Assenagon Asset Management S.A. The projects varied greatly in size and ranged from pricing a single options class to preparing complete software libraries for pricing exotic options. The latter was of a scale to incorporate all the above-mentioned research and development aspects and draw upon the complete spectrum of applied financial mathematics; it also led to publications in top-flight journals (see e.g., [6, 8, 26, 35], or [9]).

In order to remain continuously abreast of scientific developments in the field of option pricing, a multitude of PhDs were completed during this same time period relating to its various aspects. Often, algorithmic aspects were paramount, in order to make a satisfactory model fit for service in the first place. Consequently, in [11], the numerical pricing of so-called barrier options was investigated; in [12], new tree methods for pricing exotic options in the field of interest rates were developed; in [13], tree methods for option pricing in the Heston model were derived (see Sect. 6 also); and in [10], Monte Carlo methods for special multi-asset barrier options were examined.

## 4 The Foundations of Stock Price Modeling and Option Pricing

Modeling the development of the basic processes underlying the financial markets represents the foundation of financial mathematics. Depending on the market segment, these might be stock prices, indices (such as the DAX or Dow Jones), interest rates, exchange rates, or other indicators. In this section, we initially restrict ourselves to modeling only stock prices as stochastic processes and assume all other influencing quantities to be constant. For the technical bases, we refer the reader to [4].

### 4.1 Modeling Stock Prices

We wish to examine a financial market in which trading takes place in continuous time, any desired division of shares is permissible, and ancillary costs (broker and transaction fees, etc.) do not exist. As basis investment opportunities, we take the investment in a fixed-term deposit account and in (initially) one stock (or stock index).

We assume that a fixed-term deposit $B(t)$ accrues interest continuously at a constant rate $r$, which leads to the temporal development

$$B(t) = B_0 e^{rt} \tag{1}$$

for an initial deposit of $B_0$ at initial time-point $t = 0$.

The significant ingredient for modeling the stock price is the selection of the stochastic process. Motivated by the central limit theorem, according to which the (centered and standardized) sum of many independent, identically-distributed random variables is asymptot-

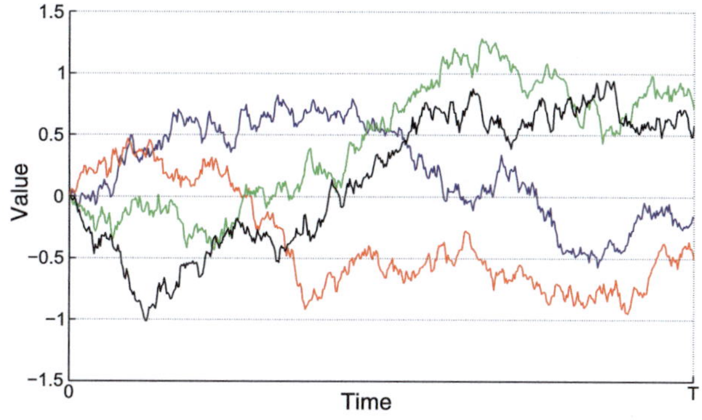


**Fig. 3** Four simulated paths of a Brownian motion with $n = 500$ and $T = 1$

ically normally distributed, we select a Brownian motion as the random driver of the stock price.

**Definition 1** Let $(\Omega, \mathcal{F}, P)$ be a complete probability space. A Brownian motion $\{(W(t), \mathcal{F}_t), t \in [0, \infty)\}$ is a real-valued stochastic process with $W(0) = W_0 = 0$ and continuous paths with stationary and independent increments, that is

$$W(t) - W(s) \sim W(t-s) \quad \forall t > s \geq 0,$$

$$W(t) - W(s) \quad \text{is independent of } \mathcal{F}_u \text{ for } t \geq s \geq u \geq 0.$$

Here, $\{\mathcal{F}_t, t \in [0, \infty)\}$ is a right-continuous filtration for which $\mathcal{F}_0$ already contains all $P$-null sets. The filtration is called natural filtration if it is the filtration generated by the Brownian motion.

*Remark 2* It can be shown that the requirements placed on the Brownian motion in Definition 1 already yield $W(t) \sim \mathcal{N}(0, t)$. From this, an algorithm follows directly to (approximately) simulate a Brownian motion. To this end, for $n \in \mathbf{N}$ and $T > 0$, let $0 = t_0 < t_1 < \cdots < t_n = T$ define a separation of the interval $[0, T]$. We then proceed as follows:

1. Set $W(0) = 0$.
2. Generate $n$ independent $\mathcal{N}(0, 1)$-distributed random numbers $Z_1, \ldots, Z_n$.
3. For $i = 1, \ldots, n$, set

$$W(t_i) = W(t_{i-1}) + \sqrt{t_i - t_{i-1}} Z_i$$

and interpolate linearly between $W(t_i)$ and $W(t_{i-1})$.

Several of the (discretized) paths generated for $T = 1, n = 500$, and $t_j = j/n$ according to the above algorithm are shown in Fig. 3. Here, another characteristic of the paths of Brownian motion can also be detected; they are nowhere differentiable with respect to time. This is very significant for modeling a stock price as a function $S_t = f(W_t)$, since

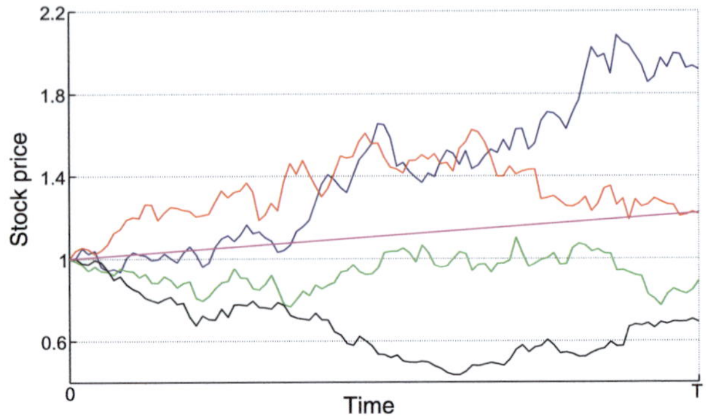


**Fig. 4** Simulated stock price paths and mean $\mathbf{E}(S(t))$ for $b = 0.2$ and $\sigma = 0.4$

this is then also not differentiable with respect to time. This characteristic is indispensable from a modeling perspective, if one sticks to the continuity of the price over time. Were the stock price to be differentiable with respect to time, then it would also be locally predictable, and, as a result, no trading would take place.

With the help of the Brownian motion, the stock price $S(t)$ is modeled as a *geometric Brownian motion*

$$S(t) = S_0 e^{(b-\frac{1}{2}\sigma^2)t+\sigma W(t)}, \tag{2}$$

that is, its logarithmized increments are assumed to be normally distributed. Here, $b$ and $\sigma$ are real numbers that describe the *mean rate of return* and the *volatility* of the stock price. Furthermore,

$$\begin{aligned} \mathbf{E}\big(S(t)\big) &= S_0 e^{bt}, \\ \mathbf{Var}\big(S(t)\big) &= S_0^2 e^{2bt}\big(e^{\sigma^2 t} - 1\big), \\ \ln\left(\frac{S(t)}{S_0}\right) &\sim \mathcal{N}\left(\left(b - \frac{1}{2}\sigma^2\right)t, \sigma^2 t\right). \end{aligned}$$

Figure 4 shows the price paths $S(t)$ associated with the paths of the Brownian motion for the parameters $b = 0.2$, $\sigma = 0.4$, and $S_0 = 1$, along with the course of the mean $\mathbf{E}(S(t))$.

*Remark 3* To formulate the model thus derived for multiple stocks, one introduces, for $d$ stocks, an $n$-dimensional ($n \geq d$) Brownian motion $W(t) := (W^{(1)}(t), \ldots, W^{(n)}(t))$, whose components are each independent, one-dimensional Brownian motions according to Definition 1, and models the price of the $j$-th stock $S^{(j)}(t)$ as

$$S^{(j)}(t) = S_0^{(j)} e^{(b^{(j)}-\frac{1}{2}\sum_{k=1}^{n}\sigma_{j,k}^2)t+\sum_{k=1}^{n}\sigma_{j,k}W^{(k)}(t)}, \quad j = 1, \ldots, d,$$

where $b^{(j)}$, $\sigma_{j,k}$, and $S_0^{(j)}$ are suitable constants. Due to the characteristics of the normal distribution, the stock prices remain log-normally distributed, and the expectations and variances can also be determined analogously.

As our next ingredient, we introduce the investors by means of the trading strategy, where the information structure of the investors is given by the filtration $\{\mathcal{F}_t\}_{t\in[0,T]}$ corresponding to the Brownian motion. Here, a trading strategy is a two-dimensional stochastic process, whose components specify the number of units of each security being held.

**Definition 2**

(a) A *trading strategy* $\varphi$ is an $\mathbf{R}^2$-valued process $\varphi(t) := (\varphi_0(t), \varphi_1(t))'$ that is progressively measurable with regard to $\{\mathcal{F}_t\}_{t\in[0,T]}$. Moreover, we require

$$\int_0^T |\varphi_0(t)|\mathrm{d}t < \infty \quad P\text{-almost surely,}$$

$$\int_0^T (\varphi_1(t)S(t))^2\mathrm{d}t < \infty \quad P\text{-almost surely.}$$

The value $x := \varphi_0(0)B_0 + \varphi_1(0)S_0$ is called initial value of $\varphi$.

(b) Let $\varphi$ be a trading strategy with initial value $x \geq 0$. The process

$$X(t) := \varphi_0(t)B(t) + \varphi_1(t)S(t)$$

is then called the *wealth process* corresponding to $\varphi$ with *initial wealth* $X(0) = x$.

(c) A trading strategy $\varphi$ is called *self-financing* if, for the associated wealth process $X(t)$, $t \in [0, T]$,

$$X(t) = x + \int_0^t \varphi_0(s)\mathrm{d}B(s) + \int_0^t \varphi_1(s)\mathrm{d}S(s)$$

$P$-almost surely, that is, the current wealth is yielded by the sum of the initial wealth and the profits/losses from investments in the time period $[0, t]$. It is then called *admissible* when its associated wealth process is non-negative.

Note that the requirement of progressive measurability of the strategy means that the investor has no information about the future development of the stock price. The economically natural requirement that the investor behaves in a self-financing way is a genuine requirement and does not result mathematically from parts (a) and (b) of the definition. For information on the analogous definition of a trading strategy in more generalized markets (see above), we refer the reader to [4].

## 4.2 Option Pricing and the Arbitrage Principle

With the mathematical market model developed in the previous section, we are now in a position to tackle the problem of option pricing. The essential idea behind option pricing is, first, that an option is a derived security (derivative) having no existence of its own independent from its underlying asset; the movement of the asset price also determines the price of the option. The second central principle that applies in option pricing is that of *absence of arbitrage*. Here, one considers an arbitrage opportunity to be a transaction involving the possibility of a profit without the risk of a loss, where none of the investor's own money must be used.

A typical example of such an arbitrage opportunity is a free ticket in a lottery. Here, although one may seldom win, neither must one put up one's own money to acquire the ticket.

An arbitrage opportunity of this type is such a good "deal" that every market participant would instantly take advantage of it. The resulting infinite demand would immediately trigger a corresponding price adjustment on the market and the arbitrage opportunity would disappear. Therefore, for theoretical deliberations, one only considers financial market models that are free from arbitrage opportunities. This assumption alone makes it possible in any arbitrary financial market model to set lower and upper bounds for option prices (as a function of each option type). See, for example, Chap. 3 in [4].

**Definition 3** An arbitrage opportunity is an admissible trading strategy $\varphi$ whose associated wealth process $X(t)$ fulfills the conditions

$$X(0)=0, \qquad X(T)\geq 0 \quad P\text{-almost surely}, \qquad P\big(X(T)>0\big)>0.$$

In the market we are considering here, with prices according to Eqs. (1) and (2), an even stronger variant of the absence-of-arbitrage principle can be shown, namely, the validity of the *replication principle*. This stipulates that two investments with identical future cash flows must have the identical price today. If this were not so, one could buy the cheaper of the two alternatives today and simultaneously sell the more expensive. The future cash flows arising from this transaction neutralize one another, but one has already accrued today an increase in wealth from the price difference, which could then simply be invested in a money market account. Because one needs no starting capital to pursue this strategy, but the final wealth is strictly positive, this represents an arbitrage opportunity.

The following is a central theorem in the theory of option pricing.

**Theorem 1** (Completeness of the market) *Using the notation* $\theta=(b-r)/\sigma$, *then* $H(t):=\exp(-(r+\theta^2/2)t-\theta W(t))$.

(a) *Let* $x\geq 0$. *For an admissible trading strategy* $\varphi$ *with wealth process* $X(t)$, *we have*

$$\mathbf{E}\big(H(t)X(t)\big)\leq x \quad \forall t\in[0,T].$$

(b) *Let* $Y\geq 0$ *be a* $\mathcal{F}_T$*-measurable random variable with*

$$\tilde{x}:=\mathbf{E}\big(H(T)Y\big)<\infty. \tag{3}$$

*Then, there is an admissible trading strategy $\varphi$ with initial value $\tilde{x}$, wealth process $X(t)$ and*

$$X(T) = Y \quad P\text{-almost surely.}$$

The complete market theorem seems unspectacular at first blush, but it is extremely significant for option pricing. Part (b) says that a non-negative final payment $Y$ (that fulfills condition (3)), which can be secured via possession of an option, for example, can be synthetically generated by pursuing a suitable trading strategy $\varphi$ in the money market account and the stock. Thus, from the perspective of the final payment, it is irrelevant whether one physically possesses the option or whether it is synthetically replicated. If there is no arbitrage opportunity in the market, then Eq. (3) yields the option price $\tilde{x}$. The non-existence of an arbitrage opportunity follows immediately from part (a) of the theorem, however (see [4]).

**Corollary 1** (Absence of arbitrage) *In the market model under consideration here, there is no arbitrage opportunity.*

Consequently, Theorem 1 and Corollary 1 together lead directly to the main result for option pricing in this market model.

**Corollary 2** (Fair price) *In the market model under consideration here, the fair price of an option with final payment $Y$, which is compatible with the arbitrage principle, is given by*

$$x_Y := \mathbf{E}\big(H(T)Y\big) \tag{4}$$

*when this value is finite.*

*Remark 4* (Option price, risk-neutral pricing, equivalent martingale measure) If one considers $H(t)$ to be a discount factor process, with which one discounts future payments, then Eq. (4) says that one obtains the price of the option with final payment $Y$ by calculating the expectation of the final payment discounted by $H(T)$. This means, first, that the net present value principle, under which the price is defined as the future payment discounted to today, is valid here with a stochastic discount factor. It can also be shown, however, that

$$\mathbf{E}\big(H(T)Y\big) = \mathbf{E}_Q\big(\exp(-rT)Y\big) \tag{5}$$

holds true, where the second expectation with regard to the (unique) probability measure $Q$ is formed in the probability space being considered, for which

$$\mathbf{E}_Q\big(\exp(-rt)S(t)\big) = S_0$$

holds true. Because, as a consequence, $S(t)/S_0$ and $B(t)/B_0$ possess the same expectation under $Q$, regardless of whether one is dealing with a risky or a risk-free investment, $Q$ is also called the *risk-neutral measure*, and one speaks of *risk-neutral pricing*, since the option price is given by the right side of Eq. (5). The existence of the risk-neutral measure $Q$ follows from the Girsanov theorem (see [4], Chap. 3), which also states that, under $Q$, the process

$$W_Q(t) = W(t) + \theta t,$$

with $\theta$ from Theorem 1, is a Brownian motion. If one inserts this in the stock price equation (2), then one obtains

$$S(t) = S_0 \exp\left(\left(r - \frac{1}{2}\sigma^2\right)t + \sigma W_Q(t)\right),$$

from which it once again follows, with the defining characteristics of the Brownian motion, that the discounted stock price process $S(t)/B(t)$ is a martingale. Since it is also true that the measure $Q$ is equivalent to $P$ (i.e., both measures possess the same null sets), one also refers to $Q$ as the *equivalent martingale measure*. The relationship between the existence of such equivalent martingale measures and the absence of arbitrage in a market model is also referred to as the first fundamental theorem of option pricing (see [25]). In general, it can be shown in an elementary fashion for analogous, arbitrage-free financial market models that establishing an option price by the right side of Eq. (5) does not lead to arbitrage opportunities when $Q$ is an equivalent martingale measure.

### 4.3 The Black–Scholes Formula: Nobel Prize for Mathematics

For the special case of the European call option, one can explicitly calculate the expectation that determines the option price. This then yields the famous *Black–Scholes formula* (see [20] or [4]).

**Theorem 2** (Black–Scholes formula) *In the market model given by the price equations* (1) *and* (2), *the price* $X_{\text{call}}(t, S(t), K, T)$ *of a European call option at time* $t \in [0, T]$ *with maturity* $T$ *and strike* $K > 0$ *is given by*

$$X_{\text{call}}\big(t, S(t), K, T\big) = S(t)\Phi\big(d_1(t)\big) - Ke^{-r(T-t)}\Phi\big(d_2(t)\big) \tag{6}$$

*where* $\Phi(.)$ *is the distribution function of the standard normal distribution and where we use the abbreviations*

$$d_1(t) = \frac{\ln(S(t)/K) + (r + \sigma^2/2)(T-t)}{\sigma\sqrt{T-t}}, \qquad d_2(t) = d_1(t) - \sigma\sqrt{T-t}.$$

*Using the same notation, the price of the corresponding European put is given by*

$$X_{\text{put}}\big(t, S(t), K, T\big) = Ke^{-r(T-t)}\Phi\big(-d_2(t)\big) - S(t)\Phi\big(-d_1(t)\big).$$

*Remark 5* (Characteristics, applications, and consequences of the Black–Scholes formula)

(a) The outstanding quality of the Black–Scholes formula is not simply that it allows a closed analytical form of the price of the European call, but that it allows this price to be independent from $b$, the stock's mean rate of return. Because this parameter is far more critical for estimating the future development of the stock price, which can only be poorly estimated from past stock prices (although one can efficiently estimate the volatility $\sigma$, at least as $\sigma^2$, from historical data), it is precisely the absence of $b$ that is one of the main reasons for the market's acceptance of the Black–Scholes formula—aside from its elegant and convincing mathematical derivation. Its standing was further underscored in 1997 with the awarding of the Nobel Prize for Economics to Robert C. Merton and Myron Scholes for their work on it. Fischer Black had already died in 1995 and could therefore no longer be honored for his contribution.
(b) In the market, the Black–Scholes formula is not generally used to calculate call prices, but rather in a manner indicating that the market does not fully believe in the Black–Scholes model. Closed pricing formulas, including those in other models, are frequently used for *parameter calibration*, that is, the input parameters for each model are defined so that the associated model prices for each derivative coincide as well as possible with the prices observed in the market. In the case of the Black–Scholes formula, this is taken a step further, in that a positive volatility is defined for all calls with different maturities and different strikes on the same stock, such that the price observed in the market coincides exactly with the model price (see Sect. 4.3.1). This volatility is called the *implied volatility* of the particular call. If one joins the resulting points by means of a suitable interpolation procedure, one then obtains a so-called *implied volatility surface*. If the Black–Scholes model corresponded exactly to the market data, then all of the implied volatilities would have to be (at least) virtually identical. In the following section, we make clear that this is not so by defining in detail the implied volatility and implied volatility surface and illustrating them with an example.

Other weaknesses of the Black–Scholes model regarding characteristic empirical properties of stock and option prices (so-called *stylized facts*) are treated in Sect. 4.6.

### 4.3.1 Implied Volatility

According to Theorem 2, the price of a European call is a function with six arguments, all of which are observable except for the volatility. Consequently, if the volatility is known, the option price can be calculated. Conversely, if the option price is known, one can easily show that the volatility can be uniquely determined under the assumption that it is positive. Therefore, it is possible to determine the implied volatility $\sigma_{\text{imp}}$ from the option prices quoted in the market.

For $i = 1, \ldots, N$, we let $X_{\text{call}}^{\text{market}}(K_i, T_i)$ denote the market price of a European call with exercise price $K_i$ and maturity $T_i$, where the same strike may very well be paired with different maturities and vice versa. If one sets these market prices equal to the theoretical

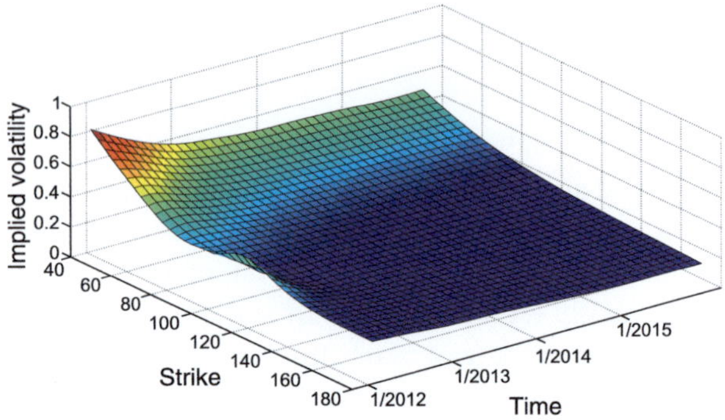


**Fig. 5** Implied volatility surface on 14 December 2011, from European calls on the stock of Allianz SE

prices of the corresponding call options in the Black–Scholes model, then $\sigma_{\text{imp}}$ can be uniquely determined from

$$X_{\text{call}}\big(t, S(t), K_i, T_i\big) \stackrel{!}{=} X_{\text{call}}^{\text{market}}(K_i, T_i),$$

for $i = 1, \dots, N$. Because the market prices are dependent on the exercise prices and maturities, the implied volatilities are dependent on them also.

**Definition 4** The representation of the implied volatility $\sigma_{\text{imp}}$ as a function of the exercise price $K$ and the maturity $T$ is referred to as the *implied volatility surface*.

Figure 5 shows the implied volatility surface on 14 December 2011, as obtained from European calls on the stock of Allianz SE. As the graphic shows, contrary to the assumption in the Black–Scholes model, options having different execution prices and maturities possess different implied volatilities.

## 4.4 Alternative Stock Price Models: Theoretical Aspects

There are several ways to redress the deficits of the Black–Scholes model, and these are often resorted to when modeling problems. Among others, they are:

- Moving from a linear to nonlinear stochastic dynamics, as introduced, for example, in Sect. 4.4.1.
- Introducing further stochastic components, such as a stochastic, rather than constant, volatility (see Sect. 4.4.2 and, particularly, Sects. 5 and 6).
- Considering a more general class of stochastic processes for modeling the uncertainty in the stock price process, such as the class of Lévy models in Sect. 4.4.3.

### 4.4.1 Local Volatility Models

Local volatility models utilize the first of the above-mentioned ideas. To avoid the problem of non-constant volatility, the volatility of the stock price is permitted to be time and location dependent. As before, a simple one-dimensional Brownian motion $W(t)$ is used as the underlying stochastic process. This is done in the hope of thereby maintaining the completeness of the market. The replication principle of option pricing would then remain valid. In point of fact, an astounding result is attained in this regard, which we will present in Theorem 3.

We consider a market model consisting of the usual money market account with interest rate $r$ (see Eq. (1)) and a stock whose price is modeled with the aid of the stochastic differential equation

$$\mathrm{d}S(t) = rS(t)\mathrm{d}t + \sigma\big(S(t), t\big)S(t)\mathrm{d}W(t), \quad S(0) = S_0. \tag{7}$$

Here, we let $\sigma(x, t)$ be a non-negative, real-valued function of such a form that Eq. (7) possesses a unique (non-negative) solution. One sees immediately that, for the constant function $\sigma(x, t) \equiv \sigma$, one obtains the Black–Scholes model.

Now, instead of prescribing a parametric form of the volatility function, Dupire [27] takes an entirely different approach. Motivated by the terminology of the implied volatility surface, he looks for a volatility function that ensures, for a specified set of call prices, that the associated theoretical option prices (calculated as the discounted expectation of the final payoff under the unique equivalent martingale measure) coincide with the given market prices. And this is precisely the assertion of the following theorem.

**Theorem 3** ([27]) *Let today's market prices $X_{\text{call}}^{\text{market}}(0, S, K, T)$ of European calls for all possible choices of strikes $K \geq 0$ and maturities $T \geq 0$ be known, be once differentiable as functions of the maturity, and be twice differentiable as functions of the strike. With the choice of the volatility function $\sigma(x, t)$ via*

$$\sigma(K, T) = \frac{1}{K}\sqrt{\frac{2\frac{\partial X_{\text{call}}^{\text{market}}}{\partial T} + rK\frac{\partial X_{\text{call}}^{\text{market}}}{\partial K}}{K^2\frac{\partial^2 X_{\text{call}}^{\text{market}}}{\partial K^2}}}, \tag{8}$$

*the market prices coincide with the theoretical call prices obtained in the corresponding local volatility model according to*

$$X_{\text{call}}(0, S, K, T) = \mathbf{E}\big(e^{-rT}\big(S(T) - K\big)^+\big) \quad \forall (T, K) \in [0, \infty)^2.$$

*Here, it is implicitly assumed that the call prices are furnished in such a way that all expressions appearing in Eq. (8) are defined.*

Theorem 3 presents exactly the desired result. Consequently, there exists for any given set of market prices for European calls, a volatility function $\sigma(x, t)$ that generates them. Thus, one has found a model in which the theoretical model prices coincide with the given market prices for simple options. It is therefore plausible to use this model for calculating the prices of more complicated options for which there are no market prices. The problem with the theorem, however, is its practical applicability; some prerequisites and assumptions that enter the result cannot be verified and/or can hardly be implemented in practice:

- To design the volatility function, one needs a continuous set of market prices. Due to the discreteness of the set of strikes and maturities, however, there is none. Therefore, the volatility function must be obtained with the help of interpolation and extrapolation methods, but is then dependent on the method being used and, in particular, is no longer unique.
- In a local volatility model generated in this fashion, there are generally no closed, analytical price formulas, even for simple standard options.
- The form of the local volatility function has no intuitive economical interpretation or motivation, but is based purely on data.

For further general aspects, we refer the reader to [27].

A popular parametrical model, which, however, represents no substantial improvement over the Black–Scholes model, is the CEV model (*C*onstant-*E*lasticity-of-*V*ariance model), for which the stock price equation is given as

$$\mathrm{d}S(t) = rS(t)\mathrm{d}t + \sigma S(t)^{\alpha}\mathrm{d}W(t), \quad S(0) = S_0$$

with $\alpha \in [0, 1]$ and $r, \sigma \in \mathbf{R}$. For the special choice of $\alpha = 0$ and $\alpha = 1$, it admits explicit solutions:

- For $\alpha = 1$, one then obtains the already familiar geometric Brownian motion (Black–Scholes case), that is, log-normally distributed stock prices.
- For $\alpha = 0$, one obtains

  $$S(t) = S_0 \exp(rt) + \sigma \int_0^t \exp\big(r(t-u)\big)\mathrm{d}W(u),$$

  from which follows that the stock price is normally distributed with

  $$\mathbf{E}\big(S(t)\big) = S_0 \exp(rt), \qquad \mathbf{Var}\big(S(t)\big) = \frac{\sigma^2}{2r}\big(\exp(2rt) - 1\big).$$

For all values $\alpha \in [0, 1)$, the CEV model admits a quite complicated, albeit closed, formula for the price of a European call (see [5]), which we will not reproduce here. The additional parameter $\alpha$ does indeed yield, in comparison with the Black–Scholes model, a somewhat better fit to option market prices, but one that is still far from perfect. Moreover, for $0 < \alpha < 1$, the model is numerically difficult to manage. For these reasons, we do not recommend it for practical application.

### 4.4.2 Stochastic Volatility Models

The economic idea behind stochastic volatility models is that price fluctuations are determined by supply and demand and, depending on the trading intensity, may be stronger or weaker. Since the intensity of the price fluctuations in the Black–Scholes model is determined by the value of the constant volatility $\sigma$, one assumes here a trading intensity that is (on average) constant.

If, on the other hand, one wishes to model a non-constant trading intensity whose variability cannot be predicted, then it makes sense to model the volatility by a stochastic process also. Such a *stochastic volatility model* is then given by price and variance process equations having the form

$$dS(t) = bS(t)dt + \sqrt{\nu(t)}S(t)dW_1(t), \quad S(0) = S_0, \tag{9}$$

$$d\nu(t) = \alpha(t)dt + \beta(t)dW_2(t), \quad \nu(0) = \nu_0, \tag{10}$$

where $\alpha(t)$ and $\beta(t)$ are suitable stochastic or deterministic processes that are progressively measurable relative to the filtration generated from the two-dimensional Brownian motion $(W_1(t), W_2(t))$. Furthermore, we let $\nu_0$ be the initial value of the variance process and $\rho \in [-1, 1]$ be the correlation of the Brownian motions $W_1(t)$ and $W_2(t)$,

$$\mathbf{Corr}\big(W_1(t), W_2(t)\big) = \rho. \tag{11}$$

Analogously to $\sigma$ in the Black–Scholes model, we call $\sqrt{\nu(t)}$ the *volatility process*. Moreover, all of the processes described above should be selected so that the coupled stochastic differential equations (9) and (10) possess a unique solution.

In practice, one tends to be less interested in the economic motivation behind stochastic volatility models. The decisive factors are the free parameters and/or processes arising from the introduction of the stochastic differential equation (10), with whose help one hopes to obtain a model that can much more accurately replicate the option prices observed in the market.

Among the various choices found in the literature for modeling the volatility process, the choice of Heston (see [29]) has proved especially effective in practice and has, in many fields, replaced the Black–Scholes model as the standard. At the ITWM, we have already successfully applied the Heston model in several industrial projects. In Sects. 5 and 6, we offer an extensive theoretical description of the model and take a closer look at the details of its application for modeling variants and pricing algorithms.

### 4.4.3 Lévy Models

In the class of Lévy models, a *Lévy process* $Z(t)$ essentially takes over the role of the Brownian motion $W(t)$ from the Black–Scholes model. A Lévy process is a stochastic process with independent and stationary increments that starts with $Z(0) = Z_0 = 0$ and possesses paths that are almost surely continuous. Thus, a Brownian motion is also a Lévy process, but a significant majority of Lévy processes possess paths exhibiting jumps. Lévy models are determined by their characteristics. They typically exhibit a large number of parameters and their distributions, in comparison with a normal distribution, possess markedly sharper densities with heavier tails. These can therefore explain even extreme stock price movements, for which the Black–Scholes model has no explanation (or only an explanation such as: "In the credit crisis, we have observed $10\sigma$-events").

For an overview of the application of Lévy processes in financial mathematics, we refer the reader to the monographs [23] and [40]. Other models known in the theory that have also been applied to market data include the hyperbolic model (see [28]), the variance gamma model (see [37]), and the NIG model (see [17]). To date, however, the Lévy models have been unable to make large-scale breakthroughs in practical application, since the extensive parameterization is connected with greater estimation effort and larger estimation error.

## 4.5 Further Application Aspects

The given application is a crucial factor in choosing a stock price model. A simple model, such as the Black–Scholes, often suffices to price relatively simple derivatives. For complicated, strongly path-dependent exotic options, however, the Black–Scholes model is generally inadequate. It is somewhat paradoxical, then, that when pricing options based on multiple assets (so-called *basket options*), one often resorts to the Black–Scholes model again in its multi-dimensional variant. The explanation here is that there are no suitable multi-dimensional variants of the above-mentioned, more realistic models, or none that would be numerically and statistically manageable.

Finally, the computation time required to determine an individual option price is another crucial argument. Banks often carry out *sensitivity analyses* when selling large amounts of a particular derivative. This involves varying all possible input parameters, which can quickly lead to an exponentially increasing number of different scenarios, for which the option prices must then be calculated. Hence, research into faster algorithms and new hardware concepts, such as the use of graphic cards or so-called FPGA as computational accelerators, remains an active field.

## 4.6 Effects with Real Data: Stylized Facts as an Argument Against the Black–Scholes Model

Stock prices, interest rates, exchange rates, and many other financial time series exhibit typical empirical characteristics that distinguish them from other time series. These characteristics are referred to as stylized facts. In the following analysis, we will present those characteristics in particular which suggest that the assumption of constant volatility in the Black–Scholes model is too restrictive. Here, we take the term *discrete time series* to mean an ordered sequence of observations at discrete time-points, such as exists with stock prices, for example.

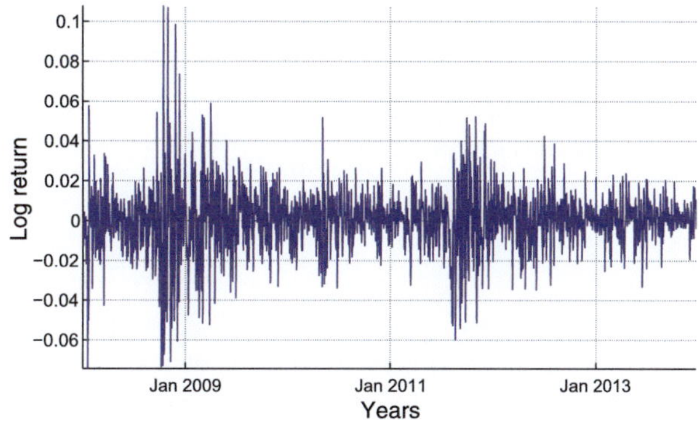


**Fig. 6** Daily log returns of the DAX between January 2008 and December 2013

**Definition 5** Let $S(t)$ be the price of a stock. We define the return $R(s,t)$ between the time points $s$ and $t > s$ as

$$R(s,t) := \frac{S(t) - S(s)}{S(s)},$$

and the logarithmized return (log return) $r(s,t)$ as

$$r(s,t) := \ln\left(\frac{S(t)}{S(s)}\right).$$

With regard to a stock's daily return, we define $r(n) := r(n-1, n)$ for $n \in \mathbf{N}$.

*Remark 6* In the following discussion, we present the typical characteristics of the daily, and thus discrete, time series of the log return $r(n)$, $n \in \mathbf{N}$. For small price changes, as are the norm with stock data, the log returns are a good approximation of the returns. The time series relevant to the investigation relate to the daily closing prices between January 2008 and December 2013.

Let the sample mean $\hat{\mu}_N$ and the sample variance $\widehat{\mathbf{Var}}_N$ of the log return be defined as

$$\hat{\mu}_N = \frac{1}{N}\sum_{n=1}^{N} r(n), \qquad \widehat{\mathbf{Var}}_N = \frac{1}{N-1}\sum_{n=1}^{N}\left(r(n) - \hat{\mu}_N\right)^2.$$

### 4.6.1 Volatility Clustering

Figure 6 shows the daily log returns of the DAX for the relevant time period. The graphic clearly illustrates that there are phases with both large and small price changes, which alternate with each other. This phenomenon is referred to as volatility clustering.

### 4.6.2 The Leverage Effect

Empirical data shows that, for returns on stocks, negative reports in the form of higher losses have a stronger impact on the perception of risk (and thus of volatility) than positive reports in the form of higher profits. The volatility thus reacts asymmetrically to the

**Table 1** Sample skewness $\hat{\gamma}_{N=1527}$ of the DAX and various DAX stocks for log returns from January 2008 through December 2013

| Sample skewness | |
|---|---|
| DAX | 0.1028 |
| BASF | 0.0594 |
| BMW | 0.0861 |
| Deutsche Bank | 0.3028 |
| Deutsche Telekom | −0.0379 |

signs of shocks. This phenomenon is known as the leverage effect. In 1976, Fischer Black commented on this as follows: *"A drop in the value of the firm will cause a negative return on its stock, and will usually increase the leverage of the stock. [...] That rise in the debt-equity ratio will surely mean a rise in the volatility of the stock."* Therefore, price and volatility changes are usually negatively correlated.

### 4.6.3 The Skewness—A Measure for the Symmetry of a Distribution

The empirical distribution of logarithmized stock price returns is often asymmetric. One measure for this asymmetry is the skewness of a random variable.

**Definition 6** Let $X$ be a real-valued random variable with $\mathbf{E}(X^3) < \infty$. The *skewness* $\gamma(X)$ of $X$ is defined as

$$\gamma(X) := \frac{\mathbf{E}((X - \mathbf{E}(X))^3)}{(\mathbf{Var}(X))^{3/2}}.$$

*Remark 7* For the discrete log returns, the skewness is estimated by means of the sample skewness

$$\hat{\gamma}_N = \frac{1}{\widehat{\mathbf{Var}}_N^{3/2}} \frac{1}{N} \sum_{n=1}^{N} \left(r(n) - \hat{\mu}_N\right)^3.$$

The sample skewness of a normally distributed random variable is equal to zero. The more $\hat{\gamma}_N$ deviates from zero, the more asymmetric is the empirical distribution of the data. If $\hat{\gamma}_N < 0$ (left-skewed), the left tail of the distribution is heavier than the right. Conversely, for $\hat{\gamma}_N > 0$ (right-skewed), the right tail is heavier than the left.

Table 1 shows the sample skewness of the DAX and some of its individual components. All observed values are non-zero and the associated time series are accordingly asymmetric. This in turn suggests considering alternative stock price models that do not assume a normal distribution.

### 4.6.4 Kurtosis—Emphasized Peaks and Tails

Figure 7 shows the histogram of the log returns of the DAX for the relevant time period, along with the density of the adjusted normal distribution. As the graphic indicates, the density of the log returns has a higher peak in the middle and heavier tails than the density

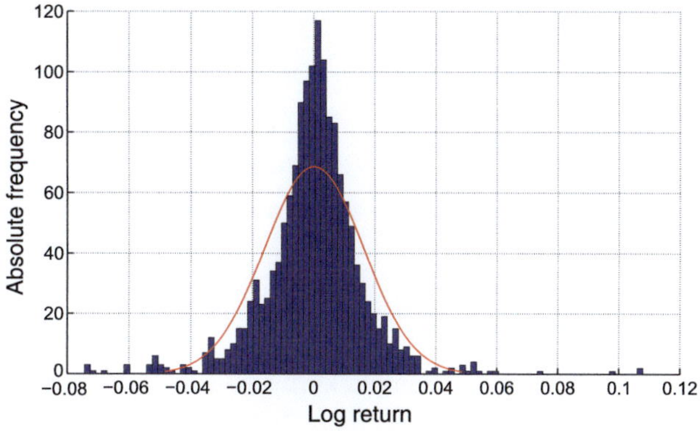


**Fig. 7** Empirical distribution of the DAX log returns and density of the fitted normal distribution

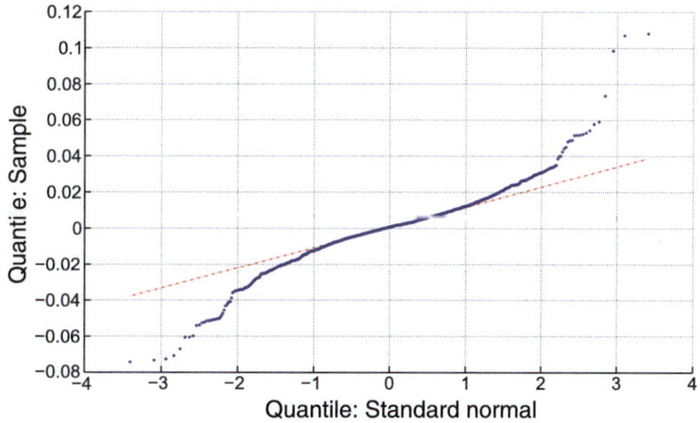


**Fig. 8** Q-Q plot of the log returns for the DAX

of the normal distribution. The quantile-quantile diagram (Q-Q plot) in Fig. 8 makes clear how heavy the tails of the empirical distribution are in comparison to the normal distribution. If the historical data had been normally distributed, it would lie on the dashed red line.

**Definition 7** Let $X$ be a real-valued random variable with $\mathbf{E}(X^4) < \infty$. The *kurtosis* $\kappa(X)$ of $X$ is defined as

$$\kappa(X) := \frac{\mathbf{E}((X - \mathbf{E}(X))^4)}{\mathbf{Var}(X)^2}.$$

*Remark 8* The kurtosis for the discrete log returns is estimated on the basis of the sample kurtosis

$$\hat{\kappa}_N = \frac{1}{\widehat{\mathbf{Var}}^2} \frac{1}{N} \sum_{n=1}^{N} \left(r(n) - \hat{\mu}_N\right)^4.$$

Normally distributed random variables have a kurtosis of 3. If the kurtosis is larger, then the distribution of the associated random variable is leptokurtic. The distribution then has a narrower peak than that of a normal distribution.

**Table 2** Sample kurtosis $\hat{\kappa}_{N=1527}$ for the DAX and various DAX stocks for the log returns from January 2008 through December 2013

| Sample kurtosis | |
|---|---|
| DAX | 8.6948 |
| BASF | 10.1601 |
| BMW | 6.5074 |
| Deutsche Bank | 9.3832 |
| Deutsche Telekom | 12.8838 |

Table 2 shows the sample kurtosis of the DAX and various stocks for the time period January 2008 through December 2013. All observed values are significantly larger than 3; the associated time series thus exhibit pronounced tails and high peaks. These characteristics are typical for mixtures of distributions with different variances. Therefore, these results also indicate that the assumption of constant volatility is not appropriate.

### 4.6.5 The Volatility Reverts to Its Mean

Another empirical characteristic of the volatility is that it reverts to its mean. To investigate this behavior, we consider the historical standard deviation of the log returns. This is referred to as the historical volatility.

**Definition 8** The *historical N-days volatility* $\sigma_{\text{hist}}$ is defined as the annualized standard deviation

$$\sigma_{\text{hist}}(N) := \sqrt{\frac{D}{N-1} \sum_{n=1}^{N} \left(r(n) - \hat{\mu}_N\right)^2}.$$

Here, $D$ stands in general for a days convention, which specifies the number of days used to approximate a year, since weekends and holidays cause the exact number to fluctuate. In practice, $D = 252$ is often used.

In order to study the historical volatility, we consider the rolling historical volatility over a longer time period.

**Definition 9** For $l \in \mathbf{Z}$, one takes the *rolling historical N-days volatility* to be the time series

$$\sigma_{\text{hist}}(N, l) := \sqrt{\frac{D}{N-1} \sum_{n=l+1}^{l+N} \left(r(n) - \hat{\mu}_N(l)\right)^2},$$

where the sample mean $\hat{\mu}_N(l)$ on the basis of $N$ is calculated for the observed data points, starting at 1, and then slides over the data.

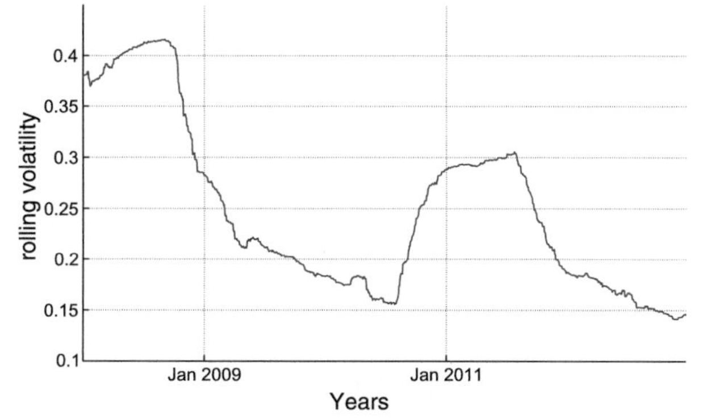


**Fig. 9** Historical rolling one-year volatility for the DAX

Figure 9 shows the rolling historical volatility on a one-year basis $\sigma_{\text{hist}}(252, l)$ for the DAX from January 2008 through January 2013. One can observe that the historical volatility, after reaching high (low) values, tends to fall (climb). Empirically, the volatility reverts to its mean.

In summary, one can state that both the stylized facts and the implied volatility observed in the market (see Sect. 4.3.1) militate against the assumption of constant volatility in the model. Instead, the volatility itself should be modeled as a random variable that is correlated with the stock price. One model that does so is Heston's stochastic volatility model (cf. [29]); this will be analyzed in depth in the following discussion, along with its variants—some of which we have put to use in industrial projects.

## 5 Theoretical Foundations of the Heston Model

The Heston model is a stochastic volatility model in which the functions $\alpha(t)$ and $\beta(t)$ from Eq. (10) possess a special form. Here, the stock price and the variance both follow the stochastic differential equations

$$dS(t) = bS(t)dt + \sqrt{\nu(t)}S(t)dW_1(t), \qquad S(0) = S_0, \tag{12}$$

$$d\nu(t) = \kappa\big[\theta - \nu(t)\big]dt + \sigma\sqrt{\nu(t)}dW_2(t), \quad \nu(0) = \nu_0. \tag{13}$$

As in Eq. (11), the Brownian motions $W_1(t)$ and $W_2(t)$ have a correlation of $\rho$. Moreover, $b$ denotes the stock drift; $\kappa$, the reversion speed of the variance to the mean reversion level $\theta > 0$; and $\sigma$, the volatility of the variance. The process $\nu(t)$ from Eq. (13) is called the square root diffusion process, or Cox-Ingersoll-Ross (CIR) process. It is the pathwise unique, weak solution of Eq. (13) and is almost surely non-negative. It is not given explicitly, but has a non-central chi-square distribution and, in particular, is finite. If the Feller condition

$$2\kappa\theta \geq \sigma^2 \tag{14}$$

also holds, then the process is strictly positive, that is, $P(\nu(t) > 0) = 1 = Q(\nu(t) > 0)$ for all $t \geq 0$. Furthermore, the variance process reverts to its mean reversion level $\theta$, which—as described in Sect. 4.6—is an empirical characteristic of the volatility. The correlation of the Brownian motions is in a position to replicate the leverage effect described earlier, and is thus generally negative (sometimes even very close to $-1$!). All in all, the Heston model thus models all the characteristics of the volatility that were described as stylized facts.

As does $W_1(t)$, the Brownian motion $W_2(t)$ also represents a source of uncertainty. However, because the volatility is not an asset that can be traded in the market, the replication principle—which is based on the completeness of the market (see Theorem 1)—can no longer be applied. In such an incomplete market, the risk-neutral pricing measure $Q$ is no longer unique. Moreover, there are infinitely many equivalent martingale measures (see [19] or [18]).

Up to this point, the Heston model has been considered under the physical measure $P$, which is supposed to describe the price movements in the real market. The dynamics under an equivalent martingale measure $Q$ can be derived from the dynamics (12) and (13). For a positive constant $\lambda$, the risk-neutral parameters

$$\kappa^\star = \kappa + \lambda, \qquad \theta^\star = \frac{\kappa\theta}{\kappa + \lambda},$$

and the Girsanov transformations

$$\begin{aligned} \mathrm{d}W_1^Q(t) &= \mathrm{d}W_1(t) + (b - r)\int_0^t \frac{1}{\sqrt{\nu(s)}}\mathrm{d}s, \\ \mathrm{d}W_2^Q(t) &= \mathrm{d}W_2(t) + \frac{\lambda}{\sigma}\int_0^t \sqrt{\nu(s)}\mathrm{d}s, \end{aligned}$$

can be used to define the risk-neutral form of the Heston model as follows:

$$\mathrm{d}S(t) = rS(t)\mathrm{d}t + \sqrt{\nu(t)}S(t)\mathrm{d}W_1^Q(t), \qquad S(0) = S_0, \tag{15}$$

$$\begin{aligned} \mathrm{d}\nu(t) &= \left[\kappa\left(\theta - \nu(t)\right) - \lambda\nu(t)\right]\mathrm{d}t + \sigma\sqrt{\nu(t)}\mathrm{d}W_2^Q(t) \\ &= \kappa^\star\left[\theta^\star - \nu(t)\right]\mathrm{d}t + \sigma\sqrt{\nu(t)}\mathrm{d}W_2^Q(t), \quad \nu(0) = \nu_0. \end{aligned} \tag{16}$$

Here, $W_1^Q(t)$ and $W_2^Q(t)$ denote $Q$-Brownian motions with correlation $\rho$.

*Remark 9* In Heston's original work (cf. [29]), the term $\lambda\nu(t)$ is referred to as the *market price of the volatility risk* $\Phi$. This (and therefore the associated Girsanov transformation, also) can be *a priori* freely selected. Both economic and mathematical arguments militate for modeling proportional to variance $\nu(t)$; only for this choice is there a known semi-closed formula for the price of European calls and puts.

In closing, we want to point out that the choice of a particular equivalent martingale measure equates to the choice of a market price for the volatility risk, which is ultimately determined by the choice of the positive constant $\lambda$. Consequently, we must also pose the question of which measure is to be used in the specific application. The answer to this question is revealed in Sect. 6.1.

## 5.1 Closed Form Solution for the Price of European Calls

One of the main reasons for the success of the Heston model in practice is a semi-closed price formula for European calls and puts that allows one to efficiently determine the model parameters from market prices, and thus, to calibrate the model (see Sect. 6.1). Using classical arbitrage arguments, one obtains the following partial differential equation for determining the price of a European call $X_{\text{call}}(t, S, K, T)$:

$$\begin{aligned} 0 = & \frac{\partial X_{\text{call}}}{\partial t} + \frac{\nu S^2}{2}\frac{\partial^2 X_{\text{call}}}{\partial S^2} + \rho\sigma\nu S\frac{\partial^2 X_{\text{call}}}{\partial \nu \partial S} + \frac{\sigma^2 \nu}{2}\frac{\partial^2 X_{\text{call}}}{\partial \nu^2} + rS\frac{\partial X_{\text{call}}}{\partial S} \\ & - rX_{\text{call}} + \big[\kappa(\theta - \nu) - \lambda\nu\big]\frac{\partial X_{\text{call}}}{\partial \nu}, \end{aligned} \tag{17}$$

where it is assumed that the market price of the volatility risk is proportional to the variance, according to the relationship $\Phi = \lambda\nu(t)$. There is no known explicit solution for the partial differential equation (17). However, Heston found a way to express the solution with the aid of characteristic functions. Analogously to the Black–Scholes formula (6), he chooses the approach

$$\begin{aligned} X_{\text{call}}\big(t, S(t), K, T\big) = & \, S(t)P_1\big(S(t), \nu(t), t, \ln(K)\big) \\ & - Ke^{-r(T-t)}P_2\big(S(t), \nu(t), t, \ln(K)\big) \end{aligned}$$

for the solution, where $P_1(S(t), \nu(t), t, \ln(K))$ and $P_2(S(t), \nu(t), t, \ln(K))$ describe the probabilities that the stock finishes above the strike. Both probabilities fulfill the partial differential equation. If the characteristic functions $\varphi_1(S(t), \nu(t), t, u)$ and $\varphi_2(S(t), \nu(t), t, u)$ belonging to the probabilities exist, then $P_1(S(t), \nu(t), t, \ln(K))$ and $P_2(S(t), \nu(t), t, \ln(K))$ are given by their inverse Fourier transforms

$$P_j\big(S(t), \nu(t), t, \ln(K)\big) = \frac{1}{2} + \frac{1}{\pi}\int_0^\infty \Re\left[\frac{e^{-iu\ln(K)}\varphi_j(S(t), \nu(t), t, u)}{iu}\right]\mathrm{d}u \tag{18}$$

for $j = 1, 2$, where $\Re(.)$ denotes the real part. The linearity of the coefficients then suggests the approach

$$\varphi_j\big(S(t), \nu(t), t, u\big) = \exp\big(C_j(\tau, u) + \nu D_j(\tau, u) + iu\ln\big(S(t)\big)\big), \tag{19}$$

for $j = 1, 2$ and $\tau := T - t$ for the characteristic functions. Utilizing $\varphi_1(S(t), \nu(t), t, u)$ and $\varphi_2(S(t), \nu(t), t, u)$ in Eq. (17) then delivers the following system of linear differential equations

$$0 = -\frac{u^2}{2} + \rho\sigma ui D_j + \frac{\sigma^2}{2} D_j^2 + u_j ui - b_j D_j - \frac{\partial D_j}{\partial \tau}, \tag{20}$$

$$0 = rui + aD_j - \frac{\partial C_j}{\partial \tau} \tag{21}$$

for the unknowns $C_j(\tau, u)$ and $D_j(\tau, u)$ with initial conditions

$$C_j(0, u) = 0, \qquad D_j(0, u) = 0 \tag{22}$$

and

$$u_1 = \frac{1}{2}, \qquad u_2 = -\frac{1}{2}, \qquad a = \kappa\theta, \qquad b_1 = \kappa + \lambda - \rho\sigma, \qquad b_2 = \kappa + \lambda. \tag{23}$$

The solution of the system (20), (21) and (22) is given by

$$\begin{aligned} C_j(\tau, u) &= rui\tau + \frac{a}{\sigma^2}\left[(b_j - \rho\sigma ui + d_j)\tau - 2\ln\left[\frac{1 - g_j e^{d_j\tau}}{1 - g_j}\right]\right], \\ D_j(\tau, u) &= \frac{b_j - \rho\sigma ui + d_j}{\sigma^2}\left[\frac{1 - e^{d_j\tau}}{1 - g_j e^{d_j\tau}}\right] \end{aligned} \tag{24}$$

with

$$g_j = \frac{b_j - \rho\sigma ui + d_j}{b_j - \rho\sigma ui - d_j}, \qquad d_j = \sqrt{(\rho\sigma ui - b_j)^2 - \sigma^2(2u_j ui - u^2)}. \tag{25}$$

The following theorem summarizes the results.

**Theorem 4** (Heston's price formula) *Let the market price of the volatility risk be given by $\Phi = \lambda\nu(t)$. Then, in the Heston model, which is specified by Eqs.* (12), (13), *and* (11), *the arbitrage-free price of a European call is given by*

$$\begin{aligned} X_{\text{call}}(t, S(t), K, T) = {} & S(t)P_1(S(t), \nu(t), t, \ln(K)) \\ & - Ke^{-r(T-t)}P_2(S(t), \nu(t), t, \ln(K)). \end{aligned}$$

*The probabilities $P_j(S(t), \nu(t), t, \ln(k))$ and the associated characteristic functions $\varphi_j(S(t), \nu(t), t, u)$ are given by Eqs.* (18) *and* (19). *The further quantities are defined in Eqs.* (23), (24), *and* (25).

## 5.2 Variants of the Heston Model—Requirements Arising from Practice

On the basis of the acceptance and popularity of the Heston model in practice, the Financial Mathematics Department of the Fraunhofer ITWM received numerous research commissions from the financial and insurance industries, whose goals were the model's theoretical generalization and algorithmic implementation. In the wake of these projects, new and innovative variants of the closed formula from Theorem 4 were developed and implemented. In this section, we treat several of these variants—particularly those that resulted in publications in relevant journals.

### 5.2.1 The Heston Model with Time-Dependent Coefficients

The partial differential equation (20) is a nonlinear differential equation of the Riccati type. Therefore, generalizing the Heston model for non-constant parameters is non-trivial. The work of Mikhailov and Nögel (cf. [38]) considers diverse variants for treating time-dependent coefficients. For example, since Eq. (20) is not dependent on the mean reversion level $\theta$, a general solution for a time-dependent enhancement $\theta(t)$ can be found. Other special cases include solutions with the help of hyper-geometric functions, for cases in which the reversion speed is modeled as $\kappa(t) = at + b$ or $\kappa(t) = ae^{-\alpha t}$. Strictly speaking, however, one must resort to other techniques. By numerically solving Eqs. (20) and (21), the model's application can be extended with relative ease to the situation of time-dependent parameters. Here, Runge–Kutta algorithms are good candidates. The use of semi-closed price formulas arises for the algorithmic implementation—especially for calibrating the model.

**Asymptotic Expansion** Because an analytical solution for the partial differential equation (20) can only be found for a few special cases, it seems appropriate to apply asymptotic methods. We therefore assume that $\rho(t)$ results from a superposition of time-dependent functions and, for small variations $\epsilon$, possesses a potential series expansion around a constant value $\rho_0$:

$$\rho(t) = \rho_0 + \epsilon\rho_1(t) + \epsilon^2\rho_2(t) + \cdots .$$

Using the approach

$$D_j(t) = D_{j,0}(t) + \epsilon D_{j,1}(t) + \epsilon^2 D_{j,2}(t) + \cdots$$

the first order approximation delivers a linear equation with time-dependent coefficients, whose solution is given by

$$\begin{aligned} D_{j,1}(t) = -\sigma u_j i \int_0^t \rho_1(\tau) D_{j,0}(\tau) \exp\left( \int_0^\tau D_{j,0}(\xi)\mathrm{d}\xi - (-\rho_0\sigma u_j i + b_j)\tau \right) \mathrm{d}\tau \\ \times \exp\left( - \int_0^t D_{j,0}(\tau)\mathrm{d}\tau + (-\rho_0\sigma u_j i + b_j)t \right). \end{aligned}$$

As an alternative to the above asymptotic approach, one could perform an asymptotic analysis of the system with slowly changing parameters.

**Piece-Wise Constant Parameter** If one sub-divides the time interval $[t, T]$ into $n$ sub-intervals $[t, t_1], \dots, [t_i, t_j], \dots, [t_{n-1}, T]$ and defines the model parameters to be constant in each sub-interval, then a closed solution can be found for Eq. (20), even for different parameters in different sub-intervals. With the help of the time inversion $\tau_k = T - t_{n-k}$, $k = 1, \dots, n-1$, the initial condition for the first sub-interval $[0, \tau_1]$ is exactly zero. For this interval, one can then use the solution (24) of the Heston model. For the second sub-interval, we need solutions for the differential equations (20) and (21) with arbitrary initial conditions

$$C_j(0,u) = C_j^0, \qquad D_j(0,u) = D_j^0, \tag{26}$$

which are given by

$$\begin{aligned} C_j(\tau,u) &= rui\tau + \frac{a}{\sigma^2}\left[(b_j - \rho\sigma ui + d_j)\tau - 2\ln\left[\frac{1 - g_j e^{d_j\tau}}{1 - g_j}\right]\right], \\ D_j(\tau,u) &= \frac{b_j - \rho\sigma ui + d_j - (b_j - \rho\sigma ui + d_j) g_j e^{d_j\tau}}{\sigma^2(1 - g_j e^{d_j\tau})} \end{aligned} \tag{27}$$

with

$$g_j = \frac{b_j - \rho\sigma ui + d_j - D_j^0\sigma^2}{b_j - \rho\sigma ui - d_j - D_j^0\sigma^2}, \qquad d_j = \sqrt{(\rho\sigma ui - b_j)^2 - \sigma^2(2u_j ui - u^2)} \tag{28}$$

and (23). The continuity requirement for the functions $C_j(\tau, u)$ and $D_j(\tau, u)$ at the intersection of the first and the second sub-interval $\tau_1$ delivers the initial conditions for the second sub-interval as

$$C_j(0,u) = C_j^0 = C_j^H(\tau_1,u), \qquad D_j(0,u) = D_j^0 = D_j^H(\tau_1,u), \tag{29}$$

where $C_j^H(\tau_1, u)$ and $D_j^H(\tau_1, u)$ refer to the Heston solution with the initial conditions (22). If one solves the above equations relative to the initial conditions $C_j^0$ and $D_j^0$, one obtains the initial conditions for the second sub-interval. The procedure is then repeated for each jump point of the parameters $\tau_k$, for $k = 2, \dots, n-1$. Summarizing, the calculation of the option price in the Heston model with piece-wise constant parameters consists of 2 phases:

1. Determine the initial conditions for each sub-interval with the aid of the formulas in (29).
2. Determine the functions $C_j(\tau, u)$ and $D_j(\tau, u)$ using the solutions (27) and (28) with the initial conditions (26).

### 5.2.2 Forward Starting Options in the Heston Model

For pricing many exotic options in the Heston model, one must often resort to numerical methods, such as a Monte Carlo simulation or tree method (cf. Sect. 6). There are also instances, however, where closed formulas have been derived for complex derivatives. One example is the so-called *forward starting option*, which is treated in the work of Kruse and Nögel (cf. [35]).

A forward starting option is one whose exercise price is not completely determined until a time-point $t^\star$. This time-point lies between the issuing date and the option's maturity, and is referred to as the starting point. Here, one can see that the forward starting option belongs to the class of path-dependent options. The payoff function for this option is given by

$$Y_{\text{fso}} = \big(S(T) - kS(t^\star)\big)^+, \tag{30}$$

where $k \in [0, 1]$ denotes a percentage.

Using the principle of risk-neutral pricing, a semi-closed pricing formula can be obtained for the option. The derivation goes beyond this discussion, however, so that we refer the interested reader to [35] for more information and present only the result here.

**Theorem 5** (Forward starting option in the Heston model) *Let $\kappa \geq \rho\sigma$ and $0 \leq t < t^\star < T$. If the stock price and the variance fulfill the risk-neutral dynamics* (15) *and* (16), *and if the Feller condition* (14) *also holds, then the price of a forward starting option at time t with payoff* (30) *is given by*

$$X_{\text{fso}}\big(t, S(t), K, T\big) = S(t)\hat{P}_1(t) - ke^{-r(T-t^\star)}S(t)\hat{P}_2(t), \tag{31}$$

*where*

$$\hat{P}_j(t) := \int_0^\infty P_j\big(1, \xi, t^\star, k\big)p\big(\xi, v(t)\big)\mathrm{d}\xi$$

*and the probabilities $P_j$ are given in Eq.* (18). *Moreover,*

$$p\big(\xi, v(t)\big) = \frac{B}{2}e^{-(B\xi+\Lambda)/2}\left(\frac{B\xi}{\Lambda}\right)^{(R/2-1)/2} I_{R/2-1}(\sqrt{\Lambda B\xi})1_{\{\xi>0\}},$$

$$\Lambda = Be^{-(\kappa-\rho\sigma)(t^\star-t)}v(t), \tag{32}$$

$$B = \frac{4(\kappa-\rho\sigma)}{\sigma^2}\big(1 - e^{-(\kappa-\rho\sigma)(t^\star-t)}\big)^{-1}, \tag{33}$$

*and*

$$R = \frac{4\kappa\theta}{\sigma^2},$$

*where $I_{R/2-1}(.)$ denotes the modified Bessel function of the first kind.*

For a forward starting option on the return of a stock with payoff function

$$Y_{\text{rfso}} = \left( \frac{S(T)}{S(t^\star)} - K \right)^+, \tag{34}$$

a corresponding variant of the option price formula (31) can be specified on the basis of

$$\left( \frac{S(T)}{S(t^\star)} - K \right)^+ = \frac{(S(T) - KS(t^\star))^+}{S(t^\star)}.$$

In the Heston model, the option price belonging to the payoff (34) is given by

$$X_{\text{rfso}}\big(t, S(t), K, T\big) = e^{-r(t^\star - t)} \big( \hat{P}_1(t) - K e^{-r(T-t^\star)} \hat{P}_2(t) \big), \tag{35}$$

where the expression $\kappa - \rho\sigma$ is replaced by $\kappa$ in Eqs. (32) and (33).

*Remark 10* For the numerical implementation of the option price formulas (31) and (35), we refer in particular to the calculation of the modified Bessel function of the first kind, which can be approximated by the following series expansion:

$$I_{R/2-1}\big(\sqrt{\Lambda B \xi(t^\star)}\big) \approx \sum_{n=0}^{N} \frac{(\Lambda B \xi(t^\star))^n}{2^{2n} n! \Gamma(n + R/2)}.$$

For practical applications, it turns out that the series converges with sufficient speed, so that even relatively small values of $N$ are acceptable.

With the aid of the closed formulas (31) and (35), we have the efficient tools we need in order to price forward starting options.

### 5.2.3 A Sparsely Parameterized Multi-Asset Heston Model

In order to price options based on several underlying assets, a multi-dimensional version of the Heston model was developed at the Fraunhofer ITWM by Dimitroff, Lorenz, and Szimayer (cf. [26]). We now wish to present this work. We first point out that, with the help of the Cholesky decomposition, the risk-neutral dynamics of the Heston model (15) and (16) can be represented as follows:

$$\mathrm{d}S(t) = rS(t)\mathrm{d}t + \sqrt{v(t)}S(t)\mathrm{d}W(t), \qquad S(t) = S_0,$$

$$\mathrm{d}v(t) = \kappa\big[\theta - v(t)\big]\mathrm{d}t + \sigma\sqrt{v(t)}\big[\rho\mathrm{d}W(t) + \sqrt{1-\rho^2}\mathrm{d}\widetilde{W}(t)\big], \quad v(t) = v_0,$$

where, for simplicity's sake, we dispense with the notation $\star$ and $Q$, and start directly with the risk-neutral parameterization relevant for the pricing.

**Multi-Dimensional Generalization** In the following treatment, we describe a parsimonious, multi-dimensional extension of the one-dimensional Heston model, in which each one-dimensional sub-model is a classical one-dimensional Heston model, although the price processes may exhibit correlations to each other. Consequently, the model is parsimonious in the sense that for a $d$-dimensional model, only $d(d-1)/2$ correlations between the risky securities are needed. For $i = 1, \dots, d$,

$$\begin{aligned}\begin{pmatrix} \mathrm{d}S_i(t) \\ \mathrm{d}\nu_i(t) \end{pmatrix} &= \begin{pmatrix} rS_i(t) \\ \kappa_i(\theta_i - \nu_i(t)) \end{pmatrix} \mathrm{d}t \\ &\quad + \begin{pmatrix} \sqrt{\nu_i(t)}S_i(t) & 0 \\ 0 & \sigma_i\sqrt{\nu_i(t)} \end{pmatrix} \begin{pmatrix} 1 & 0 \\ \rho_i & \sqrt{1-\rho_i^2} \end{pmatrix} \begin{pmatrix} \mathrm{d}W_i(t) \\ \mathrm{d}\widetilde{W}_i(t) \end{pmatrix} \end{aligned} \tag{36}$$

denotes the Heston model in vectorized form, where $W_i(t)$ and $\widetilde{W}_i(t)$ describe the uncorrelated Brownian motions.

The model is thus defined, except for its dependency structure. Let $W(t) = (W_1(t), \dots, W_d(t))$ and $\widetilde{W}(t) = (\widetilde{W}_1(t), \dots, \widetilde{W}_d(t))$ now be $d$-dimensional Brownian motions. For $i = 1, \dots, d$ and $j = 1, \dots, d$, we assume that $W(t)$ and $\widetilde{W}(t)$ are described by the following dependency structure:

1. $W(t)$ has the correlation matrix $\Sigma^S = (\rho_{i,j})$, i.e., $\langle W_i(t), W_j(t)\rangle = \rho_{i,j}$,
2. $\widetilde{W}(t)$ has the correlation matrix $I_d$, i.e., $\langle \widetilde{W}_i(t), \widetilde{W}_j(t)\rangle = \delta_{i,j}$,
3. $W(t)$ and $\widetilde{W}(t)$ are independent.

The complete correlation matrix of $(W(t), \widetilde{W}(t))$ is thus given by

$$\Sigma = \Sigma^{(W,\widetilde{W})} = \begin{pmatrix} \Sigma^S & 0 \\ 0 & I_d \end{pmatrix}. \tag{37}$$

The first assumption allows for an arbitrary correlation structure between the risky securities. In contrast, the second and third assumptions stipulate that the dependency structure of the variance processes is determined by the corresponding correlations of the Brownian motions, which are transferred to the variance processes by the parameters $\rho_i$ and $\rho_j$.

The model specification (36) and the assumed form of the correlation matrix (37) thus define the following correlation structure:

$$\begin{aligned} \frac{\mathrm{d}S_i(t)\mathrm{d}S_j(t)}{\sqrt{(\mathrm{d}S_i(t))^2(\mathrm{d}S_j(t))^2}} &= \rho_{i,j}, \\ \frac{\mathrm{d}S_i(t)\mathrm{d}\nu_j(t)}{\sqrt{(\mathrm{d}S_i(t))^2(\mathrm{d}\nu_j(t))^2}} &= \rho_{i,j}\rho_j, \\ \frac{\mathrm{d}\nu_i(t)\mathrm{d}\nu_j(t)}{\sqrt{(\mathrm{d}\nu_i(t))^2(\mathrm{d}\nu_j(t))^2}} &= \begin{cases} \rho_{i,j}\rho_i\rho_j, & \text{for } i \neq j, \\ 1, & \text{for } i = j. \end{cases} \end{aligned}$$

*Remark 11* The one-dimensional models presented here $(S_i(t), \nu_i(t))$ are affine with the corresponding closed formulas, according to Theorem 4. However, the multi-dimensional generalization is not affine, and as a consequence, its characteristic function cannot be simply determined. Therefore, Monte Carlo methods and tree methods—as described in Sect. 6—are generally required for pricing options with multiple underlying investment assets.

**Empirical Correlations and Correlation Adjustment** Under the assumption that the parameters of the one-dimensional sub-models are known, there are additional $(d-1)d/2$ free parameters from the matrix $\Sigma^S$ that must be determined in order to correlate the risky securities. If there is sufficient data available, this is accomplished with the help of the implied correlations of multi-asset options. If this data is not available, the empirical correlations $\widehat{\Sigma}^{\mathrm{emp}}$ from the time series of the risky securities can be estimated and adjusted to the model correlations $\Sigma^S$. Here, it is known that $\widehat{\Sigma}^{\mathrm{emp}}$ is an unbiased estimator for the correlation matrix $\Sigma^{\mathrm{emp}}$ of the investment assets, which is evidently strongly dependent on the non-observed quantity $\Sigma^S$.

The idea is now to adjust the correlation matrix $\Sigma^S$ so that it fits $\Sigma^{\mathrm{emp}}$, which, in turn, is estimated by $\widehat{\Sigma}^{\mathrm{emp}}$. Here, it is important to point out that $\Sigma^S$ describes the infinitesimal correlation of the Brownian motion $W(t)$ and $\Sigma^{\mathrm{emp}}$ describes the correlation of the log returns. We refer to the adjustment of $\Sigma^S$ to $\Sigma^{\mathrm{emp}}$ as the *correlation adjustment*. In the following treatment, we now formally define the estimator $\widehat{\Sigma}^{\mathrm{emp}}$. Let $r_i(k)$ for $k = 1, \ldots, K$ be discrete log returns of the $i$-th stock. Moreover, let

$$\hat{\nu}^{\mathrm{emp}}_{i,j,T,K}(\Sigma) = \frac{1}{K-1} \sum_{k=1}^{K} \big(r_i(k) - \hat{\mu}^i_K\big)\big(r_j(k) - \hat{\mu}^j_K\big). \tag{38}$$

Then, the empirical correlation matrix of the log returns is defined as

$$\widehat{\Sigma}^{\mathrm{emp}}_{T,K}(\Sigma) = \big(\hat{\rho}^{\mathrm{emp}}_{i,j,T,K}(\Sigma)\big)_{1\le i,j\le d}$$

and its elements, as

$$\hat{\rho}^{\mathrm{emp}}_{i,j,T,K}(\Sigma) = \frac{\hat{\nu}^{\mathrm{emp}}_{i,j,T,K}(\Sigma)}{\sqrt{\hat{\nu}^{\mathrm{emp}}_{i,i,T,K}(\Sigma)\hat{\nu}^{\mathrm{emp}}_{j,j,T,K}(\Sigma)}}. \tag{39}$$

It can now be shown that the entries $\hat{\rho}^{\mathrm{emp}}_{i,j,T,K}(\Sigma)$ of the empirical correlation matrix converge suitably to the entries $\rho_{i,j}$ of the model correlation matrix $\Sigma^S$. That is, the model correlations $\Sigma$ can be determined by calculating the historical, empirical correlations $\widehat{\Sigma}^{\mathrm{emp}}_{T,K}(\Sigma)$ using (38) and (39). This makes it possible to develop a procedure for estimating the unknown correlations. If we assume that the empirical correlations are observed under the risk-neutral measure $Q$, then, if $T$ and $K$ are large, the observed sample correlations are good approximations for the expected sample correlations, given the true correlation structure of the Brownian motions; that is,

$$\widehat{\Sigma}^{\mathrm{emp}}_{T,K} \approx \mathbf{E}^{Q}\,\widehat{\Sigma}^{\mathrm{emp}}_{T,K}\big(\Sigma^{\mathrm{true}}\big) =: \Sigma^{Q}\big(\Sigma^{\mathrm{true}}\big).$$

The unknown correlations can thus be determined by means of a minimization problem:

$$\min_{\Sigma \in \mathrm{Cor}(d)} \big\| \Sigma^{Q}(\Sigma) - \widehat{\Sigma}^{\mathrm{emp}}_{T,K} \big\|, \tag{40}$$

where $\mathrm{Cor}(d)$ denotes the space of the $d \times d$-dimensional correlation matrices and $\|.\|$, a suitable matrix norm. The solution of the minimization problem (40) is not trivial; however, it can be solved using standard software. We denote the solution as $\Sigma^{\star}$.

**Generating an Admissible Correlation Matrix** It is possible that the correlations estimated with the above algorithm may not lead to a valid (positive semi-definite) correlation matrix. In this case, a transformation is required. One possible algorithm that generates a genuine correlation matrix from an estimated one is the following (see [32] also):

1. Determine an eigenvalue decomposition of $\Sigma^{\star}$ as $\Sigma^{\star} = S \Lambda S^{T}$, where $\Lambda = \mathrm{diag}(\lambda_i)$.
2. Define the diagonal matrix $\tilde{\Lambda}$ with entries

$$\tilde{\lambda}_i = \begin{cases} \lambda_i & \text{if } \lambda_i \geq 0 \\ 0 & \text{if } \lambda_i < 0. \end{cases}$$

3. Generate the diagonal matrix $T$ with entries

$$t_i := \Big( \sum_{m} s_{im}^{2} \tilde{\lambda}_m \Big)^{-1}.$$

4. Define $B := \sqrt{T} S \sqrt{\tilde{\Lambda}}$ and obtain a new positive semi-definite correlation matrix as $\hat{\Sigma}^{\star} := B B^{T}$ with $\hat{\Sigma}^{\star}_{ii} = 1$.

For other relevant algorithms, we refer to [39], for example. Finally, then, with the generation of the correlation matrix, the sparsely parameterized multi-asset Heston model is completely defined.

## 6 The Heston Model in Action—Algorithmic Implementation

In this section, we turn to the questions that are relevant for implementing the Heston model.

### 6.1 Problems of Calibration

As previously shown, in a complete, arbitrage-free market, a derivative can be uniquely replicated by other investments available in the market. Therefore, in the theory of financial mathematics, the equivalent martingale measure and/or the market price of risk, is uniquely given by the model. As a consequence, the price of the derivative is also uniquely determined.

Because Heston's stochastic volatility model defines an incomplete financial market, the absence of arbitrage alone here does not suffice to uniquely determine a price; there are infinitely many equivalent martingale measures that define infinitely many arbitrage-free product prices. So-called lower and upper arbitrage bounds can then be specified for a financial product and, ultimately, all prices lying within these bounds are correct—according to financial mathematics theory.

In practice, these price bounds are insufficient. For the specific pricing of products, a single equivalent martingale measure must be chosen, which raises the following interesting question:

*"Who determines the martingale measure?"*

The short and amazing answer is (cf. [19]):

*"The market does!"*

The implication of the answer is simple: in determining the measure, one should include information available in the market in the form of traded products. This process, known as *model calibration*, uses the option prices observed in the market as input parameters. The goal is to use them to determine the model parameters so that the model prices correspond as closely as possible to the observed market prices.

However, since the number of traded products typically exceeds the number of model parameters by a wide margin, it frequently happens that not all market prices can be replicated exactly. The following algorithm uses the least squares method to calibrate the model for European calls.

For $i = 1, \dots, N$, let $X_{\text{call}}^{\text{market}}(K_i, T_i)$ be the prices of $N$ European calls observed in the market for various exercise prices $K_i$ and maturities $T_i$, and let $\omega_1, \dots, \omega_N$ be positive weights that add up to 1. We then obtain the simple *calibration algorithm* for the parameters $(\nu_0, \kappa, \theta, \sigma, \rho)$ describing the Heston model:

1. Solve the minimization problem

$$\min_{(\nu_0,\kappa,\theta,\sigma,\rho)} \sum_{i=1}^{N} \omega_i \left(X_{\text{call}}^{\text{market}}(K_i, T_i) - X_{\text{call}}\big(t, S(t), K_i, T_i\big)\right)^2.$$

The (calibrated) parameter set found here offers the best possible explanation for the observed market situation.

Here, one sees the decisive advantage of the Heston model. Since there are semi-closed calculations for the prices of European options, the required model prices do not have to be determined by means of laborious methods. In each iteration of the minimization algorithm, the $N$ model prices can thus be obtained very quickly.

*Remark 12*

(a) Because the above minimization problem is highly nonlinear, one needs methods of nonlinear optimization to find a solution. Here, one must take particular care that the solution algorithm for the global optimization problem can terminate in a local minimum. This makes it absolutely essential to check the resulting parameters for plausibility and, if needed, to start the optimization again using different initial values or different minimization algorithms.
(b) There are both deterministic and stochastic algorithms available for solving the optimization problem, and each type has specific advantages and disadvantages. For example, deterministic algorithms lend themselves to situations in which good initial values for the calibration exist. Based on the initial solution, these then attempt to minimize the target function by locally changing the parameters. As a result, the deterministic methods often converge very quickly, but do not leave the neighborhood of a local optimum. In contrast here, the stochastic optimization methods offer the possibility of abandoning an already discovered local minimum and continuing the search for a better solution. Implementing these algorithms is typically more laborious, but the calibration results are often superior to those obtained via deterministic methods.
(c) In addition to the option prices observed in the market, the market prices of other products can be used for calibration purposes. If these products do not have closed form solutions in the model, however, laborious numerical simulations are needed to determine the prices, and these are frequently very time-consuming. Therefore, the market prices of derivatives for which analytical solutions exist in the model form the basis for a satisfactory model calibration.
(d) For practically relevant applications, the prices observed in the market exert differing influences on the calibration. This might be a function of the product-specific bid/ask spread, for example, which is a sign of a product's liquidity. For this reason, when calibrating, practitioners often use various weights $\omega_i$ to weight the individual input prices, in order to emphasize relevant situations or reduce the influence of less significant ones.

For realistic applications, the calibrated parameters typically vary over time. This means that it may be necessary to re-calibrate the model repeatedly within a short time period (within a single day, for example). For these applications, the calibrated parameters are often used as the new initial values for the re-calibration.

## 6.2 Pricing Complicated Products; Aspects of Numerical Simulation

In practice, pricing simple products such as European calls and puts is generally not a problem. In the following section, we consider the pricing of more complicated derivatives using numerical methods such as Monte Carlo simulations and tree approximations. While the Monte Carlo simulation for determining option prices is based on the strong law of large numbers, tree pricing relies on the central limit theorem. Each method has its advantages and disadvantages, so that, in practice, it has proved to be effective to implement both methods. In addition to these methods, there are other numerical methods, such as those for solving partial differential equations or Fourier techniques. We will not discuss these further here, but instead, refer the interested reader to [4] or [22] for more information.

### 6.2.1 Variants of the Euler Discretization

In order to price complex products traded in non-liquid markets, it is necessary to simulate the stock and variance paths of the Heston model.

Although the variance does not have a closed solution, its distribution—the non-central chi-squared distribution—is known. Thus, a promising approach might be to exactly simulate variance values $\nu(t)$ directly with the aid of the distribution. Such an approach is presented in [21]. With the exactly simulated variance process, the stock price process can then be determined using a suitable discretization method.

These methods function well for independent and therefore uncorrelated Brownian motions. However, problems arise in the generalized case for high absolute values of the correlation. For this case, an unbiased method is described in [21] that includes an inverted Fourier transformation. However, this method is much more time intensive than simpler discretization methods (see [36]).

The following algorithm introduces a naive discretization method suited to the Heston model that is based on the Euler–Maruyama method for the numerical solution of stochastic differential equations.

1. Initialize the variance and stock price approximation by $\nu(0) = \nu_0$ and $S(0) = S_0$.
2. Define $\Delta = T/n$, where $T$ denotes the product maturity and $n$ the number of discretization steps.
3. Repeat for $j = 1, \ldots, n$:
   (a) Simulate independent random variables $Z_1, Z_2 \sim \mathcal{N}(0, 1)$.
   (b) Define $Z_3 = \rho Z_1 + \sqrt{1-\rho^2} Z_2$.
   (c) Discretize the stochastic differential equation of the variance and iterate

$$\nu(j\Delta) = \nu((j-1)\Delta) + \kappa(\theta - \nu((j-1)\Delta))\Delta + \sigma\sqrt{\nu((j-1)\Delta)\Delta} Z_3.$$

(d) Discretize the stochastic differential equation for the logarithmized stock price $\mathcal{X}(t) = \ln(S(t))$ and iterate

$$\mathcal{X}(j\Delta) = \mathcal{X}\big((j-1)\Delta\big) + \left(r - \frac{\nu((j-1)\Delta)}{2}\right)\Delta + \sqrt{\nu\big((j-1)\Delta\big)\Delta}Z_1.$$

4. Determine the path $\mathcal{X}(t)$ as a linear approximation between the discrete time-points $\mathcal{X}(j\Delta)$ for $j = 0, 1, \ldots, n$. Then, $S(t) = \exp(\mathcal{X}(t))$ is the stock price path.

Although the continuous-time solution of the variance process assumes only non-negative values, the approximation can indeed generate negative values. Here, however, the root terms that must be determined in steps 3(c) and 3(d) are complex and unusable for the next iteration.

Various methods are described in the literature that compensate for this obvious weakness. For a systematic investigation of the methods presented below, we refer the reader to [36], which is based on empirical results.

1. Absorption (A): Use the positive part of the predecessor of the variance iteration $\nu((j-1)\Delta)^+$ to approximate the variance

$$\nu(j\Delta) = \nu\big((j-1)\Delta\big)^+ + \kappa\big(\theta - \nu\big((j-1)\Delta\big)^+\big)\Delta + \sigma\sqrt{\nu\big((j-1)\Delta\big)^+\Delta}Z_3$$

and to determine $\mathcal{X}(j\Delta)$ in the simulation step.
2. Reflection (R): Use the absolute amount of the predecessor of the variance iteration, that is,

$$\nu(j\Delta) = \big|\nu\big((j-1)\Delta\big)\big| + \kappa\big(\theta - \big|\nu\big((j-1)\Delta\big)\big|\big)\Delta + \sigma\sqrt{\big|\nu\big((j-1)\Delta\big)\big|\Delta}Z_3$$

to determine the variance value. Use the absolute amount also for $\mathcal{X}(j\Delta)$ in the simulation step.
3. Higham and Mao (HM): Use the absolute amount $|\nu((j-1)\Delta)|$ only in the root terms, that is, once each for calculating the succeeding value of the variance and the stock price. The other expressions of $\nu((j-1)\Delta)$ remain unchanged (see [30]).
4. Partial truncation (PT): Use the positive part $\nu((j-1)\Delta)^+$ of the preceding value of the variance approximation only in the root terms to calculate the succeeding value of the variance and the stock price. The other incidences of $\nu((j-1)\Delta)$ remain unchanged (see [24]).
5. Full truncation (FT): Use the positive part of the predecessor of the variance iteration in the drift and diffusion component of the variance approximation, that is,

$$\nu(j\Delta) = \nu\big((j-1)\Delta\big) + \kappa\big(\theta - \nu\big((j-1)\Delta\big)^+\big)\Delta + \sigma\sqrt{\nu\big((j-1)\Delta\big)^+\Delta}Z_3$$

and for $\mathcal{X}(j\Delta)$ in the simulation step (see [36]).

**Table 3** Simulated prices and standard deviations (in parentheses) for a European call in the Heston model with $S_0 = K = 100$, $T = 1$, $v_0 = \theta$, $r = 0.05$, and $\rho = -0.9$. Moreover, $n = 100\,000$. The exact value corresponds to the analytical price

| $\kappa$ | $\theta$ | $\sigma$ | exact | A | R | HM | PT | FT |
|---|---|---|---|---|---|---|---|---|
| 2.0 | 0.04 | 0.2 | 9.349 | 9.328 (0.037) | 9.328 (0.037) | 9.328 (0.037) | 9.328 (0.037) | 9.328 (0.037) |
| | | 0.5 | 8.881 | 8.890 (0.029) | 8.919 (0.029) | 8.871 (0.029) | 8.890 (0.029) | 8.859 (0.029) |
| | 0.01 | 0.2 | 5.594 | 5.584 (0.017) | 5.585 (0.017) | 5.582 (0.017) | 5.584 (0.017) | 5.582 (0.017) |
| | | 0.5 | 5.156 | 5.512 (0.014) | 5.862 (0.016) | 5.593 (0.019) | 5.516 (0.014) | 5.149 (0.012) |
| 0.5 | 0.04 | 0.2 | 9.278 | 9.255 (0.034) | 9.255 (0.034) | 9.255 (0.034) | 9.255 (0.034) | 9.255 (0.034) |
| | | 0.5 | 8.317 | 8.467 (0.024) | 8.633 (0.025) | 8.534 (0.027) | 8.468 (0.024) | 8.307 (0.023) |
| | 0.01 | 0.2 | 5.507 | 5.515 (0.014) | 5.533 (0.015) | 5.515 (0.014) | 5.515 (0.014) | 5.496 (0.014) |
| | | 0.5 | 4.723 | 5.352 (0.012) | 6.041 (0.016) | 5.852 (0.027) | 5.355 (0.012) | 4.743 (0.009) |

The example in Table 3 confirms the result in [36]; namely, of all the methods described above, full truncation functions best. Here, we consider a European call with a residual term of one year, and use $n = 100\,000$ paths for the Monte Carlo simulation. The remaining parameters are chosen so that, for falling $\kappa$ and $\theta$ and rising $\sigma$, the discretized variance process becomes more frequently negative and the various truncation methods must be applied. In the table, we present the analytical value, the simulated option price, and, in parentheses, the standard deviation of the option price estimator.

One notices here—especially in cases where the variance process must be modified frequently—considerable price differences for similarly small and therefore unremarkable standard deviations. Thus, the danger for practical application is that incorrect option prices having small standard deviations might mistakenly be considered good. All told, we can propose the following simple procedure:

1. Repeat for $i = 1, \ldots, N$:
   (a) Simulate one path each of the Heston price process $S(t)$ and the variance process $v(t)$, $t \in [0, T]$ as described above, using the Euler–Maruyama scheme and the FT variant for the variance process.
   (b) Calculate the corresponding option payoff $Y^{(i)}$.
2. Estimate the option price $X_Y$ as

$$X_Y := e^{-rT} \frac{1}{N} \sum_{i=1}^{N} Y^{(i)}.$$

### 6.2.2 Tree Approaches

The Monte Carlo simulation technique, which simulates stock price paths successively, is especially well suited for pricing path-dependent derivatives.

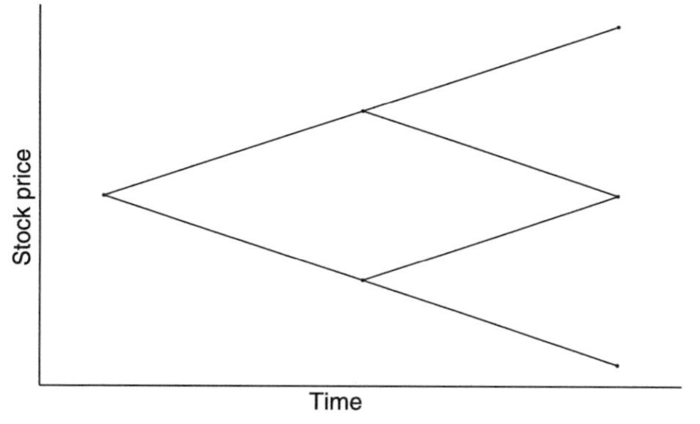


**Fig. 10** Example of a two-step binomial approximation

When pricing products that allow for multiple exercise times or even for a permanent possibility to exercise the option—so-called Bermuda or American options—tree methods offer simpler and more efficient approaches than Monte Carlo methods. Here, for each time increment, one assumes several possible developments—the next points or next nodes—and assigns them transition probabilities. The next nodes therefore represent possible future stock prices, each of which has a different probability. To determine the option prices, the nodes are then processed from the leaves toward the root using backward induction. Analogously to the algorithm of the Euler–Maruyama method for path generation, $\Delta = T/n$.

To approximate efficiently, it is crucial to be able to calculate both the nodes of the tree and the transition probabilities before the actual backward induction. Moreover, if the probabilities are chosen so that the first two moments of the price increments of the continuous and the approximated models coincide, then, according to Donsker's theorem, the tree approximation converges to the continuous process. For a detailed examination of both the standard approximation methods in the Black–Scholes model and the theory of convergence, we refer the reader to [4]. Figure 10 shows a two-period binomial tree.

We now present an algorithm developed at the ITWM by Ruckdeschel, Sayer, and Szimayer (see [9]) that achieves an efficient tree approximation in the Heston model. The method's fundamental idea is to model the variance and stock price processes as separate trees and to incorporate the correlation of the Brownian motions via a modification of the resulting transition probabilities.

However, because the variance process is mean stationary and its diffusion component depends on the current value $v(t)$, a naive approximation of the process leads to difficulties during implementation. The tendency to revert to the mean causes the process drift to become larger as the process moves further away from $\theta$. For large trends, however, one sees negative and thus non-admissible transition probabilities. On the other hand, the dependency of the diffusion component on the current state leads to jump heights that depend on the starting level.

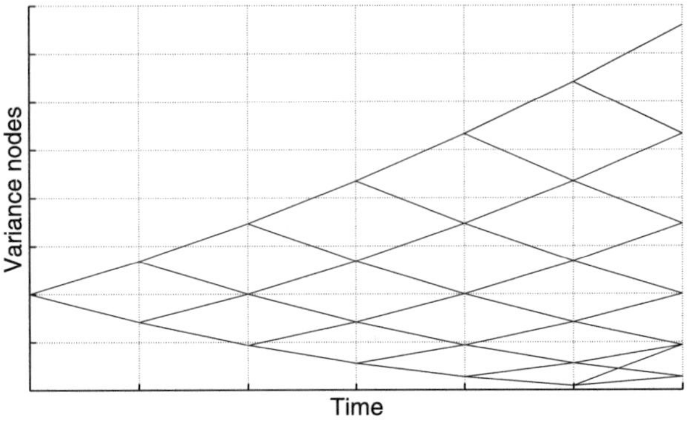


**Fig. 11** Binomial approximation of the variance process

For such a tree approximation, the number of nodes increases exponentially, the computational effort increases, and the tree becomes inefficient, that is, useless for practical application. Here, the Itô transformation

$$R(t) = \frac{2\sqrt{\nu(t)}}{\sigma},$$

offers a remedy, since the variance of the resulting process

$$\mathrm{d}R(t) = \left(\left(\frac{2\kappa\theta}{\sigma^2} - \frac{1}{2}\right)\frac{1}{R(t)} - \frac{\kappa}{2}R(t)\right)\mathrm{d}t + \mathrm{d}W_2(t), \quad R(0) = \frac{2\sqrt{\nu_0}}{\sigma}$$

is constant and a binomial approximation re-combines, since all approximation nodes exhibit the distance $\sqrt{\Delta}$. Inversion of the transformation allows one to then determine the variance values for the detected nodes. If, for each of these nodes, one now chooses successors that surround the drift, one can ensure that the transition probabilities are positive and add up to one, and that the approximation converges to the continuous model.

Figure 11 shows a variance approximation. Note, first, that the state-dependent diffusion causes the node intervals to increase as one moves upward and, second, that one sees irregular jumps—that is, jumps with multiple jump heights—for small variances, due to the tendency to revert to the mean.

To approximate the stock price, [9] uses a trinomial tree. Although this increases the computational effort, it also improves the accuracy of the approximation. Analogously to the variance approximation, the diffusion component of the logarithmized stock price process is not constant, but depends instead on the current variance value, that is, on the current node of the variance approximation. Therefore, a naive approximation also leads here to a non-efficient (from a numerical perspective) tree.

One possible way around this problem is to define a constant $\tilde{\nu}$, which describes the smallest variance unit allowed for the approximation. Possible approximation nodes then exhibit the distance $\sqrt{\tilde{\nu}\Delta}$. If one also defines all needed stock jumps as integer multiples of this unit, then the nodes of the stock price approximation lie on a uniform grid and the approximation re-combines. In order to ensure convergence, one determines the transition probabilities in the model such that the first two moments in the continuous and approximated models coincide.

In summary, with this approach, one has determined the tree approximation for the variance and stock price processes, since the node set, each successor node, and the transition probabilities are known.

The next step is to combine both separate approximations into one tree model. Here, one must determine the successor nodes and transition probabilities for each possible combination of the two node sets.

The successor nodes for a node combination of the stock and variance approximations are given by the six combinations of each successor node of the separate approximations. For uncorrelated Brownian motions, each transition probability is calculated as the product of the separate probabilities. For a non-zero correlation, the authors of [9] introduce an adjustment of the product probabilities that retains the marginal moments already determined in the course of preparing the separate trees. Because the adjustment of the probabilities can be determined before the actual backward induction and the tree approximation is re-combining, the resulting approximation method is fast and accurate, even for high correlation values.

[13] presents an application of the algorithm described here. In this application, the author prices employee stock options having permanent exercise rights and specific execution hurdles.

## 6.3 The Complex Logarithm—An Important Detail for Implementation

In financial mathematics, the use of characteristic functions for product pricing is based, in particular, on the very generally applicable price formula from [22], which is, in turn, based on a fast Fourier transformation. This representation also forms the theoretical basis of Theorem 4 for analytical solutions in the Heston model. Implementing and numerically evaluating this semi-closed formula requires the use of complex values, which for our purposes, are incorrectly treated under some circumstances.

In order to permit a detailed investigation of the problem in the following discussion, we waive the case distinction from Theorem 4 by defining

$$\varphi(u) = \varphi_2\big(S(t), \nu(t), t, u\big)$$

and using the relationship

$$\varphi_1\big(S(t), \nu(t), t, u\big) = \frac{e^{-r(T-t)}}{S(t)}\varphi(u-i).$$

The characteristic function then becomes

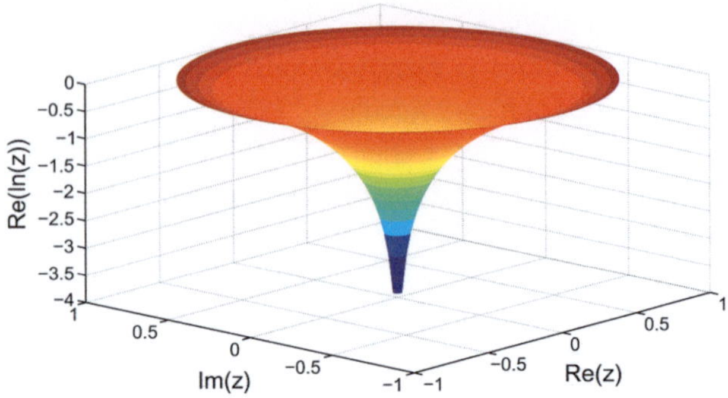


**Fig. 12** Real part of the complex logarithm for a complex number $z$

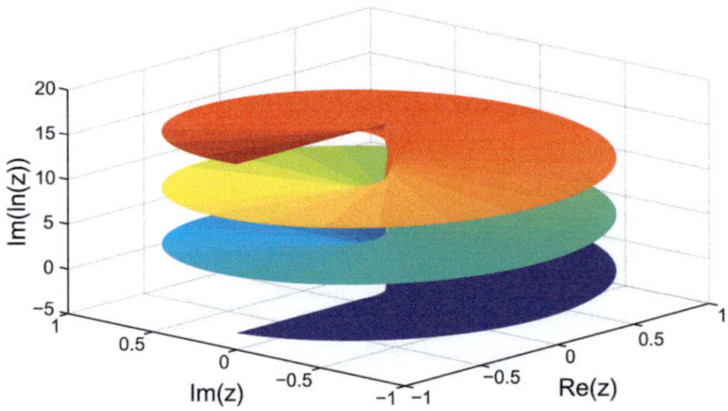


**Fig. 13** Imaginary part of the complex logarithm for a complex number $z$

$$\begin{aligned}\varphi(u) &= \exp\bigl(iu\bigl(\ln\bigl(S(t)\bigr) + r\tau\bigr)\bigr)\\ &\quad\times \exp\left(\frac{\kappa\theta}{\sigma^2}\left((\kappa - \rho\sigma ui + d)\tau - 2\ln\left(\frac{1 - ge^{d\tau}}{1 - g}\right)\right)\right)\\ &\quad\times \exp\left(\frac{\nu(t)}{\sigma^2}(\kappa - \rho\sigma ui + d)\frac{1 - e^{d\tau}}{1 - ge^{d\tau}}\right)\end{aligned}$$

with

$$\tau = T - t, \qquad g = \frac{\kappa - \rho\sigma ui + d}{\kappa - \rho\sigma ui - d}$$

and

$$d = \sqrt{(\rho\sigma ui - \kappa)^2 + \sigma^2\bigl(ui + u^2\bigr)}. \tag{41}$$

A significant problem with the implementation is the complex logarithm, which, in contrast to a real logarithm, is not unique. The standard software systems used for pricing financial products typically implement the principal value of the complex logarithm. Figures 12 and 13 show the real and imaginary parts of the complex logarithm for different branches.

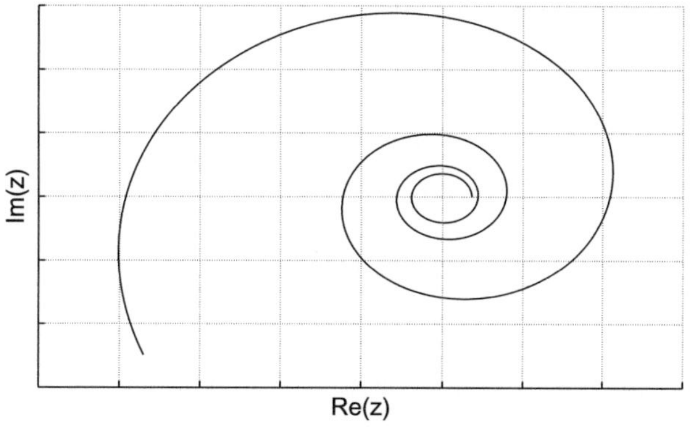


**Fig. 14** Trajectory of $(1 - ge^{d\tau})/(1 - g)$ in the complex plane

Due to the non-continuity described earlier, the integration of the characteristic function—which must be performed to determine the price in Theorem 4—is not stable starting at a certain residual time-to-maturity $\tau$. Frequently, the problem of the ambiguity leads to large price differences that are hard to locate as numerical difficulties. Numerical problems automatically arise for sufficiently large residual time-to-maturity if the Heston parameters are chosen such that $\kappa\theta \neq m\sigma^2$ for an integer $m$ (see [14]). This is because the trajectory of $(1 - ge^{d\tau})/(1 - g)$ describes a spiral around the origin with an exponentially increasing radius (see Fig. 14).

If the residual time-to-maturity is large enough, the trajectory inevitably crosses the negative real axis, thus producing a discontinuity. One remedy is to add $2\pi$ to the imaginary part of the result for each crossing of the negative real axis. A more elegant variant is to modify the characteristic function. To do so, one takes

$$\begin{aligned}\tilde{\varphi}(u) &= \exp\big(iu\big(\ln\big(S(t)\big) + r\tau\big)\big)\\ &\times \exp\left(\frac{\kappa\theta}{\sigma^2}\left((\kappa - \rho\sigma ui - d)\tau - 2\ln\left(\frac{1 - \tilde{g}e^{-d\tau}}{1 - \tilde{g}}\right)\right)\right)\\ &\times \exp\left(\frac{v(t)}{\sigma^2}(\kappa - \rho\sigma ui - d)\frac{1 - e^{-d\tau}}{1 - \tilde{g}e^{-d\tau}}\right)\end{aligned}$$

with

$$\tilde{g} = \frac{\kappa - \rho\sigma ui - d}{\kappa - \rho\sigma ui + d} = \frac{1}{g}$$

as the modified characteristic function. The only difference between $\tilde{\varphi}$ and $\varphi$ is the negative sign of $d$, that is, the choice of the negative root in Eq. (41). Since

$$\begin{aligned}d\tau - 2\ln\left(\frac{1 - ge^{d\tau}}{1 - g}\right) &= d\tau - 2\ln(e^{d\tau}) - 2\ln\left(\frac{1 - e^{-d\tau}/g}{1 - 1/g}\right)\\ &= -d\tau - 2\ln\left(\frac{1 - \tilde{g}e^{-d\tau}}{1 - \tilde{g}}\right)\end{aligned}$$

**Table 4** Initial values of the calibration and calibrated parameters

| | $\nu_0$ | $\kappa$ | $\theta$ | $\sigma$ | $\rho$ |
|---|---|---|---|---|---|
| Initial value | 0.12 | 3.00 | 0.09 | 0.10 | −0.95 |
| Calibration | 0.28 | 1.01 | 0.21 | 1.50 | −0.79 |

and

$$\frac{d(1-e^{d\tau})}{1-ge^{d\tau}} = \frac{d(1-e^{-d\tau})}{g-e^{-d\tau}} = \frac{-d(1-e^{-d\tau})}{1-\tilde{g}e^{-d\tau}}$$

are valid, $\tilde{\varphi}$ is equivalent to $\varphi$. The trajectory of $(1-\tilde{g}e^{-d\tau})/(1-\tilde{g})$, however, does not cross the real negative axis and the modification $\tilde{\varphi}$ is thus more stable numerically. To implement the analytical solution, it is therefore advisable to use the characteristic function $\tilde{\varphi}$.

### 6.4 Empirical Quality of the Heston Model

In this section, we want to illustrate the empirical quality of the Heston model, that is, its ability to replicate reality, by calibrating a real volatility surface. The stock of Allianz SE will serve as our example.

The corresponding volatility surface from 14 December 2011, obtained from the implied volatilities of European calls, is shown in Fig. 5 in Sect. 4.3.1. The shape of the surface is characteristic for volatility surfaces in general. Thus, for a fixed maturity, an option's implicit volatility is typically lower, the closer the strike lies to the current stock price. One also observes that, for a fixed exercise price, the implied volatility declines as the term of the option increases.

The freely chosen initial values for the calibration of the Heston parameters and the calibrated results obtained by applying a deterministic minimization algorithm are listed in Table 4. Figure 15 shows the calibrated surface that results when the implicit volatilities for given maturities and execution prices are calculated and presented with the help of the calibrated Heston parameters. Typically, the calibrated surface is considerably smoother than the original, but the characteristics of the real volatility surface are retained.

## 7 Mathematical Modeling and Algorithmic Implementation in the Financial Market—A Few Closing Remarks

This example of option pricing in connection with the Heston model is but one of many similar research and implementation projects that have been successfully dealt with by the Financial Mathematics Department of the Fraunhofer ITWM in cooperation with partners from the financial and insurance industries. Some examples of other projects involving innovative in-house developments and algorithmic implementations are:

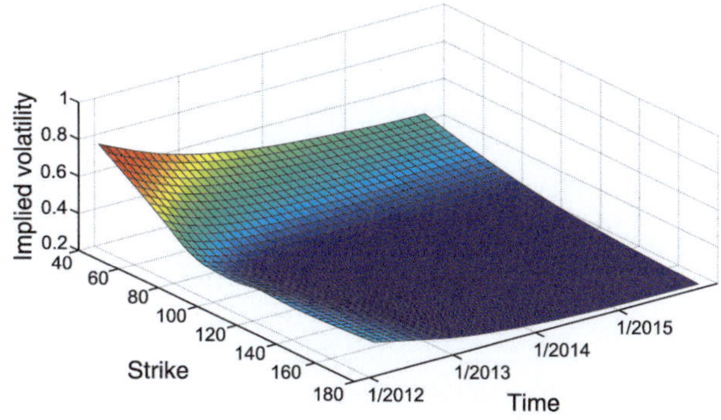


**Fig. 15** Calibrated implied volatility surface

- development of a new stock price model based on the explicit modeling of future dividend payments, in cooperation with the University of Cambridge (see [8]);
- development of a dynamic mortality model for evaluating longevity bonds, together with the Hypovereinsbank (see [7]);
- algorithmic implementation of robust statistics in the field of operational risk (see [31]), honored with a "best paper award";
- development of a completely new approach for efficient, multi-dimensional binomial trees (see [6]).

In addition, many of the algorithms used in the daily work of the ITWM are described extensively in [5].

There are several components common to all these projects and developments that are typical for implementations in the financial and insurance markets:

1. In general, the methods used are based on continuous-time stochastic processes and require thorough training in the fields of Itô calculus, martingale theory, and stochastic processes.
2. The client's wish for the best possible explanation of observed market prices leads to a wish for the generalization of existing models. Here, one must always make sure that the introduction of further parameters (e.g., by replacing a constant with a deterministic function) does not lead to numerical or statistical instability.
3. The use of a variety of numerical methods (e.g., Monte Carlo simulation, tree methods, Fourier transformation) is necessary in order to be able to calculate the prices of the diverse (exotic) options. Here, the character of the option determines the choice of the numerical algorithm. There is no universal, standard algorithm that performs well for all option types.
4. Calibration of the parameters plays a very significant role. While it's true that no spectacular theoretical results can be achieved in this domain, reliably calibrated parameters form the basis of all mathematical modeling and calculation that succeeds in the market.

5. Theoretical understanding of the models is indispensable if one is to calculate those values that are actually desired. The lack of understanding in shifting between the risk-neutral and the physical model worlds, in particular, is a frequent source of error.

Finally, it is important to emphasize the responsibility of the financial mathematician to help ensure a correct—and above all, wise—application of his models. Particularly with a view toward the financial crisis of these recent years, the financial mathematician must

- warn against mistaking the model for the reality,
- point out the inability of most models to predict, and
- avoid bringing excessive complexity into derivative products.

It was precisely the successful mathematical treatment of ever newer and more complex problems in the financial market that encouraged product designers to offer ever more complexly structured products—products whose effects were, in large measure, incomprehensible to customers but were bought anyway, despite this lack of understanding. Here too, the financial mathematician has a responsibility to warn against such dangerous developments.

## References

### Publications of the Authors

1. Korn, R.: Optimal Portfolios. World Scientific, Singapore (1997)
2. Korn, R.: Elementare Finanzmathematik. Berichte Fraunhofer ITWM **39**, 1–89 (2002)
3. Korn, R.: Optimal portfolios—New variations of an old theme. Comput. Manag. Sci. **5**, 289–304 (2008)
4. Korn, R., Korn, E.: Optionsbewertung und Portfolio-Optimierung. Vieweg, Wiesbaden (2001)
5. Korn, R., Korn, E., Kroisandt, G.: Monte Carlo Methods and Models in Finance and Insurance. Chapman & Hall/CRC Financial Mathematics Series. CRC Press, London (2010)
6. Korn, R., Müller, S.: The decoupling approach to binomial pricing of multi-asset options. J. Comput. Finance **12**, 1–30 (2009)
7. Korn, R., Natcheva, K., Zipperer, J.: Langlebigkeitsbonds: Bewertung, Modellierung und Anwendung auf deutsche Daten. Blätter der DGVFM **27**(3), 397–418 (2006)
8. Korn, R., Rogers, L.: Stocks paying discrete dividends: modelling and option pricing. J. Deriv. **13**(2), 44–49 (2005)
9. Ruckdeschel, P., Sayer, T., Szimayer, A.: Pricing American options in the Heston model: a close look at incorporating correlation. J. Deriv. **20**(3), 9–29 (2013)

### Dissertations in the Area at Fraunhofer ITWM

10. Horsky, R.: Barrier Option Pricing and CPPI-Optimization. Ph.D. thesis, TU, Kaiserslautern (2012)

11. Krekel, M.: Some New Aspects of Optimal Portfolios and Option Pricing. Ph.D. thesis, TU, Kaiserslautern (2003)
12. Natcheva, K.: On Numerical Pricing Methods of Innovative Financial Products. Ph.D. thesis, TU, Kaiserslautern (2006)
13. Sayer, T.: Valuation of American-style derivatives within the stochastic volatility model of Heston. Ph.D. thesis, TU, Kaiserslautern (2012)

## Further Literature

14. Albrecher, H., Mayer, P., Schoutens, W., Tistaert, J.: The little Heston trap. Wilmott Magazine, 83–92 (2007)
15. Artzner, P., Delbean, F., Eber, J.M., Heath, D.: Coherent measures of risk. Math. Finance **9**(3), 203–228 (1999)
16. Bachelier, L.F.: Théorie de la spéculation. Ann. Sci. Éc. Norm. Super. **17**, 21–86 (1900)
17. Barndorff-Nielsen, O., Shephard, N.: Non-Gaussian Ornstein–Uhlenbeck based models and some of their uses in financial economics. J. R. Stat. Soc. B **63**, 167–241 (2001)
18. Bingham, N., Kiesel, R.: Risk-Neutral Valuation: Pricing and Hedging of Financial Derivatives. Springer, Berlin (1998)
19. Björk, T.: Arbitrage Theory in Continuous Time, 2nd edn. Oxford University Press, Oxford (2004)
20. Black, F., Scholes, M.: The pricing of options and corporate liabilities. J. Polit. Econ. **81**, 637–654 (1973)
21. Broadie, M., Kaya, Ö.: Exact simulation of stochastic volatility and other affine jump diffusion processes. Oper. Res. **54**(2), 217–231 (2006)
22. Carr, P., Madan, D.: Option valuation using the fast Fourier transform. J. Comput. Finance **2**, 61–73 (1999)
23. Cont, R., Tankov, P.: Financial Modelling with Jump Processes. Financial Mathematics Series. Chapman & Hall, Boca Raton (2003)
24. Deelstra, G., Delbaen, F.: Convergence of discretized stochastic (interest rate) processes with stochastic drift term. Appl. Stoch. Models Data Anal. **14**(1), 77–84 (1998)
25. Delbaen, F., Schachermayer, W.: The Mathematics of Arbitrage. Springer, Berlin (2006)
26. Dimitroff, G., Lorenz, S., Szimayer, A.: A parsimonious multi-asset Heston model: calibration and derivative pricing. Int. J. Theor. Appl. Finance **14**(8), 1299–1333 (2011)
27. Dupire, B.: Pricing and hedging with smiles. In: Dempster, M.A., Pliska, S.R. (eds.) Mathematics of Derivative Securities, pp. 103–111. Cambridge University Press, Cambridge (1997)
28. Eberlein, E., Keller, U.: Hyperbolic distributions in finance. Bernoulli **1**, 281–299 (1995)
29. Heston, S.L.: A closed-form solution for options with stochastic volatility with applications to bond and currency options. Rev. Financ. Stud. **6**(2), 327–343 (1993)
30. Higham, D.J., Mao, X.: Convergence of Monte Carlo simulations involving the mean-reverting square root process. J. Comput. Finance **8**(3), 35–61 (2005)
31. Horbenko, N., Ruckdeschel, P., Bae, T.: Robust estimation of operational risk. J. Oper. Risk **6**, 3–30 (2011)
32. Jäckel, P.: Monte Carlo Methods in Finance. Wiley, West Sussex (2002)
33. Karatzas, I., Shreve, S.E.: Brownian Motion and Stochastic Calculus, 2nd edn. Springer, Berlin (1991)
34. Kloeden, P.E., Platen, E.: Numerical Solution of Stochastic Differential Equations. Springer, Berlin (1999)

35. Kruse, S., Nögel, U.: On the pricing of forward starting options in Heston's model on stochastic volatility. Finance Stoch. **9**, 233–250 (2005)
36. Lord, R., Koekkoek, R., van Dijk, D.: A comparison of biased simulation schemes for stochastic volatility models. Quant. Finance **2**, 177–194 (2010)
37. Madan, D.B., Seneta, E.: The variance gamma model for share market returns. J. Bus. **63**, 511–524 (1990)
38. Mikhailov, S., Nögel, U.: Heston's Stochastic Volatility Model Implementation, Calibration and Some Extensions. Wilmott Magazine (2003)
39. Mishra, S.: Optimal Solution of the nearest Correlation Matrix Problem by Minimization of the Maximum Norm (2004). http://mpra.ub.uni-muenchen.de/1783/
40. Schoutens, W.: Lévy Processes in Finance: Pricing Financial Derivatives. Wiley, New York (2003)

# Part IV
# Education

# Applied School Mathematics—Made in Kaiserslautern

Wolfgang Bock and Martin Bracke

## 1 Why Applied School Mathematics?

The mathematical modeling week as an event format was introduced in 1993 by Helmut Neunzert in Kaiserslautern. Its direct successor, the Felix Klein Modeling Week, is thus the most venerable event of its type in Germany. In the interim, this successful event has inspired others far beyond the borders of Rheinland-Pfalz and led to modeling weeks in Aachen, Graz, Hamburg, and a number of other German cities. In addition to sponsoring the Felix Klein Modeling Week, the Felix Klein Center for Mathematics in Kaiserslautern also acts as supporting partner for a modeling week in Tramin, Italy.

The demand for modeling events in schools reflects the need for more applied school mathematics, that is, mathematics that deals with visible, authentic, concrete problems. This demand has been the driving force behind more than 20 years' worth of modeling activities aimed at young learners.[1] The success of this concept is verified by booked-out events with waiting lists and by the receipt of the "School Meets Science" prize, received from the Robert Bosch Foundation in 2011.

[1]For the sake of readability, we will dispense with the simultaneous use of both masculine and feminine singular pronouns; in general, the masculine form will be used. All references to persons, however, may be understood to apply to both genders.

W. Bock · M. Bracke (✉)
Fachbereich Mathematik, Technische Universität Kaiserslautern, Gottlieb-Daimler Straße 48, 67653 Kaiserslautern, Germany
e-mail: bracke@mathematik.uni-kl.de



H. Neunzert, D. Prätzel-Wolters (eds.), *Currents in Industrial Mathematics*,
DOI 10.1007/978-3-662-48258-2_11

**The Felix Klein Center for Mathematics, e.V, in Kaiserslautern** In conjunction with the State of Rheinland-Pfalz's "Mathematics Initiative," the Felix Klein Center for Mathematics was founded in late 2008. It is named after the important mathematician and scientific promoter Felix Klein (1849–1925) and establishes an institutional connection between the Mathematics Department of the TU Kaiserslautern and the Fraunhofer Institute for Industrial Mathematics ITWM, two organizations who have worked together closely and successfully for many years.

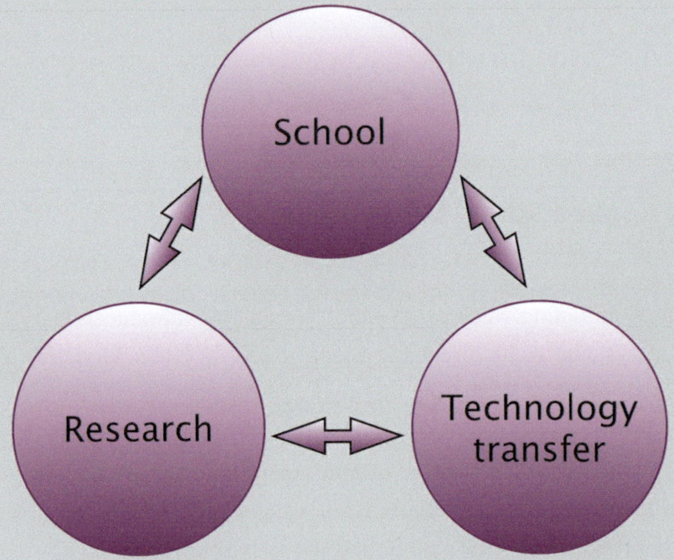


The importance of each of the Felix Klein Center's three cornerstones—"Research," "School," and "Technology transfer"—is underscored by the diverse activities offered to students. All three are necessary to allow them to gain a genuine look inside the working methods and worldview of modern applied mathematics.

An important reason for the increased interest in applied mathematics and its impact on everyday life is that, over the past four decades, mathematics has found itself playing an entirely new role in society. The causes for its new status are closely related to the rapid development of the computer. Today's computers are some 10 000 times faster than their counterparts of 20 years ago. Moreover, powerful algorithms can deliver solutions for such applications as large equation systems—another factor of even greater significance. Today, this combination of computer and algorithm allows one to solve typical problems many millions of times faster than was possible two decades ago. This has made it possible to evaluate even complex mathematical models for industrial products and manufacturing procedures, as well as for scientific, economic, and medical processes, with acceptably high precision and speed. As a consequence, one can simulate the behavior of an automobile engine, the progress of a bypass operation, and the weather for tomorrow by preparing the appropriate mathematical models, developing the corresponding algorithms, and evaluating them on the suitable computers (cf. The Concepts—Modeling). Today, virtually every industrial development and every prediction is based on these mathematical activities. In other words, mathematics has evolved into a key technology.

At the same time, this new and crucial role that mathematics plays throughout society appears not to be fully appreciated in our schools. The linking of the MINT[1] subjects is not sufficiently evident in either the teaching curriculum or in practical lesson designs. One reason for this is the structure of the teacher training programs. Here, applied mathematics still plays a quite modest role. Neither modeling nor work with algorithms—which, in practice, have largely supplanted the use of complicated formulas—are given a position in schools commensurate with their significance. Nor is interdisciplinary cooperation between mathematics, computer science, the natural sciences, and technology adequately demonstrated or trained.

Obviously, with the means available to even an advanced high-school mathematics class, it is not possible to generate models or algorithms suitable for simulating a bypass operation. But one can develop the same abilities using other "authentic problems" that are indeed amenable to school mathematics. For example, one can measure the quality of fleeces, optimize simple radiation therapy plans, and identify turtles on the basis of their shell patterns. However, these foundational activities—modeling, calculating, and interdisciplinary work—are hardly practiced in schools. There are, of course, so-called word problems. But these very seldom reflect "authentic problems," whose relevance is clear and interesting to the students. To practice ratios, for example, one poses the following problem: "Six bulldozers need 24 hours to level a building site. After six hours, three more bulldozers come to help. How long does it take to level the building site?" In posing the problem, it is assumed that one is to work only with ratios to find a solution. However, it would be much more interesting to ask whether it even makes sense to add three more dozers, since the three extra machines might result in a "traffic jam" that actually slows down the work. It is clear to anyone who stops to think about it that arbitrarily increasing the number of dozers on the building site won't necessarily lead to a reduction in the time required for leveling.

Algorithms do appear in school, but hardly any like those developed in recent years to handle large-scale problems. An applied mathematician cannot really understand why school computers are only allocated to computer science classes. Most students have computers at home, with Internet access, and use them for playing video games, for chatting, for writing, and for surfing—for everything except the mathematics that the computer actually represents.

[1] MINT (Mathematics, Information technology, Natural sciences and Technology) is the abbreviation used in Germany. In English speaking countries, the common equivalent is STEM (Science, Technology, Engineering, and Mathematics).

Although algorithms can be taught, instructed, and learned from books, they must be implemented and tested to be truly understood. Modeling cannot be taught at all; the only valid approach is learning by doing. The goal, here, is "meta-strategic knowledge," erected on a solid foundation of mathematics expertise. Above all, however, one must learn how to structure problems from the real world, separate the important data from the trivial, and then transform it all into mathematics. Interdisciplinary thinking and working must also be learned in the course of practical activity; knowledge alone is no guarantee that information and methods from the fields of science and technology can be profitably applied.

The adoption of authentic MINT projects into mathematics instruction currently fails on two accounts. First, most teachers simply have no experience with it and therefore don't trust themselves to tackle such authentic, open-ended problems with their students. This cannot be done in the traditional lecture format, that is, "face to face," but only in a cooperative format, that is, "side by side." Often, the transition from face to face to side by side is somewhat tentative and uncertain. Second, finding suitable, authentic problems is difficult at the start, before teachers have attained "problem identification competence."

Mathematical modeling, calculating, and interdisciplinary work expose young people to new experiences and, above all, convey a more realistic and lively picture of the subjects involved. In the end, this promotes a better and deeper understanding of mathematics, science, and technology. The basic prerequisite for successfully combating the shortage of teachers trained in the MINT subjects—and this includes engineers—is a sufficiently large pool of students who are motivated to grapple with these disciplines. Solid technical skills, when coupled with the joy that arises naturally from successfully managing complex and interesting problems, will lead automatically to a greater interest in the subject areas that make such successes possible in the first place.

## 2 Modeling and the Search for Authentic Problems

To kick off all modeling events at the Felix Klein Center, a real problem is introduced. After years of experience with mathematical modeling weeks, modeling days, and other similar events, we can safely conclude that students are more interested in problems stemming from their real-life environments. We have thus arrived at the following definition (cf. [3]):

**Definition 1** (Authentic problem) An authentic problem is one posed by a customer who is looking for a solution that can be applied to meet his needs. The problem is neither filtered nor reduced, but posed without manipulation in complete generality (i.e., posed exactly as perceived). A real-world problem (a.k.a. a realistic problem) is an authentic problem having elements that relate to the student's reality.

With this definition, we have identified the problem—the object, so to speak, upon which we want to practice "modeling." But what is "modeling" really? What must one do? There is extensive literature on mathematical modeling describing so-called modeling cycles, which have been continuously extended over the course of time (cf. [2] and the references it contains). Comparable cycles have also been applied in operations research (cf. [1]) since the 1950s. From the perspective of a problem solver, one can lay out a simple modeling procedure.

**The four phases of modeling**

1. Discuss the problem with the end-user (e.g., the customer)
2. Define the problem exactly
3. Design a mathematical model
   a. Analyze the problem
   b. Describe the problem mathematically
   c. Search for a suitable mathematical method
   d. Solve the mathematical problem
4. Interpret the solution in light of the end-user's original application

**Define the Problem Exactly** With Definition 1, we have already introduced the problem of the end-user, that is, the authentic problem. According to our definition, this problem description should be as unfiltered as possible. Naturally, this complicates Point 2. Defining the problem exactly represents a difficult challenge for students, for many working applied mathematicians, and for the customer alike. Plainly put, the real world is large and has many degrees of freedom, which allows a measure of latitude in defining any problem. The problem definition is also very dependent on the customer or problem supplier. It may even come to pass that the initial problem is formulated in such a way that it points toward a particular solution approach that later proves to be unsuited for solving the problem. Therefore, when defining the problem exactly, one must examine it carefully from all sides and, where needed, extract more information from the problem supplier. This allows one to clarify misunderstandings early on and to compensate for any missing data sets or specifications by means of skillful assumptions. For the latter, in particular, it is useful to ask oneself: What does the customer want to do with the solution? Is she perhaps also interested in something else, perhaps some issue or concern that lies hidden beneath the problem?

**Design a Mathematical Model** Designing a mathematical model, or mathematizing and solving the problem, surely represents the core of the mathematical modeling process. First, however, one must acknowledge that it is not easy in practice to draw a sharp boundary between "defining the problem exactly" and "designing a mathematical model." Both students and teachers, because they have purposefully enrolled in a mathematical modeling event, often get an idea immediately about which of the mathematical methods they are familiar with ought to be applied in a particular case. It is the same with mathematicians, who have many different mathematical models and techniques "in stock" in their heads and believe they recognize the appropriate ones. The expediency of such models or techniques is only revealed, however, when they are actually applied to the problem at hand.

Designing a mathematical model can be viewed as building a bridge from the real world to the mathematical world, where problems can be solved with the aid of mathematical techniques (cf. The Concepts—Modeling). However, depending on the type of problem, a model often requires more than just mathematics to be effective. A multitude of other disciplines may also be needed, such as the natural sciences, computer science, technology, geography, sports, economics, and many others. This can once again be traced back to the authenticity of the problem. For the "prediction of finish times in mountain races," for example, in which the finish time of an athlete for a race with many ascents and descents is to be predicted on the basis of his finish times in flat-land races, models from the field of biomechanics and physics may be incorporated, depending on the modeling approach. These then have to be skillfully assembled into a suitable, calculable mathematical model. This example makes clear one basic characteristic of mathematical modeling: it is often interdisciplinary.

It is especially important at the start of the modeling process to keep in mind the above-mentioned calculability of the mathematized problem using a suitable model. Here, it is essential to simplify the problem by means of intelligent assumptions so as to attain an easily calculable basis model (cf. The Concepts—Modeling). In practice, this means concentrating initially on the essentials and trying to design a model "under laboratory conditions." Naturally, such a model will have to be subsequently refined.

Once the model has been prepared and the problem translated into the language of mathematics, one can then commence to solve the mathematical problem. For authentic problems, this is a particular challenge for students. Either the techniques known from classroom instruction must be modified so as to be applicable to the problem or new techniques must be developed—on the basis of information gained in literature searches, for example. In the case of the latter, experience shows that students often cope better with the application of algorithms than teachers anticipate. Naturally, one cannot expect 12th graders to penetrate deeply into the mysteries of differential equation theory, for example. They can, however, learn to implement a Euler method.

The mathematical solution of the problem is often arrived at with the help of a computer—for example, when either very large data sets or suitable numerical methods are used. Of course, one also finds geometrical and analytical solutions on occasion, but we must confess that, today, this is more the exception than the rule. The same holds true, however, for applied mathematics at the university level and in industrial and research settings.

Thus, we sometimes hear the complaint from both students and teachers that modeling weeks have everything to do with computer science and nothing to do with mathematics. We, the authors, beg to differ. Certainly, one needs programming knowledge to implement the methods. However, those who express this complaint often overlook the fact that the method itself, as applied to a mathematical model that one has laboriously assembled, is also mathematics. If one wishes to use a computer to solve a problem, then one must also speak the language of the computer. And this language is—programming languages aside—mathematics.

**Interpret the Solution in Light of the End-User's Original Application** In order to re-cross the bridge from the world of mathematics and re-enter the real world, it is important to verify the solution's fitness for everyday use. This can be done by posing some simple questions: Does the solution I have just calculated make sense? How exact does it appear to be? Can it be improved in this regard?

At this point, the modeling procedure described at the start of this section develops into a cycle. In most cases, a comparison of the solution with reality reveals hints as to how to optimize the model or account for new influencing variables. Then, the model design process starts again from the top. If the initial model works crudely, it can perhaps be improved upon by making new assumptions, for example, or by incorporating new data sets, or by integrating previously ignored effects. It happens all too often, however, that the solution attained simply doesn't reflect reality—or doesn't reflect it accurately enough to be useful. Here, it may be that the approach selected simply will not lead to the desired goal. In this case, one must go back to the drawing board and design a new model.

The modeling sequence therefore illustrates only a single iteration of the modeling activities. These are repeated, as needed, until a satisfactory solution can be presented. One also observes, especially in student groups, that modelers don't necessarily follow the modeling sequence from top to bottom. Instead, they tend to jump back and forth from one phase to another—and this is often a quite reasonable approach. It can, in fact, be helpful during the model design phase to give some thought to the subsequent real interpretation, in order to avoid wasting time calculating unrealistic solutions, for example.

For longer modeling events in particular, one can—depending on the problem being addressed—add two more points to the four listed at the start of this section in our description of the modeling process. The extended list then takes the following form:

**Extension to the four phases of modeling**

1. Discuss the problem with the end-user (e.g., the customer)
2. Define the problem exactly
3. Design a mathematical model
    a. Analyze the problem
    b. Describe the problem mathematically
    c. Search for a suitable mathematical method
    d. Solve the mathematical problem

4. Interpret the solution in light of the end-user's original application
5. **Translate the solution into the language of the end-user**
6. **Generate a product prototype**

Both of these additional points could certainly be subsumed under Point 4 of our modeling recipe. Indeed, the time available for the modeling event ultimately determines whether Points 5 and 6 are feasible. Both the Felix Klein Modeling Week and the modeling days end with student presentations, which are delivered in the language of the end-user—that is, of the problem supplier or customer. One example of a product prototype, as mentioned in Point 6, might be a computer calculation program. At any rate, the presenters should strive to speak the language of the end-user—that is, one containing as little mathematics as possible—and to demonstrate in that language an applicable solution. Roughly speaking, the student groups explain to the end-user what he can do to manage his problem without his needing to understand the underlying "deeper mathematics". Experience shows that this is a difficult task for both the students and the teachers in the groups.

## 3 Modeling—The Attempt of a Didactic Classification

We saw in Sect. 2 what the term "modeling activities" encompasses. Namely, an authentic problem, taken whenever possible from the student's everyday life, is solved with the aid of the appropriate model design, mathematical terms, and methods, and the solution is then translated into the language of the end-user for purposes of comparing solution and reality. Here, wherever possible, the students should work out their own solutions; teachers and facilitators consciously keep a low profile. The problem suppliers are admonished to answer the students' questions exclusively from the perspective of the customers they are pretending to be. Depending on how much time is available, the facilitators can, however, attempt to help the student groups avoid getting lost in details by skillfully posing questions of their own. This gentle guidance is needed above all for groups with little or no previous experience in modeling. The goal of the modeling event for the students is to present the solution or product developed by their group to an audience consisting of the problem supplier and the students from the other groups.

In these modeling events, we find four of the five features of Jank and Meyer's activity oriented instruction (cf. [6]):

**Orientation on Interests** The active, hands-on investigation of various topics—a characteristic of activity oriented instruction—allows students to recognize, critically reflect upon, and further develop their own interests.

The same principle is applied at the modeling events sponsored by the Felix Klein Center for Mathematics. The problems posed during a modeling week are designed to represent a broad spectrum of topics, from which the students can then choose the ones that most appeal to their own interests.

**Self-Initiative and Guidance** In activity oriented instruction, the students receive as few guidelines as possible. They should remain free to explore, discover, and plan for themselves. This self-initiative poses the risk for the students, however, that the instruction degenerates into mere action and fun. Action and fun are well and good, but something sensible must come of the instruction as well. Here, one needs a dialectic comprising self-initiative and guidance.

As already indicated, much importance is attached to the student groups' generating their own solutions during modeling events. The primary goal here is that the students identify with and take ownership of their problem. By virtue of their own investigations, the students develop their own expertise and can be justifiably proud of their own accomplishments. The guidance provided by the facilitators or the teachers in attendance should come in the form of skillful questioning, done in such a way that the students critically reflect on their work and can recognize mistakes on their own. Experience shows that the student groups themselves take over guidance and leadership functions, since they know that they must later present their work in front of their peers.

**Linking Head and Hand** "The students' mental and manual labors achieve a dynamic equilibrium in the teaching-learning process" ([6]).

This means that, step by step, a culture of learning develops in which the material activities of the students are viewed as the expression of human development. Although genuine manual labor also takes place occasionally in the modeling events, in the form of building a functioning prototype, for example, this is certainly not the rule. In our opinion, however, creating simulation software, testing parameter sets with this same software, or conducting real experiments can also be considered examples of material activities. Since these very same activities have been carried out by almost all student groups in recent years, one might also identify this feature of activity oriented instruction in the modeling events, albeit in a less pronounced form.

**Practicing Solidarity** Drawing on Jörg Habermas's distinction between communicative and instrumental activity, Jank and Meyer define two forms of activity: first, linguistic-argumentative deliberation about the meaning and significance of activities and, second, purposeful, goal-oriented work. They supplement these with a third, namely, demonstrating solidarity in behavior, which is contained in both communicative and instrumental activity. Linguistic deliberation belongs to communicative activity. It consists of deliberation among the participants in the teaching-learning process about activities and possible approaches. Purposeful, goal-oriented work consists of students carrying out the activities. Solidarity in behavior arises when the first two forms are in balance with each other. It is oriented on mutual rather than individual benefit and relies on teamwork and cooperative teaching-learning forms.

Mathematical modeling—at least in the events sponsored by the Felix Klein Center for Mathematics—is a group process. Group dynamics have an enormous influence on the course of the modeling event. The degree to which the group can identify with its given (or chosen) problem also plays a role.

**Orientation on the Product** Activity oriented instruction is also product oriented. Students and teachers agree upon a product that is to be the goal of the students' efforts. The students can then identify themselves with this product and it can serve as a basis for criticism and for the students to evaluate the lesson period.

Here, there is a basic difference between activity oriented instruction and mathematical modeling. Only in modeling events of longer duration, such as the Felix Klein Center sponsored Junior Engineer Academy or the Fraunhofer MINT-EC Math Talents, do the participants work toward a product that has been decided upon in advance. Here, precisely because of the long duration and the planning latitude given to the students, the goal is not defined in detail in advance and can be altered as needed in the course of the event. The focus for these events, however, is on the process of modeling or, more exactly, on all the processes that belong to modeling. This having been said, most events do indeed conclude with project presentations.

Mathematical modeling is thus in accord with activity oriented instruction in many respects, perhaps even in accord with project instruction. This includes, of course, both the advantageous and the disadvantageous. We have already highlighted the advantages of this approach, and we would now like to examine the disadvantages.

Mathematical modeling places two demands on teachers that, under some circumstances, may be difficult to reconcile. First, the teachers must ensure that the guidelines for the solutions are appropriate to both the subject matter and the developmental stage of the students. Second, they must keep an eye on the presumptive motives of the students and offer them as much latitude as possible to play with and implement their (i.e., the students') own ideas. Moreover, they must also try to weave together the instructional goals and the articulated goals of the activity. As in activity oriented instruction, this can be accomplished in two ways: "Either one succeeds in getting the students to embrace the instructional goals as the goals of their activity (which becomes more likely as the instructional situation more closely involves problems, topics, and tasks that are of significance to the students), or one offers space to discuss differing interests, to argue out opposing opinions, and to develop and pursue mutual activity goals" (cf. [5]).

Schools themselves still evidence some resistance to mathematical modeling. For one thing, it is difficult to fit such an instructional unit into a 45-minute class period. Nor do the curricular pressures, compartmentalization of subjects, or often standardized classrooms necessarily have a conducive effect on activity oriented instruction.

In light of these obstacles, the introduction of this instructional concept requires at least those conditions, "that are possible with goodwill and without too much effort" (cf. [5]). It is often possible, for example, to schedule the class periods of related subjects (such as biology and physics or French and English) consecutively, thus gaining more time for activity oriented lessons. It is also advisable to consult with the other teachers, since the time-consuming and laborious project phases may indeed draw students' energies away from their other subjects—whether they like it or not—and students may behave differently in the class periods following a modeling lesson, due to its differing didactical framework. Students and teachers should also discuss the important topic of grades. Now and again, a project founders; students should be made aware of this and also that such an outcome will not be penalized with a poor grade.

## 4 The Felix Klein Modeling Weeks

As briefly described at the outset, the first modeling week for students and teachers was staged in 1993, and has been held every year since. The concept has been modified only slightly over the ensuing years. In this section, we will first discuss the organizational aspects and then present several projects from previous modeling week events. This should serve to give the reader an impression of the complexity of the problems being treated and also illustrate the very wide variety of topics that can be covered in this format.

### 4.1 The Event Concept

Since 2009, the Felix Klein Center for Mathematics has staged two modeling week events per year for some 40–48 students and 16 teachers in Rheinland-Pfalz. Once the event has been announced, teachers apply for participation, each with 2–4 students from grades 11–13. The recommendation is to select students who have very good math skills and are interested in interdisciplinary work and the use of computers. Programming skills are not required, but certainly do not hurt. We'll take a closer look at this point later. For the teachers, the event is accredited by the Ministry of Education, Science, Continuing-Education, and Culture as professional training.

The event is usually held at a youth hostel, where all participants and teachers lodge for the duration. On the one hand, this increases expenses. However, it also offers plenty of advantages over other possible venues, such as the University, for example, where the students come in the morning and go home in the afternoon, after putting in their prescribed hours.

In any case, the schedule is always the same: Participants arrive on Sunday evening and, after dinner, are welcomed officially and offered a brief introduction to the event. Then, a total of eight projects are introduced by the project tutors, who are normally lecturers, employees, or PhD students from the TU Kaiserslautern, the Fraunhofer Institute of Industrial Mathematics, or the University of Koblenz–Landau—with whom we have been collaborating for a number of years. The project introductions typically last 5–10 minutes each and, as described in Sect. 2, are formulated in the language of the end-user—that is, they contain virtually no mathematical terminology (Several examples of these project introductions follow this discussion of the event's organizational concept). After each introduction, the participants have a chance to ask questions. At this point, there is typically a conspicuous lack of curiosity; equally conspicuous, however, is how consistently this changes in the course of the week.

Once the introductions are finished, the participants can state their preferences regarding the eight projects. These are then logged into a software program developed expressly for this purpose.[2] Here, participants can choose up to three favorites and also register one project on which they definitely do not want to work. Once the preferences have all been logged, the software divides the participants into eight project groups of 5–7 students and two teachers. The most important criterion, of course, is the student preferences, but several other constraints are also taken into account:

- Participants from the same school are placed into different groups whenever possible. This practice is not usually greeted by the students with thunderous applause, since one's own schoolmates and teachers are generally the only people one knows before the event begins. However, in our experience, this approach greatly facilitates teamwork. For one thing, it prevents automatic sub-group formation within the project groups. Moreover, the participants communicate more openly with one another, since they don't know their fellow team members very well and therefore rarely make (possibly false) assumptions about strengths and weakness—a dynamic that tends to hinder progress.

  Because the participants very quickly transition to a first-name basis (and the familiar pronoun "Du"), it also makes life less complicated for the teachers if they are not working directly with the students they supervise daily in school.
- Programming skills are desirable for most projects. Therefore, when the groups are assembled, at least one person with such skills is assigned to each group, if possible. Depending on the requirements of the particular project, the project's tutor may request a larger number of programmers or students with other special skills. This might be familiarity with databases, for example, or experience with graphic programming or special expertise in a particular field of science.

After the project teams have been established, the first evening is concluded by a brief informational meeting with the project tutors and the teachers, in which the plan for the week is discussed and the teachers are informed about the role they play in the event (cf. Sect. 4.2).

On Monday, the project work begins. Each team is assigned its own room containing a flipchart or whiteboard. The program is easy to follow, since the work sessions are same each day, from 09:00–12:30 and 14:30 to 18:30, with fixed tea & coffee breaks during both the morning and afternoon sessions. Normally, however, the teams arrange their breaks to meet their own needs and often have to be reminded at break and lunch time that a change

[2]Interestingly, the genesis of this software was a modeling week project in 2009.

of pace now and then can also promote progress and that human beings need to eat and drink occasionally!

On Wednesday afternoon, we interrupt the routine for a joint field trip consisting of visits to the Fraunhofer Institute for Industrial Mathematics ITWM and the TU Kaiserslautern—provided the event location is close enough to permit this. If not, a visit to a business in the vicinity or a recreational activity, such as geo-coaching or rafting on the Rhine, serve as substitutes. This field trip is intended as an enforced time-out from work on one's own modeling project. It allows the participants to "power down" so to speak, and gives them a chance to exchange notes with members of the other groups. As a consequence of this exchange, Wednesday evening, after the participants' return, is often used for another work session, since students are eager to try out new ideas that have come to them during the day.

Officially, there are no work sessions scheduled for the evenings. As the week advances, however, the students begin to invest their supposed free time in their project work also. In future events, we plan to offer optional mini-workshops on such topics as *Using the scripting system LaTeX* or *Introduction to MATLAB*. On Thursday evening—their final evening together—the participants all work late. Since modeling week ends on Friday with presentations of the results from all project groups, and since a short written documentation of the work is expected, the first teams usually head off to bed around midnight. For some, the work continues deep into the night. Here, dry runs of the presentations—sometimes repeated dry runs—are also part of the (voluntary) program.

Friday revolves around the presentations of group results. Here, the students from each team present the solution they have developed. These presentations last 20–25 minutes; each is then followed by a brief discussion and concluding remarks from the project's combined tutor and "end-user." It is very important that each presentation be delivered in the language of the end-user—since it is designed to suit his needs—and that mathematical details are dispensed with as much as possible. Naturally, these details can be included in the written documentation. This non-technical approach allows all participants to understand the results and promotes lively discussion with active participation and many questions. One can observe here, first, that the students in the audience have the self-possession to ask the questions that are really on their minds and, second, that the presenters really have become experts in their respective project fields and can in almost all instances respond with cogent and intelligent answers.

Guests attending the final presentations, who have had no chance to observe the work in the individual groups, confirm this observation, which is a very important outcome of the modeling week and other similar events involving students. At the end, all participants are awarded certificates of participation, of course, and the event closes with a joint "coffee and cake" buffet.

One further organizational note involves computers. In the event advertisement and invitation, participants are encouraged to bring their own laptops or tablets with them. There is also a pool of laptops available, so that each team can be allocated at least one powerful device. In the early days of the modeling week, a central computing room was usually set up housing eight desktop computers with Internet access and printers. Today, however, the laptops in the eight group workrooms suffice. Internet access comes via WLAN, and there is even the chance to exchange data and execute printing jobs over a separate network. Whereas the eight desktop computers in the computing room used to be the only devices present, today, it is not unusual to find three or more laptops in each group.

As far as software is concerned, we make no specifications, merely offers: the allocated devices run standard software, such as Office, along with tools for graphics processing and various programming environments (usually Java, Free Pascal/Lazarus, Python, and C/C++). In addition, we also offer participants the use of the scientific programming environment MATLAB. This is commercial software, but the company Mathworks issues us temporary licenses for the duration of the events, which can also be used on the participants' own computers. MATLAB has the tremendous advantage that its comprehensive documentation and function catalog make it possible even for programming neophytes to implement their mathematical models and algorithms with the computer in their search for solutions. On the basis of excellent results in other school projects, we also plan to introduce LaTeX, since the students generally get up to speed quickly with it and are proud when they can produce professional-looking documents containing mathematical and scientific content.

## 4.2 The Role of the Teachers

As described above, the teachers work with the students in their groups during the modeling week. Here, they take on various roles. First and foremost, on the basis of their legal capacity, they serve as chaperones for the students. On the other hand, they are also to take part in the modeling process. Often, this is also the first time the teachers have attempted to solve an everyday problem with the aid of mathematical modeling. The important thing, however, is that teachers don't force their own solutions onto the students. Rather, they should assume the role of just another modeler in the group and, working "side by side" with their fellow modelers, discuss the project. In every instance, it should be the students themselves that chart the course the work is to take.

Nonetheless, if the students are grappling with the implementation of unknown techniques, teachers can serve as sounding boards and provide them with the needed support.

We emphasize the importance of explaining to the teachers in advance that "failure" is also part of the modeling process. Teachers are advised to let the students fail, if it comes to that. Thereafter, they can support the students with their troubleshooting and in the search for new ideas or workarounds.

## 4.3 Selected Modeling Projects

This section presents five very different projects that have been undertaken in connection with prior modeling weeks. The descriptions are composed in the language of each end-user and represent written drafts of the project presentations that the project tutors give at the start of an event. When possible, the oral presentations are supplemented with short video sequences or practical demonstrations.

Our goal here is not to give a detailed description of the selected projects with various possible solutions. Instead, we want to give the reader an impression of the complexity and diversity of the questions that have been addressed (and which therefore *can* be addressed) in modeling events. After each of the descriptions, there are several remarks about the general solution strategies developed by the working groups, as well as the data and software used.

*Example 1* (Is the penguin's waddle energy efficient?) Penguins are very agile in the water. Their top underwater speed of more than 10 km/h is ample evidence of their skill as swimmers. With their torpedo-shaped bodies and very short legs, however, they don't seem to be designed for walking. The style of locomotion resulting from their anatomy is the familiar waddle. It is known that penguins require more energy for walking than other terrestrial animals of the same weight. For this reason, one might assume that penguins do not use energy very efficiently for walking. This would appear strange, however, in light of the fact that some penguin species travel great distances over land to reach their breeding and nesting grounds. And, in fact, some research results suggest that penguins have developed the ideal walking style for their body shape, and that their waddle even saves energy! This conclusion is supported by the research results of Rodger Kram, which he presents in his article "Penguin waddling is not wasteful":

http://spot.colorado.edu/~kram/penguins.pdf

Your task is to get to the bottom of this question, from a mathematical perspective, by finding a way to describe the waddle and assess the amount of energy it requires. Can you confirm the biologist's research conclusions?

In 2000, Rodger Kram published his research results in the renowned scientific journal *Nature*. He conducted experiments to verify that the penguin's waddle uses energy very efficiently for terrestrial locomotion, given that the penguin's anatomy is optimized for the world of water.

The first thought of the applied mathematician is of course, is it possible to verify the efficiency of these creatures' strange-looking gait without conducting experiments with them? In this project, which has already been carried out twice in modeling week events, it is not easy to develop a suitable model. The significant biomechanical idiosyncrasies of the penguin must be duplicated, but excessive detail can lead very quickly to too much complexity—at least for a further analysis with the means available in school mathematics (cf. Fig. 1).

Nonetheless, one arrives very quickly at a model based on differential equations that can be subsequently solved numerically. Obtaining realistic data for the dimensions and mass distribution of penguins is not easy, but with the appropriate values, one can in fact also show in the simulation that land-going penguins, in comparison to other animals, use their energy quite well.

**Fig. 1** Humboldt penguins and physical model (Photo, Drawing©Andreas Roth)

One might note here that the students' attempts to imitate the penguin's waddle—so as to conduct their own experiments and gain valuable insights for preparing a good model—provided first rate entertainment!

*Example 2* (Judging the Laufladen Cup fairly) In 2006, the *Laufladen Cup* was inaugurated in the West Pfalz, and it has been a yearly event ever since. In 2013, the series consisted of the following 10 amateur races:

| Date | Event | Length |
|---|---|---|
| 24.03. | The Laufladen Half-marathon of the TSG Kaiserslautern | 21.1 km |
| 07.04. | Trail Run Rodenbach | 10 km |
| 21.04. | Fun Run *Auf der Platte* (Pirmasens) | 10 km |
| 03.05. | 41st Intern. Fun Run Höheinöd | 10 km |
| 31.05. | Mölschbacher Wildsaulauf | 10 km |
| 07.06. | Intern. Evening Fun Run (Frankenstein) | 15.6 km |
| 08.06. | 9th Kuseler Hutmacherlauf | 10 km |
| 04.08. | Udo-Bölts-Lindenparkfestlauf (Heltersberg) | 12.8 km |
| 10.08. | Residenzfestlauf Kirchheimbolanden | 10 km |
| 31.12. | 41st Intern. New Year's Eve Road Race (Kottweiter) | 10 km |

At the end of the year, each participant who has competed in at least 6 of the 10 events receives a combined score. For every competition, each participant was assigned a point score according to the following formula:

$$P_i = 550 - 250 \cdot \frac{z_i}{z_s}, \quad P_i : \text{ point score for runner } i, \tag{1}$$

$$z_i : \text{ time for runner } i \text{ in seconds } z_s : \text{ winner's time in seconds} \tag{2}$$

For foot and bicycle races that are conducted in stages, it is typical that all participants compete in each stage; the times are then simply added together to obtain a combined result. This means of scoring is simple and seems fair. In contrast, however, several questions arise for the method of combined scoring described above:

- Do the above point values accurately reflect the differences in performance within a race?
- Does a total score consisting of the sum of the best 6 results yield an objective ranking?
- Is it possible for a racer to optimize his own point total by cleverly choosing his events? If so, how must the system be changed to eliminate this possibility?

The questions posed here seem quite unremarkable and, if one looks only superficially at the solutions presented at the end of the week, one might get the idea that they don't amount to much. This impression is deceptive, however. One need only consider the new point awarding system introduced for ski jumping competitions before the 2009–2010 season, in which the influence of the wind and the length of the approach are incorporated in a complicated manner. A complex model is used here that renders the judging of the events difficult for both athletes and spectators to understand.

In view of this somewhat dubious "standard" from the world of winter sports, it was important from the start to create an improved point awarding system that the athletes could understand and, yet, was still relatively easy to calculate. There were many discussions about what constitutes a *fair scoring system* which is of course essential for developing a model.

One recognizes quite early on that always awarding 300 points to the winner of a race is the heart of the problem: If the winner posts a comparably slow time, then all other racers receive more points for their performance then they normally would. Since only 6 of 10 events are considered for the overall ranking, participating in events with relatively high point awards offers a significant advantage over direct competitors who were not able to compete in such events.

But how can one redress this problem without making the point awarding system unnecessarily complicated at the same time? The project group that worked on this problem found a very convincing solution that was implemented in 2014 as the official judging system for the Laufladen Cup, and which is not much more difficult to calculate than the old system. Moreover, one of the participating students developed a web-based software that will permit automatic evaluation of future Cups and offer many possibilities for racers to make comparisons and predictions. This was an example where the modeling process yielded a product that was actually implemented by the end-user—an especially strong confirmation of the modelers' capabilities.

*Example 3* (Optimal spread pattern for road salting vehicles) The moment snow and icy roads are forecast, road salting vehicles take to the streets and highways. Within city limits, where the vehicles travel at an average of 40 km/h, they represent no great hindrance to other traffic. On the expressways, however, where their average speed is 60 km/h, they slow down other drivers considerably. According to traffic regulations, on-duty salting vehicles have right of way and may only be passed when road conditions safely permit such a maneuver. Why, then, don't salting vehicles travel faster?

It is essential for road maintenance departments to know whether they need to buy special vehicles to service expressways, or whether the city vehicles are also suitable for use at the higher speeds. To plan the use of the vehicles effectively, road maintenance departments and vehicle manufacturers must know exactly which factors influence the distribution of salt on the road surface:

- What is the optimal speed for spreading salt on roads?
- What are the effects of changes in the throwing angle?
- How can winter road maintenance vehicles achieve an optimal spread pattern?

In dealing with these questions, one must be very careful not to get tangled up in the details: What effect exactly does salt have on the roads? How densely must it be spread to achieve the desired effect? What are the environmental impacts? Which chemical reactions—with which complex interrelationships—must be considered? To further exacerbate the problem, empirical data from actual practice is very rare.

Another important influencing factor is the method with which the salt is dispensed from the vehicle via a rotating disc (cf. Fig. 2). Here, one must investigate how rotational speed, plate inclination, frictional forces, and external factors, such as wind and turbulence behind the vehicle, impact salt distribution.

If one can find a good compromise between adequate simplicity and the needed details, then the simulations offer revealing relationships between the speed of the vehicle, the frequency of the rotating plate, and the amount of salt distributed per unit time in relation to the width of the street.

*Example 4* (Optimizing the quality of fleece fabrics) In the textile industry, needling technology is used to convert fluffy, light fleece into inexpensive, durable, and tear-resistant fabric that can then be used in carpeting, lining, and insulating materials, for example (cf. Fig. 3).

During the needling process, fleece material on a conveyor belt is fed slowly under a needle board, which is pressed on the material for a fixed period of time (cf. Fig. 4). In this manner, the loose fleece material is converted into a stable fabric. The holes punched by the needles produce a pattern in the fabric that depends on the needle distribution and the belt feed rate. These patterns determine the applicability and quality of the final product. Because stripes, grids, and variations in hole density equate to diminished quality in the market, our client wishes to produce an especially homogeneous distribution of needle holes. How can he perfect his needling technology toward this end?

**Fig. 2** Salting vehicle on duty (Foto: glasseyes view (flikr), Creative-Commons License)

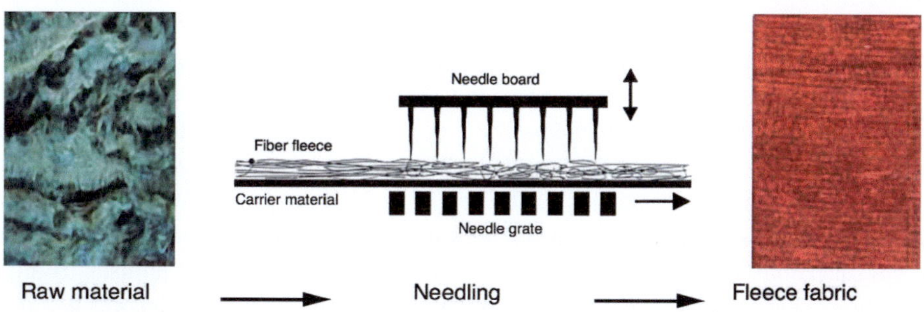


**Fig. 3** Schematic depiction of the needling process (Source: Fraunhofer ITWM)

This problem confronts the students with the challenge of using a pre-designed needle board to generate a fleece needling pattern that is as random as possible. The parameters that can be varied are the frequency of the needle board pressing and the (constant) speed at which the raw fleece passes under the needle board.

An essential task is to find a suitable definition for a "non-perceivable" needling pattern. Intense structures, such as lines and regularly repeating areas of higher needling density can be easily detected. But how can one establish a continuous measure for the degree of patterning or regularity that can then be used to assess the quality of alternative configura-

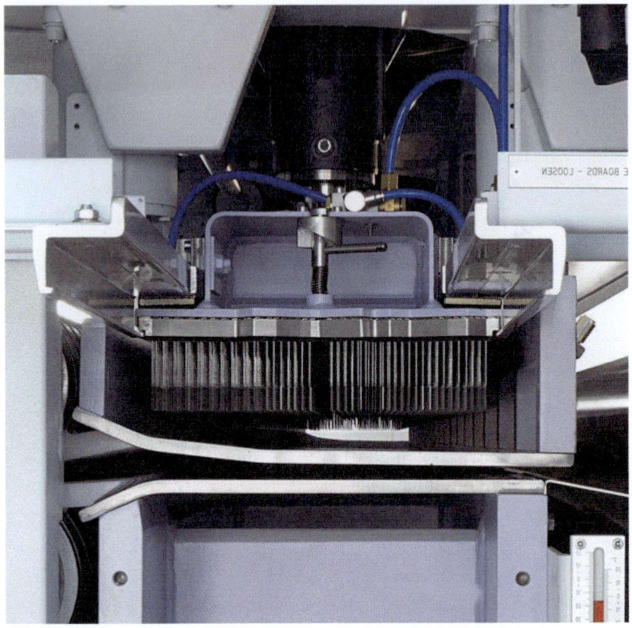

**Fig. 4** Machine for needling fleeces (Photo: AUTEFA SOLUTIONS GERMANY GMBH)

tions? One approach used by some students is to first specify a fixed needle arrangement on the board and then use simulations to study the homogeneity and randomness of the resulting patterns when the two available parameters are varied. Here, however, defining a random pattern presents the students with a problem. Moreover, they must find a way to systematically investigate and appraise various parameter combinations.

To date, this challenge has inspired a wide variety of approaches: Some student groups invested a great deal of time in mathematically describing randomness and producing genuine random numbers. Others experimented with mechanical alterations to the machinery that allowed horizontal movements of the needle board to supplement the constant frequency vertical movements. On occasion, the students' attempts stimulated some very unexpected questions. One group of 10th graders wanted to place needles equidistantly along a sine curve. The group's teachers suddenly found themselves trying to answer the question of how to define and calculate the arc length of a curve—a process that necessitated explaining to their charges the previously unfamiliar derivative terms...

Along with the mathematical approach, this problem also lends itself to the use of experimentation in the search for a solution. One might build one's own needle board out of cork or Styrofoam and thin needles, for example, and then use it to simulate the real process by punching holes in a steadily moving sheet of wallpaper. The resulting hole pattern can then be analyzed and evaluated. This experimental approach allows even quite young students to delve into the underlying problem without the use of computer simulations. When this kind of experimentation succeeds in awakening the students' interest, they are often motivated to learn programming skills so as to transport the laborious real experiments into the virtual world, where new and exciting possibilities then present themselves.

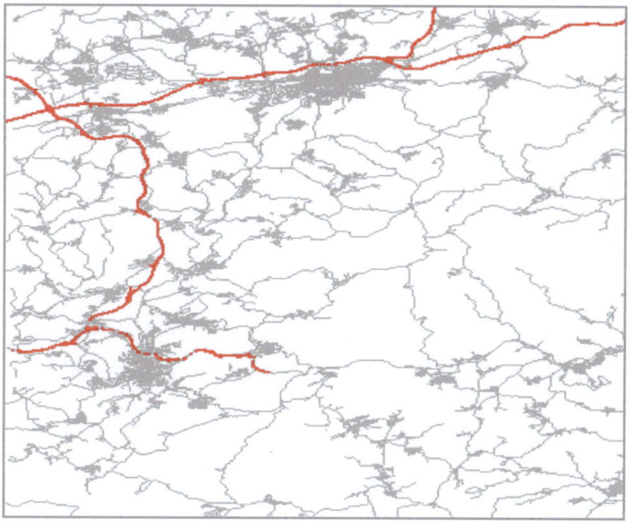

**Fig. 5** Street map of Kaiserslautern and environs (OpenStreetMap data—visualized by MATLAB®)

*Example 5* (Street evaluation based on route data) In this project, one is to investigate how existing route data can be used to evaluate an individual street or an entire street network. The students use a dataset for Kaiserslautern, which was provided by the *Mathematical Methods in Dynamics and Durability* am Fraunhofer-Institut für Techno- und Wirtschaftsmathematik ITWM was provided by the *Mathematical Methods in Dynamics and Durability* Department of the Fraunhofer Institute for Industrial Mathematics ITWM. This data is based on the open source database *OpenStreetMap*.

Characterizing a street network allows one to predict typical stresses for the chassis, brakes, clutches, and transmissions of motorized vehicles, such as trucks. This renders time-consuming test drives partially or entirely unnecessary, since information for areas with similar characteristics is already known and can be transferred.

Some of the project goals are:

- characterizing individual streets on the basis of individual data points,
- describing a street network,
- comparing streets or entire street networks on the basis of characteristic features.

At first glance, this project seems rather unremarkable, and it is not clear just how mathematical methods might be helpful. The students find challenges in several areas, however. For example, interpreting and processing *OpenStreetMap* data about the streets within a relatively small region of some $50 \times 50$ km (see Fig. 5) proves unexpectedly laborious. The dataset contains information about individual intersections, geographic data (longitude and latitude must first be converted prior to creating the graphical presentation), altitude, street category (highway, expressway, etc.), speed limits, and much more. A non-trivial problem, for example, is to identify from the almost 30 000 intersections in the map excerpt, those streets and junctions that must be known for the remaining processing steps.

Subsequently, various street network characteristics were discussed in the search for ways to ascertain corresponding values from the existing data. The first approaches involved the junction density of an area, the (maximal) incline of a street section, or its *curviness*. After several reasonable features were selected, the next goal was to rate a region on the basis of these features.

The selected features were then used to define a rating index that can be determined for a given street network.

The students quickly recognized, however, that average values alone are not enough for characterization purposes; they must also account for the possibility of driving around zones exhibiting poor characteristics. Thus, the more general idea arose of producing a navigation algorithm that investigates, on the basis of self-selected street network features, those connections between a region's various important points having the best possible rating. The distribution of these indices for a given region's important connections could then serve as a measure for rating and comparing different regions. Within the framework of a five-day modeling week, however, this idea could not be brought to fruition. It will be pursued as an extracurricular project over a longer time period of several months.[3]

## 5 Modeling Days

Modeling days are a short-format offshoot of modeling weeks. Every year, the Felix Klein Center for Mathematics stages up to 10 such events in schools. In this section, we describe the event concept in more detail. We also discuss the evolution of modeling days from the perspective of teacher training.

### 5.1 The Event Concept

Upon request, the Felix Klein Center for Mathematics stages several modeling days each year in schools in Rheinland-Pfalz. Twenty to thirty students, as well as members of the technical teaching staff (i.e. teachers of MINT subjects), participate in such an event. The inquiries come from the schools themselves. Because the Felix Klein Center's mission is to establish mathematical modeling in schools, the selection process considers whether a particular school already has modeling experience—either through its own modeling activities or previous participation in modeling days—and wishes to expand on its own offerings or gain more modeling experience. The modeling teams frequently comprise 11th and 12th graders in advanced college prep courses, but modeling days are also held

[3] At the time this chapter was written, the extracurricular continuation of the modeling week project was in the planning stage.

for 9th and 10th graders with equal regularity. Moreover, modeling days are also staged for other student groups, even in elementary schools. The duration of the modeling days generally depends on the schedules of the schools and varies from one to three days.

Depending on the school and the age of the participants, four to eight class periods per day are dedicated to the modeling event. Here, along with the student groups, teachers are to be present during each class period. Ideally, the teachers run the modeling days themselves, with the support of Felix Klein Center employees. The schedule is like that of a modeling week (cf. Sect. 4) in compact form: First, the students receive a short introduction to the work of applied mathematicians and mathematical modeling. Then, depending on how many students are participating, various themes are introduced for investigation in the project groups. As described in Sect. 3, the themes are introduced using the language of the end-user. Because modeling days follow the same principles as modeling weeks, the same problems are generally suitable for both types of events. In our experience, the problem sponsors need not worry about the feasibility of solving the problems within the shorter time period. Problems involving very large datasets, that is, problems for which students must invest a good deal of time in preparing the data, should be avoided or modified, however. Here, it can be useful for the problem sponsor to reduce the size of the data set in advance, or to subject it to some preliminary processing.

Directly after establishing the project groups, the members begin collecting ideas for treating their respective problems. Due to the shorter work period, it makes sense for problem sponsors to offer their groups a bit more guidance than is necessary during a modeling week. Intervening with critical questioning or encouraging the students to critically examine their own approach or model can shorten or even eliminate periods of stagnation. It is important, however, not to intervene so much that the students' motivation for solving the problem suffers.

On the final day, the student groups share their results via short presentations. This can be done at different scales. One can form an audience merely from the teachers, the problem sponsors, and the other students. More commonly, however, the audience also includes other students from the same grade. Consequently, it is important here also for presenters to use the language of the end-user, so as to make the results comprehensible to as wide an audience as possible. This comprehension then forms the basis for a further scientific discussion.

When planning a modeling day, there are a few organizational issues to bear in mind. One should try to have a computer room available or at least a laptop trolley. One should also provide Internet access. Should the above facilities not be available, however, one can also take advantage of the Felix Klein Center's laptop pool—just as in the modeling week events. Because the modeling days' relatively short time frame makes introductory programming sessions impracticable, the participants should rely on familiar programming languages and tabular calculation programs. It also makes good sense, before the modeling day commences, to get a feel for how much programming expertise the participants possess.

## 5.2 Modeling Days as Training Platforms for Teachers

In our view, one must practice and experience modeling oneself in order to learn it. During modeling weeks, one repeatedly hears teachers say that the concept is good, but implementing it in schools is hardly possible, due to time pressure and curricular demands, and also due to the lack of experience among the teachers. To counteract this situation, teacher training during modeling days was introduced, a concept designed to prepare teachers for integrating mathematical modeling into their own lesson plans. Here, along with mathematical modeling, some time is also devoted to practical issues, such as dealing with the 45-minute cycle time in modeling events or coping with the lack of suitable rooms.

At the moment, these training measures are in an experimental stage. If they are positively evaluated, they will then be offered to a wider audience. Training part or all of a school's technical teaching staff means that modeling teams can then be formed in which the teachers help one another in planning and carrying out modeling in the school. Experience shows that modeling is indeed possible in a school; in order to institute it on a broader basis, one needs several teachers to initiate it and serve as champions.

# 6 Teacher Trainings

Again and again, we see great willingness among teachers to take a chance on "Modeling and MINT at school." There is often great confusion as well, however, especially with regard to finding suitable test cases and designing lessons around real problems, which are often interdisciplinary in nature. The fact that modeling problems and implementing solutions can only be learned by doing places the question of finding appropriate instructional material in an unusually critical light. It is not possible to deliver complete solutions to teachers, since there is always a multiplicity of solution paths, which the students (and the teachers) themselves are supposed to discover. In practice, the teachers must often react to unforeseeable approaches, as well as to difficulties that suddenly arise in the search for solutions. This cannot be done with prepared materials. Instead, it requires intensive training to be able to respond appropriately and spontaneously. The design of training seminars for teachers should take this into account. Here it is not enough for teachers to listen passively to how a specially selected problem should be solved. They must actively sharpen their own problem-solving skills—by solving problems. This represents meta-strategic knowledge, which one can indeed learn, but which is quite difficult to teach. Naturally, teachers must also be given the appropriate tools and resources. Their knowledge and experience base must be extended so that they are capable of finding suitable problems, modeling them, and calculating approximate solutions. They must also learn this by doing—together with students, for example, "side by side"—in the course of a modeling week, or in intensive teaching sessions. Recently conducted in-house training events for several technical teaching staffs—conducted in the style of a modeling week—have proven effective. These trainings take into account that teachers, of course, have a deeper mathematical understanding than students. However, they usually cannot put this knowledge to work in a suitable fashion when confronted with complex, real-life problems. Our initial impression is that having the chance to use fellow teachers as a sounding board makes it noticeably easier to later implement modeling in the classroom. One such training concept might be for a teacher to first participate in a modeling week, then conduct project days in his own school (with the support of the Felix Klein Center), and finally, design and support his own project during a modeling week. Teachers trained in this manner are then in a position to pass along their knowledge and experience to their own colleagues, as well as to teachers from neighboring schools. In this way, local centers of problem solving—and also problem finding—competence are slowly built up, which require less and less support from outsiders as time goes by. Eventually, the impulse to bring modeling into schools reaches a critical mass and becomes self sustaining. For some time, teacher training opportunities have been evolving in Kaiserslautern: continuing education seminars at the Pädagogische Landesinstitut (formerly, IFB Speyer); the Felix Klein modeling weeks (twice per year in Rheinland-Pfalz and once per year in South Tyrol, together with the German school board in Bozen); in-house trainings in modeling for the technical teaching staffs of various schools; and the Junior Engineer Academy at the Heinrich Heine High School in Kaiserslautern.

## 7 Junior Engineer Academy

Since 2010, the Felix Klein Center for Mathematics and the Heinrich Heine High School in Kaiserslautern have jointly conducted a Junior Engineer Academy (JEA). From year to year, the JEA concept has been enhanced and refined. In this section, we will take a closer look at this developmental process and discuss the special features of this instructional form.

### 7.1 The Event Concept

The Junior Engineer Academy consists of a three-year project, in which an interdisciplinary topic encompassing the subjects of mathematics, computer science, and at least one natural science is offered as an alternative compulsory subject in the form of a weekly, three-period lesson.

The goal of this instructional form is not only to offer a new concept for mathematics and physics lessons, but also to help establish new organizational structures in schools. The Heinrich Heine High School, which has a gifted-and-talented curriculum, was seen as a promising partner for the JEA, since they had already introduced an alternative compulsory MINT course for students in the 7th grade and upward.

In this predecessor course, the subject was sub-divided into three sections: In the first year, students were instructed in computer science; in the second year, mathematics lessons were added; and in the third year, there was instruction in one of the natural sciences, that is, biology, chemistry, or physics (in rotating order). During this 3-year course, the class had three lesson periods of MINT per week. It is important to emphasize that the lesson contents were not supposed to overlap with the normal curriculum of the 7th to 10th grades. In selecting Heinrich Heine as partner, the main criterion was not the school's gifted-and-talented curriculum, but its already existing organizational structures. Similar projects had already been conducted by the TheoPrax® Center in Pfinztal for normal classes in grades 8–10 (cf. [4]). The teaching-learning method TheoPrax® was developed in the 1990s by Peter Eyerer, Bernd Hefer, and Dörthe Krause at the Fraunhofer Institute for Chemical Technology ICT. The goal is to use activity and practice oriented instructional concepts—implemented in cooperation with external partners in industry, research, and the services sector—to create an "interface between the school and the marketplace."

The school administrators and MINT teachers developed the new MINT course together. Naturally, it remained an alternative compulsory class with three lesson periods per week in the 7th, 8th, and 10th grades (The 9th grade was skipped, due to the BEGYS program—Gifted-and-Talented Support in Academic High Schools). However, there is now a common topic for the entire three-year cycle, along with weekly instruction in mathematics, computer science, and a natural science. In the first round, which ran from 2010 to 2013, the common topic was *location planning for wind farms*, and physics joined mathematics and computer science as third subject. The second round, begun in 2011, addressed the topic of *batteries, power packs, and fuel cells: the search for the super storage device*, and included chemistry as the natural science component. The round begun in 2012 tackled the topic *bioacoustics: automatic recognition of birdsongs*, and added biology to the trio of underlying subjects. The recently commenced Junior Engineer Academy has taken on the challenge of re-designing one of Kaiserslautern's former industrial parks, and is learning geography in the process. What makes this project special is that the topic is being shaped into its final form during the course of the Academy, which gives individual students a chance to pursue their particular interests. In the meanwhile, the group has decided to address questions from the very current field of electric mobility.

Some readers, simply from scanning these brief topic descriptions, might conclude that these are very ambitious projects for this age group, perhaps even impossible—especially for the 7th and 8th graders. Our idea was to work on real-world MINT problems that capture the interest of the students. The course is not simply about learning concepts and solving problems, and the teachers are not simply project leaders, but project partners exploring the MINT topic together with the students. For each round, there is a team of regular teachers and external teachers. The external teachers are computer scientists (for computer science instruction in rounds 1 and 2) and mathematicians (for mathematics instruction in round 2 or for mathematics and computer science instruction in rounds 3 and 4). In addition, each subject has an expert from the TU Kaiserslautern or the Fraunhofer Institute for Industrial Mathematics ITWM. The Academy includes the regular 3 hours per week of classroom instruction plus additional field trips and workshops, which are embedded in the 3-year course. Field trips include visits to the University's laboratories (not just for sight-seeing, but also for doing experiments) or other institutes or businesses. The workshops cover such subjects as *team building*, *time and project management*, *creativity training*, and *conflict management*. Financing the pilot project is another important consideration. For the first round, the Felix Klein Center and the Heinrich Heine High School were awarded a so-called Junior Engineer Academy grant from the Deutsche Telekom Foundation. This covered the costs of the field trips and workshops, as well as materials and equipment not included in the regular school budget. Because of clear successes visible in the first year, the two partners decided to extend the program to at least three rounds, that is, five years. The fourth round was begun in the 2013/14 school year, and the JEA has in the interim become a regular feature of the Heinrich Heine High School.

## 7.2 Features of the Junior Engineer Academy

Self-initiated planning increases motivation and a sense of ownership in the results. The above-mentioned workshop *time and project management* is designed to enable students to plan as many phases of the instruction for themselves as possible. Here, small projects and/or lesson content are to be assigned as often as possible by the class itself. This requires becoming conscious of where one is missing knowledge and includes bringing in expertise from outside, where needed. The practice of incorporating student ideas makes changes in lesson plans unavoidable. As a result, the planning horizon is much shorter than for regular instruction. A compensating virtue, however, is that students identify more strongly with the project and are more highly motivated to grapple with the lesson material. It is permissible—perhaps even desirable—for students to make one plunge down a blind alley, that is, to think a bad idea through to its logical conclusion.

**Students Should Be Given Time to Gather as Much Experience for Themselves as Possible** During the MINT lessons, students should be allowed enough time to investigate the topic from a variety of angles. One way to do this is to arrange for as much small-group work as possible. Here, the group members take on various tasks, such as timekeeper, materials supervisor, group speaker, and group secretary. Many experiments are also carried out, either with the computer or, when necessary, with physical equipment. From the very beginning, emphasis is placed on regularly presenting and discussing results. Starting with the 8th graders in round three of the JEA, the class was divided into "expert teams" so as to take better account of the students' individual interests and strengths. Granted, this brings with it the extra challenge of suitably integrating the various teams, so as to keep the focus on the overall goal.

**The Path to the Product Is the Goal** The problems addressed in the JEA stem from the real word and, therefore, a real product is expected at the end of the project. Working towards a significant goal serves to chart the course through the three years. It motivates the students to keep trying when they find themselves at a dead-end. When the whole group considers the question "Where do we want to go from here?" new aspects and perspectives often come to light. The end product is also important for strengthening group identification with the project.

**Use of Technology** In round three, during the 2013/14 school year, tablets were introduced into the class. Put briefly, their subsequent use can be divided into two categories: First, they were used for such traditional purposes as documentation, communication, presentation, and, of course, calculation. In addition, they were also used as "black box" tools to replace challenging (mathematical) techniques with the appropriate computer programs. For example, the computer can perform a Fourier analysis of an audio signal. Or, an oscilloscope can be used to illustrate that noise consists of vibrations, which one can recognize and process again as an audio signal.

## 7.3 Setting and Development

Even before the Heinrich Heine High School began its collaboration with the Felix Klein Center for Mathematics, it had an alternative compulsory MINT subject in its gifted-and-talented curriculum for junior high school students. Here, students had the choice between a 3-year MINT course with three class periods per week or a course in Japanese. Because the class is categorized as an alternative compulsory subject, its mandate was to cover MINT material that supplemented the regular instruction material, that is, it was to avoid, where possible, treating material included in the regular curriculum of the 7th, 8th, and 10th graders (9th grade is skipped). This was designed to prevent the MINT students from gaining an advantage over the Japanese students—an important point, since positive grades in alternative compulsory subjects count towards graduation.

### 7.3.1 Evolution of the JEA (Since the 2012/13 School Year)

**Time Allocation** IFor the first two rounds of the Junior Engineer Academy, the three subjects (mathematics, computer science, and a natural science) were each taught one period per week. Given the course's project format, this allocation was not especially helpful. Starting in the 2012/13 school year, it was possible to reorganize the class scheduling; subsequently, the natural science lesson in the 7th grade was still given as a single period lesson, but mathematics and computer science were team taught as a double period. Since the start of the 2013/14 school year, the biology class in round 3 is taught by a teacher who also participates in the mathematics and computer science teaching team. This flexible allocation provides the freedom needed for the project with respect to scheduling and content. The joint team teaching approach greatly improves coordination among the teachers and facilitates lesson planning and reflection.

**Small-Group Instruction** As mentioned previously, students involved in project work have a product as their goal. This focus on the product serves as the leitmotiv for the 3-year project. Within the project, however, students also work on small or mini-projects in order to acquire the knowledge and skills needed for the large goal at the project's end. Here, the students themselves should dictate which mini-projects should be worked on and what knowledge is needed. Since the start of the 2013/14 school year, the students in round 3 have been working in expert teams on various sub-areas of the project. The project goal of this round is to program an app that automatically recognizes birdsongs. Along with the algorithmic implementation, mathematical modeling, technical, and even design aspects must be considered. Accordingly, the small groups work in these sub-areas and confer with one another as often as possible, to ensure that they all keep their eyes on the common project goal. This approach is intended to foster the development of individual skills, but it also takes into account the interests of the various students. Moreover, the required coordination of the various groups forces the specialists of a given group to transmit their knowledge as clearly as possible to the non-specialists of the other groups. Initially, this was an unaccustomed challenge. Regular, short presentations starting in the 7th grade are designed to strengthen these communication skills. As a result, one observes that the class has become very critical of poor presentations and unintelligible explanations.

### 7.3.2 The Role of the Teachers—Team Teaching

During class periods, the teachers serve more frequently as moderators, and where needed, as experts. Discussions often extend beyond the instructional material, and it is therefore necessary for teachers to sometimes admit that they have no answer to the question posed and must seek out expertise and advice elsewhere. Managing experiments and software frequently takes more time than has been planned for. After all, the students are supposed to experiment and try things out as much as possible. Teaching in teams has proven to be very helpful in dealing with these circumstances. Team teaching makes it possible to individually address any one student's specific problem without abandoning the remainder of the class. The presence of two teachers for a double period also allows one to divide up the class—to learn programming skills, for example. This technique has proven very effective. For example, one part of the class can conduct physical experiments, while the other part carries out programming tasks on the computers. During the second half of the double period, the roles are then reversed. Another advantage of team teaching is that it makes lesson planning easier, especially in view of the project's interdisciplinary nature.

### 7.3.3 Challenges and Ideas

The open structure reduces the time available for lesson preparation. The goal of adopting the ideas of the students, where feasible, and giving them as much leeway as possible to do their own planning means that lesson contents are often not foreseeable or predictable. The introduction of tablets has improved access to information, but resulted in the problem of how to deal with the abundance of available information. Students often have trouble critically examining the quality of the information they find, working it into usable form, and summarizing it. The interdisciplinary nature and project format seem to deter many teachers from MINT project instruction. Team teaching surely offers a way around this problem, although the demands of teachers' day to day instruction routines often makes it difficult to convince colleagues in other subjects to join project teams as experts. This is a particular problem in non-MINT subjects. Team teaching and the formation of MINT teams in schools could strengthen communication channels between teachers in different subject areas. This is especially important for lesson planning and for dealing with questions that straddle the boundaries of the individual subjects.

# 8 Fraunhofer MINT-EC Math Talents

In the Autumn of 2011, the Fraunhofer MINT-EC Talents program was initiated as a cooperative venture between the Fraunhofer Gesellschaft and the MINT-EC Association. The program focuses on *mathematics* and *chemistry*, the latter under the direction of the Fraunhofer Institute for Production Technology and Applied Material Research, in Bremen. Following a selection process involving 10th (12-year high schools) and 11th (13-year high schools) graders from various MINT-EC schools throughout Germany, 12 students were chosen to participate in the two-year program. This began with a joint kickoff event for all participants in January of 2012, and the Felix Klein Center for Mathematics is providing support and consultation for the selected students.

## 8.1 *Math Talents* Projects

For the second round of the *Math Talents* program, the selection process was changed somewhat. Out of the group of applicants from MINT-EC schools throughout Germany, approximately 40 students were invited to participate in a Fraunhofer Talent School for *mathematics & athletics*, which took place in January of 2014. This event was closely modeled on the Felix Klein modeling week, although shortened to three days. The Talent School served, first, to determine the suitability of the candidates for, and their interest in, a two-year MINT program and, second, to provide them with a glimpse into the world of mathematical modeling in the context of MINT projects. This was valuable experience, even for those who did not make the final selection.

Due to the fine performance of many applicants, the number of participants in the second round of the *Math Talents* was increased to 24. The plan is for these students to work on four different MINT projects and convene for six multi-day workshops by the time they graduate from high school. In the intervening time, they can exchange intermediate results via an Internet platform and confer with the project sponsors.

In contrast to round one of the program, the participants were much more actively involved in selecting their projects. They were initially presented with only very general fields of investigation. In April 2014, in the course of the first workshop, the students then defined a total of four projects, which we will present briefly in this section. The projects chosen by the students are not necessarily from the field of "mathematics and athletics" and only partially overlap with the projects from the Fraunhofer Talent School.

*Example 6* (Computer-supported analysis of pocket billiard strategies) In this project, students consider how to analyze and assess game situations in pocket billiards in order to derive a long-term strategy or, at least, recommend which shot makes the most sense at

**Fig. 6** Game situation in pocket billiards (GNU General Public License)

any given point in the game (cf. Fig. 6). Due to the complexity of game situations and the abundance of factors to consider, the data analysis is performed via computer. Thus, the challenge is to develop software that recommends one (or more) shots to the player that are either easy to pocket or make strategic sense. One might also account for the experience of the player regarding shots of differing degrees of difficulty (such as banked shots or shots with English) or even the personal tastes of the player. The latter would naturally require the development of player profiles.

The project members have established the following preliminary goals:

- Use photo/video to map the geometry of the table and the position of the balls; identify which balls are playable.
- Develop a suitable model to assess the difficulty of the possible shots (angles and distances are already captured in the model; further factors are to follow) and present the player with a top-3 list of suggested shots.
- Analyze the shot with respect to outcome (made/missed/fouled) via the video, so as to document the course of the game.
- Create a computer program or smartphone app with an appropriate user interface.

*Example 7* (Self-piloting quadcopter) Small flying devices outfitted with cameras and sensor technology, such as quadcopters, are today readily available to everyone and have a potentially wide field of application. Their use for automatic delivery service in the logistics branch or for cheaply and effectively monitoring large public events both lie within the realm of possibility (cf. Fig. 7).

**Fig. 7** Self-piloted flight with object recognition and a virtual swarm in MATLAB®

In order to realize these possibilities, however, the quadcopter must be able to pilot itself autonomously to a certain degree. Aside from the legal difficulties associated with such a procedure (Who is liable when a technical failure causes the device to fall on someone's head?), there are numerous technical challenges to overcome.

In this project, a quadcopter is to be first made capable of autonomous flight, that is, the device must be able to carry out a search within a specified area. Based on the data from its on-board camera, the quadcopter must identify color-coded objects and execute the flight maneuvers needed to reach them.

This is to be accomplished via a wireless connection between the drone and a guidance computer that manages the image processing and flight control functions. This permits the use of a light, programmable platform for the test phase. It would also be interesting to establish the coordinated flight of multiple drones with limited sensor technology by enabling them to inform one another about their surroundings. This goal of squadron flying is also to be pursued, as soon as the automated control system for a single drone functions as desired.

*Example 8* (Optimized fitness training with electric mountain bikes) Electric bikes are gaining an ever-increasing share of the bicycle market. Their great advantage is that they enable riders to enjoy longer and more difficult tours than their fitness levels would normally permit. This trend is also visible in the mountain biking sector.

The main aim of this project is to generate a control system to optimize fitness training with an electric bike. To do so, one needs to simulate effects such as headwinds and slipstreams. Another possible feature is simulating special training tours on one's standard route by using the electric motor in either support or re-charge (i.e., braking) mode, as appropriate. Integrating performance data, such as pulse and pedaling frequency, also helps provide the rider with an optimized training program.

The required hardware and control system are both to be developed by the project team. The base hardware consists of a Hardtail mountain bike (see Fig. 8), over-sized disc brakes (the motor adds weight), and a hub motor for either the back wheel or the bottom bracket. The control system is to be based on a *Raspberry Pi* or an *Arduino* micro-controller, although there are also plans to design a smartphone based system.

**Fig. 8** Standard mountainbike (Foto: Martin Bracke)

*Example 9* (A self-navigating outdoor robot) The goal of this project is to develop a self-propelled, "all-terrain" robot that can independently find a designated target and advance over open land to reach it. The robot is to plot its own course to the target, while optimizing energy consumption and range, and recognize and evade both stationary and non-stationary obstacles.

The current design plan calls for a sturdy chassis with an electric drive system, and will possibly take the form of a tracked vehicle. The control system will use a *Raspberry Pi* mini-computer, and there will be several sensors, including a camera and a distance sensor.

The first challenges recognized by the students were technical in nature: How can one make the chassis robust enough to survive the rigors of traversing open land? How can sensor data be evaluated and the motors controlled via the *Raspberry Pi* One must also answer the question of how to use the sensor data to help the robot orient itself in its surroundings, so as to then set a course for the given target.

For the duration of the program, the participants will continue to develop their projects as independently as possible, although they can draw on the support of the Felix Klein Center as needed. Mini-workshops have been planned or requested on the topics of *Raspberry Pi* programming, mathematical image processing, simulation and hardware control with MATLAB/Simulink, and app programming for smartphones.

Because the development of a specific product plays an important role in all the projects, the Math Talents participants will be assisted at the end of the program, should they wish to enter their projects in the *Jugend forscht* (Young Researchers) competition. In this case, additional soft skills courses—*presentation techniques*, *project planning*, and *time management*—will be offered. Along these lines, all the participants were also introduced to LaTeX, at the start of the program, a typesetting and composition system very widely used in mathematics and the natural sciences. Thus, they had a powerful tool for documenting and designing the layout of their results at their disposal throughout the duration of the project.

We would also like to mention that the projects described in this section were not at all pre-defined; estimates of the challenge level and difficulty of implementation could thus only be made once the projects were underway. This resulted in a certain measure of uncertainty regarding their practicability, since the goals set by the students for themselves were very high—so high, in fact, that some aspects even touch upon current state-of-the-art research! This lack of pre-definition also presented a great opportunity, however: that of drawing on student capabilities that were hidden at the start and only developed over the course of time. The key, here, as is often the case, is for all participants to manage their expectations about the results. If very specific goals or even quantitative targets are set, this will quite likely lead to frustration and failure. If, instead, the work is viewed as a research project with an uncertain outcome (which, in fact, it is), then winding up in a blind alley does not automatically mean the end of the project. Rather, it means rethinking one's previous work and, if necessary, re-aligning one's goals. As with the countless successful projects from modeling weeks and days, here too, each group will ultimately have produced a functioning product in the form of hardware and/or software—although the originally defined features may be partially scaled-back or perhaps, as also happens, even extended.

## 9 Final Remarks

To conclude this chapter, in which we have presented an overview of the possibilities for implementing mathematical modeling with real-world applications in student projects, we would like to offer some encouragement to our readers.

Dare to enter the exciting world of application problems waiting to be discovered all around by those whose eyes are open to see them. Your point of entry for an intriguing modeling project might come in the form of a claim in a newspaper article that has no convincing evidence to back it up. Or it could be a conversation with a friend or acquaintance about a tricky problem he is facing in his company, for which there is no standard solution. Or a simple observation of a generally accepted procedure that suddenly triggers the thought, why does it have to be done exactly that way? Is there not perhaps some better alternative?

When you find yourself fascinated with such a problem and are able to dive into models and simulations of your own in the effort to solve it, then the optimal conditions have been established for you to also capture the fancy of young learners. And if the students themselves can then develop their own projects on the basis of your inspiring example, then it is almost certain that they too will be carried away on their own exciting journeys of invention, creation, and learning.[4]

[4]Information about current school projects at the Felix Klein Center for Mathematics in Kaiserslautern is available at the website of the *Kompetenzzentrum für Mathematische Modellierung in* MINT-*Projekten* in der Schule (KOMMS), http://komms.uni-kl.de.

**Acknowledgements** The authors wish to express their appreciation to many colleagues at the Felix Klein Center for the years of successful collaboration in carrying out school projects. We are especially grateful to all the students and teachers who, as the main actors, have enriched and shaped the project work with their creative ideas and concentrated efforts, and who have evidenced much joy in everything they undertook to accomplish.

## References

1. Ackoff, C., Arnoff, R., Churchman, E.: Introduction to Operations Research. Wiley, New York (1957)
2. Blum, W., Borromeo Ferri, R., Maaß, K.: Mathematikunterricht im Kontext von Realität, Kultur und Lehrerprofessionalität. Springer, New York (2012)
3. Bock, W., Bracke, M.: Project teaching and mathematical modelling in stem subjects: a design based research study. In: Ubuz, B., Haser, C., Mariotti, M. (eds.) Proceedings of CERME 8, Lecture Notes in Computational Science and Engineering, pp. 1010–1020. Middle East Technical University, Ankara (2013)
4. Eyerer, P.: TheoPrax®-Projektarbeit in Aus- und Weiterbildung. Bausteine für Lernende Organisationen. Klett-Cotta, Stuttgart (2000)
5. Gudjons, H.: Handlungsorientiert lehren und lernen: Projektunterricht und Schüleraktivität. Klinkhardt, Bad Heilbrunn (1986)
6. Jank, W., Meyer, H.: Didaktische Modelle. Cornelsen Scriptor, Berlin (2003)

MIX
Papier aus verantwortungsvollen Quellen
Paper from responsible sources
FSC® C105338

FSC
www.fsc.org

If you have any concerns about our products,
you can contact us on
**ProductSafety@springernature.com**

In case Publisher is established outside the EU,
the EU authorized representative is:
**Springer Nature Customer Service Center GmbH**
**Europaplatz 3, 69115 Heidelberg, Germany**

Printed by Libri Plureos GmbH
in Hamburg, Germany